2013 9th International Workshop on Electromagnetic Compatibility of Integrated Circuits

(EMC Compo 2013)

Nara, Japan
15-18 December 2013

IEEE Catalog Number: CFP1348P-POD
ISBN: 978-1-4799-5004-1

Copyright © 2013 by the Institute of Electrical and Electronic Engineers, Inc
All Rights Reserved

Copyright and Reprint Permissions: Abstracting is permitted with credit to the source. Libraries are permitted to photocopy beyond the limit of U.S. copyright law for private use of patrons those articles in this volume that carry a code at the bottom of the first page, provided the per-copy fee indicated in the code is paid through Copyright Clearance Center, 222 Rosewood Drive, Danvers, MA 01923.

For other copying, reprint or republication permission, write to IEEE Copyrights Manager, IEEE Service Center, 445 Hoes Lane, Piscataway, NJ 08854. All rights reserved.

***This publication is a representation of what appears in the IEEE Digital Libraries. Some format issues inherent in the e-media version may also appear in this print version.

IEEE Catalog Number: CFP1348P-POD
ISBN 13:   978-1-4799-5004-1

Additional Copies of This Publication Are Available From:

Curran Associates, Inc
57 Morehouse Lane
Red Hook, NY 12571 USA
Phone: (845) 758-0400
Fax: (845) 758-2633
E-mail: curran@proceedings.com
Web: www.proceedings.com

Session Program

Date: Sunday, 15/Dec/2013

JW1: Joint Workshop 1
Time: Sunday, 15/Dec/2013: 9:00am - 10:30am ·   *Location:* Kinsho Hall

JW1-1: Devices, Circuits, Packages and Systems Understanding their Interplay for Managing Signal, Power and Thermal Integrity

Madhavan Swaminathan
Georgia Institute of Technology, Atlanta, USA

JW1-2: The Pathways to Cost-Effective Power/Signal Integrity Designs for High-Performance Mobile Systems

Woong Hwan Ryu
Samsung Electronics, Korea, Republic of (South Korea)

JW2: Joint Workshop 2
Time: Sunday, 15/Dec/2013: 10:50am - 12:10pm ·   *Location:* Kinsho Hall

JW2-1: Transition from 2D to 3D IC design: Advantage, Challenge and Solutions

William Wu Shen
TSMC Corp., Taiwan

JW2-2: Introduction of JEITA LPB WG

Yoshinori Fukuba
JEITA EDA technical committee LSI-Package-Board Interoperable Design Working Group, Japan

TU1: LPB:
Time: Sunday, 15/Dec/2013: 2:00pm - 3:00pm ·   *Location:* Kinsho Hall

TU1: JEITA LSI-Package-Board (LPB) Standard Format

Yoshinori Fukuba
JEITA EDA technical committee LSI-Package-Board Interoperable Design Working Group, Japan

TU2: SEISME:
Time: Sunday, 15/Dec/2013: 3:20pm - 4:20pm ·   *Location:* Kinsho Hall

TU2: Simulation of Emission and Immunity of Systems and Modules Electronics
Christian MAROT[1], Andre DURIER[2], Etienne SICARD[3], Alain SAUVAGE[4], Zouhair RIAH[5], Olivier Crepel[6], Olivier MAURICE[7], Genevieve DUCHAMP[8]
[1]EADS IW, France; [2]CONTINENTAL Corporation; [3]INSA Toulouse; [4]AIRBUS SAS; [5]ESIGELEC; [6]EADS IW, France; [7]GERAC; [8]IMS-Bordeaux, France

Date: Monday, 16/Dec/2013

SI: Signal Integrity
Time: Monday, 16/Dec/2013: 8:50am - 10:10am ·   *Location:* Kinsho Hall

SI-1: A Technique for Estimating Signal Waveforms at Inaccessible Points in High Speed Digital Circuits   pp. 1-4
Tomohiro Kinoshita, Shoichi Hara, Eiji Takahashi, Kazuhide Uriu, Panasonic Corporation, Japan

SI-2: Design and Measurement of a Compact On-interposer Passive Equalizer for Chip-to-chip High-speed Differential Signaling   pp. 5-9
Heegon Kim[1], Jonghyun Cho[1], Daniel Hyunsuk Jung[1], Jonghoon Jay Kim[1], Sumin Choi[1], Junho Lee[2], Kunwoo Park[2], Joungho Kim[1]
[1]KAIST, Korea, Republic of (South Korea); [2]SK Hynix Semiconductor Inc., Korea, Republic of (South Korea)

Session Program

SI-3: Signal Integrity and EMC Performance Enhancement using 3D Integrated Circuits – A Case Study   pp. 10-14

Etienne SICARD[1], Jianfei WU[2], Jian-cheng LI[2]

[1]INSA, France; [2]National University of Defense Technology, China

SI-4: Kron Simulation of Field-to-line Coupling using a Meshed and a Modified Taylor Cell   pp. 15-20

Sjoerd Op 't Land[1], Richard Perdriau[1], Mohamed Ramdani[1], Olivier Maurice[2], M'hamed Drissi[3]

[1]Groupe ESEO, France; [2]GERAC, France; [3]INSA de Rennes, France

WS: APD and Near-field Measurement   *Time:* Monday, 16/Dec/2013: 10:40am - 12:00pm ·   *Location:* Kinsho Hall

WS-1: Measurement of amplitude probability distribution of an EM noise for analyzing its impact on wireless systems

Yasushi Matsumoto

National Institute of Information and Communications Technology, Japan

WS-2: Developing a Multi-channel Near-field Measurement System to Detect Electromagnetic Noise on Electronic Devices Using APD Measurement Method

Hiroshi Tsutagaya

Peritec Corporation, Japan

WS-3: Binocular analysis for RF Cross Domain Analyzer

Yasunori Hirai

Advantest Corporation, Japan

KS: Keynote Speech   *Time:* Monday, 16/Dec/2013: 1:50pm - 2:35pm ·   *Location:* Kinsho Hall

KS: Semiconductor Innovation for Smart Society and Its EMC Solutions

Toru Shimizu

Renesas Electronics Corp., Japan_

EMI: Emission Measurement and Control   *Time:* Monday, 16/Dec/2013: 2:35pm - 3:35pm ·   *Location:* Kinsho Hall

EMI-1: Extraction of Deterministic and Random LSI Noise Models with the Printed Reverberation Board   pp. 21-26

Umberto Paoletti, Takashi Suga

Hitachi, Ltd., Yokohama Research Laboratory, Japan

EMI-2: Broadband Detection of Radiating Moments using the TEM-cell and a Phase-calibrated Oscilloscope   pp. 27-32

Renaud Gillon[1], Niko Bako[2], Adrijan Baric[2]

[1]ON Semiconductor Belgium BVBA, Belgium; [2]Faculty of Electrical Engineering, University of Zagreb, Croatia

EMI-3: Substrate Noise Reduction Based on Impedance Balance using Tunable Resistances   pp. 33-36

Atsushi Nakamura[1], Masaaki Maeda[2], Tohlu Matsushima[2], Osami Wada[2]

[1]Renesas Electronics; [2]Kyoto University, Japan

SS: [Special Session] Tackling EMC in Real IC Chips   *Time:* Monday, 16/Dec/2013: 4:00pm - 6:00pm ·   *Location:* Kinsho Hall

SSI-1: Different IC Radiated-emission Models to Analyze Far-field Radiation of a Test PCB

Jingnan Pan, Yaojiang Zhang, Jun Fan

Missouri University of Science and Technology, United States of America

Session Program

SSI-2: An Integrated Simulation Approach for Addressing the Chip-Package-Board Power Noise Challenge
Dian Yang
Apache Design Inc. Subsidiary of ANSYS, Inc.

SS-3: Measurement-based Diagnosis of Wireless Communication Performance in the Presence of In-band Interferers in RF ICs pp. 37-41
Makoto Nagata[1], Shunsuke Shimazaki[1], Naoya Azuma[1], Satoru Takahashi[2], Motoki Murakami[3], Kazuaki Hori[3], Satoshi Tanaka[4], Masahiro Yamaguchi[4]
[1]Kobe-univ., Japan; [2]Renesas Mobile Corp, Japan; [3]Renesas Electronics Corp., Japan; [4]Tohoku Univ., Japan

SS-4: Measurements and Simulation of Substrate Noise Coupling in RF ICs with CMOS Digital Noise Emulator pp. 42-46
Naoya Azuma[1], Shunsuke Shimazaki[1], Noriyuki Miura[1], Makoto Nagata[1], Tomomitsu Kitamura[2], Satoru Takahashi[2], Motoki Murakami[3], Kazuaki Hori[3], Atsushi Nakamura[3], Kenta Tsukamoto[4], Mizuki Iwanami[4], Eiji Hankui[4], Sho Muroga[5], Yasushi Endo[5], Satoshi Tanaka[5], Masahiro Yamaguchi[5]
[1]Kobe-univ., Japan; [2]Renesas Mobile Corp, Japan; [3]Renesas Electronics Corp., Japan; [4]NEC Corporation, Japan; [5]Tohoku Univ., Japan

SS-5: In-Band Spurious Attenuation in LTE-Class RFIC Chip using a Soft Magnetic Thin Film pp. 47-52
Sho Muroga[1], Yutaka Shimada[1], Yasushi Endo[1], Satoshi Tanaka[1], Naoya Azuma[2], Makoto Nagata[2], Motoki Murakami[3], Satoru Takahashi[4], Kazuaki Hori[3], Masahiro Yamaguchi[1]
[1]Tohoku University, Japan; [2]Kobe Universiy, Japan; [3]Renesas Electronics Corporation, Japan; [4]Renesas Mobile Corporation, Japan

Date: Tuesday, 17/Dec/2013

STD: Standards for Semiconductor EMC *Time:* Tuesday, 17/Dec/2013: 8:30am - 10:10am · *Location:* Kinsho Hall

STD-1: Conducted immunity of three Op-Amps using the DPI measurement technique and VHDL-AMS modeling pp. 53-58
Siham HAIROUD AIRIEAU, Tristan DUBOIS, Angelique TETELIN, Geneviève DUCHAMP
Univ. Bordeaux, IMS, UMR 5218, France

STD-2: Improvement of Reproducibility of DPI Method to Quantify RF Conducted Immunity of LDO Regulator pp. 59-62
Tohlu Matsushima, Nobuaki Ikehara, Takashi Hisakado, Osami Wada
Kyoto University, Japan

STD-3: A Generalized Accurate Modelling Method for Automotive Bulk Current Injection (BCI) Test Setups up to 1 GHz pp. 63-68
Sergey Miropolsky, Alexander Sapadinsky, Stephan Frei
AG Bordsysteme, Technische Universität Dortmund, Dortmund, Germany

STD-4: IC-Stripline Design Optimization using Response Surface Methodology pp. 69-73
Tvrtko Mandic[1], Renaud Gillon[2], Adrijan Baric[1]
[1]University of Zagreb, Faculty of Electrical Engineering and Computing, Croatia; [2]ON Semiconductors BVBA, Belguim

STD-5: Study of radiated immunity of an electronic system in a reverberating chamber pp. 74-77
Laurent GUIBERT[1], Patrick MILLOT[1], Xavier FERRIÈRES[1], Etienne SICARD[2]
[1]ONERA Centre de Toulouse, France; [2]INSA Toulouse, France

DSN: EMC-aware IC Design *Time:* Tuesday, 17/Dec/2013: 10:40am - 12:00pm · *Location:* Kinsho Hall

DSN-1: Transient Analysis of EM Radiation Associated with Information Leakage from Cryptographic ICs pp. 78-82
Yuichi Hayashi[1], Naofumi Homma[1], Takafumi Aoki[1], Yuichiro Okugawa[2], Yoshiharu Akiyama[2]
[1]Tohoku University, Japan; [2]Nippon Telegraph and Telephone Corporation, Japan

DSN-2: Noise-immune Design of Schmitt Trigger Logic Gate using DTMOS for Sub-threshold Circuits pp. 83-88
KyungSoo Kim, Wansoo Nah, SoYoung Kim
SungKyunKwan University, Korea, Republic of (South Korea)

Session Program

DSN-3: Reliability Analysis of an On-chip Watchdog for Embedded Systems Exposed to Radiation and EMI — pp. 89-94

Chri'stofer Oliveira[1], Juliano Benfica[1], Leti'cia BolzaniPoehls[1], Fabian Vargas[1], Jose' Lipovetzky[2], Ariel Lutenberg[2], Edmundo Gatti[3], Fernando Hernandez[4], Alexandre Boyer[5]

[1]Catholic University - PUCRS, Brazil; [2]Universidad de Buenos Aires, Buenos Aires, Argentina; [3]Instituto Nacional de Tecnologia Industrial - INTI, Buenos Aires, Argentina; [4]Universidad ORT, Montevideo, Uruguay; [5]LAAS-CNRS / Universite' de Toulouse, Toulouse, France

DSN-4: An Optimizing Technique to Lower Both Phase Noise and Susceptibility of a Voltage Controlled Oscillator — pp. 95-100

Jeremy Raoult, Amable Blain, Sylvie Jarrix
IES, France

PDE: Power Device EMC
Time: Tuesday, 17/Dec/2013: 2:00pm - 3:00pm ·   *Location:* Kinsho Hall

PDE-1: EMC Analysis of Current Source Gate Drivers — pp. 101-106

Alexis Schindler[1], Bernhard Wicht[1], Benno Koeppl[2]
[1]Reutlingen University, Germany; [2]Infineon Technologies AG

PDE-2: EMI Resisting LDO Voltage Regulator with Integrated Current Monitor — pp. 107-112

Philipp Schröter[1], Stefan Jahn[1], Frank Klotz[1], Fabio Ballarin[2], Fabio Gini[2], Marco Piselli[2]
[1]Infineon Technologies AG, Germany; [2]Infineon Technologies Italia

PDE-3: A Study on Gate Voltage Fluctuation of MOSFET Induced by Switching Operation of Adjacent MOSFET in High Voltage Power Conversion Circuit — pp. 113-118

Tsuyoshi Funaki
Osaka University, Japan

Poster Session
Time: Tuesday, 17/Dec/2013: 3:20pm - 5:00pm ·   *Location:* Small Hall

PP-01: Active Magnetic Field Canceling System — pp. 119-122

Wei-li Sun[1], Feng-Chang Chuang[2,3], Yu-Lin Song[4], Chwen Yu[5], Tzyh-Ghuang Ma[1], Tzong-Lin Wu[2], Luh-Maan Chang[3]
[1]Communication and Electromagnetic Engineering, National Taiwan University of Science and Technology, Taipei, Taiwan; [2]Department of Electrical Engineering, National Chung Hsing University, Taichung, Taiwan.; [3]Department of Civil Engineering, National Taiwan University, Taipei, Taiwan; [4]High Technology Research Center, Yen Tjing Ling Industrial Research Institute, Taipei, Taiwan; [5]Taiwan Semiconductor Manufacturing Company, Ltd., Hsinchu, Taiwan.

PP-02: Spread Spectrum Clocking for Emission Reduction of Charge Pump Applications — pp. 123-128

Bernd Deutschmann
Infineon Technologies, Germany

PP-03: Evaluating the Impact of Substrate Noise on Conducted EMI in Automotive Microcontrollers — pp. 129-133

Marco Cazzaniga[1,2], Patrice Joubert Doriol[1], Aurora Sanna[1], Emmanuel Blanc[3], Valentino Liberali[2], Davide Pandini[1]
[1]Central CAD and Design Solutions, STMicroelectronics, Agrate Brianza, Italy; [2]Dipartimento di Fisica, Università degli Studi di Milano, Milano, Italy; [3]Apache Design Inc., Grenoble, France

PP-04: Impedance Balance Control for Suppression of Fluctuation on Ground Voltage in LSI Package — pp. 134-137

Masaaki Maeda, Tohlu Matsushima, Osami Wada
Kyoto University, Japan

PP-05: Automatic Conducted-EMI Microcontroller Model Building — pp. 138-141

Shih-Yi Yuan, Shry-Sann Liao
Feng Chia University, Taiwan, Republic of China

Session Program

PP-06: Evaluation of PDN Impedance and Power Supply Noise for Different On-chip Decoupling Structures   pp. 142-146
Haruya Fujita, Hiroki Takatani, Yosuke Tanaka, Masaomi Sato, Shohei Kawaguchi, Toshio Sudo
Shibaura Institute of Technology, Japan

PP-07: Characterization of Conducted Emission at High Frequency under Different Temperature   pp. 147-151
Néstor Berbel, Raúl Fernández-García, Ignacio Gil
Universitat Politècnica de Catalunya (UPC), Spain

PP-08: Using the EM Simulation Tools to Predict the Conducted Emissions Level of a DC/DC Boost Converter : Introducing EBEM-CE Model   pp. 152-157
Andre Durier[1], Olivier Crepel[2], Christian Marot[3]
[1]CONTINENTAL AUTOMOTIVE FRANCE, France; [2]EADS IW, France; [3]EADS IW, France

PP-09: Design of Contactless Wafer-level TSV Connectivity Testing Structure using Capacitive Coupling   pp. 158-162
Jonghoon J. Kim[1], Heegon Kim[1], Sukjin Kim[1], Bumhee Bae[1], Daniel H. Jung[1], Sunkyu Kong[1], Junho Lee[2], Kunwoo Park[2], Joungho Kim[1]
[1]KAIST, Korea, Republic of (South Korea); [2]SK Hynix Semiconductor Inc., Republic of (South Korea)

PP-10: Modeling and Analysis of Open Defect in Through Silicon Via (TSV) Channel   pp. 163-166
Daniel H. Jung[1], Heegon Kim[1], Jonghoon J. Kim[1], Hyun-Cheol Bae[2], Kwang-Seong Choi[2], Joungho Kim[1]
[1]KAIST, Korea, Republic of (South Korea); [2]ETRI, Korea, Republic of (South Korea)

PP-11: The Direct RF Power Injection Method up to 18 GHz for Investigating IC's Susceptibility   pp. 167-170
Yin-Cheng Chang[1], Shawn S. H. Hsu[2], Yen-Tang Chang[3], Chiu-Kuo Chen[3], Hsu-Chen Cheng[1], Da-Chiang Chang[1]
[1]National Chip Implementation Center, National Applied Research Laboratories, Hsinchu, Taiwan; [2]Institute of Electronics Engineering, National Tsing Hua University, Hsinchu, Taiwan; [3]Bureau of Standards, Metrology and Inspection, M.O.E.A, Taipei, Taiwan

Date: Wednesday, 18/Dec/2013

MDL: Power Integrity and Conducted Emission Modeling   *Time:* Wednesday, 18/Dec/2013: 8:50am - 10:30am ·   *Location:* Kinsho Hall

MDL-01: Anti-resonance Peak Frequency Control by Variable On-die Capacitance   pp. 171-174
Wataru Ichimura, Sho Kiyoshige, Masahiro Terasaki, Ryota Kobayashi, Genki Kubo, Hiroki Otsuka, Toshio Sudo
Shibaura Institute of Technology, Japan

MDL-02: Estimation of Data-dependent Power Voltage Variations of FPGA by Equivalent Circuit Modeling from On-board Measurements   pp. 175-179
Kengo Iokibe, Yoshitaka Toyota
Okayama Univ, Japan

MDL-03: Microcontroller Emission Simulation based on Power Consumption and Clock System   pp. 180-185
Thomas Steinecke
Infineon Technologies AG, Germany

MDL-04: A Microcontroller Conducted EMI Model Building for Software-level Effect   pp. 186-189
Shih-Yi Yuan
Feng Chia University, Taiwan, Republic of China

MDL-05: Characterization and Modeling of Electrical Stresses on Digital Integrated Circuits Power Integrity and Conducted Emission   pp. 190-195
Alexandre BOYER, Sonia BEN DHIA
LAAS-CNRS, France

Session Program

ERE: ESD and Robustness Evaluation in IC-level
Time: Wednesday, 18/Dec/2013: 11:00am - 12:00pm ·   *Location:* Kinsho Hall

ERE-1: System-ESD Validation of a Microcontroller with External RC-Filter   pp. 196-201
Thomas Steinecke[1], Markus Unger[1], Stanislav Scheier[2], Stephan Frei[2], Josip Bačmaga[3], Adrijan Barić[3]
[1]Infineon Technologies AG, Germany; [2]Technical University of Dortmund, Germany; [3]University of Zagreb, Croatia

ERE-2: Automatic Verification of EMC Immunity by Simulation   pp. 202-207
Bertrand Vrignon[1], Pascal Caunegre[1], John Shepherd[1], Jianfei Wu[2]
[1]Freescale, France; [2]National University of Defense Technology, China

ERE-3: Electro-Magnetic Robustness of Integrated Circuits: from statement to prediction   pp. 208-213
Sonia Ben Dhia, Alexandre Boyer
INSA - LAAS, France

IM1: Chip Level Immunity
Time: Wednesday, 18/Dec/2013: 2:00pm - 3:20pm ·   *Location:* Kinsho Hall

IM1-1: EMC Immunity of Integrated Smart Power Transistors in a non-50Ω Environment   pp. 214-219
Hermann Nzalli[1], Wolfgang Wilkening[1], Rolf H. Jansen[2]
[1]Robert Bosch GmbH, Germany; [2]Chair of EM Theory (RWTH Aachen University), Germany

IM1-2: Discrete Low-frequency Transistors Subjected to High-frequency CW and Pulse-modulated Sine Signals   pp. 220-225
Sylvie JARRIX[1], Jeremy RAOULT[1], Adrien DORIDANT[1], Clovis POUANT[2], Patrick HOFFMANN[2]
[1]Institut d 'Electronique du Sud, France; [2]CEA Gramat, France

IM1-3: Noise Analysis using On-chip Waveform Monitor in Bandgap Voltage References   pp. 226-231
Akitaka Murata[1], Shuji Agatsuma[1], Daisaku Ikoma[1], Kouji Ichikawa[1], Takahiro Tsuda[1], Makoto Nagata[2], Kumpei Yoshikawa[2], Yuuki Araga[2], Yuji Harada[2]
[1]DENSO CORPORATION, Japan; [2]Kobe University, Japan

IM1-4: Immunity Evaluation of Inverter Chains against RF Power on Power Delivery Network   pp. 232-237
Kumpei Yoshikawa[1], Yuji Harada[1], Noriyuki Miura[1], Noriaki Takeda[2], Yoshiyuki Saito[2], Makoto Nagata[1,3]
[1]Kobe University, Japan; [2]Panasonic Corporation, Japan; [3]CREST, JST

IM2: Automotive Immunity
Time: Wednesday, 18/Dec/2013: 3:50pm - 4:50pm ·   *Location:* Kinsho Hall

IM2-1: Magnetic Field Coupling on Analog-to-digital Converter from Wireless Power Transfer System in Automotive Environment   pp. 238-242
Bumhee Bae, Sunkyu Kong, Joonghoon J Kim, Sukjin Kim, Jongho Kim
KAIST, Korea, Republic of (South Korea)

IM2-2: Immunity Simulation Method for Automotive Power Module using Electromagnetic Analysis   pp. 243-248
Yosuke Kondo, Kei Tsunada, Norimasa Oka, Masato Izumichi
DENSO CORPORATION, Japan

IM2-3: Translation of Automotive Module RF Immunity Test Limits into Equivalent IC Test Limits using S-parameter IC Models   pp. 249-253
Hugo Pues[1], Ben Briké[1], Celina Gazda[1], André Durier[2], Dries Vande Ginste[3], Peter Teichmann[1], Kristof Stijnen[1], Christian Peeters[1]
[1]Melexis Technologies, Tessenderlo, Belgium; [2]Continental Automotive France, Toulouse, France; [3]Ghent University, INTEC, Gent, Belgium

Chair's Welcome

Welcome to the 9th International Workshop on Electromagnetic Compatibility of Integrated Circuits – EMC Compo 2013 in the "Historic Monuments of Ancient Nara", a UNESCO World Heritage Site. EMC Compo 2013 is the first workshop to be held outside Europe, and it's indeed a great honor for me as a General Chair to have this workshop in Japan with the cooperation of many people.

Another important feature of EMC Compo 2013 is that it is prepared and organized as a series Symposium/Workshop with the "2013 IEEE Electrical Design of Advanced Packaging & Systems Symposium (EDAPS 2013)", which is one of the IEEE CPMT Society Symposiums, chaired by Prof. Hideki Asai of Shizuoka University. Technical topics in EMC and Advanced Packaging are closely related and we share common issues and challenges to create new frontiers. I hope the attendees of both EDAPS and EMC Compo will have fruitful communications and discussions here at Todai-ji, Nara.

We appreciate very much the support of the technical and financial sponsors, and also thank the authors and participants, and all the committee members. Please enjoy Nara.

Osami Wada, *General Chair of EMC Compo 2013*

Program Overview:

The technical program starts with a **Joint Workshop** together with **EDAPS 2013**, and we have three guest speakers, Prof. Madhavan Swaminathan from the Georgia Institute of Technology, Dr. Woong Hwan Ryu from Samsung Electronics, and Dr. William Wu Shen from TSMC. The topics cover Signal/Power/Thermal Integrity, high-performance mobile systems, and 3D IC design. An introduction to a new activity in Japan, the LPB interoperable design process working group (LPB-WG), will also be presented. LPB stands for the entire set of LSI, Packages and Boards, which will be co-designed for high-performance electronic systems.

After the Joint Workshop, EMC Compo 2013 will start with two **tutorials**, and a **Keynote Speech** by Dr. Toru Shimizu from Renesas Electronics. Forty eight **regular papers** in nine Regular Sessions, one Special Session, and a Poster Session including eleven poster presentations will follow. In the **Special Session**, "Tackling EMC in Real IC Chips", two guests, Prof. Jun Fan and Dr. Dian Yang, will give talks. A **workshop** on new EM noise evaluation methods with APD measurements will also be presented.

15-18
DEC
2013

Proceedings of the 9TH International Workshop on
Electromagnetic Compatibility of Integrated Circuits

EMC COMPO 2013 *NARA, JAPAN*

Organized by

Graduate School of Engineering,
Kyoto University, Japan

Technically cosponsored by

IEEE EMC Society
IEEE EMC Society Japan Chapter
IEEE EMC Society Sendai Chapter

Editors: Yoshitaka Toyota, Kengo Iokibe, Tohlu Matsushima, Osami Wada

2013 Committee

Organizing Committee

General Chair:
Osami Wada,   Kyoto University,   Japan

TPC Co-Chairs:
Toshio Sudo,   Shibaura Institute of Technology,   Japan
Kouji Ichikawa,   DENSO CORPORATION,   Japan

Organized Session Chair:
Makoto Nagata,   Kobe University,   Japan

Tutorial Chair:
Hideki Sasaki,   Renesas Electronics,   Japan

Publication Co-Chairs:
Yoshitaka Toyota,   Okayama University,   Japan
Kengo Iokibe,   Okayama University,   Japan

Publicity Chair:
Yuichi Hayashi,   Tohoku University,   Japan

Financial Chair:
Yoshiyuki Saito,   Panasonic Corporation,   Japan

IEEE Liaison Chair:
Liuji R. Koga,   Okayama University,   Japan

Industry Liaison Chair:
Atsushi Nakamura,   Renesas Electronics   Japan

Local Arrangement Chair:
Sadahiro Tani,   ST-Lab   Japan

Secretary:
Tohlu Matsushima,   Kyoto University,   Japan

Secretary Assistant:
Akemi Iwata,   Kyoto University,   Japan

Technical Program Committee

Jean-Michel Redoute,	Monash University,	Australia
Renaud Gillon,	ON Semiconductor,	Belgium
Hugo Pues,	Melexis,	Belgium
Bernd Deutschmann,	Infineon Technologies,	Germany
Uwe Keller,	Zuken,	Germany
Frank Klotz,	Infineon Technologies,	Germany
Thomas Steinecke,	Infineon Technologies,	Germany
Wolfgang Wilkening,	Robert Bosch GmbH,	Germany
Sonia Ben Dhia,	INSA – LAAS,	France
Alexandre Boyer,	LAAS-CNRS,	France
Frederic Lafon,	Valeo,	France
Jean-Luc Levant,	Atmel,	France
Christian Marot,	EADS France,	France
Richard Perdriau,	ESEO,	France
Mohamed Ramdani,	ESEO,	France
Etienne Sicard,	INSA,	France
Bertrand Vrignon,	Freescale Semiconductor,	France
Adrijan Baric,	University of Zagreb,	Croatia
Franco Fiori,	Politecnico di Torino,	Italy
Mauro Merlo,	STMicroelectronics,	Italy
Davide Pandini,	STMicroelectronics,	Italy
Umberto Paoletti,	Hitachi, Ltd.,	Japan
Masahiro Yamaguchi,	Tohoku University,	Japan
Joung Ho Kim,	KAIST,	Korea
HarkByeong Park,	Samsung Electronics,	Korea
Hyun Ho Park,	The University of Suwon,	Korea
Mart Coenen,	EMCMCC,	Netherlands
Jan Niehof,	NXP Semiconductors,	Netherlands
ErPing Li,	A*STAR,	Singapore
Tzong-Lin Wu,	National Taiwan University,	Taiwan
Shih-Yi Yuan,	Feng Chia University,	Taiwan
Todd Hubing,	Clemson University,	U.S.A.
David Pommerenke,	Missouri S & T University,	U.S.A

A technique for estimating signal waveforms at inaccessible points in high speed digital circuits

Tomohiro Kinoshita, Shoichi Hara, Eiji Takahashi

Panasonic Corporation
R&D Division, Device Solutions Center
1006 Kadoma, Kadoma City, Osaka, Japan
kinoshita.tomohiro@jp.panasonic.com

Kazuhide Uriu

Panasonic Corporation
Automotive & Industrial Systems Company
Corporate Business Development Division
Industrial Business Development Center

Abstract— **There is a large difference between the waveforms of high speed digital signals measured on a printed circuit board and those at the receiving terminals of LSI chips. In order to ensure that the high speed digital signal is correctly transferred, one of the major issues is how to estimate the waveform at the receiving terminal, where the waveform cannot be directly measured. Therefore, we have developed a new technique for estimating the waveform at the receiving terminal from the waveform measured at an accessible test point, based on the input impedance and transmission characteristics of the traces. By comparing the estimated waveform and the measured waveform in a Blu-ray Disc recorder system, it was confirmed that it is possible to estimate the waveform with high accuracy. Furthermore, it was confirmed that the estimated waveform has good correlation with the timing margin for correct data transfer.**

Keywords—signal integrity; simulation; high speed digtal circuits; timing margin; eye diagram; jitter; LSI package; DDR3

I. INTRODUCTION

Higher performance and lower power consumption for digital equipment are ever in demand, so the transfer rate of digital interfaces is continually becoming higher while voltage amplitudes are increasingly becoming lower. Therefore, noise margins are decreasing and malfunctions occur more frequently. When a malfunction occurs, it is very important to observe the signal waveform. In reference [1][2], a technique for evaluating high speed digital signals was proposed. However, since signal frequencies are very high, waveforms observed at measurement points on a printed circuit boards (PCB) are quite different from the waveforms at the receiving terminals on Large Scale Integrated (LSI) circuit chips. Additionally, since LSI circuits are enclosed in packages, it is impossible to measure the signal waveform directly at the receiving points on an LSI chip. As a result, the time required to analyze any malfunction is fairly lengthy.

To address this problem, we have developed a method to estimate the waveform at an arbitrary point using the waveform measured on the PCB and the transmission characteristics of the traces, which include the traces in the package.

II. ALGORITHM FOR ESTIMATING THE WAVEFORM AT AN ARBITRARY POINT

Fig.1 shows a two-terminal-pair network. In Fig.1, the relationship between the voltages and currents at the two points can be represented by (1), using a fundamental matrix (F-matrix) that describes the characteristics of the two-terminal-pair network. Here, V_1 and I_1 are the voltage and current in the frequency domain at point 1, respectively. V_2 and I_2 are the equivalent voltage and current at point 2. The expression for V_2 in equation (2) is obtained from the inverse of the F-matrix, and equation (3) is obtained by substituting Z_1 into equation (2), where Z_1 ($=V_1/I_1$) is the input impedance at point 1. Assuming that points 1 and 2 are the point at which the signal is measured and the arbitrary point, respectively, equation (3) signifies that we can estimate the waveform at the arbitrary point (V_2) from the F-matrix, the input impedance (Z_1) and the measured waveform (V_1).

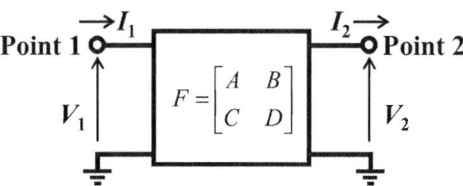

Fig. 1. Two-terminal pair network.

$$\begin{pmatrix} V_1 \\ I_1 \end{pmatrix} = \begin{bmatrix} A & B \\ C & D \end{bmatrix} \begin{pmatrix} V_2 \\ I_2 \end{pmatrix} \quad (1)$$

$$V_2 = D \cdot V_1 - B \cdot I_1 \quad (2)$$

$$V_2 = (D - B / Z_1) \cdot V_1 \quad (3)$$

The F-matrix can be calculated with high accuracy using an already established analytical method [3]. The measured waveform (V_1) can be obtained using an oscilloscope. The input impedance (Z_1) can be measured at point 1 or can be calculated using an electromagnetic analysis tool and semiconductor device model. Therefore, the waveform at the arbitrary point (V_2) can be estimated from equation (3).

III. VERIFICATION OF THE ACCURACY OF THE METHOD

To demonstrate that the technique can be used to help analyze malfunctions, we verified the accuracy of it. We applied the technique to the DDR3 memory data bus of a Blu-ray Disc (BD) recorder system. Fig.2 shows a part of the BD recorder system which includes a DDR3 memory and a system LSI (sLSI) chip. TP1 (Testpoint1) and TP2 (Testpoint2) in Fig.2 are points on the data bus that can be monitored. The

978-1-4799-5004-1/13 $31.00 © 2013 IEEE

data bus has a transfer rate of 1.333 Gbps. We estimated the waveform at TP2 from the measured waveform at TP1, and compared the estimated waveform with the measured waveform at TP2 to verify the accuracy.

First, we measured the impedance Z_{tp1} on the sLSI chip side of TP1 using a network analyzer (Agilent Technologies E5071C). Normally, the impedance of an electrical circuit including an LSI chip cannot be measured stably due to voltage fluctuations caused by the switching current of the chip. Therefore, in order to keep the switching current to a negligible level, we applied the following two conditions when the impedance was measured:

1. The data bus of the DDR3 interface with the sLSI chip was fixed in the read data mode.

2. The trace was cut off at TP1 to separate the DDR3 memory from the target data bus.

Secondly, the voltage at TP1 (V_{tp1_mes}) was measured using an oscilloscope (Agilent technologies DSA-X 92004A). Then, $V_{tp1\_mes-in-f}$ can be calculated by converting V_{tp1_mes} into the frequency domain.

Thirdly, the voltage at TP2 ($V_{tp2\_sim-in-f}$) was estimated in the frequency domain by substituting Z_{tp1} and $V_{tp1\_mes-in-f}$ for Z_1 and V_1 in equation (3), respectively.

Finally, the voltage at TP2 (V_{tp2_sim}) was obtained by converting $V_{tp2\_sim-in-f}$ into the time domain. Fig. 3 shows the estimated result as well as the measured voltages (V_{tp1_mes} and V_{tp2_mes}).

The plots in Fig.3 confirm that there is good agreement between V_{tp2_sim} and V_{tp2_mes}, demonstrating the accuracy of our proposed technique.

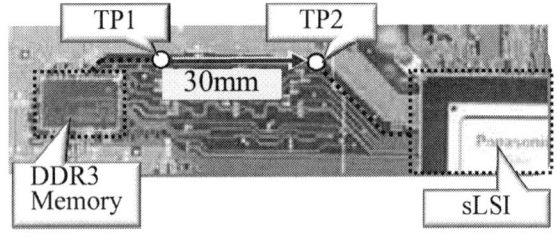

Fig. 2. Part of the BD recorder system used for verification.

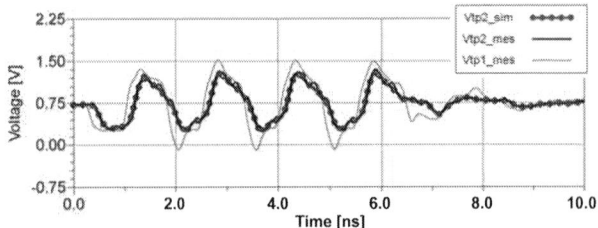

Fig. 3. Comparison of the estimated waveform (V_{tp2_sim}) and the measured waveform (V_{tp2_mes}).

IV. ESTIMATED WAVEFORM AT THE RECEIVING TERMNAL OF THE LSI CHIP

In this section, we estimate the waveform at the receiving terminal of an LSI chip using our proposed technique. However, since we cannot directly observe the waveform at the receiving terminal, we verified the accuracy of the estimated waveform by observing the relationship between the estimated waveform and the timing margin for correct data transfer. In order to observe this relationship under different conditions, we estimated the waveform and measured the timing margin while introducing distortion to the waveform by injecting RF power noise to the LSI.

A. Evaluation system and how to control the distortion of the output signal waveform

Fig.4 shows the system we used for evaluation. A sLSI circuit for media processing and a DDR3 memory are mounted on the PCB. The sLSI chip has a DDR3 interface controller with a data transfer rate of 1.333Gbps. We estimated the waveform at the receiving point from the measured waveform at the measurement point shown in Fig.4. Deliberately injecting RF power noise into the DC power supply of the DDR3 memory interferes with the waveform at the measurement point.

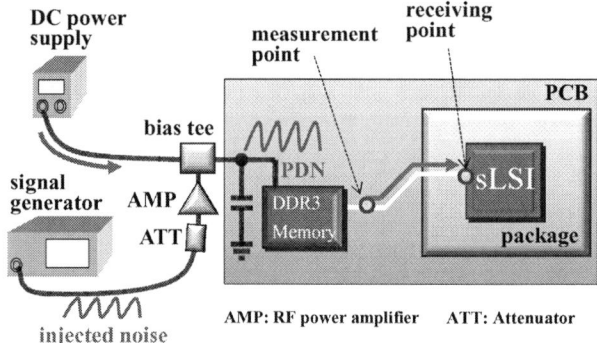

Fig. 4. Evaluation system.

The power distribution network (PDN) of the DDR3 memory is separate from the PDN of the sLSI on the PCB. DC power is supplied to the PDN of the DDR3 memory from the off-board DC power supply. Since a signal generator (Agilent Technologies, E8254A), an attenuator (-20dB) and an RF power amplifier (Mini-Circuits, ZHL-42W, 30dB) are connected to the PDN of the DDR3 memory via a bias tee, RF power noise can be intentionally injected into the PDN of the DDR3 memory. The noise source is a sinusoidal wave.

The noise injected into the PDN causes the voltage supplied to the DDR3 memory to fluctuate, producing some distortion in the waveform of the output signal from the DDR3 memory.

We observed the relationship between the RF power noise injected into the PDN and the distortion in the output signal waveform. First, we examined the relationship between the frequency of the RF power noise and the distortion in the output signal waveform. We used the jitter in the output signal as a parameter for the distortion. We measured the jitter while sweeping the frequency of the RF power noise from 10MHz to 3GHz in 10MHz steps. Despite this RF power noise, the jitter at 3GHz didn't increase. The jitter was measured on 0.75V, which is the midpoint voltage of the data bus signal of the DDR3 interface. As a result, the jitter in the output signal was at its largest when the frequency of the RF power noise was 1000 MHz. We surmised that the insertion loss from the point at which the noise is coupled in to the pads of the power pins might be low at this frequency. So, we fixed the noise frequency to 1000MHz after this.

Next, we measured the jitter while increasing the amplitude of the RF power noise. Fig.5 shows the relationship between the voltage of the noise injected in and the jitter in the output signal. In Fig.5, the x-axis shows the amplitude of the noise, which is the output voltage displayed on the monitor of the signal generator.

In Fig.5, it can be seen that the jitter increases almost monotonically with the amplitude of the noise. Thus, it is possible to control the distortion in the output signal waveform by varying the amplitude of the RF power noise. This means that we can create different conditions by injecting RF power noise into the PDN of the DDR3 memory.

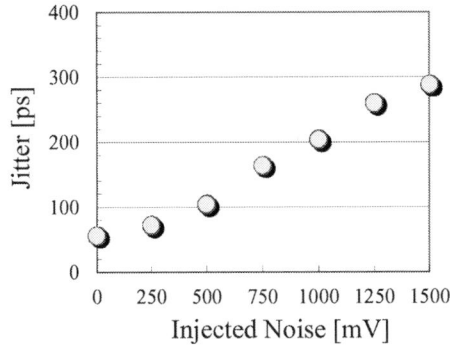

Fig. 5. Relationship between the noise amplitude and the jitter in the output signal.

B. Relationship between the timing margin and the jitter in the output signal

We measured the timing margin for data transfer in the DDR3 interface while injecting RF power noise into the PDN of the DDR3 memory. The timing margin is the period in the time window in which the data can be correctly transferred from the DDR3 memory to the sLSI chip. It is expected that jitter in the output signal decreases the timing margin. The sLSI chip has a function by which the latch timing for the data read from the DDR3 memory can be changed. By using this function, we measured the timing margin while gradually shifting the latch timing. Fig.6 shows the variation in the measured timing margin with the amplitude of the noise injected into the PDN of the DDR3 memory. Fig.6 confirms that the timing margin decreases with the RF power noise.

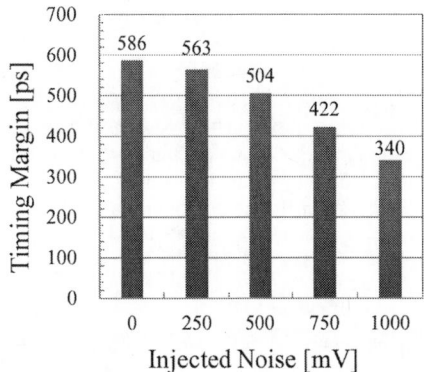

Fig. 6. Relationship between the noise amplitude and the timing margin.

C. Verification of the accuracy of the technique

In order to verify the accuracy of the proposed technique, we compared the timing margin measured in the previous section with two different kinds of eye opening. One eye opening was extracted from the eye diagram derived from waveforms measured at the measurement point. The other was extracted from the eye diagram derived from estimated waveforms at the receiving terminal using the techniques described in section III. Since the timing margin is the time window in which data can be transfered correctly, the timing margin decreases as the jitter becomes greater. Therefore, it is supposed that there is a correlation between the timing margin and the eye opening.

First, we plotted the eye diagrams of measured waveforms at the measurement point. Fig.7 (a) shows the eye diagram without the RF power noise. The eye opening is about 509ps and the eye height is about 590mV. Fig.7 (b) shows the eye diagram with 500mV noise. The eye opening is about 108ps and the eye height is about 150mV. By comparing these eye diagrams, we can confirm that there is a large amount of distortion in the data signal due to the RF power noise. On the other hand, as can be seen in Fig.6, despite the 500mV noise, the timing margin is not much reduced. The timing margin without the RF power noise is about 586ps and with 500mV noise it is about 504ps. Thus, the eye openings are quite different from the timing margins in Fig.6. In particular, the timing margin (504ps) is more than four times greater than the eye opening (108ps) with 500mV noise.

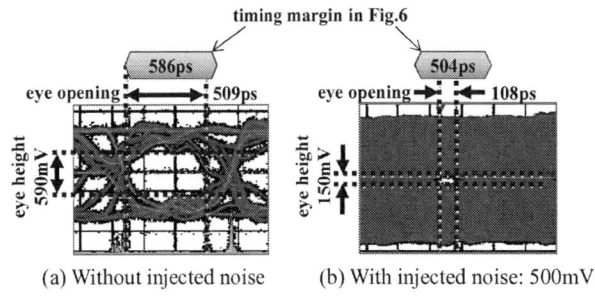

(a) Without injected noise   (b) With injected noise: 500mV

Fig. 7. Eye diagrams from the measured waveforms at the measurement point.

Next, we plotted the eye diagrams of the estimated waveforms at the receiving terminal. Fig.8 (a) shows the eye diagram without the RF power noise. The eye opening is about 605ps and the eye height is 1000 mV. Fig.8 (b) shows the eye diagram with 500mV noise. The eye opening is about 489ps and the eye height is 660mV. This shows that, using the proposed estimation technique, the timing margins (586ps and 504ps) are quite similar to the eye openings (605ps and 489ps).

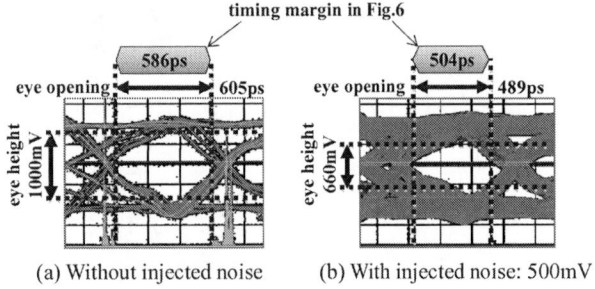

(a) Without injected noise   (b) With injected noise: 500mV

Fig. 8. Eye diagrams from the estimated waveforms at the receiving terminal using the proposed estimation technique.

Eye openings were extracted from eye diagrams in the same way described above while changing the noise from 0mv to 1000mV in 250mV steps. Fig.9 shows a comparison of the timing margin, the eye openings extracted from the eye diagrams from the estimated waveforms, and the eye openings extracted from the eye diagrams based on the measured waveforms at the measurement point.

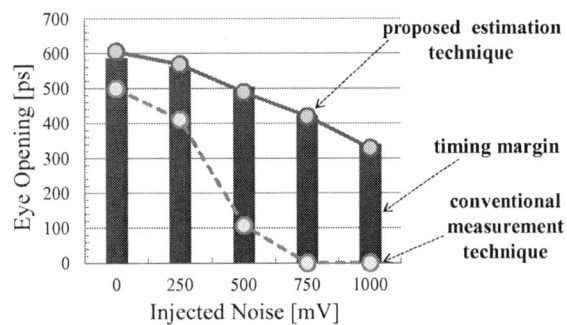

Fig. 9. Comparison between the eye openings and the timing margin.

In Fig.9, both the timing margin and the eye openings decrease as the noise increases. However, in the case of the eye openings (broken line) based on the measured waveforms, the reduction is quite different from that of the timing margin. On the other hand, the eye openings (solid line) based on the estimated waveforms are very similar to the timing margins. This demonstrates that the waveform at the receiving point can be estimated with high accuracy using our proposed technique.

V. CONCLUSION

We have developed a technique for estimating the waveform at an arbitrary point. The technique is based on the measured waveform on a PCB and the transmission characteristics of the traces. In this study, we verified the accuracy of this technique by applying it to the DDR3 memory data bus of a BD recorder system.

Furthermore, we evaluated the accuracy of the estimated waveforms by confirming the close relationship between the estimated waveform and the timing margin for data transfer. In order to confirm the close relationship between these under different conditions, we introduced some distortion into the waveform by injecting some RF power noise into the power distribution network of the DDR3 memory. As a result, while the eye openings extracted from the measured eye diagram were quite different from the timing margins, the eye openings based on the estimated waveforms were nearly equal to the timing margins. Therefore, we are able to confirm that a highly accurate estimation of the waveform at an arbitrary point can be obtained using our technique. Even if the arbitrary point is inaccessible, such as a bonding pad on an LSI chip, the waveform can be estimated. Therefore, the time required to analyze a malfunction can be considerably shortened.

REFERENCES

[1] Y. Uematsu, H. Osaka, Y. Nishio and S. Hatano, " A Method for Measuring Vref Noise Tolerance of DDR2-SDRAM on Test Board That Simulates an Actual Memory Module," pp.65-69, IEICE Technical Report CPM2007-139, 2008.

[2] J. Fan, Xiaoning Ye, Jingook Kim, Archambeault B, Orlandi A, "Signal Integrity Design for High-Speed Digital Circuits: Progress and Directions," pp.392-400, IEEE Transactions, Electromagnetic Compatibility, 52(2), 2010.

[3] M. Yamaoka, K. Uriu and T. Yamada, "Design and Analysis Methodology of Electronic Devices using a Novel Electromagnetic Tool 'MomCACE'," pp3-4, Proceedings of JIEP Annual Meeting, 2009.

Design and Measurement of a Compact On-interposer Passive Equalizer for Chip-to-chip High-speed Differential Signaling

Heegon Kim, Jonghyun Cho, Daniel H. Jung,
Jonghoon J. Kim, Sumin Choi and Joungho Kim
Korea Advanced Institute of Science and Technology
TERA Lab
Daejeon, South Korea
heegon87@eeinfo.kaist.ac.kr

Junho Lee and Kunwoo Park
SK Hynix Semiconductor Inc.
Advanced Design Team
Icheon, South Korea

Abstract—In this paper, a compact on-interposer passive equalizer for chip-to-chip high-speed differential signaling was proposed and experimentally verified. By using the parasitic resistance and inductance of the coil-shaped on-interposer metal line, the proposed on-interposer passive equalizer achieves not only the wide-band equalization but also the compact size. Moreover, the symmetric structure of the proposed equalizer maintains the balance between the differential signals. The remarkable performance of the proposed on-interposer passive equalizer for differential signaling was successfully verified by a frequency- and time-domain measurement of up to 10 Gbps.

Keywords – Inter-symbol Interference (ISI), On-interposer passive equalizer, Silicon-interposer, Differential data transmission

I. INTRODUCTION

Recently, the required system bandwidth for high-speed application has dramatically increased. To satisfy the industrial demands, the TSV-based interposer that enables not only the large number of I/Os but also the wide channel bandwidth by reducing chip-to-chip interconnect length gradually replaces the conventional package. Especially, the silicon interposer has been widely developed because of its outstanding advantages, such as the low-cost of fabrication, the high thermal conductivity, and the fine pitch patterning [1-2]. Though the silicon interposer enables the fine pitch patterning, the number of I/Os should be limited by the routing issues, resulting in the requirement of the wider bandwidth per channel. The differential signaling, that transmits two complementary signals with the opposite phases, effectively improves the channel bandwidth by rejecting the common-mode noise. However, the inter-symbol interference (ISI) of the channel is hard to compensate by the differential signaling. As the data rate of the transmitted signal increases, the inherent frequency-dependent loss of the silicon substrate causes the considerable ISI, resulting in the degradation of the on-interposer channel bandwidth. An active equalizer is the general and the most conventional way to counteract the ISI problem. However, it has several decisive problems, such as limited bandwidth and excessive I/O power consumption. A passive equalizer overcomes those problems and enables the wide bandwidth equalization and low power consumption. Though the passive

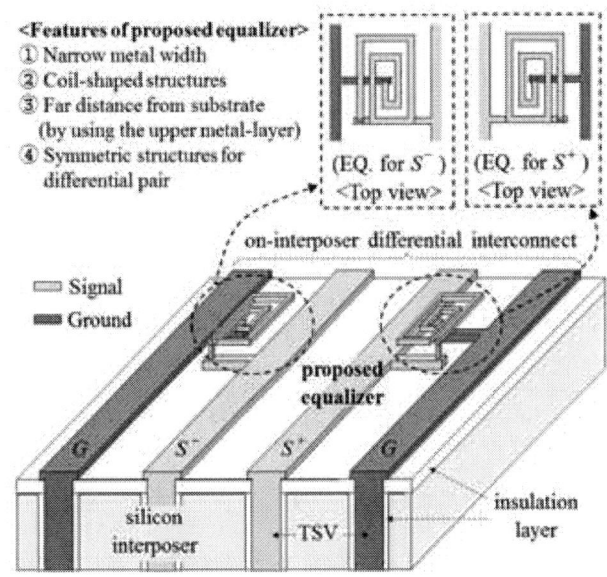

Figure 1. Structure of the proposed on-interposer passive equalizer for chip-to-chip high-speed differential signaling.

equalizer is suitable more than the active one for the higher on-interposer channel bandwidth, it consumes the relatively large area. Several passive equalizer designs that can be implemented on the interposer have also been studied [3-5]. However, a passive equalizer described in [3] requires significant process support and the TSV equalizer described in [4] has low adjustability for various channel loss compensation. Moreover, the on-interposer passive equalizer describes in [5] only covers the single-ended channel and the performance of the equalizer is not experimentally verified. Therefore, a new passive equalizer design for chip-to-chip high-speed differential signaling for the on-interposer channel and its experimental verification are highly desirable.

In this paper, a compact on-interposer passive equalizer design for chip-to-chip high-speed differential signaling is proposed and experimentally verified. By using the parasitics of the coil-shaped on-interposer metal line with a narrow width, the proposed on-interposer passive equalizer achieves the

---

This work was supported by the Smart IT Convergence System Research Center funded by the Ministry of Education, Science and Technology as Global Frontier Project (STRC-2011-0031863) and by the R&D program of ISTK [Development of an image-based, real-time inspection and isolation system for hyperfine faults].

978-1-4799-5004-1/13 $31.00 © 2013 IEEE

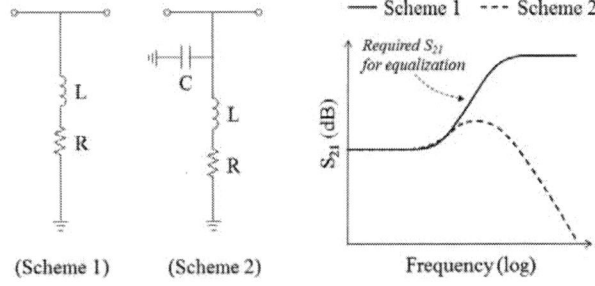

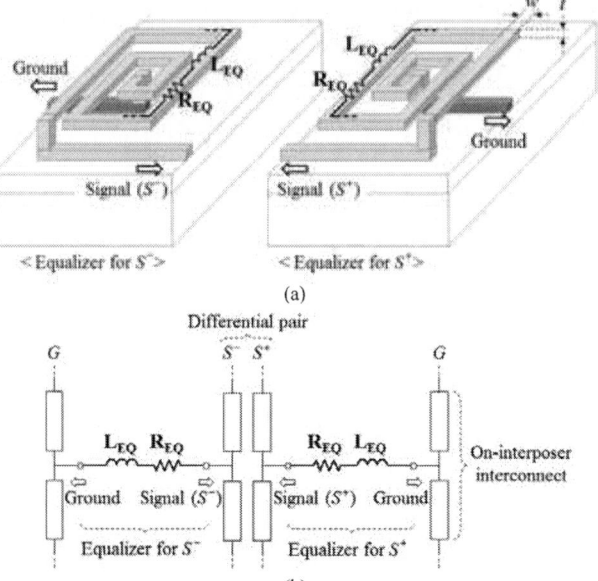

Figure 2. Simplified schematics and S_{21} of shunt metal lines with and without the capacitance model. S_{21} of Scheme 1 is required for equalization.

wide-band equalization and small area consumption. In addition, the coil-shaped structure in the proposed equalizer enables the high adjustability by varying the turn number of coil. The symmetric structure of the proposed equalizer maintains the balance between the differential signals, resulting in the suitable common-mode rejection. For experimental verification of the proposed equalizer, the several test vehicles were fabricated and measured in frequency- and time-domains for data rates of up to 10 Gbps. By adding the proposed equalizer to the on-interposer channel, the quality of the transmitted differential signals on interposer was remarkably improved. The proposed on-interposer passive equalizer design enables the very wide channel bandwidth for the differential lines in a interposer-based 2.5-D or 3-D IC system.

II. STRUCTURE OF PROPOSED ON-INTERPOSER PASSIVE EQUALIZER FOR DIFFERENTIAL SIGNALING

The structure of the proposed on-interposer passive equalizer for chip-to-chip high-speed differential signaling is shown in Fig. 1. Target on-interposer differential interconnect is located on an M1 metal layer. The signaling type of the target on-interposer interconnect is Ground-Signal-Signal-Ground (GSSG). The differential signals (S^+ and S^-) have the ground lines on either sides as the guard traces. The proposed on-interposer passive equalizer consists of the two compact coil-shaped structures with the M2 metal layers. Each coil-shaped structures are embedded on the areas between the each signal and ground lines as shown in Fig. 1. The proposed on-interposer passive equalizer structures connect the signal lines to the near ground lines.

The proposed on-interposer passive equalizer is based on a high-pass filter that uses the parasitic resistance and inductance of the shunt on-interposer metal line. Simplified schematics of the shunt metal lines and their frequency responses (S_{21}) are shown in Fig. 2. Schemes 1 and 2 in Fig. 2 show the schematics without and with the parasitic capacitance model, respectively. In the case of Scheme 1, the S_{21} represents that of a high-pass filter. At low frequencies, the S_{21} level is reduced depending on the amount of the parasitic resistance. As the frequency increases, the impedance of the parasitic inductance becomes considerable, resulting in the increased S_{21} level. However, the parasitic capacitance of the shunt metal lines should be considered, as shown in Scheme 2, for a realistic case [6]. When considering the parasitic capacitance, the response level in the high frequency range decreases. This is because the low impedance of the parasitic capacitance at high

Figure 3. (a) Parasitic resistances and inductances of the proposed structure and (b) the simplified equivalent-circuit model of the proposed on-interposer passive equalizer for differential signaling.

frequencies reduces the impedance of the shunt metal line. If the shunt metal line is designed to reduce the parasitic capacitance, the S_{21} of the shunt metal line is closer to that of Scheme 1 and therefore it acts like an equalizer.

The structure of the proposed on-interposer passive equalizer for differential signaling has four features, as shown in Fig. 1: a narrow metal line width, a coil-shaped structures, a far distance from the silicon substrate by using the upper metal-layer, and symmetric structures for differential pair. The narrow metal width of the proposed equalizer structure leads to the considerably high parasitic resistance and inductance. Moreover, the coil-shaped structure more enhance the amount of the parasitic inductance of the proposed structure. The narrow metal width and the far distance from the silicon substrate of the proposed equalizer structure achieve the negligible parasitic capacitance. Therefore, the proposed structure in Fig. 1 is simply modeled as the Scheme 1 in Fig. 2 and it acts like a passive equalizer. In order to balance the impacts of the proposed equalizer to the differential signals, the proposed on-interposer passive equalizer has the symmetric structures for differential pair. The embedded position of the two symmetric coil-shaped structures in the proposed equalizer should also be symmetrical because the different embedded position can lead to the critical problems such as the DC voltage imbalance between the differential signals.

$$R_{EQ} \approx \rho_{metal} \times \frac{l}{t \times w} \quad (1)$$

The equivalent-circuit model of the proposed on-interposer passive equalizer for differential signaling is shown in Fig. 3. The parasitic resistances and inductances of the proposed equalizer structure are shown in Fig. 3(a) and the simplified equivalent-circuit model using those parasitics is shown in Fig. 3(b). The parasitic capacitance is ignored in the simplified equivalent-circuit model due to its negligible amount. The

978-1-4799-5004-1/13 $31.00 © 2013 IEEE

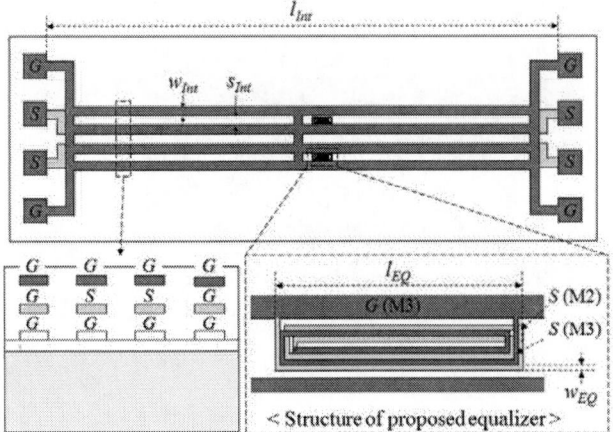

Figure 4. Structure of designed test vehicle including the proposed on-interposer passive equalizer for differential signaling. Lengths of the target on-interposer differential interconnect (l_{int}) and the turn number of the coil in the proposed structure are varied.

TABLE I. PHYSICAL DIMENSIONS OF DESIGNED TEST VEHICLE

Symbol	Value	Symbol	Value
l_{Int}	4500 μm	l_{EQ}	220 μm (@ 1-turn coil)
	450 μm		110 μm (@ 2-turn coil)
w_{Int}	10 μm		63 μm (@ 4-turn coil)
s_{Int}	10 μm	w_{EQ}	1 μm

amount of the parasitic resistance (R_{EQ}) is simply calculated by using (1); l indicates the total length of the coil-shaped on-interposer metal line. Due to the complicated coil-shaped structure of the proposed equalizer, the amount of the parasitic inductance (L_{EQ}) is obtained by using the 3D full-wave simulation; the simulation is performed by *Q3D Extractor 8* from *Ansoft*. The noticeable property of the proposed on-interposer passive equalizer structure is that the amount of L_{EQ} can be adjusted regardless of the amount of R_{EQ} by varying the turn number of the coil. It enhances the adjustability of the proposed equalizer, resulting in the suitable channel equalization for the different loss of the various on-interposer differential channels.

III. EXPERIMENTAL VERIFICATION

To validate the proposed on-interposer passive equalizer, a test vehicle including the proposed structure was designed. By measuring the fabricated test vehicle in frequency- and time-domains, the performance of the proposed on-interposer passive equalizer structure for differential signaling was experimentally verified.

A. Structure of designed test vehicle

The structure of the designed test vehicle including the proposed on-interposer passive equalizer for differential signaling is shown in Fig. 4. The differential signal lines were designed with M2 metal layer and ten ground lines surrounded those signal lines for higher loss of the on-interposer channel. The ten ground lines were connected each other at the edge and center of the on-interposer interconnect. The proposed

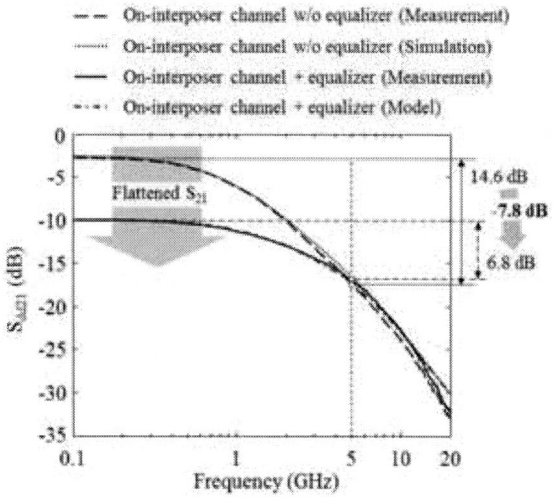

Figure 5. Measured S_{dd21} of an on-interposer differential channel without and with the proposed on-interposer passive equalizer structure (with 4-turn coil). By using the proposed equalizer, the S_{dd21} of a target on-interposer channel is successfully flattened by 7.8 dB from DC to 5 GHz.

equalizer structure was designed at the center of the on-interposer interconnect with M2 and M3 metal layers. The proposed equalizer in Fig. 4 has four-turn coil-shaped structure and those with one- and two-turned coils were also designed and fabricated.

The physical dimensions of the designed test vehicle is shown in Table I. The target on-interposer differential interconnect for equalization has a metal width (w_{Int}) of 10 μm, a space (s_{Int}) of 10 μm and a length (l_{Int}) of 4500 μm. The test vehicle with the short on-interposer interconnect of 450 μm was also designed to compare the S_{dd21} of the proposed equalizer with various turn number of coil. The proposed equalizer structure has a narrow metal width (w_{EQ}) of 1 μm. The length of proposed equalizer structure (l_{EQ}) is varied depending on the turn number of the coil-shaped structure as shown in Table I. In the case of the four-turn coil, the proposed on-interposer passive equalizer structure achieves the compact size of 63 μm by 10 μm.

B. Measurement results

The measured S_{dd21} of an on-interposer differential channel without and with the proposed on-interposer passive equalizer structure is shown in Fig. 5. The l_{Int} was 4500 μm and the turn number of the coil was four. The vector network analyzer (VNA), an N2530 from *Agilent Technologies* whose bandwidth is from 300 kHz to 20 GHz, was employed for the S_{dd21} measurement. Although the DC level is decreased, the proposed on-interposer passive equalizer successfully flattens the S_{dd21} of the target on-interposer differential channel by 7.8 dB from DC to 5 GHz. The flattened S_{dd21} leads to the reduced rise time, resulting in the improved quality of the transmitted differential signal.

The simulated S_{dd21} of the on-interposer differential channel by using the equivalent-circuit model of the proposed passive equalizer is also shown in Fig. 5. Despite of the simplified equivalent-circuit model that only includes the parasitic

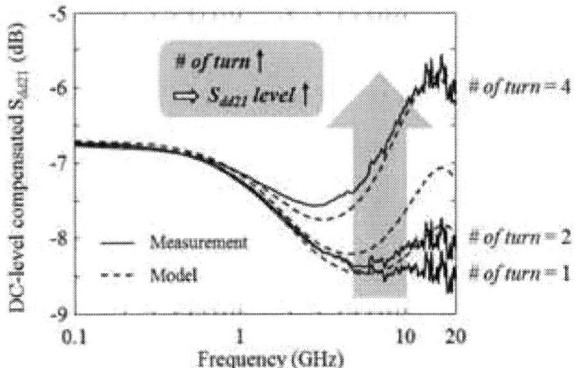

Figure 6. Measured S_{dd21} of the short on-interposer channels (l_{Int} = 450 μm) for different turn number of coils. As the turn number of coil increases, the S_{dd21} level at high frequency is increased due to the increased parasitic inductance.

TABLE II. EQUIVALENT-CIRCUIT MODEL PARAMETERS OF PROPOSED EQUALIZER WITH VARIOUS COIL TURN NUMBERS

Symbol	# of turn	Value	Symbol	# of turn	Value
R_{EQ}	1	29.4 Ω	L_{EQ}	1	0.18 nH
	2	29.5 Ω		2	0.3 nH
	4	29.7 Ω		4	0.49 nH

resistance and inductance of the on-interposer metal line, the model and the measurement result show good correlation because the amount of the parasitic capacitance is sufficiently small in the proposed structure. Thus, the validity of the equivalent-circuit model of the proposed on-interposer passive equalizer in Fig. 3 is experimentally verified.

The measured S_{dd21} of the fabricated test vehicles with the short on-interposer channel (l_{Int} = 450 μm) for different turn number of coils are shown in Fig. 6. The low loss of the on-interposer interconnect enables the suitable comparison between the S_{dd21} of the proposed on-interposer passive equalizer for different turn number of coils. The different DC-levels that are caused by the contact resistance variation during the measurement and the slightly different total length of the coil metal line are compensated for proper comparison.

As the turn number of the coil increases, the parasitic inductance of the proposed on-interposer passive equalizer is increased, resulting in the higher S_{dd21} level at the high frequency range. In contrast, the parasitic resistance is independent to the turn number of the coil, resulting in the almost constant S_{dd21} level from DC to the low frequency range. The equivalent-circuit model parameters of the proposed on-interposer passive equalizer with various turn number of the coil are shown in Table II. The R_{EQ} is almost constant when the turn number of the coil varies from one to four, whereas the L_{EQ} is considerably increased by a factor of 2.7. It verifies the noticeable adjustability of the proposed on-interposer passive equalizer structure that enables the suitable channel equalization for the different loss of various on-interposer differential channel.

The measured eye-diagrams of the fabricated test vehicles without and with the proposed on-interposer passive equalizer

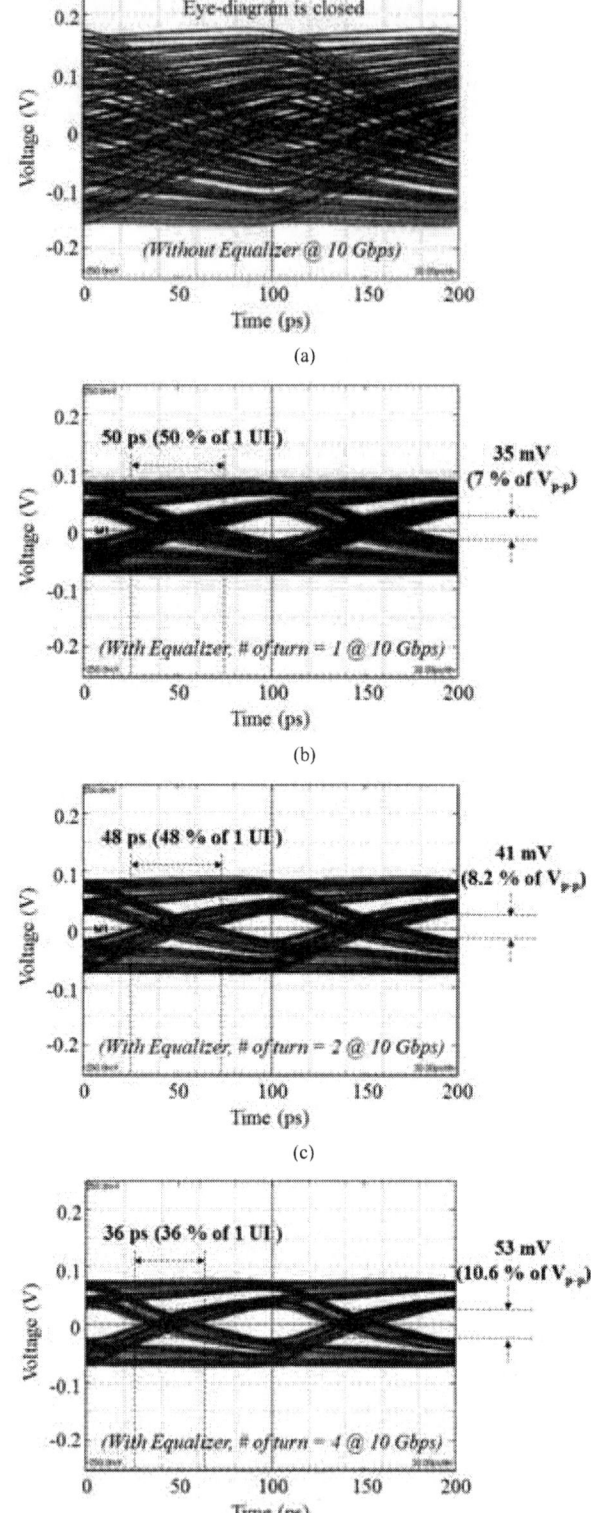

Figure 7. Measured eye-diagrams of test vehicles at a data rate of 10 Gbps (a) without an equalizer (b) with the proposed on-interposer passive equalizer for differential signaling with 1-turn coil, (b) 2-turn coil, and (d) 4-turn coil.

for differential signaling are shown in Fig. 7. The measured eye-diagram in Fig. 7(a) is the on-interposer channel without the equalizer and those in Fig. 7(b)-(d) are the cases with the proposed on-interposer passive equalizer with one-, two-, and four-turn coils, respectively. The input signals were the differential pseudo-random bit sequences of 2^{11}-1 with a rise- and fall-time of 30 ps. The amplitude and the data rate of the differential input signal were 0.5 V and 10 Gbps, respectively. The digital sampling oscilloscope, an TDS8000B from *Tektronix* whose bandwidth is 20 GHz, was employed to measure the eye-diagram of the output waveform.

The measured eye-diagram of the non-equalized on-interposer differential channel was absolutely closed at the data rate of 10 Gbps, whereas those of the equalized channels by using the proposed on-interposer passive equalizer were remarkably improved. Despite of the DC-level attenuation, the reduced rise time by the proposed passive equalizer improved the quality of the transmitted differential signal, resulting in the open eye-diagram. By using the proposed on-interposer passive equalizer with one-, two-, and four-turn coils, the eye-opening voltages were improved by 7 %, 8.2 % and 10.6 % of the peak-to-peak voltage, respectively, and the timing jitter were improved by 50 %, 52 % and 64 % of the one unit-interval, respectively. As the turn number of the coil increases, the loss compensation level is also increased as shown in Fig. 6, resulting in the better eye-diagram with the higher eye-opening voltage and the lower timing jitter. From the measurement results, the equalization performance and the adjustability of the proposed on-interposer passive equalizer for differential signaling were successfully demonstrated for data rates of up to 10 Gbps.

IV. CONCLUSION

In this paper, a compact on-interposer passive equalizer for chip-to-chip high-speed differential signaling was proposed and experimentally verified. As the silicon interposer gradually replaces the conventional package, the signal integrity of the channel on the lossy silicon substrate becomes significant. Though the differential signaling can enhance the channel bandwidth by rejecting the common-mode noise, the ISI still degrades the quality of the transmitted differential signal. For chip-to-chip high-speed differential signaling on the silicon-based on-interposer channel, the proposed passive equalizer structure compensates for channel loss up to very high frequencies. Because the proposed equalizer employs the parasitic resistance and inductance of the coil-shaped on-interposer metal line with a narrow width, it achieves not only wide-band equalization but also the compact size. In addition, the coil-shaped structure enables the suitable equalization for the different loss of the various on-interposer channels.

The performance of the proposed equalizer were experimentally verified by the frequency- and time-domain measurements for data rates of up to 10 Gbps. The proposed on-interposer passive equalizer successfully flattens the S_{dd21} of the target on-interposer differential interconnect with the length of 4500 µm by 7.8 dB from DC to 5 GHz. The eye-diagram of the target on-interposer differential channel without an equalizer was absolutely closed, whereas that with the proposed equalizer became considerably improved. Moreover,

the adjustability of the proposed on-interposer passive equalizer was also experimentally verified. The measured S_{dd21} and the eye-diagram were significantly varied depending on the turn number of coil in the proposed equalizer structure. The proposed on-interposer passive equalizer for differential signaling enables the very wide channel bandwidth in a 2.5-D or 3-D IC system.

REFERENCES

[1] Q. Chen, T. Bandyopadhyay, Y. Suzuki, F. Liu, V. Sundaram, R. Pucha, M. Swaminathan, and R. Tummala, "Design and demonstration of low cost, panel-based polycrystalline silicon interposer with through-package-vias (TPVs)," in Proc. *ECTC 2011*, June 2011, pp. 855-860

[2] S. Masahiro, T. Takayuki, K. Takashi and H. Mitsutoshi, "Silicon interposer with TSVs (through silicon vias) and fine multilayer wiring," in Proc. *ECTC 2008*, May 2008, pp. 847-852

[3] R.-B. Sun, C.-Y. Wen and R.-B. Wu, "Passive equalizer design for through silicon vias with perfect compensation," *IEEE Trans. Component, Packaging and Manufacturing Technology*, vol. 1, no. 11, pp. 1815-1822, Nov, 2011

[4] J. Kim, E. Song. J. Cho, J. Pak, J. Lee, H. Lee, K. Park and J. Kim, "Through silicon via (TSV) equalizer," in *Proc. IEEE 18th Electrical Performance of Electronic Packaging and Systems*, pp. 13-16, Oct. 2009

[5] H. Kim, J. Cho, J. Kim, K. Kim, S. Choi, J. Kim and J. S. Pak," A compact on-interposer passive equalizer for chip-to-chip high-speed data transmission," in Proc. *EPEPS 2012*, Oct 2012, pp. 95-98

[6] G. Luo, W. Yin, K. Kang, J. Shi, J. Yue, and J. Mao, "Wideband circuit model of silicon-based interconnects up to 50 GHz," in Proc. Asia-Pacific Microwave Conference 2007, pp. 1-4, Dec. 2007

978-1-4799-5004-1/13 $31.00 © 2013 IEEE

Signal integrity and EMC performance enhancement using 3D Integrated Circuits – A Case Study

Etienne Sicard[1], Wu Jian-fei[2], Li Jian-cheng[2]

[1] INSA, University of Toulouse, 135 av. de Rangueil, 31077 Toulouse, France

[2] School of Electronic Science and Engineering, National University of Defense Technology, Changsha, Hunan, China, 410073

Contacts: (1) Etienne.sicard@insa-toulouse.fr, (2) wujianfei990243@126.com

Abstract—In this paper, the signal integrity (SI) and Electromagnetic Compatibility (EMC) performance of a microcontroller and memory are simulated in 2D and 3D assembly versions. Three types of configurations are investigated: conventional 2D routing on printed-circuit-board, stacked dies with wire bonding and stacked dies with Through-Silicon-Via (TSV). The study addresses signal integrity of the memory bus and the conducted emission of the microcontroller. An equivalent bus model is presented for order reduction and improved simulation efficiency. The benefits of 3D integration are highlighted, in terms of improved eye diagram and one decade reduction in parasitic emission.

Keywords- Signal integrity; EMC; Microcontroller and memory; 3D ICs; Equivalent bus model; 1 Ω/150 Ω; IC Strip line; Emission

I. INTRODUCTION

Three-dimensional integration allowing the stacking of different types of devices within the same 3D chip is offering exciting new possibilities in terms of system integration. Academic and industrial researchers have been very active the past recent years to provide ways to exploit 3D integration for commercial applications such as smartphones, memory-based hard disks, memory-processor stacks for workstations or miniature cameras [1][2]. Wire-bonded 3D stacking, package-on-package (PoP) and Through-Silicon-Vias (TSV) [3] have recently entered the mass-production market.

3D stacking leads to higher density, shorter interconnects, reduced propagation delays, with significant benefits in terms of operating frequencies and power efficiency. In [2], Patti describes a stacking of R8051 and memory which leads 5 times improvement in exchange bandwidth together with a reduction by 10 of the power consumption, as compared to the PCB implementation with 2 conventional packages. Power savings close to a factor of 100 may be achieved with 10-μm thinned dies, micro-range TSV such as described in [2] [8], together with high-pitch Direct Bonding Interconnects (BDI).

3D integration may also reduce resonance effects and simultaneous switching noise. However, crosstalk noise between adjacent lines may remain significant due to decreased interconnect pitch in 3D structures, as well as vertical die-to-die couplings, specifically in TSV technologies [7][8]. Despite these drawbacks, some improvements in signal integrity may be expected, as well as reduced parasitic emission and improved immunity to radio-frequency interference. However, not many case studies have been reported in the literature to compare 2D and 3D interconnects approaches using identical dies.

In this paper, the signal integrity and EMC performances of a Microchip DSPIC 33F microcontroller and a 1MB SRAM memory are simulated in 2D and 3D versions. Three types of configurations are investigated, as illustrated in Figure 1: conventional 2D routing on printed-circuit-board (Fig. 1-a), stacked dies with wire bonding (Fig. 1-b) and stacked dies with Through-Silicon-Via (TSV, Fig. 1-c).

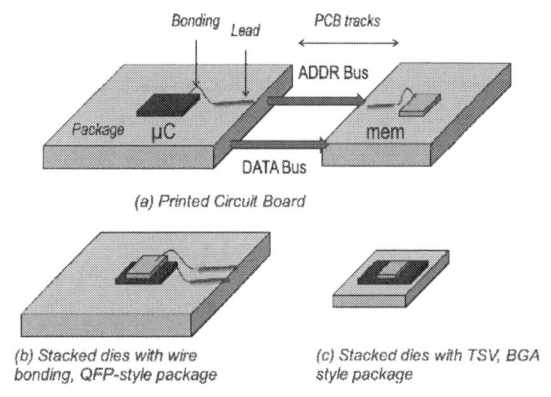

Figure 1. Three types of microcontroller-memory interconnection investigated in this paper (a) 2D Printed-Circuit-Board interconnects, (b) 3D with wire bonding and die stacking (c) 3D with TSV

The case study is described in section 2, with details on the model construction. In Section 3, we compare the signal integrity performances of the memory bus using the eye diagram simulation. Section 4 is focused on conducted emission prediction, followed by a conclusion.

II. DESCRIPTION OF THE EMC CASE STUDY

A 16 bits microcontroller DsPIC33FJ128GP706 [4] from Microchip is associated to a 1MB SRAM memory from Brilliance BS62LV8001EIP55 [5], implemented for the 2D version in a TEM-compatible 10 x 10 cm, 6 layers board. The 2D board is shown in Fig. 2. The voltage regulator, microcontroller and memory are routed on the inner side of the PCB.

978-1-4799-5004-1/13 $31.00 © 2013 IEEE

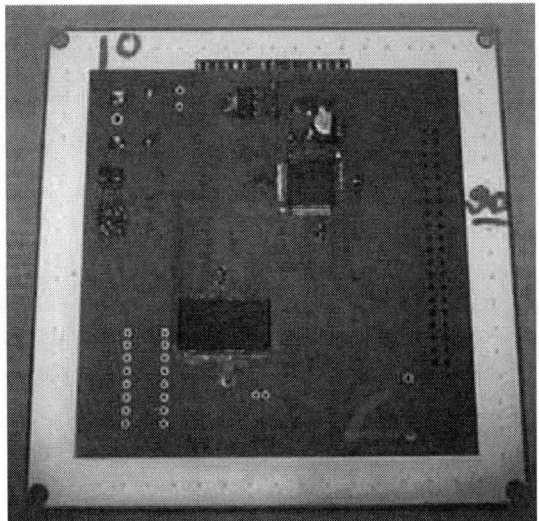

Figure 2. Inner view of the TEM-compatible 2D implementation of the microcontroler and memory, which serve as reference for EMC performance

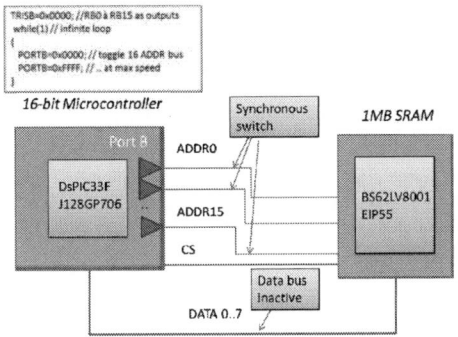

Figure 3. Details of the microcontroller-memory link. The test software toggles the 16 ADDR lines synchronously at 5.7 MHz rate.

The communication bus between the microcontroller and the memory is made of a 16-bit address bus and 8-bit data bus, routed on the inner side to characterize the global system emission. The measured emission and immunity levels of the 2D implementation serve as reference levels for later comparison with 3D implementation. The IBIS models of the two components are available from the foundries, as well as PCB track models extracted from routing information.

III. SIGNAL INTEGRITY EVALUATION

We consider a synchronous switching of the 16-bit address bus, together with the nominal activity of the microcontroller. The software code toggles the 16 address lines at a 5.7 MHz rate, while the code clock is around 40 MHz.

A. 2D and 3D case studies

We compare the SI and EMC performances between 2D and 3D variants of microcontroller/memory sub-system, as shown in Tab. I. The average characteristics of interconnects are reported, with an evaluation of the corresponding characteristic impedance Z0. The die stacking considers wire

bonding while the TSV technology considers interconnects routed internally, through the die of the ICs. We use TSV dimensions close to the Terrazon process [2] opened to academics: 1 x 1 μm size, 5 μm pitch and 10-μm thinned, with 100nm oxide thickness, which may stack 3-8 dies by combining TSV and DBI.

TABLE I.   DETAILS ON THE 2D AND 3D IC CONFIGURATIONS

Config.	Description	IC Packages	Interconnects
1	2D IC with PCB	μC: QFP 64, pitch 0.5mm Mem: TSOP 44, pitch 0.8mm	5-cm PCB, Z0 75-100 Ω
2	3D IC with wire bonding	μC: QFP 64, pitch 0.5mm Mem: stacked die bond wired	1-3 mm bonding wires, Z0 135-275 Ω
3	3D IC with TSV	μC: BGA 64, pitch 0.5mm Mem: thinned die with TSV	1 x 1 x 10 μm TSV, 2 x 2 μm DBI, Z0 25 Ω

The 2D aspect of the board is given in Fig. 4 with associated physical and electrical characteristics, computed using IC-EMC [5]. The upper layer (signal 1) is twice thicker than signal 2, and situated far above the ground plane, leading to higher characteristic impedance. The 3D aspect of the stacked dies (microcontroller on the bottom, memory on the top) with double-deck wire bonding is reported in Fig. 5. We notice that the characteristic impedance of the bonding reaches 275 Ω, which is twice higher than the lead. This is mainly due to the high inductance per unit length. Also notice the skin effect at 1 GHz which raises the serial resistance to 0.1 Ω/m.

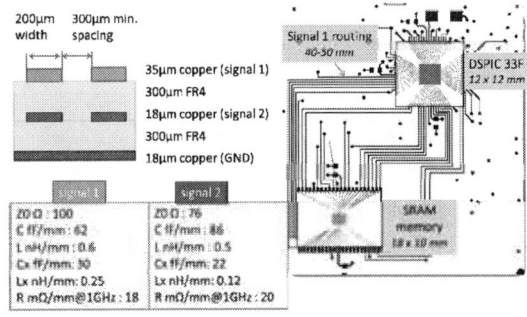

Figure 4. Interconnect characteristics in the 2D PCB implementation

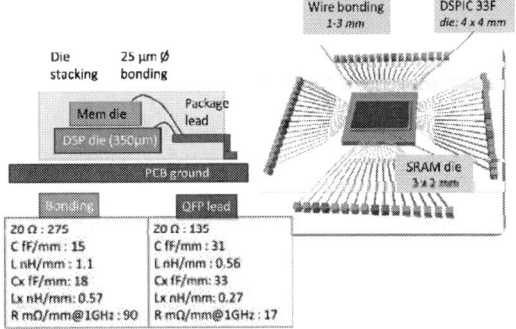

Figure 5. Interconnect characteristics in the 3D wire-bonded implementation with wire bonding and stacked dies

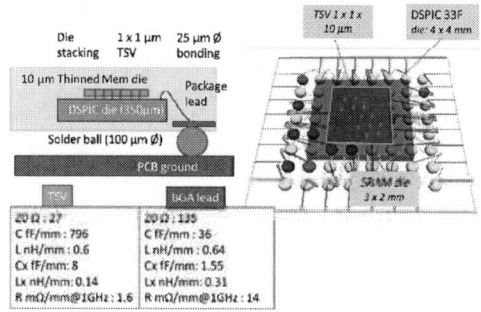

Figure 6. Interconnect characteristics in the 3D TSV implementation and 64-BGA package

B. A focus on Terrazon 3D technology

The die stacking using 1 x 1 μm TSV is illustrated in Fig. 6. A zoom at the die-to-die interconnection is reported in Fig. 7. The dimensions of the TSV and upper die thickness are derived from the Terrazon 3D process which has recently been made available for academic research [2]. Many different TSV processes and 3D stacking options exist, which may lead to significantly different electrical values [6], specifically to the capacitance which may rise to several pF per TSV. The modelling of TSVs has been the focus of many research papers such as [8] (2 x 2 x 400 μm case study) and the book of Er-Ping Li [7] (20 x 20 and 100 x 100 μm TSVs). The R,L,C values computed by IC-EMC [10] should be used with care as they are frequency and technology dependent, due to the semiconductor nature of the silicon die.

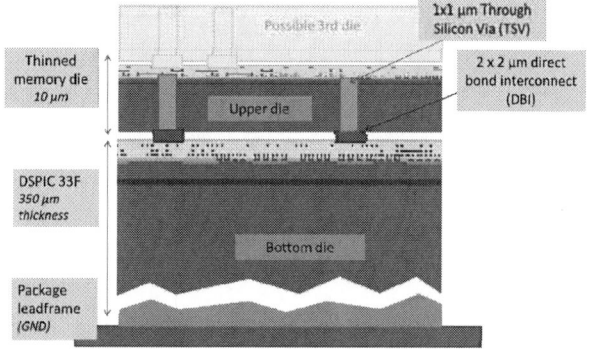

Figure 7. Aspect of the two stacked dies in 3D TSV technology inspired from Terrazon process [2]

However, the thinned memory substrate down to 10 μm combined with very small TSV section (1 x 1 μm) leads to C=8 fF, L = 10 pH, which are negligible, while R_{DC}=150 mΩ remains significant. The TSV modeling proposed in IC-EMC (Tool "Interconnect parameters") is inspired from analytical models proposed in [7], with an estimation of R,L,C and frequency dependency.

C. Signal integrity simulations

The signal integrity analysis focuses on communication lines between the microcontroller and memory.

We consider the switching of the address bus using the 3 types of configurations with the same IBIS I/V information for the pull-up and pull-down structures, as well as the die capacitance. Depending on the configuration, the package and interconnect models are evaluated in order to match the physical characteristics of the DsPIC/memory implementation. We use the memory IBIS information to model the clamp and die capacitance of the interconnect loads.

To evaluate the signal quality, we used the eye diagram function implemented in IC-EMC [10]. At first, we generate a random bit stream using the PWL source generator tool that is used for bus stimulation. The data rate is 100 Mb/s, and the signal amplitude is 3.3 V (Fig. 8).

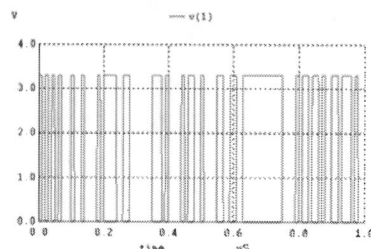

Figure 8. Random pulse stimulation for SI eye diagram simulation

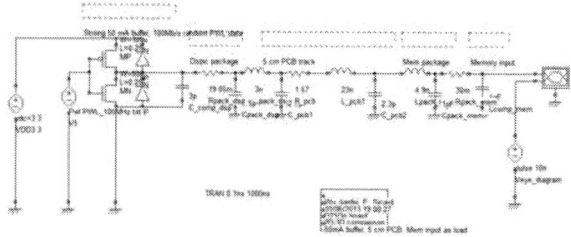

Figure 9. Equivalent bus model for SI simulation in configuration 1 (see table I)

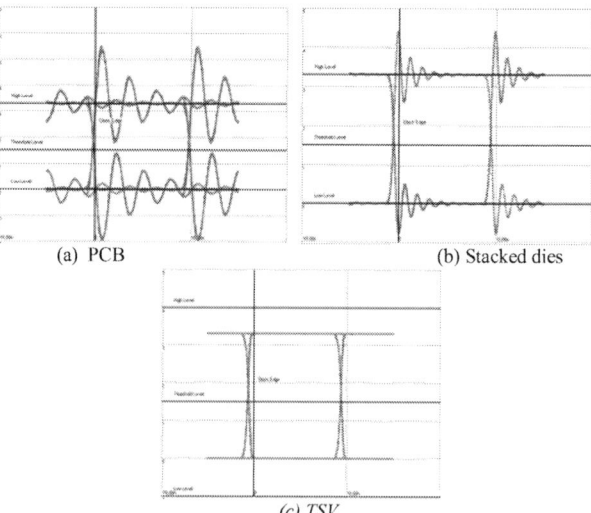

Figure 10. Eye diagram simulation results for config. 1 (a), config. 2 (b) and config.3 (c)

Then, we inject the PWL as a command of the buffer that drives the memory through different types of packages and interconnects, depending on the configurations described in Table I. The schematic diagram used for simulation of configuration 1 is reported in Fig. 9. It consists of a 50 mA buffer activated by the stimulation source at 100 Mb/s rate of Fig. 8, the QFP package, 5-cm PCB, TSSOP memory package and memory input load. The eye diagram symbol is controlled by a 100 MHz clock synchronized with the input data.

In the model reported in Fig. 9, we consider ideal IO supply voltages. We also neglect the couplings between adjacent wires. Therefore, simulations of Fig. 10 are a first-order approximation only. It can be seen that the eye diagram is strongly improved by 3D stacking, and is close to ideal using TSV.

IV. CONDUCTED EMISSION ANALYSIS

In the emission simulations, the 1 Ω direct coupling method [11] is used to compare conducted emission between the 3 configurations. In accordance to ICEM Conducted Emission Modelling approach [13], the model of the DSPIC includes the core noise and its decoupling circuit, tuned with measured I(t) and Z(f) (Fig.11), a substrate coupling resistance (IBC) and the ADDR bus model reused from previous signal integrity simulation. We apply a model reduction technique on the address bus in order to replace the 16 wire by an equivalent model which has identical switching performances and conducted noise spectrum.

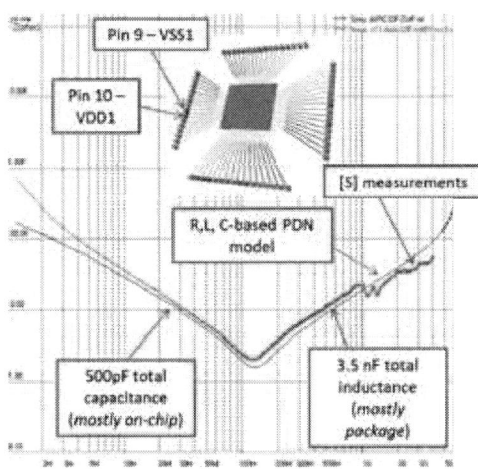

Figure 11. The PDN model is extracted from [S] measurements

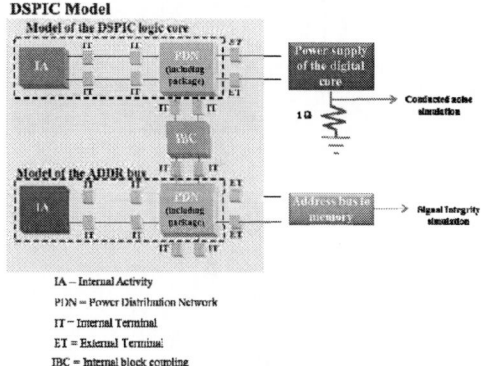

Figure 12. DSPIC model for EMC emission simulation

The predicted conducted emission in configuration 1 can be compared with the measurement performed on the EMC test board (Fig. 2). The matching is quite good for both the core activity (40 MHz harmonics) and the IO switching (5.7 MHz harmonics). Configurations 2 and 3 are only based on predictive simulations as no experimental die stacking is available for measurements.

The model used for conducted emission simulation in IC-EMC [10] is closely matched with ICEM-CE strategy, with the PDN and IA of the DSPIC core, an IBC resistance and PDN of the DSPIC IOs (Fig. 15). The MOS devices represent the Internal Activity of the IO block. The 16 ADDR lines are merged into a single buffer, as well as the TSV and Memory models.

The same core and buffer models are used, while the package and interconnect models are changed according to Table I and figs. 4-6. As may be seen in the result summary of Table II, a 13dB reduction is obtained with config. 2 and upto 20 dB in config. 3 at fADDR (5.7 MHz) while the peak emission at fCPU (40 MHz) is only reduced by 7 dB.

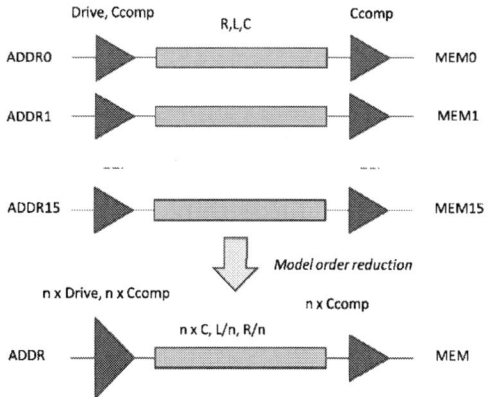

Figure 13. Model-order-reduction applied to the ADDR bus to speed up conducted emission simulations

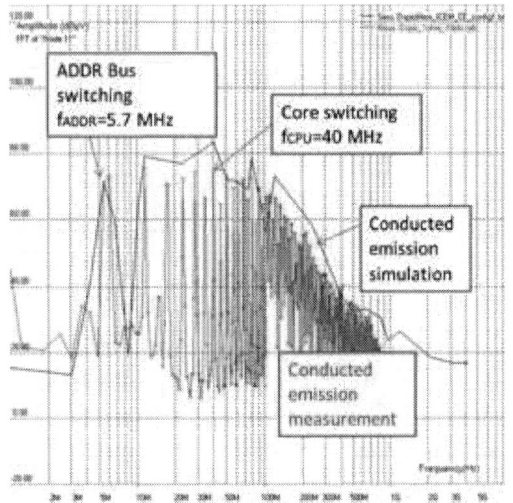

Figure 14. Conducted emission in Config. 1 : comparison between measurements (red curve) and simulation (black envelope)

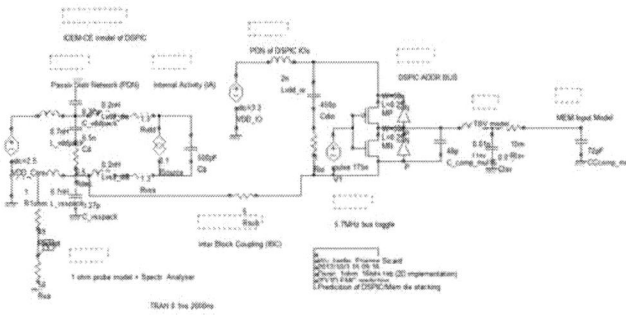

Figure 15. Conducted emission simulation in IC-EMC including the DSPIC and TSV model (Config. 3)

TABLE II.   CONDUCTED EMISSION COMPARISON BETWEEN 2D AND 3D IC CONFIGURATIONS

Configuration	Description	Simulated Peak Emission	Comments
1	2D IC with PCB	82 dB@ 40 MHz 73dB@5.7 MHz	See Fig. 14
2	3D IC with wire bonding	78 dB@ 40 MHz 60dB@5.7 MHz	13 dB reduction at f_{ADDR}
3	3D IC with TSV	75dB@ 40 MHz 54dB@5.7 MHz	20dB reduction at f_{ADDR}

V.   CONCLUSION

Based on an existing EMC test board implementing a DSPIC microcontroller and a 1Mb SRAM memory, we have investigated the benefits of a 3D die stacking using either the conventional double wire bonding or the more advanced Through-Silicon-Via technology. Significant gains in signal integrity and one decade of parasitic emission reduction at bus switching frequency have been forecast. Three-dimensional integration allowing the stacking of heterogeneous dies or packages should enhance multi-giga-bit data links, reduce parasitic emissions thanks to very significant benefits in signal propagation and switching power reduction, and offer new exciting possibilities of efficient shielding and filtering, reduced parasitic emission and improved immunity to radio-frequency perturbations.

REFERENCES

[1] G. H. Loh, Y. Xie, "3D Stacked Microprocessor: Are We There Yet?" IEEE Micro, May-June 2012, pp. 60-63

[2] R. Patti, "3D Integration: New Opportunities for Speed, Power and Performance", Invited talk, IMAPS, 2012, http://www.terrazon.com

[3] R. Huemoeller, "OSAT positioning in the emerging Mid-End: Fan Out, 3D ICS and 2.5D multi die interposers", 3D Packaging Magazine; Nov. 2012, pp. 20-22

[4] Data sheet "DsPIC33FJXXXGPX06/X08/X10, High-Performance, 16-Bit Digital Signal Controllers", Microchip Technology Inc., Ref. DS70286C, 2009, http://www.microchip.com

[5] Data Sheet "Very Low Power CMOS SRAM, 1M X 8 bit", Brilliance Semiconductor, Inc, Ref. R0201-BS62LV8001, 2008

[6] Jian-Qiang Lu, "3-D Hyperintegration and Packaging Technologies for Micro-Nano Systems", Proceedings of the IEEE, Vol. 97, No. 1, Jan. 2009, pp. 18-30

[7] Er-Ping Li "Electrical Modeling and Design for 3D Integration", Wiley, 2012, IEEE Press.

[8] G. Katti, M. Stucchi, K. De Meyer, W. Dehaene, "Electrical Modeling and Characterization of Through Silicon via for Three-Dimensional ICs", IEEE Transactions On Electron Devices, Vol. 57, No. 1, January 2010

[9] R. Yarema, "Review of 3D Related Technologies for High-Energy Physics (HEP)", LHC-ILC Workshop on 3D Integration Techniques, Paris, Nov. 2007

[10] E. Sicard, A. Boyer IC-EMC v2.5 User's Manual, Oct. 2011, 260 pp, ISBN 978-2-87649-061-1, www.ic-emc.org

[11] IEC 61967: "Integrated Circuits, Measurement of Electromagnetic Emission – 150 KHz to 1 GHz", www.iec.ch

[12] IEC 62132: "Integrated Circuits, Measurement of Electromagnetic Immunity – 150 KHz to 1 GHz", www.iec.ch

[13] IEC 62433: "Models of Integrated Circuits for EMI behavioral simulation", www.iec.ch

[14] E. Sicard, Wu Jianfei, L. Guibert, S. Serpaud "Modelling the emission of the DSPIC-Mem Board", IC-EMC application notes; on-line at www.ic-emc.org

Kron Simulation of Field-to-line Coupling Using a Meshed and a Modified Taylor Cell

Sjoerd Op 't Land[*], Richard Perdriau[*], Mohamed Ramdani[*], Olivier Maurice[†], M'hamed Drissi[‡]

[*]Department of Electronics, Ecole Superieure d'Electronique de Ouest (ESEO), Angers, France
Email: {sjoerd.optland, richard.perdriau, mohamed.ramdani}@eseo.fr
[†]Groupe d'Études et de Recherches Appliquées à la Compatibilité Électromagnétique (GERAC), Trappes, France
Email: olivier.maurice@gerac.fr
[‡]Université européenne de Bretagne IETR, Rennes, France
Email: mhamed.drissi@insa-rennes.fr

Abstract—**Printed Circuit Board (PCB) traces play a role in the immunity of electronic products. Contrary to Integrated Circuits (ICs), the layout of PCB traces can be changed rather late in a product's design. Therefore, it is interesting to equip the PCB designer with simple tools that predict the immunity of his PCB traces.**

In this article, we compare two simulations of field-to-long line coupling based on Taylor's model. Firstly, the line is meshed into electrically short Taylor cells and numerically simulated using Kron's method. Secondly, we use *one* modified Taylor cell, which does not need meshing and is a closed-form, analytical result.

The two simulations turn out to be equally precise on a straight microstrip line, the meshed simulation being more flexible, the simulation using a modified Taylor cell being faster.

Index Terms—**PCB, EMC, field-to-line coupling, immunity, microstrip, Kron, frequency-adaptive meshing, modified Taylor**

I. INTRODUCTION

Electromagnetic compatibility (EMC) problems can often be understood as a three-element chain: agressor-coupling path-victim [1]. In the case of unshielded, wireless electronics, the dominant coupling path can consist in the PCB traces. Therefore, the routing of PCB traces may be decisive for product compliance.

In contrast to integrated circuits (ICs), PCB layout may be changed rather easily and in a late design stage. Field-to-line coupling models could help the PCB designer to *predict* and *explain* product immunity. The prediction helps the designer to detect problems before fabricating the first prototype, the explanation helps the designer to do something about the detected problems. As electromagnetic coupling is rarely easy to explain, even by humans, we do not believe in automatic explanation of product immunity. Very fast automatic prediction, however, would allow the designer to freely play around with his design and develop intuition for the coupling mechanisms. Note that this prediction need not be very precise, as long as it faithfully reveals the influence of the designable parameters. Therefore, the focus of this article will be on fast, numerical prediction of field-to-trace coupling.

We will now define a rather simple case study to evaluate the methods. However, we keep more realistic PCB traces in mind when concluding on their performance.

We choose a microstrip, i.e. a trace above a ground plane, because it is still widely used. Moreover, with respect to coplanar waveguides (CPWs) and striplines, it is good antenna and therefore prone to create immunity problems.

Operational and harmonic frequencies of electronics keep rising, so the wavelengths keep falling. For example, the Wireless Home Digital Interface (WHDI) uses a 5 GHz carrier, or a 6 cm wavelength in free space. Back-up radars may use ultra-wideband signals up to 24 GHz, or down to 1.25 cm. PCBs still have sizes in that order of magnitude, so we may expect long-line effects. Therefore, we choose to illuminate a 5 cm trace with a frequency up to 20 GHz.

In practice, traces are never characteristically terminated, because the terminating ICs and passives have frequency dependent impedances. Neither are real-world traces uniform, because of width changes and unmitered bends. However, we believe that there is already sufficient microwave theory to incorporate these non-idealities in simulation. Here, we would like to focus on modelling of field-to-trace coupling. Therefore, we allowed ourselves to study a uniform $50\,\Omega$ trace that is characteristically terminated. According to typical technology, the $\varepsilon_r = 4.6$ substrate is $362\,\mu m$ thick.

Finally, the most constraining simplification is that of grazing incidence (cf. Figure 1). The vertically polarised plane wave is not refracted by the air-substrate interface and the incident wave is simply doubled by the ground plane. The field strength in the dielectric substrate thus amounts to:

$$H = 2H^i \tag{1}$$
$$E = 2E^i/\varepsilon_r. \tag{2}$$

For low frequencies, grazing incidence constitutes the worst case [3]. Also, it models Gigahertz Transverse ElectroMagnetic cell (GTEM-cell) measurements, which integrate a PCB in the waveguide wall. Otherwise, this is a serious limitation.

For the numerical calculation, the field generated by a standard GTEM cell will be entered: 1 V at a $50\,\Omega$ septum, separated by 42 mm from the PCB. Hence, the terminal voltage in dBV is numerically equal to the S_{21} coefficient that would be measured between GTEM input and trace terminal.

Figure 1. Grazing incidence: the incident far-field wave vector $k^i = \omega/c_0$ is tangential to the substrate.

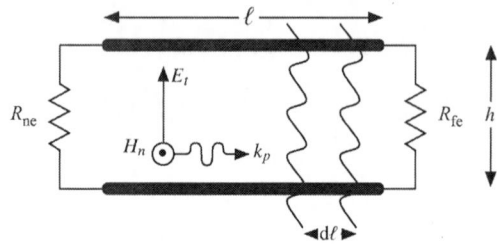

(a) Line geometry: the indices n, t and p indicate vector components normal to the plane of the wires, transversal to the line and parallel with the wires, respectively. R_{ne} and R_{fe} are the near-end and far-end resistive terminations.

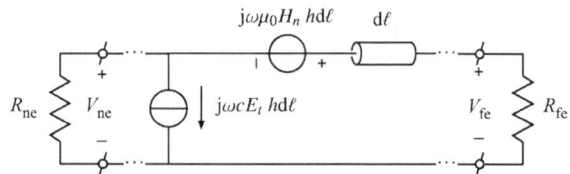

(b) Taylor cell: a current source models the electrostatic force (capacitive coupling) and a voltage source the electromotive force (emf or inductive coupling) in each line segment $\Delta\ell$. c denotes the per-unit-length p.u.l.) capacity of the line.

Figure 2. Taylor's field to line coupling model for a bifilar transmission line.

In section II, we will briefly review existing field-to-trace coupling models and pick Taylor's model to continue with. We will implement this model using frequency-adaptive meshing and Kron's method simulation in section III. Alternatively, we will explain and apply the modified Taylor's cell in section IV. Both simulations will be compared, conclusions drawn and suggestions for future research given in section VI.

II. STATE OF THE ART

Field-to-line coupling is complicated, like almost any real-life electromagnetic problem. Let us therefore start by reviewing four common simplifications.

If wavelengths are great with respect to the studied geometry, the *quasi-static* approximation may be used. The field then still changes with time, but propagates instantaneously everywhere. In that case, the illumination field is uniform. This approximation yields low-frequency asymptotes, useful for checking our model. However, as we specifically chose an electrically long line, we will not use this approximation.

In reality, everything interacts bilaterally. Indeed, the field emitted by an aggressor will couple to a guided wave in the victim line. However, this guided wave will also affect the aggressor. This, in turn, will affect the victim again, and so on. To predict what will happen, knowledge about the aggressor is needed. Because we suspect that there be only *weak coupling* and would like to avoid incorporating knowledge about the aggressor, we will only consider the unilateral interaction from field to line.

As long as the cross section of the microstrip transmission line remains small with respect to the wavelength, there is only one dominant mode: the differential Transversal ElectroMagnetic (TEM) mode. Because this assumption only gradually breaks down at several GHz for modern electronics [2], we choose to adopt this approximation. This allows us to use transmission line theory to describe the trace.

Finally, we will suppose traces to be lossless. Consequently, immunity predictions will be pessimistic with respect to reality, but only slightly [2].

There are three equivalent, weakly coupled, transmission line-based field-to-line coupling models [4]: that of Taylor et al. [5], of Agrawal et al. [7] and of Rachidi [6]. They all model the coupling of an illuminating field by means of current and/or voltage sources, distributed along the line. Agrawal and Rachidi also need sources at both terminals.

All of these models basically represent a uniform transmission line. Consequently, when modeling a piecewise non-uniform line using Agrawal or Rachidi, sources appear at the transitions. Using Taylor's model, on the other hand, only the distributed sources along the line change value, generally. That way, it is even possible to model continuously non-uniform lines. In view of the long-term goal to model non-uniform lines, we prefer the simplicity of Taylor's model and will continue with that model only.

The specialisation of Taylor's model for a two-wire transmission line in vacuum is shown in Figure 2. In each transmission line slice $d\ell$, the magnetic field normal to the plane of the wires H_n induces a voltage and the electric field in the plane and transversal to the wires E_t induces a current.

III. MESHED TAYLOR SIMULATION USING KRON

The most obvious application of Taylor's model for a non-uniform incident field, is to mesh (segment) the line in short enough *cells*, in order for the field to become approximately uniform to each cell. The passive transmission line itself must also be modeled, for example as an *rglc* telegrapher's cell. As we are considering a lossless line, we omit the dissipative elements r and g. The resulting model of a line for the case of three cells is depicted in Figure 3.

With increasing frequency, wavelength decreases and generally, the field becomes less uniform along the line. Therefore, a large number of cells may be needed to accurately model the line. In the perspective of a simple tool, we avoid manually entering this multitude of cells, because it is error-prone and time-consuming. We chose to analyse this problem in terms of Gabriel Kron's formalism [8], because of its promise to handle complex electromagnetic systems [9].

We will now first describe the basic approach of the problem in Kron's formalism. Then, we will describe the practical implementation, including performance optimisation.

978-1-4799-5004-1/13 $31.00 © 2013 IEEE

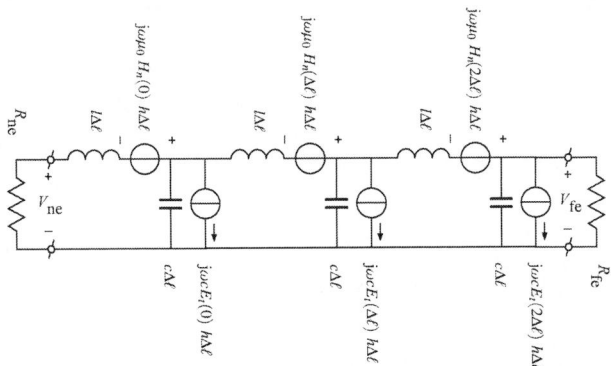

Figure 3. Lossless transmission line meshed in three cells ($\Delta\ell = \frac{1}{3}\ell$). The passive transmission line is modeled with l and c being the per-unit-length inductance and capacity, respectively.

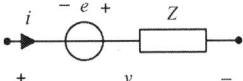

Figure 4. Simplified Kirchhoff branch. The difference of potential v across the branch and the current i through the branch are defined such, that when iv is positive, net power is dissipated in the branch (passive sign convention).

A. Basic Approach

Generally, solving a problem in Kron's formalism consists of eight steps: stating the problem, drawing the associated graph, define the topological base, entering the sources, transforming, solving in mesh space, deducing the differences of potentials and deducing other required quantities [10].

Let us draw the graph corresponding to the problem of Figure 3. In this graph, we identify *meshes* (or loops) and *nodes* (or junctions). Meshes consist of *branches* (or vertices) that each connect two nodes. We will here use simplified Kirchhoff branches, which generally consist of an impedance Z and a voltage source e as defined in Figure 4. The resulting graph is depicted in Figure 5.

Let i, v and e be column vectors in the *branch space*, that is: containing currents and voltages for every branch. The (arbitrary) branch numbers of Figure 5 define which vector component represents which voltage and current: we just defined the *topological base*. Ohm's and Kirchoff's laws then hold as in $v + e = Zi$. In our case, the impedance matrix

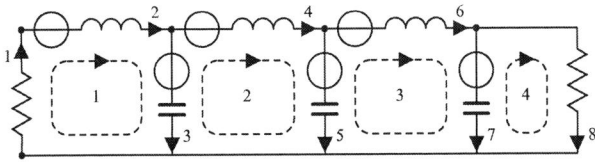

Figure 5. Graph representation of a three-cell transmission line model. Please verify that there are 4 meshes (dashed loops, numbered), 8 branches (with arrows, numbered) and 5 nodes (dots, not numbered).

Z only has entries on its main diagonal:

$$\mathrm{diag}(Z) = \Big[R_{\mathrm{ne}}, \; j\omega l\Delta\ell, \; \frac{1}{j\omega c\Delta\ell},$$
$$j\omega l\Delta\ell, \; \frac{1}{j\omega c\Delta\ell}, \; j\omega l\Delta\ell, \; \frac{1}{j\omega c\Delta\ell}, \; R_{\mathrm{fe}} \Big]. \tag{3}$$

To incorporate the current sources in the simplified Kirchhoff branch, we need to use their Thévenin equivalents $E_t h$. The source vector e stemming from the illumination electromagnetic field thus becomes:

$$e = \begin{bmatrix} 0 & 0 \\ H_n(0) & 0 \\ 0 & E_t(0) \\ H_n(\Delta\ell) & 0 \\ 0 & E_t(\Delta\ell) \\ H_n(2\Delta\ell) & 0 \\ 0 & E_t(2\Delta\ell) \\ 0 & 0 \end{bmatrix} \begin{bmatrix} j\omega\mu_0 \; h\Delta\ell \\ h \end{bmatrix}. \tag{4}$$

To solve for the mesh currents, we need to transform our equations to another topological base: that of the *mesh space*. At the same time, we will connect the branches together. This is done by means of the *connectivity* matrix L, which links the branches (rows) with the meshes (columns). In our example,

$$L = \begin{bmatrix} 1 & 0 & 0 & 0 \\ 1 & 0 & 0 & 0 \\ 1 & -1 & 0 & 0 \\ 0 & 1 & 0 & 0 \\ 0 & 1 & -1 & 0 \\ 0 & 0 & 1 & 0 \\ 0 & 0 & 1 & -1 \\ 0 & 0 & 0 & 1 \end{bmatrix}. \tag{5}$$

Note that a minus signs signifies a branch going against the mesh direction. We will denote tensors in mesh space with a hat, e.g.:

$$\hat{e} = L^{-1}e \qquad i = L\hat{i} \qquad \hat{v} = L^{-1}v \equiv \mathbf{0},$$

where the last vector (voltage around every mesh) is zero by Kirchhoff's mesh rule. The inverse of L can be found by its transpose, because L always is a Hadamard matrix. We can transform Kirchoff's laws to mesh space as follows:

$$L^{-1}v + L^{-1}e = L^{-1}Zi = L^{-1}ZL\,\hat{i} \tag{6}$$
$$\hat{e} = \hat{Z}\,\hat{i}. \tag{7}$$

Notice that by transforming to the lower-dimensional mesh space, we connected the branches together.

To solve the system, we use the pseudoinverse (denoted $^+$):

$$\hat{i} = \hat{Z}^+\hat{e}, \tag{8}$$

because only the sources e are given.

We are interested in the near-end and far-end voltages, which can now be found by means of the terminal impedances:

$$V_{\mathrm{ne}} = -\hat{i}_1 R_{\mathrm{ne}} \tag{9}$$
$$V_{\mathrm{fe}} = \hat{i}_8 R_{\mathrm{fe}}. \tag{10}$$

As we are interested in the frequency-domain response, we need to perform this calculation for each frequency sample.

B. Implementation

In view of a simple tool, we want the user to describe the essential: the geometry of the trace and the illumination. The meshing is a repetitive task, which is a tedious and error-prone task if performed by humans. Therefore, we chose to automate it.

In order for the simulator to be easily incorporated in a PCB design tool, we preferred a scripting language that can provide an object-oriented (OO) application programming interface (API). In order to perform reproducible computational research [11], we preferred a free-to-use language. Therefore, we implemented the simulator in Python and published the code that produces the figures of this article on Github [12].

In order to mesh the transmission line, we need to decide upon the number of cells to use. For the field to be approximately uniform to each cell, we decide to take 50 cells per illumination wavelength for the highest frequency of interest. In our case study (5 cm until 20 GHz), this means 167 cells. With 301 frequency points from 20 MHz to 20 GHz, the calculation takes 24.2 s on an Intel 2.53 GHz Core 2 Duo processor.

To numerically solve (8), we use the Moore-Penrose pseudoinverse implementation of NumPy, which uses singular value decomposition (SVD). About half of the total execution time is spent on this call. This and other matrix manipulations depend heavily on the matrix size.

We recognise that for low-frequencies, we do not need a great number of cells. Therefore, we decide to re-mesh the transmission line for each frequency with a certain number of cells per wavelength. For example, with 50 cells per wavelength, the simulation now only takes 1.8 s on the same platform.

How many cells per wavelength does one need? We used the first simulation (50 cells per wavelength, non-adaptive) as reference, and calculate the error of subsequent adaptive simulations while varying the number of cells per wavelength. Then, we calculated the log-frequency weighted average error. Finally, we calculated the log-frequency average absolute deviation from this average error. Both error measures are shown in Figure 6. In our case study, a 20 cells per wavelength resolution yields an acceptable error (< 1 dB).

The simulation was run with 20 cells per wavelength adaptive meshing, which took 0.45 s. The result is displayed in Figure 9.

IV. MODIFIED TAYLOR CELL

Alternatively, we can elaborate Taylor's model analytically for the case of a grazing incident wave [2]. The result turns out to be a slightly modified Taylor cell, without the need for meshing. In the present article, the model is presented in an intuitive manner (similar to [13]); for a more rigorous underpinning of the model, please refer to [2].

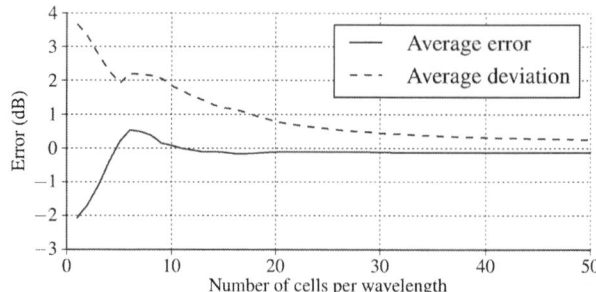

Figure 6. Sensibility of the adaptively-meshed simulation result for number of cells per wavelength.

We start with the low-frequency case, where a single Taylor cell suffices. Then we will try to imagine what happens with rising frequency. Finally, we will postulate an analytical modification on the low-frequency case to take into account high-frequency effects.

A. Low-frequency Case

Let us consider the low-frequency case using the quasi-static approximation. Because the illumination wavelength is long with respect to the line length, the field can be considered uniform along the line, and we can lump the line as a single cell ($\Delta \ell = \ell$). Because the wavelength in the transmission line is long with respect to the line length, we can ignore the phase shift introduced by the transmission line. In our case of characteristic loads ($R_{\text{ne}} = R_{\text{fe}} = Z_c$) we can find the either-end terminal voltages by inspecting Figure 2b [1]:

$$V_{\text{LF}} = -\frac{1}{2}j\omega\; cE_t Z_c\; h\ell \mp \frac{1}{2}j\omega\; \mu_0 H_n\; h\ell, \quad (11)$$

where c is the per-unit-length (p.u.l.) capacitance of the line. Unless otherwise noted, we simultaneously present the near-end and far-end results; $\mp$ means minus for the near end and plus for the far end.

B. Thought experiment

Let us perform a thought experiment on the lossless, characteristically terminated line of Figure 2a, illuminated from the near-end side ($k_p = +\|k\|$). The illuminating field has a normalised amplitude i which is just a phase lag:

$$i(z) = e^{-jk_p z}; \quad k_p = \frac{\omega}{c_0}, \quad (12)$$

where z is the coordinate along the line. Let us look at the far-end induced voltage, caused by a forward traveling wave on the line: the forward *eigenwave*. Its normalised amplitude w also is a phase lag:

$$w(z) = e^{-j\beta z}; \quad \beta = \frac{\omega}{v}, \quad (13)$$

where v is the phase speed of a wave on the transmission line.

We start at an illumination frequency where the error of a single Taylor cell is negligible, and let the frequency increase little by little. Using Figure 7, we try to imagine what happens.

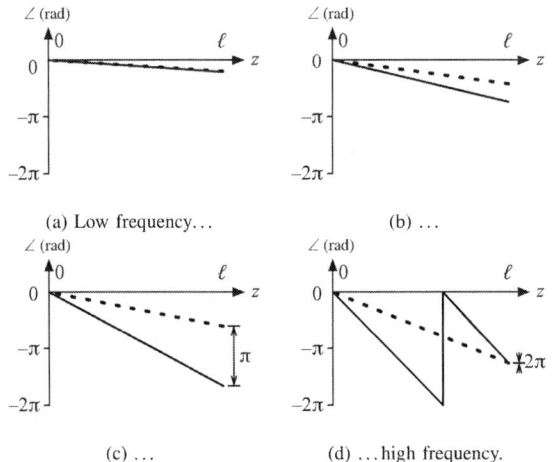

(a) Low frequency...　　　　(b) ...

(c) ...　　　　(d) ...high frequency.

Figure 7. Phase along the transmission line of the line's eigenwave $\angle w$ (solid line) and illuminating plane wave $\angle i$ (dashed line), for increasing frequency.

For low frequencies (Figure 7a), the incident field remains the same along the line, so modelling the line as one Taylor cell is legitimate.

As the frequency increases, the wavelength decreases (Figure 7b). When the wavelength is in the order of the line length, we see a propagating wave, both in free space and in the transmission line. Yet, this does not immediately invalidate the model. Indeed, the field is no longer uniform along the line, but the forward eigenwave of the line and the free space plane wave travel in the same direction. That means that, for every line slice, the free space wave and the eigenwave have approximately the same phase. Therefore, it is still reasonable to model the line as one cell.

Let the frequency increase further (Figure 7c). Now the phase difference between the forward eigenwave and the incident plane wave becomes significant; in the example shown, the phase *difference* goes from 0 at $z = 0$ to π at $z = \ell$. On average, both waves are still cross-correlated, but less so than for low frequencies.

In an extreme case (Figure 7d), the phase difference goes all the way from 0 at $z = 0$ to 2π at $z = \ell$. On average, the two waves are no longer cross-correlated and we expect no coupling.

C. Modification

So, the low frequency coupling (as predicted by a single Taylor cell) must be corrected by a measure for the length-average cross-correlation between the line's eigenwave and the incident wave. This measure should be unity for low frequencies, as not to modify the low frequency coupling. This measure should amount to zero when the phase difference along the line goes all the way from 0 to 2π. Let us call this unitless measure K.

The cross-correlation of the incident field and the line's eigenwave amplitudes is given by the complex conjugated

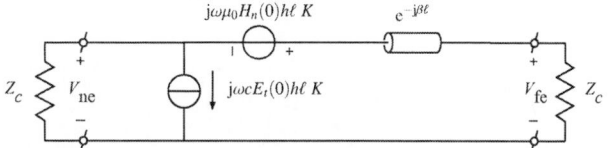

Figure 8. Modified Taylor cell, taking into account long-line effects. Note that K must be selected to predict the coupling to either the forward- or the backward travelling eigenwave.

product iw^*. K is then found by averaging along the line [2]:

$$K = \frac{1}{\ell} \int_0^\ell i(z) \cdot w^*(z) \, \mathrm{d}z = \frac{1}{\mathrm{j}(k_p \mp \beta)\ell} \left(1 - \mathrm{e}^{\mathrm{j}(k_p \mp \beta)\ell} \right). \quad (14)$$

To calculate the near-end induced voltage, the backward travelling eigenwave $w = \mathrm{e}^{+\mathrm{j}\beta z}$ was used. The resulting, modified Taylor cell is depicted in Figure 8.

This closed-form analytical solution was evaluated using a Python script [12] in 0.7 ms.

V. RESULTS

The numerical results of both simulations are compared in Figure 9. The meshed results differ by -0.4 dB on average from the modified Taylor cell, with an average absolute deviation of 1.0 dB from this error. The two can be made to approach slightly, by more cells in meshed simulation, at the expense of greater execution time (cf. Table I).

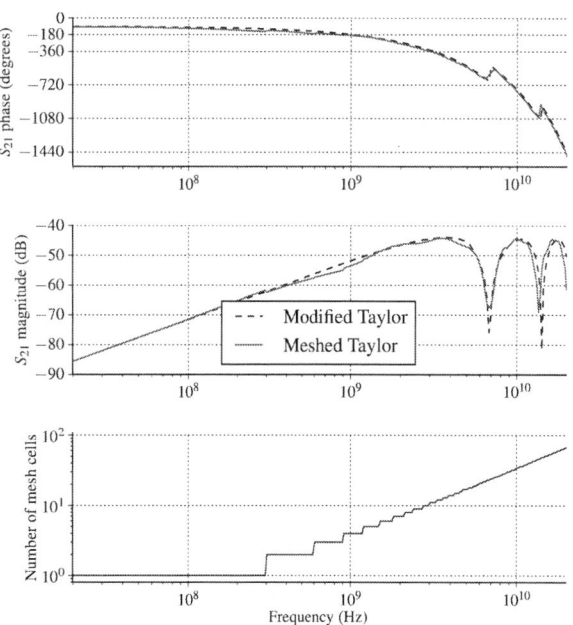

Figure 9. Simulation of the coupling from a GTEM cell input to the far end of a 5 cm microstrip trace. On the one hand the modified Taylor cell: the analytical solution of Figure 8 and (14). On the other hand a meshed Taylor cell, solved using Kron: the transmission line was meshed adaptively in 20 cells per free-space-wavelength (the 50- and 100-cell curves are indistinguishable).

Table I
CORRELATION OF MESHED AND MODIFIED TAYLOR CELL SIMULATIONS

#Cells/ wave-length	Adaptive	Meshed Execution Time	Average Meshed – Modified	Deviation Meshed – Modified
20	Yes	0.5 s	−0.4 dB	1.0 dB
50	Yes	1.8 s	−0.4 dB	0.7 dB
100	Yes	9.9 s	−0.3 dB	0.6 dB
100	No	91.1 s	−0.3 dB	0.7 dB

VI. CONCLUSIONS AND RECOMMENDATIONS

This paper presented two simulations of the coupling of a grazing-incident, vertically polarized plane wave to a characteristically terminated microstrip PCB trace. Both simulations are based on Taylor's model, which uses distributed voltage and current sources along the line. The first simulation automatically meshes the line in 20 cells per wavelength and solves the resulting circuit using Kron's formalism. The second simulation uses one modified Taylor cell, that does not need to be meshed to predict long-line effects.

The first simulation executes in 0.5 s on an 2.53 GHz Intel Core 2 Duo processor. Potentially, real world non-idealities, like non-characteristic, frequency-dependent termination impedances and excess capacitances along the line, could easily be added to the circuit. The second method executes in 0.7 ms on the same platform. The difference between the first and the second method amounts −0.4 dB on average, which can be slightly improved by increasing the number of cells.

To sum up: both simulations yield the same results, while a meshed Taylor cell is more flexible and a modified Taylor's cell is faster.

Future work on both methods seems interesting. As for the meshed Taylor cell: originally, we would have liked to use Branin's cells to represent the transmission line. However, it seemed that an otherwise uniform transmission line cannot be meshed with impunity into Branin's cells. Apparently, a cut in the model *must* correspond to some non-uniformity in the modeled line. This suggestion led Casagrande and Maurice to discover the modified Branin cell.

We here elaborated the matrices and vectors for our particular problem. Translating a circuit to a representation in Kron's formalism is a recurring and error-prone task that could be automated. One could imagine an open-source library that allows connecting circuit elements together in an object-oriented fashion. Using open libraries for symbolic calculation like sympy, symbolic simulation results may be given to the user. Apart from standard elements like resistances and capacities, there may also be circuit elements that are adaptively meshed 'under the hood' (hidden for the user).

As for the modified Taylor's cell: there is first some analytical work to be done, to take into account non-grazing incidence angles. Worst-case analysis must probably be employed to keep the solution closed-form. Moreover, it should be joined to existing microwave theory, to allow for arbitrary terminal impedances and trace discontinuities.

The speed of the closed-form calculation opens up practical possibilities. For example, the angle of incidence could be swept to produce an antenna diagram within a second. A designer could click on a PCB trace or net and almost immediately see its associated antenna diagram, instead of performing this simulation in an external full-wave solver.

By reciprocity, the far field emissions can be calculated too, if the signal levels are known by the PCB design tool.

ACKNOWLEDGEMENTS

This bit of research was made possible by the French national project SEISME (simulation of emissions and immunity of electronic systems).

REFERENCES

[1] C. R. Paul, *Introduction to Electromagnetic Compatibility*. Wiley, 2006.
[2] S. T. Op 't Land, M. Ramdani, R. Perdriau, M. Leone, and M. Drissi, "Simple, Taylor-based worst-case model for field-to-line coupling," *JPIER*, vol. 140, pp. 297–311, 2013.
[3] M. Leone and H. L. Singer, "On the coupling of an external electromagnetic field to a printed circuit board trace," *Electromagnetic Compatibility, IEEE Transactions on*, vol. 41, no. 4, pp. 418–424, 11 1999.
[4] C. A. Nucci, F. Rachidi, and M. Rubinstein, "An overview of field-to-transmission line interaction," *Applied Computational Electromagnetics Society Newsletter*, vol. 22, no. 1, pp. 9–27, 2007.
[5] C. D. Taylor, R. Satterwhite, and C. W. Harrison, Jr., "The response of a terminated two-wire transmission line excited by a nonuniform electromagnetic field," *Antennas and Propagation, IEEE Transactions on*, vol. 13, no. 6, pp. 987 – 989, nov 1965.
[6] F. Rachidi, "Formulation of the field-to-transmission line coupling equations in terms of magnetic excitation field," *Electromagnetic Compatibility, IEEE Transactions on*, vol. 35, no. 3, pp. 404 –407, aug 1993.
[7] A. K. Agrawal, H. J. Price, and S. H. Gurbaxani, "Transient response of multiconductor transmission lines excited by a nonuniform electromagnetic field," *Electromagnetic Compatibility, IEEE Transactions on*, vol. EMC-22, no. 2, pp. 119–129, 5 1980.
[8] G. Kron, *Tensor Analysis of Networks*. John Wiley and Sons, Inc., 1939.
[9] O. Maurice, "Theoretical application of the tensorial analysis of network for EMC at the system level," PSA Peugeot-Citroen, Tech. Rep., 2007. [Online]. Available: http://hal.archives-ouvertes.fr/hal-00166215
[10] ——, "xTan par la pratique." [Online]. Available: http://olivier.maurice. pagesperso-orange.fr/topologieAppliquee/ExTAN_for_Students_bat.pdf
[11] V. Stodden, C. Hurlin, and C. Perignon, "Runmycode.org: A novel dissemination and collaboration platform for executing published computational results," in *E-Science (e-Science), 2012 IEEE 8th International Conference on*, 2012, pp. 1–8.
[12] Field-to-line coupling models and measurement data. [Online]. Available: https://github.com/eseo-emc/field2line
[13] S. T. Op 't Land, T. Mandić, M. Ramdani, A. Barić, R. Perdriau, and B. Nauwelaers, "Comparison of field-to-line coupling models: Coupled transmission lines model versus single-cell corrected taylor model," in *EMC Europe 2013*, accepted.

Extraction of Deterministic and Random LSI Noise Models with the Printed Reverberation Board

Umberto Paoletti and Takashi Suga
Hitachi Ltd., Yokohama Research Laboratory,
292 Yoshida-cho, Totsuka-ku, Yokohama 244-0817, Japan
Email: umberto.paoletti.ff@hitachi.com

Abstract—The main features of the printed reverberation board are revised and the methodology to extract an LSI noise model for electromagnetic radiation estimation at frequencies above 1 GHz with the printed reverberation board is presented. A test PCB has been designed and fabricated to verify the feasibility of the method. Measurement results are presented and discussed for two variations of the model: the deterministic and the random LSI models.

I. INTRODUCTION

Power supply noise of printed circuit boards (PCBs) is a source of high frequency electromagnetic interferences (EMI) that is mainly generated by simultaneous switching of large scale of integration (LSI) circuits as shown schematically in Fig. 1. Estimating the amount of power supply noise that an LSI produces is fundamental for predicting the emissions and also for the design of printed circuit boards with low EMI.

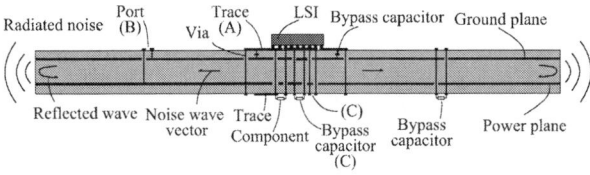

Fig. 1. Noise radiation mechanism and measurement locations (A), (B), (C)

Since an estimation based on simulations is very difficult and often not practicable, measurement of power supply noise is necessary (e.g. [1]). However, the available measurement and modeling techniques are usually limited to frequencies below 1 GHz (e.g. [2]), and their application to packages with many pins or to ball grid arrays (BGA) is problematic.

A special fixture to measure simultaneously the current in all the BGA pins can be found in [3], which however has frequency limitations at around 2-3 GHz due to the fixture itself, and is aimed at monitoring the LSI current distribution among the pins. Measurements of the board impedance on the IC pads when the LSI is unmounted have been conducted in [4] in order to indirectly extract an LSI model, but the model is limited up to 1 GHz and is valid only for the particular board where the impedance has been measured.

A model having one current source for each pin is too complicated to handle, also because a multiport LSI impedance model is required. For this reason in [5] it was proposed to use a single current source to model all the current injected by all the pins into a power plane pair. The idea of using a single noise source for a whole LSI had been proposed also

in [6], where it was shown based on numerical simulations that the model was limited up to frequencies at which the package is smaller than one sixth of the wavelength. The reason for the conclusion in [6] can be attributed to the fact that the noise emissions were observed at single points in the PCB, and it is clear that by increasing the frequency the noise emissions become less isotropic causing different observations in different positions that cannot be expressed with a single isotropic noise current source. The effect is probably further amplified by the board resonances.

However, regardless of the emission pattern, information about the total emissions of the LSI is already extremely important to decide whether to take actions to reduce the excessive noise at some frequencies, and possibly also to calculate the radiated emissions from the board with a statistical approach. With this in mind, in [5], a new approach to the problem of extracting the LSI noise source model above 1 GHz was introduced. The idea consists in making a large number of measurements at one high frequency port in the position (B) of Fig. 1, by changing the power ground plane conditions at each measurement. Measurements can be conducted with the proposed printed reverberation board (PRB) of Fig. 2, where conditions similar to a two-dimensional reverberation chamber [7] are created, and therefore the emitted noise by the LSI can be detected independently of the particular angular emission pattern.

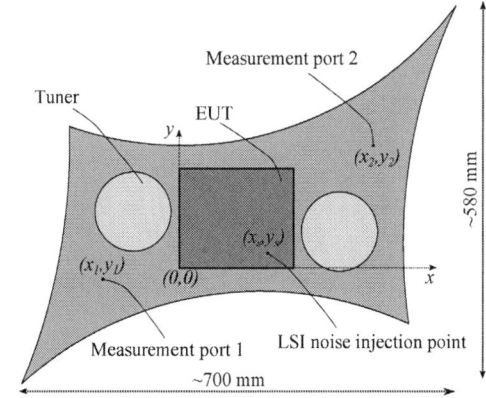

Fig. 2. PRB and coordinate system.

The present work is the continuation of the work in [5], and contains additional details of the setup and of the results with respect to [8]. In section II the main features of the

978-1-4799-5004-1/13 $31.00 © 2013 IEEE

PRB are revised. In section III the LSI model is defined by means of an equivalent port and the methodology to extract the LSI model is explained for both the usual deterministic model and the random LSI model. The model has been verified experimentally with the help of an active PCB that has been designed and fabricated to this purpose as well as for other experiments, and which is presented in section IV together with the measurement setup. The results for the extracted LSI model are discussed in section V.

II. THE PRINTED REVERBERATION BOARD

In the present realization shown in Fig. 2, the PRB is a two layer PCB with full conducting planes on the top and bottom layers. In order to measure the power supply noise, the EUT must be inserted into a hole provided in the PRB and the power supply planes of interest must be connected to the PRB all along the EUT perimeter. In this way the noise power is injected into the PRB and it is measured at the high frequency measurement ports that are provided in the PRB between power and ground planes. This requires the EUT to be a special multilayer PCB where the power and ground planes of interest are accessible from the top and bottom layers, for example by means of blind vias all around the edges of the PCB as shown in Fig. 3. It is also important that as few vias as possible cross the power ground planes, except those of interest for the measurements.

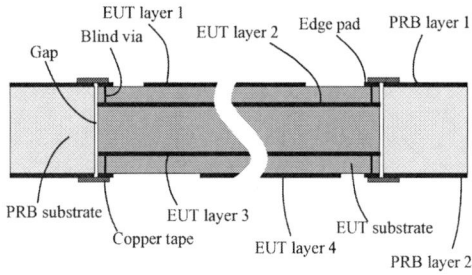

Fig. 3. Connection between EUT and PRB.

Since the main target of the present model is the radiation estimation, in the case of PCBs with more than two power supply planes it should be sufficient to connect the top and bottom planes of the test PCB to the PRB, because the radiation is determined by the total edge voltage between these two planes. However, if the target of the model is the power supply noise between two particular planes in the PCB, then those two planes must be connected to the PRB.

A suitable number of tuners are provided, in order to automatically change the boundary conditions. In the present PRB, two tuners have been realized with circular PCBs having three paddles made of vias placed along three rows and connected by traces on the top and bottom layers. A higher tuner efficiency can be obtained by emulating blocks of metals using a large number of closely spaced vias connected by portion of planes on the top and bottom layers. The tuners are rotated by means of stepping motors below the board, and are covered on the top and bottom by two copper disks, which are connected to the top and bottom layers of the PRB and do not rotate. Other type of tuners might present some practical

advantages with respect to mechanical tuners, such as electric or magnetic tuners, and might be used in the future.

The PRB must have low dielectric and conductor loss and large dimensions with respect of the wavelength at the frequencies of interest. The low-loss requirement is necessary in order to have a high quality factor, which corresponds to a large variation of the power supply impedance. On the other hand, if the quality factor is too high, one mode is expected to become dominant at its resonance frequency, and this can reduce the spatial statistical uniformity of the cavity at that frequency.

A target feature is the statistical uniformity of the PRB except for some special regions, which strictly speaking means that the statistical distribution of the outcomes of a set of experiments is the same in almost any position of the board. For example, the statistical distribution of the power supply input impedance in one set of measurements in one board position is the same as the statistical distribution in a second position, when both ports are at a distance larger than half-wavelength from the board edges.

The advantages of a statistical uniform environment, are that it simplifies the preparation of a statistical model of the power supply impedance, and it allows to make measurements in almost any position. In more practical realizations, the statistical distribution of the outcomes of a set of experiments can be more realistically described as the combination of a strictly statistically uniform component and a component that is dependent on the position.

It must be remarked that the PRB can have also other applications than the one of interest here. For example, it can be used for evaluating the immunity of one LSI to the power supply noise, or for testing how the LSI works with different conditions of the power supply impedance.

III. LSI MODEL AND CALCULATION METHOD

The equivalent circuit for the PRB is shown Fig. 4, where port 1 represents the equivalent port on the LSI side, and port 2 represents one of the measurement ports. Port 2 can be easily defined at the SMA connector, but the definition of the LSI port requires more attention. The equivalent current I_1 represents the total current injected by all the LSI pins into the power supply planes. Its position (x_s, y_s) in Fig. 2 can be conventionally set at the center of the LSI. For the equivalent LSI port, the via port definition in [9] is used, which is based on the integration of electric and magnetic field around the port perimeter. For simplicity an equivalent circular port can be used, whose radius can be freely selected for convention, such as half of the package size in the case of a BGA package. A smaller size can be also used to reduce the port dimensions with respect to the wavelength. To this purpose an extension of the port definition might be required, for example based on the total emitted power.

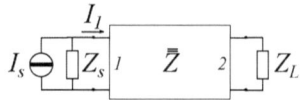

Fig. 4. Equivalent circuit of the PRB.

The source impedance Z_s represents the dependency of the port current I_1 on the board input impedance, and it is important in order to use the same LSI model with source current I_s for PCBs having different input impedances, assuming that above 1 GHz the LSI noise can be represented with a linear model. The load impedance Z_L is the impedance of the measurement equipment. A statistical model for the two-port impedance matrix $\bar{\bar{Z}}$ can be estimated according to [5] with the selected equivalent port radius. The port voltage can be now expressed in terms of the still unknown source current and impedance by introducing the transfer impedance Z_{2s}:

$$V_2 = \frac{Z_L Z_{21} Z_s}{(Z_L + Z_{22})(Z_s + Z_{11}) - Z_{21} Z_{12}} I_s = Z_{2s} I_s. \quad (1)$$

A. Deterministic LSI Model

In the deterministic current model the source impedance Z_s as well as the source current I_s are supposed to be deterministic variables. The problem is now how to calculate them from known statistical distributions of the port voltage and impedance matrix. Since in this phase of the work the calculation efficiency is not important, the most straightforward Monte Carlo method is used.

The calculation procedure is shown schematically in Fig. 5. At first the statistical distribution of the port voltage V_2 is measured for several positions of the tuners, and the corresponding cumulative distribution function (c.d.f.) is calculated. The calculation of the random model of the PRB can be done as second step. According to [5], 10,000 samples of the 2-port impedance matrices between the source position and each measurement port have been calculated. By assuming some initial values for the source current and impedance in step 3, for each sample of the impedance matrix at the frequencies of interest the transfer impedance can be calculated using Eq. (1) in step 4, and the expected statistical distribution of the voltage amplitude at one measurement port can be obtained in step 5. The resulting amplitude distribution is compared with the measured distribution, and the area of the difference between the corresponding c.d.f. is calculated similarly to [5] in step 6. This scalar real value becomes the goal of an optimization process and is used to modify the source current and complex source impedance values until a convergence criteria is reached. For the optimization the pattern search algorithm in Matlab R2011b has been used. Relatively loose lower and upper bounds have been used in the optimization, which has been repeated at each frequency and at each measurement port separately.

B. Random LSI Model

In the random LSI model it is assumed that the LSI noise itself is a stochastic signal, and its probability density function (p.d.f.) must be determined. Although the same rationale can be applied to both the source current and impedance, in order to keep the model as simple as possible, the randomness of the model will be concentrated on the current, while the source impedance will be considered as deterministic. An extension to a random model for the impedance as well is in principle possible and might be required in the future.

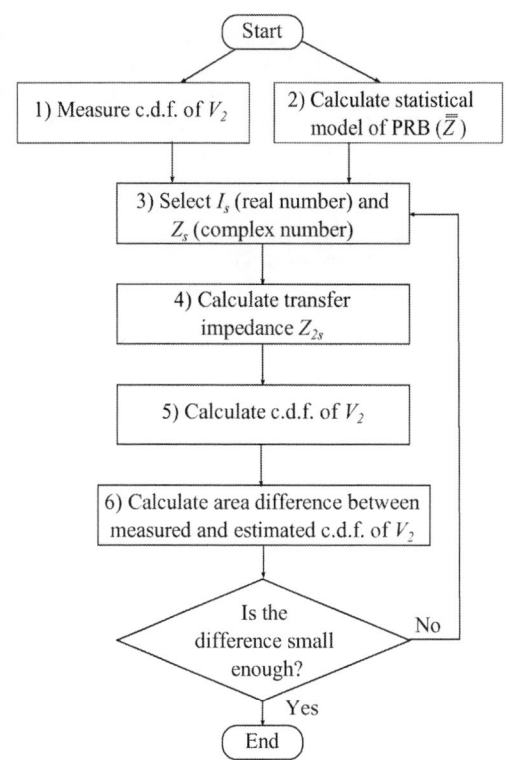

Fig. 5. Calculation procedure for deterministic LSI model.

Since the measured voltage distribution includes effects related to other noise sources and to measurement settings, the extracted LSI current distribution will be also affected. However, it is very important to understand that the highest percentiles of the current distribution can be expected to be much less dependent on the measurement settings and on the presence of other noise sources, as long as their largest contributes to the port voltages are smaller than the largest contributes of the LSI.

Therefore, after the calculation of the p.d.f. of the source current, a high percentile will be calculated, such as the 95 percentile, and the final LSI model will consist of this number, of the estimated source impedance, and of the port radius that has been selected for the calculation of the PRB model.

The calculation procedure is shown schematically in Fig. 6. The first two steps are the same as for the deterministic model. The start equation for the calculations is again Eq. (1), where this time the current as well as the impedance matrix elements must be considered as random variables. The calculation becomes simpler if the amplitudes are expressed in logarithmic scale, because the product is transformed into a sum:

$$20 \, log_{10}|V_2| = 20 \, log_{10}|Z_{2s}| + 20 \, log_{10}|I_s|. \quad (2)$$

The p.d.f. of the first term on the right-hand side can be calculated with Monte Carlo simulations in step 4 assuming an initial value for the source impedance Z_s in step 3. Assuming also an initial distribution for the p.d.f. of the current in step 3, the port voltage expressed in decibels is the sum of two

978-1-4799-5004-1/13 $31.00 © 2013 IEEE

independent random variables, and it is well known (e.g. [10]) that its p.d.f. can be calculated as the convolution of the p.d.f. of the respective random variables. From the p.d.f. of the port voltage, its c.d.f. can be calculated in step 5 and compared with the distribution of the measured port voltage in step 6. Similarly as before, the minimization of the area difference of the c.d.f can be used as goal function of an optimization process that leads to the determination of the source current p.d.f. and source impedance. The process must be repeated at each frequency and at each port separately.

The current p.d.f. in logarithmic scale in step 3 of the optimization procedure can be represented as the sum of well known functions. In the present work we used 20 Gaussian functions of unknown mean μ_i and standard deviation σ_i, leading to 40 parameters for the current in the optimization process, but alternative representations can be used.

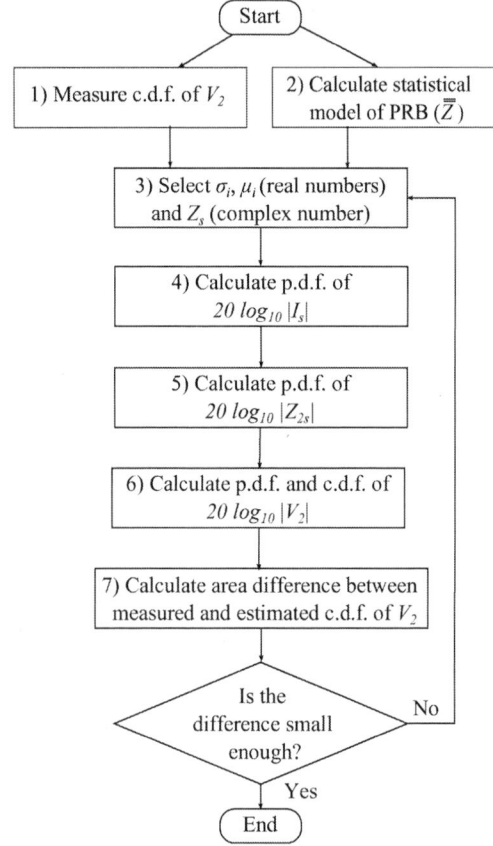

Fig. 6. Calculation procedure for random LSI model.

IV. MEASUREMENT SETUP

A. Printed reverberation board setup

The measurement setup is shown in Fig. 7. At each measurement port one SMA connector is provided, with the center conductor connected to the second layer (power plane). A DC block SMA is mounted on the SMA connector, and a spectrum analyzer is connected to it. The effect of the cable and of the DC block capacitor has been removed from the

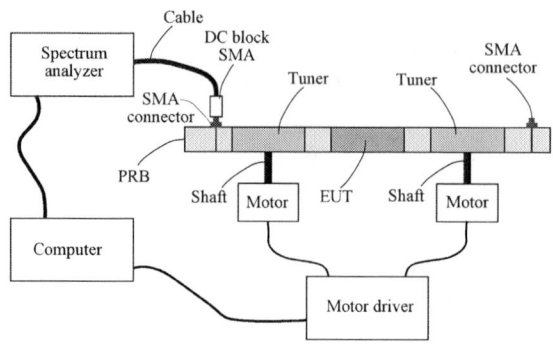

Fig. 7. Measurement setup.

measurement results. The spectrum analyzer and the stepping motors are controlled by means of a personal computer.

With the rectangular coordinate system shown in Fig. 2, having the origin in the lower left corner of the EUT, the noise source coordinates (x_s, y_s) correspond to the coordinates of the via injecting the noise signal into the power ground planes. The two measurements ports are at positions (x_1, y_1) and (x_2, y_2). The methodology for verifying the proposed technique consists in extracting two LSI models from measurements at port 1 and 2 separately, and comparing them. If the modeling technique is correct, it is independent on the measurement location, and the two models should coincide within the measurement accuracy of the proposed method.

B. Test PCB

Typically the spectrum of the power supply noise generated by an LSI above 1 GHz presents only few dominant peaks related to its clock frequency. In order to test the proposed technique it is better to have a relatively large number of harmonics in a wider frequency bandwidth, similarly to a comb generator signal. However, a comb generator is a periodic signal that lacks some of the properties of the digital noise generated by an LSI, such as its random nature or the oscillation of its clock frequency due to a spread spectrum signal that is sometimes used.

For these reasons a special digital PCB has been designed and fabricated. The basic idea is that of injecting one signal representing the LSI noise into the power supply planes in one single point. The noise signal is generated by means of a clock signal, whose spectrum is shifted above 1 GHz by means of a mixer and a sinusoidal oscillator, as shown in Fig. 8. In order to increase the bandwidth of the signal, the duty-cycle of the clock signal has been reduced with the help of a logic port with very short rise and fall times. Although the digital signal itself is deterministic, the presence of jitter that is enhanced by the noisy power supply plane gives to the spectrum of the signal some random features.

The fabricated PCB has four layers and dimensions of 180 mm × 150 mm. The substrate material is FR4. The top and bottom layers were used for signal and power supply, and their separation from the adjacent layers is 0.1 mm. In the second and third layers full ground and power planes have been prepared, and their separation is 1.2 mm. Low-voltage differential signal (LVDS) logic and current mode logic (CML) have been used.

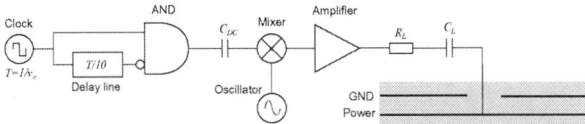

Fig. 8. Simplified schematic of EUT.

The digital clock with clock frequency ν_c of 100 MHz ($T = 1/\nu_c$) and 50% duty-cycle is delayed of 1 ns ($T/10$), inverted and combined with the original clock by means of an AND port. The resulting 10% duty-cycle pulse is multiplied with a sinusoidal signal of 5.5 GHz (ν_o), amplified and injected into the power supply planes after having removed the DC component (C_{DC}). The mixer oscillator has a separated power supply line with a voltage different from that of the power supply plane. The 50 Ω resistance R_L acts as load for the amplifier, and the capacitance C_L is required for the connection to the power plane. The capacitance C_L has been selected in order to have a self-resonance frequency of 5.5 GHz.

Along the perimeter of the PCB two rows of 5 mm separated blind vias connect pads all around the edges on the first layer to the ground plane in the second layer, and the power plane on the third layer to similar pads on the edges of the fourth layer. The pads along the perimeter of the first and fourth layer are then connected to the first and second layer of the PRB, respectively, as shown in Fig. 3. The gap between the EUT and the PRB was very small and it is expected to have a small influence in the measurement results. The variation of dielectric constant between the PRB ($\epsilon_r = 3.6$) and the test PCB ($\epsilon_r = 4.3$) is also considered to be small.

V. Measurement Results

After measuring the port voltages with the measurement setup described in the previous section and 361 different combinations of tuner positions, the deterministic and random LSI models have been calculated according to section III. For example at the frequency of 5.5 GHz, corresponding to the largest measured voltage, the measured voltage amplitude distribution and the resulting estimated c.d.f. at port 1 with the deterministic and random models are shown in Fig. 9. It is evident that the deterministic model represents only the lowest values of the voltage (and current), leading to completely wrong results for the current and to a large error in the c.d.f., which is represented by the area ΔA between the c.d.f.. On the contrary, the random model can well reconstruct the measured c.d.f. of the port voltage. This is also reflected in the huge difference in the current values of Table I, where the model parameters extracted at each port at 5.5 GHz are summarized by using the 95 percentile for the random LSI model.

The probability density function of the LSI equivalent source current I_s estimated from measurement port 1 at 5.5 GHz is shown in Fig. 10. It is important to remark that the estimated p.d.f. reflects the stochastic nature of the LSI noise current, but it is affected by other factors as well. First, it can be observed that the current values are distributed in a huge dynamic range of 50 dB. Since it is not possible in

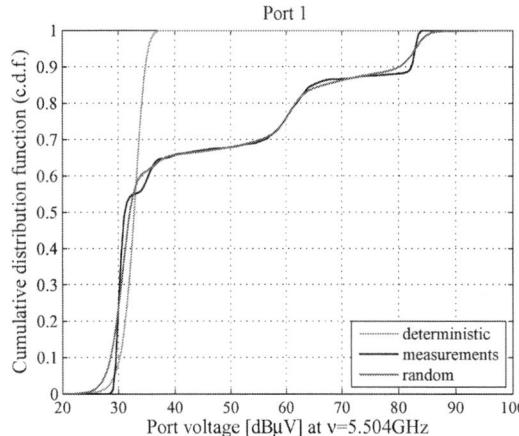

Fig. 9. Cumulative sum of measured voltages at port 1 at 5.5 GHz, and estimated cumulative distribution using deterministic and random LSI models.

TABLE I. DETERMINISTIC AND RANDOM LSI MODELS AT 5.5 GHz.

Model	Port	I_s or I_{s95}	Z_s [Ω ∠ °]	ΔA [dB]
Deterministic	Port 1	28.6 dBμA	104∠74°	12.41
	Port 2	22 dBμA	107∠69°	15.82
Random	Port 1	77 dBμA	493∠51°	0.842
	Port 2	84.5 dBμA	24∠33°	0.443

practice to have such a perfect control of the noise current in the test PCB, the lower values of the current are likely to be caused by other unknown noise sources in the test PCB and cannot be attributed to the target LSI itself. Second, the shape of the probability density function of the measured voltage, and therefore of the estimated current as well, is affected by some measurement settings of the spectrum analyzer, whereas the portion of the distribution that includes the higher percentiles is less sensitive.

Although the overall estimated probability density function cannot be completely attributed to the source current properties, in the present case we are interested only in the higher percentiles because they are going to determine the final maximum radiation levels. Therefore, it is sufficient to design the test PCB in such a way that the noise above a certain level is coming only from the target LSI.

It can be also observed that even if the lower values of the current are not reliable, the part of probability density function with the higher current values can be used to obtain some information regarding the amplitude probability distribution (APD) of the noise at one given frequency, which can be relevant for estimating the effect on an equipment that is victim of the radiation.

The frequency distributions of the 95 percentile of the random LSI model current obtained from each port is shown in Fig. 11. It can be observed that the difference between the two model is generally less than 6 dB, except at 5.5 GHz and 5.3 GHz, where it is 7 dB. The ratio of the amplitudes of the estimated impedances from each port is shown in Fig. 12. Except for the values at 5.5 GHz, 5.6 GHz and 5.8 GHz, all the remaining values differ for less than a decade, which means a reasonable accordance if the limited accuracy of the passive PRB impedance matrix model is also taken into account.

978-1-4799-5004-1/13 $31.00 © 2013 IEEE

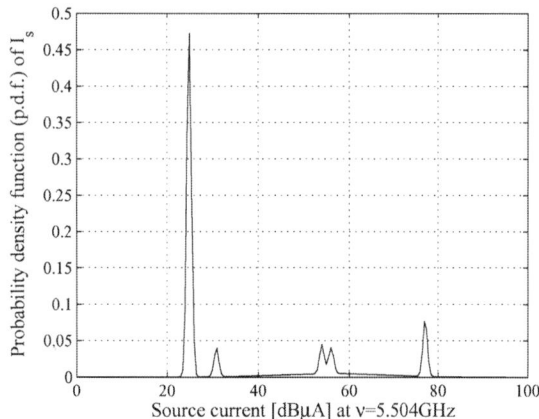

Fig. 10. Estimated probability density function of LSI equivalent port current I_s from port 1 at 5.5 GHz.

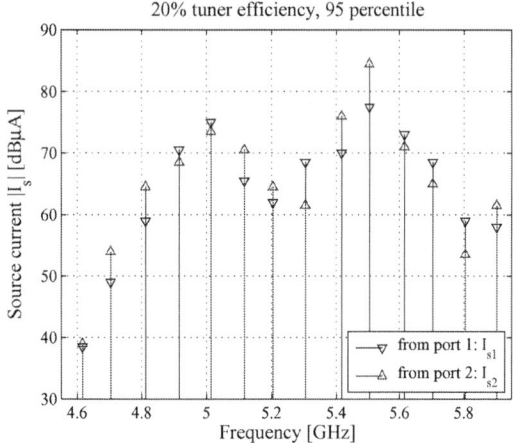

Fig. 11. Estimated LSI current obtained from measured voltages at port 1 and 2 (I_{s1} and I_{s2}).

VI. CONCLUSIONS

The methodology to extract an LSI noise model above 1 GHz using a single current source and applicable to different PCBs has been presented. An optimization method is used to determine the source parameters that fit the statistical distribution of the noise voltage measured with the PRB. The noise model consists in one frequency dependent current source, its impedance an the equivalent port radius.

Since a deterministic LSI model was not sufficient to describe the voltage distribution at the ports, a new LSI random model has been proposed. In order to verify the feasibility of the methodology, models obtained from measurements at two ports separately have been compared, and the difference is within 7 dB for the current, and a factor 20 for the impedance. Although these values may appear relatively large, considering the limited accuracy of the present PRB impedance matrix model the results confirm the feasibility of the procedure for extracting the LSI model.

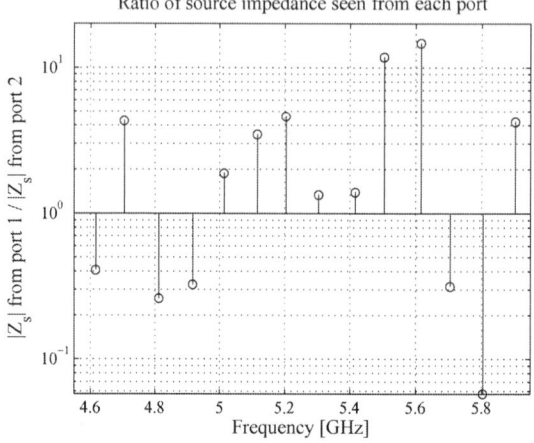

Fig. 12. Ratio of estimated LSI impedance from measurements at both ports.

It can be expected that an improvement of the PRB model and design, in particular regarding the tuner efficiency, would lead to an improvement of the estimated LSI impedance as well. Further improvements could be obtained by using also a random model for the impedance. Another possibility is to concentrate the optimization only on the higher voltage scale, because only the higher percentiles are important for the final model.

REFERENCES

[1] F. Fiori and F. Musolino, "Comparison of IC conducted emission measurement methods," *IEEE Trans. Instrum. Meas.*, vol. 52, no. 3, pp. 839–845, Jun. 2003.

[2] O. Wada, "Standardization of EMC Models of IC/LSI," in *International Zurich Symposium on Electromagnetic Compatibility*, 2006, pp. 316–319.

[3] T. Nakayama, D. Kitagawa, M. Ishii, and Y. Saito, "Proposal of a current measurement technique for each pin of a bga package," in *EMC Europe 2012 International Symposium on Electromagnetic Compatibility*, 2012.

[4] L. Li, C. Hwang, T. Wang, Y. Takita, H. Takeuchi, K. Araki, and J. Fan, "Switching-Current Measurement for Multiple ICs Sharing a Common Power Island Structure," in *IEEE EMC Symposium*, 2012, pp. 560–564.

[5] U. Paoletti and T. Suga, "The Printed Reverberation Board and its Characterization," in *EMC Europe 2012 International Symposium on Electromagnetic Compatibility*, Sep. 2012.

[6] G. Antonini, J. Drewniak, M. Leone, A. Orlandi, and V. Ricchiuti, "Statistical Approach to the EMI Modeling of Large ASICs by a Single Noise-Current Source," in *Conference on Electrical Performance of Electronic Packaging*, Oct. 2003.

[7] I. 61000-4-21, "EMC, Testing and Measurement Techniques - Reverberation Chamber Test Methods," IEC 61000-4-21, 2000.

[8] U. Paoletti and T. Suga, "LSI Noise Model Extraction with the Printed Reverberation Board," in *APEMC 2013 Asia-Pacific International Symposium on Electromagnetic Compatibility*, May 2013.

[9] Y. Zhang, G. Feng, and J. Fan, "Novel Impedance Definition of a Parallel Plate Pair," *IEEE Trans. Microwave Theory Tech.*, vol. 58, no. 12, pp. 3780–3789, Dec. 2010.

[10] A. Papoulis and S. U. Pillai, *Probability, Random Variables and Stochastic Processes*, 4th ed. McGraw-Hill, 2002.

Broadband Detection of Radiating Moments using the TEM-cell and a phase-calibrated Oscilloscope

Renaud Gillon (1), Niko Bako (2), Adrijan Baric (2)

(1) ON Semiconductor Belgium, Oudenaarde, Belgium
(2) University of Zagreb, Faculty of Electrical Engineering and Computing, Zagreb, Croatia

Abstract – The differential and common-mode signals across the TEM-cell ports are known to be correlated with magnetic and electric dipoles respectively. Historically, the absence of phase information in the measurements complicated the identification procedure for radiating moments. Using a phase-calibrated scope, this paper demonstrates a broadband and effective method for the extraction of radiating moments.

I. Introduction

Historically, most standardized EMC measurements have been based on the detection of power-levels using such equipments as power-meters or spectrum analyzers. The absence of phase information in the detected signals has complicated the problem of identifying the equivalent radiation sources from a device-under-test (DUT) using the TEM-cell. In order to deal with numerical problems in the systems of equations, Wilson introduced in [1] an algorithm based on 9 measurements allowing to determine 6 dipoles. Gerth et al. improved in [2] the procedure by introducing a least-squares method allowing to deal with the over-determined system resulting for an even large number of measurements.

Kasturi et al. proposed in [3] an interesting alternative, using a hybrid coupler to form sums and differences of signals collected at the TEM-cell ports before their detection by the spectrum analyzer or power-meter. Pan et al. showed in [4] that the method allowed them to easily determine the 3 dipoles sufficient to characterize the emissions from printed-circuit boards (PCB) with a solid ground-plane, using three measurements.

Recently, Blecic et al. in [5], replaced the usage of the hybrid coupler by recording phase information using the vector-network analyzer (VNA). Removing the hybrid coupler, the setup gains in bandwidth, the noise floor is lowered thanks to the very narrow band of the synchronous detection and parasitic losses in the cabling can be easily compensated.

However, the VNA requires a stable reference signal on which to phase-lock which drastically limits the practical application of this setup.

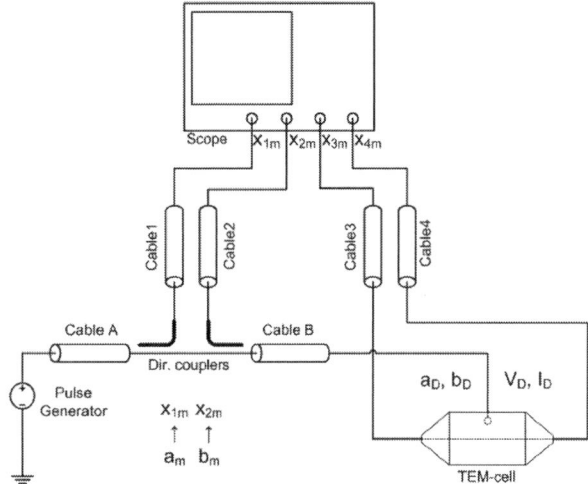

Figure 1: TEM-cell setup using the oscilloscope to monitor signals injected/collected on the DUT in the TEM-cell window and correlate to radiated waveforms.

In order to overcome this problem, the present paper introduces the usage of an oscilloscope with phase- and amplitude-calibrated measurement channels. Once distortions emanating from the cabling and the acquisition system are corrected for, one can easily compute the differential- and common-mode signals at the output of the TEM-cell which directly map to one magnetic and one electric moment respectively. One 90° rotation of the DUT PCB and an additional measurement then allows to determine the second magnetic moment.

II. Single-port Large-Signal Calibration

Figure 1 shows the set-up used to characterize radiated emissions with the TEM-cell and the oscilloscope. Scope channels 1 and 2 together with the directional coupler effectively realize a reflectometer. The pulse source can be used in active mode (sending pulses) or in passive mode (as matched load). The reflectometer is used to monitor excitation signals at the level of the DUT.

The basis of our reflectometer calibration procedure is an S-parameter calibration. S-parameter calibration techniques operate in the frequency domain and rely on the following set of linear equations which relates the raw measurements (x_{1m}, x_{2m}) to the DUT quantities (a_D, b_D) evaluated at the calibration plane :

$$
\begin{matrix} a_D(\omega) \\ b_D(\omega) \end{matrix} = \begin{bmatrix} e_{11} & e_{12} \\ e_{21} & e_{22} \end{bmatrix} \cdot \begin{matrix} x_{1m}(\omega) \\ x_{2m}(\omega) \end{matrix} \qquad (1)
$$

where the e_{ij} are the frequency-dependent error-coefficients capturing the frequency response of the measurement system. It was shown in [6] and [7] how the e_{ij} can be experimentally determined applying successively a "small-signal" and a "large-signal" calibration. The procedure consists in the following key steps :

- The small-signal Short-Open-Load calibration illustrated in Figure 3 which allows to determine 3 of the 4 complex unknowns at every frequency point;

- The power-level calibration illustrated in Figure 4, which determines the magnitude of the last complex unknown at each frequency;

- The phase calibration using a comb-generator as illustrated in Figure 5, which allows to determine the phase of the last complex unknown at each frequency.

Once the e_{ij} are known at all frequencies, the procedure described in Figure 2 can be applied to reconstruct the waveforms occurring at the DUT from those recorded on the oscilloscope channels:

1. Apply the Fourier transformation on ($x_{1m}(t)$, $x_{2m}(t)$) to obtain their spectra ($x_{1m}(\omega)$, $x_{2m}(\omega)$)

2. Perform the error correction using the e_{ij} to obtain the travelling waves spectra ($a_D(\omega)$, $b_D(\omega)$).

3. Eventually compute the voltage and current spectra ($V_D(\omega)$, $I_D(\omega)$) using the equation below :

$$
\begin{matrix} V_D(\omega) \\ I_D(\omega) \end{matrix} = \begin{bmatrix} 1 & 1 \\ 1/Z_{ref} & -1/Z_{ref} \end{bmatrix} \cdot \begin{matrix} a_D(\omega) \\ b_D(\omega) \end{matrix} \qquad (2)
$$

4. Apply the inverse Fourier transform to obtain the time-domain waveforms occurring at the DUT : ($a_D(t)$, $b_D(t)$) and ($V_D(t)$, $I_D(t)$).

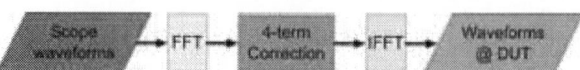

Figure 2: Main steps of the waveform processing procedure

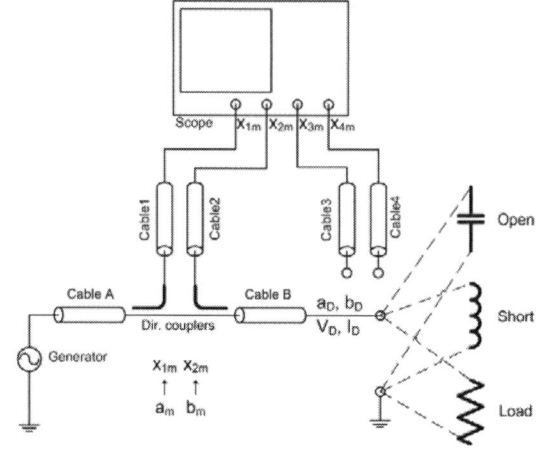

Figure 3: Single-port small-signal calibration of the analyzer consisting in the directional couplers and scope channels 1 and 2.

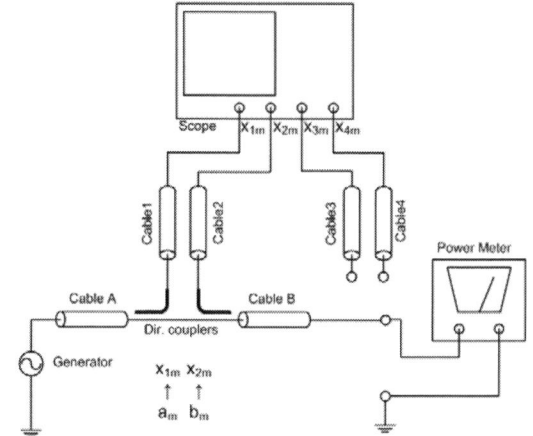

Figure 4: Power-level calibration of the analyzer consisting in the directional couplers and scope channels 1 and 2.

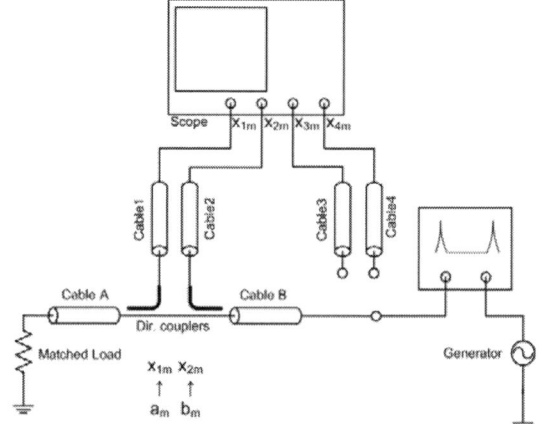

Figure 5: Phase calibration of the analyzer consisting in the directional couplers and scope channels 1 and 2.

III. Calibration Transfer

After calibrating the reflectometer consisting of first two scope channels and the directional couplers, one must now characterize the voltage transfer function of the last two channels. The procedure is described below only for channel 3, but can be easily transposed to channel 4 as well.

In order to characterize the voltage transfer function of channel 3, one must apply a known voltage at the input (DUT end of cable 3) and record the corresponding readings on the scope for a series of single tones applied at the appropriate frequencies. This can be conveniently realized by connecting the output of the reflectometer (DUT end of cable B) to the input of channel 3 (DUT end of cable 3), as illustrated in Figure 6. The resulting signals can be compared and the transfer function of the channel 3 can be extracted according to the equation below :

$$x_{3m}(\omega) = H(\omega) \cdot V_{D3}(\omega) \qquad (3)$$

Using the inverse coefficients $1/H(\omega)$ it is then possible to reconstruct the spectrum $V_{D3}(\omega)$ from the $x_{3m}(\omega)$ data recorded on the scope channel and applying the inverse Fourier transform, obtain the voltage waveform $V_{D3}(t)$. The same procedure applies to channel 4.

Figure 8 illustrates the validity of the proposed calibration procedure by comparing the waveforms obtained when measuring a power-divider, as shown in Figure 7. For the sake of the example, in order to clearly show the potential of the calibration procedure, the lengths of cables 3 and 4 where chosen to be different at 1 meter and 2 meters respectively. The additional delay on channel 4 (port 3) is clearly visible on the raw waveforms in Figure 7, whilst the calibrated waveforms show signals which are very closely in phase at both output ports of the power-divider, as expected.

IV. Detection of EMI sources

As indicated by Pan et al. in [4], from the differential and common-mode signals at the ports of the TEM-cell it is possible to reconstruct one electric moment and one magnetic moment for one position of the DUT board. One in-plane rotation of 90° (around a vertical axis) allows to determine one more magnetic moment.

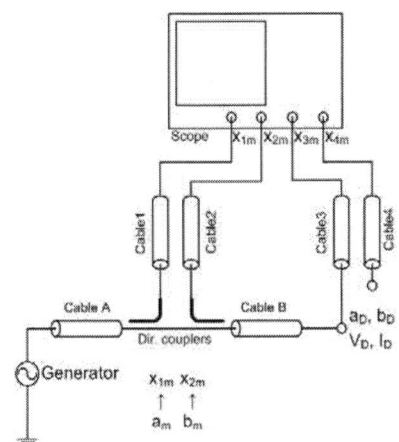

Figure 6: Setup for the calibration-transfer procedure allowing to characterize the voltage transfer function of channel 3.

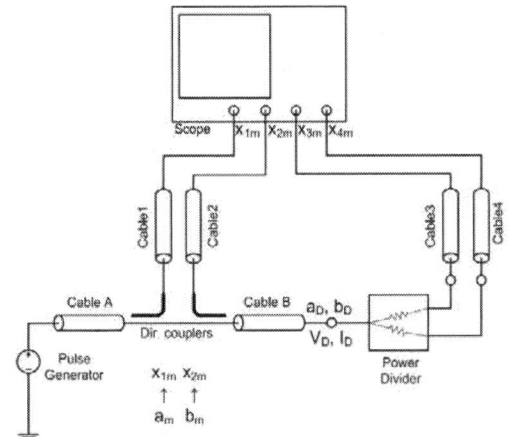

Figure 7: Validation of the setup calibration procedures using a power splitter.

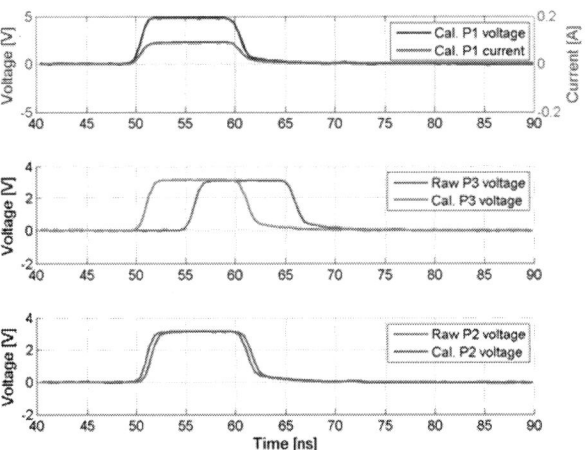

Figure 8: Comparing the raw and calibrated responses at the outputs of a power-divider to a 10ns rectangular pulse on port 1.

978-1-4799-5004-1/13 $31.00 © 2013 IEEE

With a modular board system as described by Blecic et al. in [5], using motherboards placed in the TEM-cell window with straight or right-angle bent edge connectors, it is possible to orient a daughter-board such that one more magnetic and one more electric moment can be measured. This procedure yields all 3 magnetic moments and 2 out of 3 electric moments. The last electric moment would require to rotate the PCB by 90° around a horizontal axis, which is impractical due to the use of the edge connector. Howevrer in most cases, for PCBs handling low voltages (< 100V), magnetic moments are the predominant source of radiation, which can be effectively characterized by the procedure described above.

As a result, the ability to accurately measure common- and differential-mode signals at the ports of the TEM cell is essential in order to be able to distinguish between the various radiation sources. Using the setup shown in Figure 1, with the small loop-probe shown in Figure 9 excited on one end and shorted at the other, the ability of the proposed calibration procedure to properly reconstruct the common- and differential mode signals is investigated.

Figure 10 shows the differential and common-mode signals at the output of the TEM-cell in the case the small vertical loop probe, held parallel with the septum axis. In these conditions, one can expect both a small capacitive coupling of the loop to the TEM-cell septum (electric moment) and a maximal inductive coupling (magnetic moment). Applying a 10A current pulse in the loop, one expects to see radiated signals only on the rising and falling edges, when the current and the voltage in the loop are changing. This is clearly the case for the calibrated signals, whilst the raw signals show double peaks which are uncompensated measurement artifacts due to the unbalanced delays in the cables.

In Figure 11, the vertical loop is rotated by 90° around a vertical axis, in order to zero the sensitivity of the TEM-cell to the magnetic moment of the loop. As expected, the calibrated differential signals is at a very low-level, 20dB below the common-mode signal. The raw signals, again, do not show the expected behaviour due to the unbalanced delays in the cables.

Once the differential-mode waveform at the output of the TEM-cell is accurately determined, its spectrum can be computed using the Fast Fourier Transform, and the spectrum of magnetic moment M_{MX} can be evaluated according to the formula's published ($\rightarrow$)

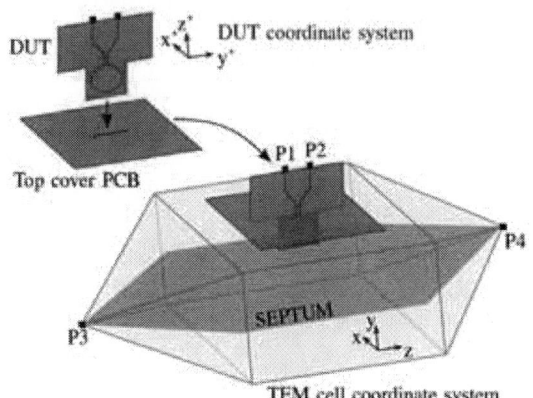

Figure 9: : Loop-probe used to excite the TEM-cell in a controlled fashion

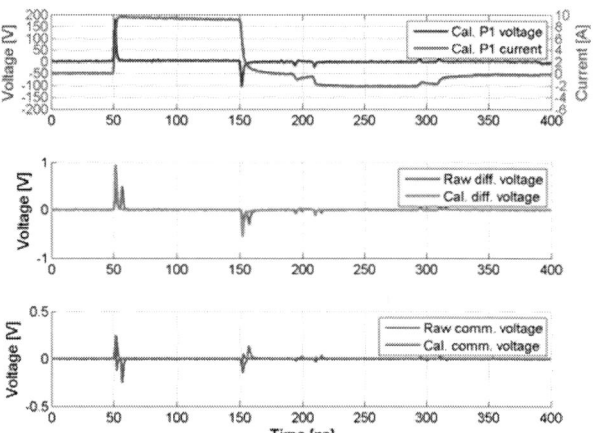

Figure 10: Differential and common-mode signals at the output of the TEM-cell in response to a 10A current pulse in a vertical loop aligned with the septum axis ..

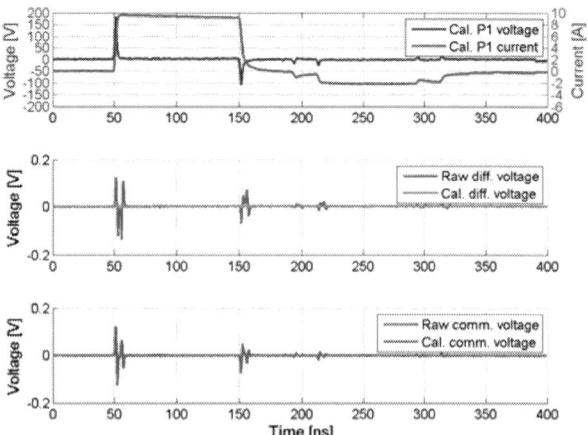

Figure 11: : Differential and common-mode signals at the output of the TEM-cell in response to a 10A current pulse in a vertical loop at right-angle with the septum axis.

by Wilson [1] or Sreenivasiah, Chang and Ma [8] :

$$M_{MX}(\omega) = \frac{B_{P4}(\omega) - B_{P3}(\omega)}{j \cdot \dfrac{\omega}{c} \cdot E_{0Y}} \qquad (4)$$

where B_{P3} and B_{P4} are the outgoing voltage waves at the TEM-cell ports; ω is the pulsation, c the speed of light in vaccum and E_{0Y} is the normalized TEM-cell field.

Figure 12 compares the spectra of the current pulse impressed on the loop shown in Figure 9 and the resulting magnetic moment computed according to the equation above. It is remarkable that the ratio between the magnitude of both quantities seems to remain constant over a wide range of frequencies. This ratio can be identified as the "effective radiating area" of the loop defined as :

$$A_{eff}(\omega) \overset{\Delta}{=} \frac{M_X(\omega)}{I(\omega)} \qquad (6)$$

Figure 13 plots the extracted effective area as a function of frequency in the case of two loops of 33 mm² geometrical area. The blue curve corresponds to the configuration shown in Figure 9, where the magnetic coupling to the TEM-cell septum is maxium. The effective area of the loop is found to match very well with the geometrical area as long as the signals remain above the noise level. The green curve corresponds to the case where the loop is rotated along a vertical axis by 90˚. In this case the magnetic coupling is expected to be zero and the residual value of the effective area measures the "isolation" level of the TEM-cell setup.

Finally, Figure 14 shows the application of the proposed method to the detection of radiated emissions from a DC-DC converter. The spectrogram of the differential signal at the TEM-cell ports was computed from the calibrated waveforms recorded at the scope and correlated with the switch-node waveform recorded on the PCB using the reflectometer channels (1 and 2) in "passive mode". This correlation clearly shows two radiating events, corresponding to the leading edge and the trailing edge resonances. A study of the common mode signals revealed that magnetic sources where the dominant mode of radiation. The data can be further processed as described by Pan et al. in [4] in order to estimate the corresponding far-field radiation spectra as well as the its evolution in the time-domain (thanks to the availability of the phase information).

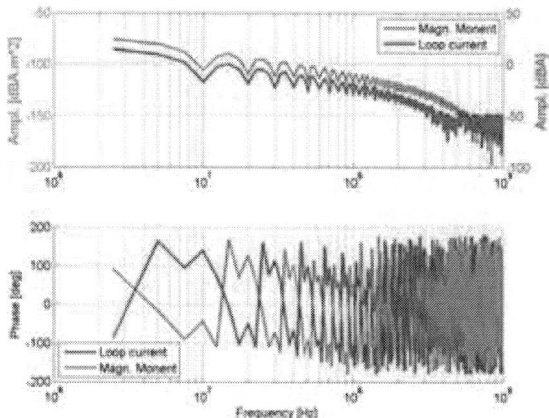

Figure 12: : Amplitude and phase of the spectral components of the current pulse impressed on the loop probe and the resulting magnetic moment deduced from the differential mode.

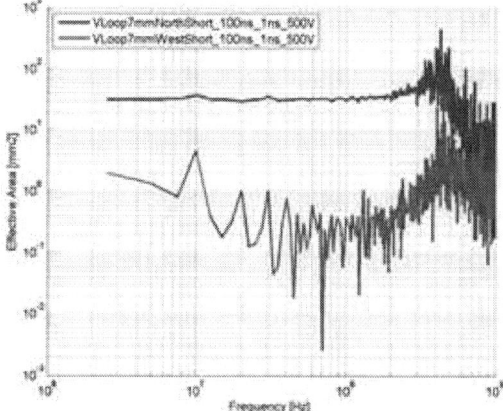

Figure 13: : Effective area of the vertical loop probe in two orientations calculated from the measured spectra. The geometrical value is 33mm².

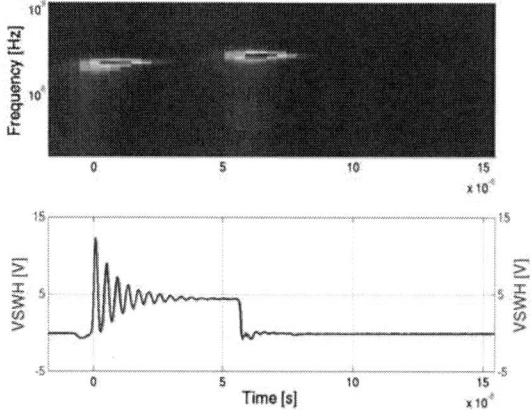

Figure 14: : Application of the proposed method to correlate magnetic-type emissions from a DC-DC converter to the switch-node voltage waveform : spectrogram of the radiated signals (top), switch-node voltage (bottom).

Conclusions

Using a phase-calibrated oscilloscope, waveforms at the TEM-cell ports can be accurately reconstructed from the recorded traces, and the corresponding common- and differential-mode signals can be evaluated, allowing to determine radiation moments with a minimum of measurements both the in the time- and the frequency domains.

The procedure has the advantage that it allows precise correlation of radiation signals with specific events recorded at the board-level in the time-domain and that it allows a very accurate compensations of losses and distortions occurring in the measurement system. The absence of a hybrid coupler in the setup allows it to operate in an intrinsicly broadband fashion.

Aknowledgements

The authors wish to acknowledge the support of Marc Vanden Bossche and Frank Verbeyst from National Instruments Belgium NV for implementing the calibration procedures in the ICE software package, [9].

References

[1] P. Wilson, "*On correlating tem cell and oats emission measurements,*" Electromagnetic Compatibility, IEEE Transactions on, vol. 37, no. 1, pp. 1 –16, Feb 1995.

[2] H. Gerth, S. Fisahn, H. Garbe, and H. Haase, "*New advances on correlating tem cell and oats emission measurements,*" Electromagnetic Compatibility, IEEE Transactions on, vol. 52, no. 1, pp. 11 –20, Feb. 2010.

[3] Kasturi, V.; Deng, S.; Hubing, T.; Beetner, D., "Quantifying electric and magnetic field coupling from integrated circuits with TEM cell measurements," *Electromagnetic Compatibility,* 2006. EMC 2006. 2006 IEEE International Symposium on , vol.2, no., pp.422,425, 14-18 Aug. 2006

[4] S. Pan, J. Kim, S. Kim, J. Park, H. Oh, and J. Fan, "*An equivalent threedipole model for ic radiated emissions based on tem cell measurements,*" in Electromagnetic Compatibility (EMC), 2010 IEEE International Symposium on, July 2010, pp. 652 –656.

[5] R. Blecic, N. Bako, R. Gillon, A. Baric, "*Broadband Measurement of the Magnetic moment Normal to the Printed Circuit Board plane using TEM Cell and Phase Information*", accepted for publication at EMC Europe 2013 Brugge, 26-30 Sept. 2013

[6] Gillon, R.; Vanden Bossche, M.; Verbeyst, F.; , "*The application of large-signal calibration techniques yields unprecedented insight during TLP and ESD testing,*" EOS/ESD Symposium, 2009 31st , pp.1-7, Aug. 30 2009-Sept. 4 2009

[7] Gillon, R.; Bossche, M.V.; Verbeyst, F.; "*Using directional couplers to overcome the bandwidth limitations of IV-probes in TLP measurements,*" EOS/ESD Symposium, 2011 33rd, p.1-7, 11-16 Sept. 2011

[8] Sreenivasiah, I.; Chang, D.C.; Ma, M.T., "*Emission Characteristics of Electrically Small Radiating Sources from Tests Inside a TEM Cell,*" Electromagnetic Compatibility, IEEE Transactions on , vol. EMC-23, no.3, pp.113,121, Aug. 1981

[9] "ICE: Integrated Component Characterization Environment" available on-line from the Network Analysis Centre of Excellence, National Instruments Belgium NV at: http://www.nmdg.be/files/ICEPresentation.pdf

Substrate Noise Reduction Based on Impedance Balance Using Tunable Resistances

Atsushi Nakamura
Renesas Electronics,
Kamimizuhonnmachi, Kodairashi, Tokyo, Japan
Email: atsushi.nakamura.fv@renesas.com

Masaaki Maeda, Tohlu Matsushima, Osami Wada
Department of Electrical Engineering,
Kyoto University
Nishikyo-ku Kyoto Daigaku Katsura, Kyoto, Japan
Email: {matsushima, wada}@kuee.kyoto-u.ac.jp

Abstract—**Substrate noise coupling from digital circuits causes degradation of performance of analog circuits on the same LSI chip. Generally large area on the chip is necessary to reduce the substrate coupling. In this paper, a proposed method eliminates the substrate coupling by extension of the impedance balance control technique which was proposed by the authors. The impedance balance condition on the LSI chip is satisfied by tunable resistances inserted into the substrate contacts. Even if the substrate coupling between two circuits is low (for example, 70 Ω,) the substrate noise coupling was reduced enough on the condition of impedance balance.**

I. INTRODUCTION

High frequency current due to LSI operation flows on power and ground conductors on the LSI. In addition, the current flows into CMOS substrate through the substrate contacts as shown in Fig. 1. In a SoC (system on a chip), analog circuits and digital circuits are integrated on the same substrate. When the logic circuits operate, the current flowing in the substrate causes voltage fluctuation of the substrate. This voltage fluctuation is transferred to the analog circuits because the circuits are on the same substrate, as shown in Fig. 2. Generally analog circuits are more sensitive to the noise than digital circuits. Therefore, an analog circuit will be affected as a victim by the voltage fluctuation through the substrate ground. In order to separate the analog circuit ground from the ground of the circuit acting as a noise source, a buffer area with highly resistivity substrate[1] is formed between the noise source area and the analog area. However, from the point of view of miniaturization, the LSI can not afford large area for the separation.

In this paper, the authors propose a new method to realize decoupling between a digital circuit and an analog circuit from the different perspectives, and verify the proposed method using a circuit simulation. To control impedance balance is important to eliminate the voltage fluctuation of the substrate [2]. According to the references [2], the authors described that the impedance balance between the impedances of the parasitic inductors of the bonding wires and the parasitic resistance in the substrate can eliminate the voltage fluctuation on the package ground and/or the bottom of the chip (voltage of the bottom side of LSI). The parasitic inductances of the bonding wires, however, depend on the type of the package the LSI on which is mounted. In this paper, we apply the impedance balance control to reduction of the substrate noise coupling on the same chip. Tunable resistances are implemented by MOSFETs and inserted between substrate contacts and on-

chip power supply networks to realize the impedance balancing network. With this new scheme, the balance can be electrically tunable when the same chip is mounted in a different package having different parasitic inductances.

Additionally extraction of the equivalent circuit model of the CMOS substrate is discussed in this paper. The extraction accuracy of the equivalent resistances of the substrate contact and substrate coupling can affect the validation of the impedance balance. In order to eliminate the voltage fluctuation of the substrate by the impedance balance control, we should know approximate values of the substrate resistances.

II. IMPEDANCE BALANCE CONTROL

A. Tunable Resistances for Impedance Balance Control

Original impedance balance [3] is represented by the ratio of the parasitic coupling on the power distribution network such as the parasitic inductances on the bonding wires and parasitic capacitances between the package and PCB. Now we apply the impedance balance to the CMOS substrate.

Using a simple CMOS inverter in Fig. 1, we explain the impedance balance control for reduction of the voltage fluctuation of the substrate. Generally, a metal electrode of a CMOS circuit is connected with the LSI package through on-chip metal interconnection and/or bonding wires. The switching currents generated by circuit operation reach to the package through this route. In addition, the current flows in the substrate below the CMOS circuit by the body contacts as shown in Fig. 1. The access to the semiconductor region via a metal contact is represented by the resistance [4]. Therefore the metal electrodes of V_{DD} and V_{SS} are connected with n-well and p-substrate by the resistive coupling, respectively. Due to the current flowing in the substrate and on the on-chip metal, the

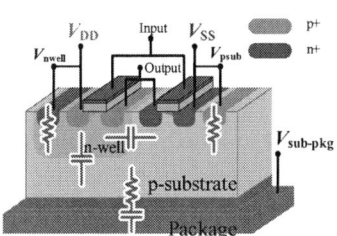

Fig. 1. Current flowing in CMOS substrate.

978-1-4799-5004-1/13 $31.00 © 2013 IEEE

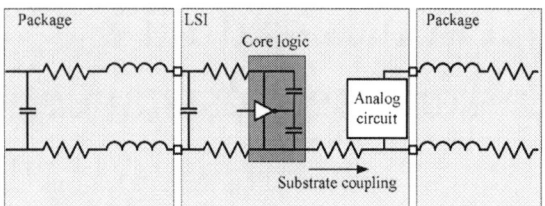

Fig. 2. Transmission of voltage fluctuation of substrate from digital circuit to analog circuit through substrate coupling.

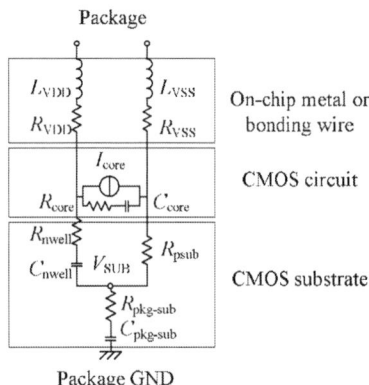

(a) Parasitic coupling around CMOS circuits.

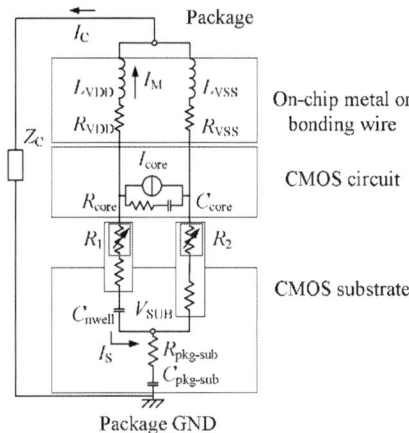

(b) Bridge circuit on assumption of low impedance bypass capacitor and tunable resistances for controlling impedance balance.

Fig. 3. Equivalent circuit model of the CMOS inverters.

voltage of VSS can fluctuate and the potential of the substrate is generally not equal to 0.

An actual digital circuit in an LSI which consists of the several CMOS circuits, such as NAND elements and Flip-Flop devices, can be represented by the equivalent circuit model illustrated in Fig. 3 from a macroscopic standpoint. The

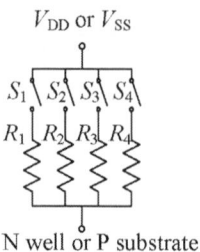

Fig. 4. Tunable resistor consisted of MOSFETs

inductances and resistances which are located in the upper part of Fig. 3 represent the parasitic coupling of the on-chip metal and bonding wires. At the middle part of the Fig. 3, LECCS model [5] are use for a noise source. The impedance model of the LSI is extract from the design information of an LSI circuit. R_{core} and C_{core} represent the ON resistance and the OFF capacitance, respectively. In addition, the current source generates the noise current due to operation of the LSI. In this paper, we use a constant current source for simplicity. Finally, the circuits elements at the bottom of the Fig. 3 show parasitic coupling in the CMOS substrate. R_{nwell} and R_{psub} mean the equivalent resistances including the series resistances, and C_{bw} represents the PN junction capacitance.

If the bypass capacitors are placed on the package and we can assume that the input impedance of the power distribution network of the package is low enough, the equivalent circuit can be illustrated as the bridge circuit in Fig. 3(b). Considering this bridge circuit, we can obtain the following condition that the voltage V_{sub} does not fluctuate,

$$\frac{R_1}{R_2} = \frac{L_{\text{VDD}}}{L_{\text{VSS}}}, \tag{1}$$

$$\frac{1}{R_2} = \frac{C_{\text{bw}}(L_{\text{VSS}}R_{\text{VDD}} - L_{\text{VDD}}R_{\text{VDD}})}{L_{\text{VSS}}^2}, \tag{2}$$

on the condition that I_{M} and I_{S} are larger than I_{C}.

However, the parasitic inductances depend on the packaging condition, such as the number of ground connection and the layout of the PCB. As a result, the impedance balance condition in Eq. (1) is changed with respect to each LSI package or PCB layout. In order to satisfy the impedance balance condition, we have to tune the impedances of the parasitic coupling composing the bridge circuit in Fig. 3. Since R_{nwell} and R_{psub} are the equivalent resistances in semiconductor material, the resistances are not controlled purposely. Then the tunable resistances R_1 and R_2 are inserted in series with the R_{nwell} and R_{psub}, respectively, in order to tune the impedance in the bride circuit. The tunable resistances consist of several MOSFETs which are connected in parallel as shown in Fig, 4, and are inserted between the V_{DD} or V_{SS} and substrate contact. After the LSI is mounted on the package and the parasitic couplings in the bonding wire and the package are fixed, appropriate resistances of the tunable resistor can realize the condition of the impedance balance formulated in Eq. (1).

B. Noise Coupling between Analog and Digital Circuit

In this subsection, the transmission of the substrate voltage from a digital circuit to an analog circuit is discussed. We

978-1-4799-5004-1/13 $31.00 © 2013 IEEE

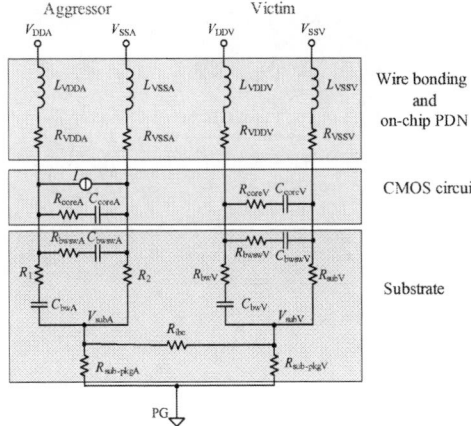

Fig. 5.   Equivalent circuit model of PDNs of digital circuit and analog circuit.

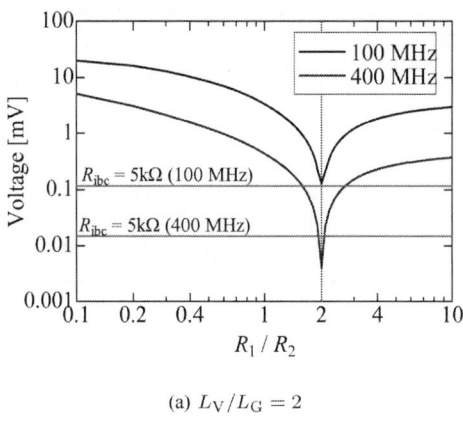

(a) $L_V/L_G = 2$

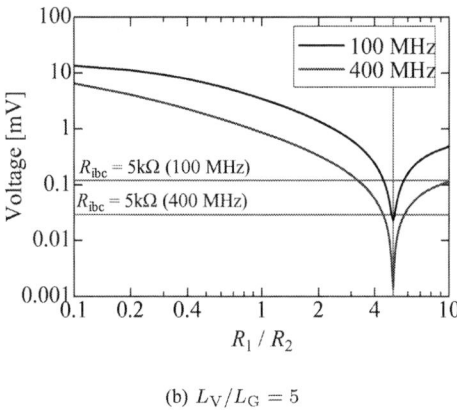

(b) $L_V/L_G = 5$

Fig. 6.   Simulation results of substrate voltage at analog circuit ($R_{ibc} = 70\Omega$)

assume an LSI which has two power domain for an digital circuit and an analog circuit. These two circuits in Fig. 5 are connected by only resistor network which represents substrate coupling. When the digital circuit operates, the substrate voltage, V_{sub}, fluctuates as a function of the noise current I, and the voltage fluctuation spread to analog circuit though the substrate resistance. Generally, the highly resistive region is laid out between the digital circuit and the analog circuit to achieve high isolation. This technique is effective to suppress the degradation of the performance of analog. However, the large isolation area is needed in the LSI.

From the standpoint of the impedance balance, reduction of the voltage fluctuation of the substrate at the digital circuit can reduce the substrate voltage at the analog circuit. According to the previous subsection, the impedance balance formulated in Eq. (1) at the digital circuit can be controlled by the tunable resistances located at the substrate contact. By achieve the impedance balance condition, we can remove the highly resistive region for isolation between the digital and the analog circuits. In the next section, reduction effect of the impedance balance will be demonstrated using circuit simulation.

III.   VERIFICATION OF SUBSTRATE NOISE REDUCTION BY CIRCUIT SIMULATION

A.   Simulation Circuit

The circuit simulation using HSPICE is carried out to verify the proposed method for reduction of transmission of the noise to the analog circuit. Simulation circuit is illustrated in Fig. 5. The left-hand side of Fig. 5 represents operation of the digital circuits as the aggressor and the right-hand side represents the analog circuit.

In the digital circuit, the parameters of these elements are obtained by the design information under the supposition that the logic circuit has about 200,000 CMOS gates with 180 nm process. Model parameters in the analog circuit are calculated as a one-five size of the logic circuit. Substrate model parameters, such as R_{nwell} and R_{psub}, are obtained by the resistor mesh network model, and the extraction method is explained in the Appendix. In addition, the amplitude of

the noise current source I is 1 A while the circuit simulation. Namely the transfer impedance from the noise current source to the voltage of the substrate in the analog circuit is calculated.

The bonding wires have generally several nano-henry inductances. These inductances influence the impedance balance as we discussed in the previous section. In order to verify the effect of these inductances to the impedance balance, two pairs of the inductances of the bonding wires connected with the digital circuit are set in this circuit simulation. We apply the tunable balance with the same chip model with two cases of different packages. One package has relatively low inductances ($L_{VDD} = 2$nH), and the other has relatively high inductance in V_{DD} path($L_{VDD} = 5$nH).

B.   Simulation Results

Before proposed method is evaluated, we discuss the transfer impedance on the condition of the high-resistive isolation between the digital circuit and the analog circuit. The horizontal lines in Fig. 6 shows the simulation results on the condition that R_{ibc} is equal to 5 kΩ as highly resistive isolation. In these simulations, R_1 and R_2 are not tunable and are fixed as 100 Ω. Of course, the substrate voltage at the analog circuit is inversely proportional to the R_{ibc}. If this isolation is not enough on the several applications, farther distance between

978-1-4799-5004-1/13 $31.00 © 2013 IEEE   35

the digital circuit and the analog circuits is necessary to isolate more.

Next, we explain the results of the proposed method. Controlling impedance balance by the tunable resistors, we obtain the V-shaped line in Fig. 6 on the condition on the low-isolation resistance which is equal to 70 Ω. The ratio of tunable resistances controls the substrate voltage of the analog circuit. When the ratio of the resistances is equal to that of the inductances, $L_{\mathrm{VDD}}/L_{\mathrm{VSS}}$, the substrate voltage takes the minimum value as indicated by the Eq. (1). Under the condition of Eq (1), the voltage fluctuation of the substrate at the analog circuit is lower in spite of low isolation resistance. These results indicate that using impedance balance substantially reduce the noise transmission to the analog circuit even if the analog circuit is located near the noisy digital circuit.

IV. CONCLUSION

In this paper, we proposed the impedance balance control for reduction of the substrate noise coupling from a digital circuit to an analog circuit on the same chip. Insertion of tunable resistances between substrate contacts and metal electrodes of V_{DD} or V_{SS} on a CMOS circuit controlled the impedance balance. Accordingly, the voltage fluctuation of the substrate at the digital circuit was eliminated, and the transmission to the analog circuit was minimized.

For verification of the proposed method, the circuit simulation was carried out using the equivalent circuit of power distribution network of LSI which had 200,000 CMOS gates for the digital circuit. Even if the substrate resistance between two circuits is 70 Ω, the substrate noise coupling was reduced enough on the condition of impedance balance.

APPENDIX

A. Extraction of Parameters of Substrate Resistance Model

In this appendix, extraction method of the substrate model is discussed. The substrate model consists of two parts. One part is bulk p-type substrate accounting for the majority of the LSI chip (we assume the substrate thickness of which is 300 um). The characteristics of the bulk substrate can be expressed by the resistances network [6]. Firstly, the substrate is separated at regular interval and the equivalent conductance of each cell is calculated using the following equation,

$$G = \sigma \left(\frac{w \times \ell}{h} \right), \quad (3)$$

where w, ℓ and h are the width, length and height of the unit cell and σ is electrical conductivity. In the simulation, the electrical conductivity is set as 10 Ωcm.

The other part is the CMOS circuit occupying the surface of the LSI chip. The substrate contacts, n-well and superior surface of the p-substrate constitute this part as shown in Fig. 7. On silicon surface, the n-well is pattered as stripe. Therefore the mesh width must be narrowed than width of the n-well because of distinction between the n-well and p-substrate. The equivalent resistor model is constructed by the same method as the bulk substrate model. In addition, substrate contact resistances are added in this model.

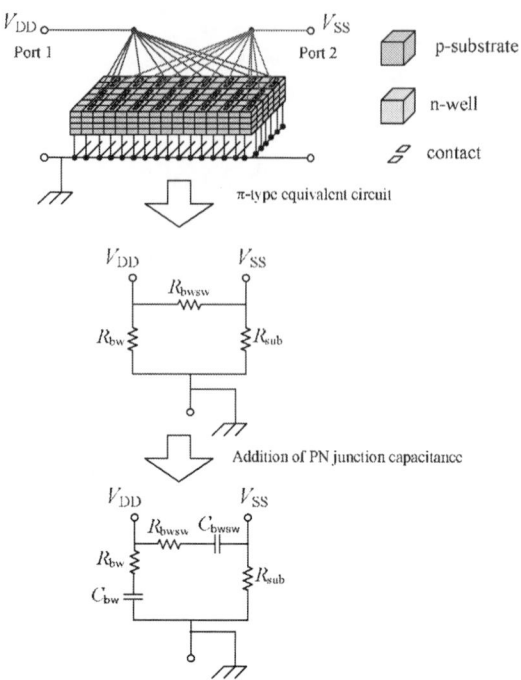

Fig. 7. Extraction of equivalent circuit model of substrate.

The equivalent resistor model of CMOS circuit is built up on the bulk substrate model. The several node of the substrate contacts are bound with respect to each V_{DD} or V_{SS}. Assuming the bottom of the substrate is set as the reference potential, we calculate the impedance matrix of the two ports circuit. We can obtain equivalent circuit model of the substrate as a middle figure in Fig. 7 using the results of calculation. Finally, the PN junction capacitors which are obtained by multiplication of the junction area and capacitor per unit area obtained from the fabrication plant.

REFERENCES

[1] S. Uemura, Y. Hiraoka, T. Kai, and S. Dosho, "Isolation techniques against substrate noise coupling utilizing through silicon via (TSV) process for RF/Mixed-Signal SoCs," *IEEE J. Solid-State Circuits*, vol. 47, no. 4, pp. 810–816, 2012.

[2] M. Maeda, T. Matsushima, T. Hisakado, and O. Wada, "Impedance balance control for suppression of substrate noise coupling in cmos ic," in *IEICE technical report, EMCJ2013-30*, vol. 113, no. 122, 2013, pp. 15–20 (in japanese).

[3] Y. Mabuchi, A. Nakamura, A. Ohmae, T. Uno, K. Ichikawa, and H. Mizuno, "Development of a low emi micro-controller package for automobile applications," in *2009 20th International Zurich Symposium on Electromagnetic Compatibility*, Zurich, Switzerland, Jan. 2009, pp. 377–380.

[4] H. H. Berger, "Contact resistance and contact resistivity," *Journal of The Electrochemical Society*, vol. 119, no. 4, pp. 507–514, 1972.

[5] O. Wada, Y. Saito, K. Nomura, Y. Sugimoto, and T. Matsushima, "Power supply current analysis of micro-controller with considering the program dependency," in *8th Workshop on Electromagnetic Compatibility of Integrated Circuits*, Dubrovnik, Croatia, Nov. 2011, pp. 93–96.

[6] S. Kumashiro, R. Rohrer, and A. Strojwas, "A new efficient method for the transient simulation of three-dimensional interconnect structures," in *Proc. IEEE Int. Electron Devices Meeting*, San Francisco, CA, USA, 1990, pp. 193–196.

Measurement-Based Diagnosis of Wireless Communication Performance in the Presence of In-Band Interferers in RF ICs

M. Nagata*, S. Shimazaki*, N. Azuma*, S. Takahashi[†], M. Murakami[‡], K. Hori[‡], S. Tanaka[¶], M. Yamaguchi[¶]

*Graduate School of System Informatics,Kobe Univ., 1-1 Rokkodai-cho, Nada-ku, Kobe, Japan
[†]Renesas Mobile Corp., 2-6-2 Nippon Bldg., Ote-machi, Chiyoda-ku, Tokyo, Japan
[‡]Renesas Electronics Corp., 1753 Shimonumabe, Nakahara-ku, Kawasaki, Japan
[¶]Tohoku Univ., 6-6-5 Aoba, Aramaki, Aoba-ku, Sendai 980-8579, Japan
nagata@cs26.scitec.kobe-u.ac.jp

Abstract—In-band interferers in wireless communication channels are due to the high order harmonics of multiple clock frequencies used by baseband digital signal processing in a single-chip solution. The impacts of in-band spurious tones on wireless performance are explored with hardware-in-the-loop simulation (HILS) of the LTE compliant systems. RF receiver circuits fabricated in a 65 nm CMOS technology are involved in the HILS, for combining circuit-level interactions at the front end and system-level digital signal processing in the back end. Experiments exhibit the sensitivity of LTE communication throughput against substrate coupling noise from a digital noise emulator to the RF receiver circuits on the same chip. The observed response is equivalently confirmed with the input referred RF sinusoidal noise components intentionally added to the input RF signal with LTE modulation. The HILS enables hierarchical diagnosis of a wireless communication system from circuit-level interactions to system-level responses against noise coupling.

Keywords—wireless communication, hardware in the loop simulation, hardware emulation, spurious noise

I. INTRODUCTION

Design complexity of RF ICs increases for the high degree of single-chip integration toward augmented functionality, a small form factor, as well as low cost, while stringently supporting performance specifications following to wireless communication standards such as long-term evolution (LTE) [1][2]. While the high sensitivity and selectivity are requested for RF chains at the front end, the high speed and large data bandwidth are strongly expected in the digital backend. Single chip integration using the latest CMOS technology is an obvious solution for meeting the requests. However, RF noise coupling from digital to RF circuits becomes very severe and potentially induces catastrophic on-chip interferers within the bandwidth of wireless channels. It is therefore needed to foresee circuit-level RF noise couplings and estimate their impacts on system-level performance of wireless communication, by looking into technology layers from device physics to modulation schemes. This is a ground challenge of integrated noise analysis technologies. On-chip noise coupling, on-board electromagnetic coupling, and in-system noise interference, are all to be covered.

Substrate noise coupling analysis has been extensively studied [3]. Chip-level simulation of noise generation in digital circuits and propagation through a silicon substrate to analog devices are accomplished for the bottom-up approach of noise managements [4]. They are very straightforward for exploring the best placements of isolation structures in a physical layout or circuit technologies of desensitization for noise coupling. On the other hand, the approach is obviously deviated from the global design optimization of wireless circuits and systems for resiliency against noise coupling and noise interference. Diagnosis techniques of RF ICs with a hybrid of hardware emulation and computer simulation will open a new paradigm for bridging the studies of circuit-level responses to environmental noises and their system-level impacts on wireless communication performance.

This paper outlines an in-system diagnosis platform of RF ICs [5] and discusses the tolerance of RF ICs against on-chip in-band interferers based on measurements. The remaining part of this paper is organized as follows. Section II describes the proposed diagnosis platform with a demonstrator implemented with a 65 nm CMOS technology. Section III then discusses measurement results. A brief conclusion of this paper will be given in Sect. IV.

II. IN-SYSTEM RFIC DIAGNOSIS PLATFORM

A. System overview

In-system diagnosis of RF ICs is embodied in a hardware-in-the-loop (HILS) based platform of Fig. 1 [5]. It consists mainly of two parts; (i) a system-level RF simulator with hardware connections with RF ICs, and (ii) a silicon emulator of on-chip interferers coupled to the RF ICs.

In the first part, an RF signal chain of a receiving path (Rx chain) is fully compliant with the LTE standard and inserted in a closed loop of simulation for in-system diagnosis. The modulated down-link RF vector signal is created by an electronic system level (ESL) simulator with LTE physical architecture (PHY) models and stored in the vector memory of a signal generator (SG). A physically modulated RF signal on a carrier sinusoid at 2.1 GHz is generated by the SG

978-1-4799-5004-1/13 $31.00 © 2013 IEEE

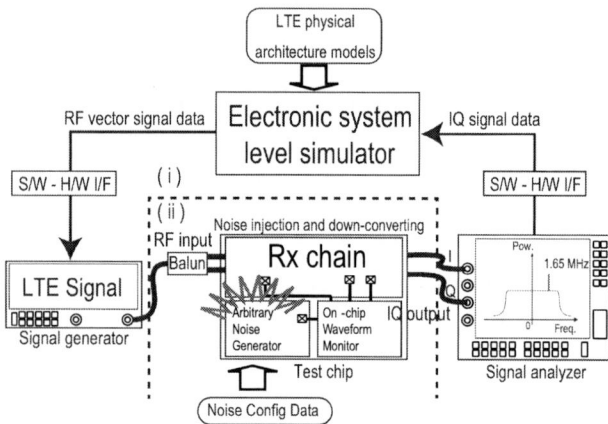

Fig. 1. In-system RFIC diagnosis platform.

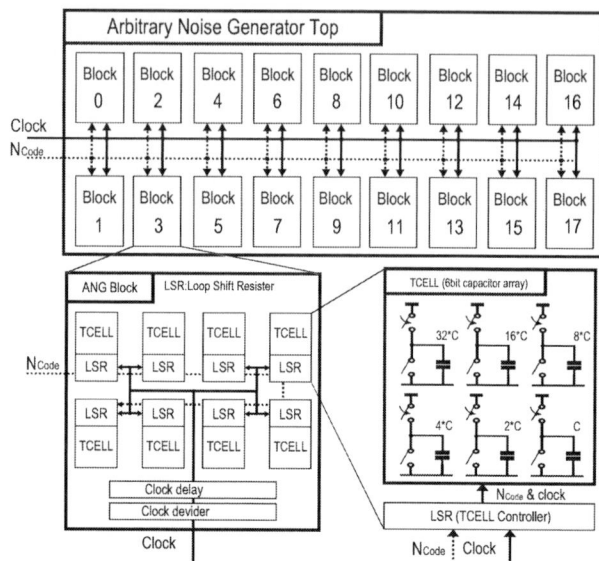

Fig. 2. Arbitrary noise generator circuit block diagram [5].

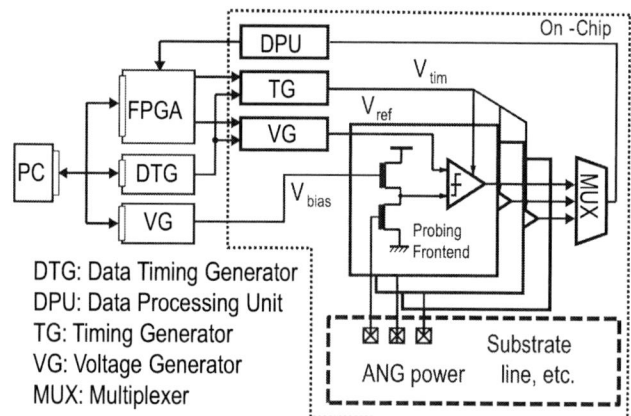

Fig. 3. On-chip monitor circuit block diagram [5].

and provided to the input of Rx chain. I/Q signals at the baseband are output from the Rx chain and captured by a signal analyzer (SA). The sampled and digitized data by the SA are then transferred to and demodulated by the simulator, in the exactly same way as in the LTE baseband digital processing of a mobile handset. The simulator can quantitatively evaluate the quality of LTE wireless channels in system-level metrics of such as the bit error rate (BER), error vector magnitude (EVM), and throughput (THP).

The second part of the platform combines an arbitrary noise generator (ANG) and an on-chip waveform monitor (OCM) for hardware emulation of RF noise coupling into the Rx chain on the same die.

The substrate coupling noise creates on-chip in-band interferers within the Rx chain and appears as spurious tones (spurs) in the frequency bandwidth. The spurs can degrade the quality of wireless channels and reduce the system-level communication performance. The in-system diagnosis platform provides the opportunities to view the response of RF ICs to the in-band spurs under real time wireless signal processing with modulated RF signals. The susceptibility of RF ICs is also quantitatively diagnosed with the function of frequency components, timing of operation, as well as the relative and electrical distance to the source of interferers. These evaluations will not be possible with a full-simulation approach.

B. Arbitral noise generator

An on-chip arbitrary noise generator (ANG) of Fig. 2 emulates power and substrate noise generation in digital circuits. The ANG represents the process of dynamic power current consumption in standard CMOS digital circuits, following to the capacitor charging model where power current consumption during switching operations among gate elements is substituted by charging of a single capacitor. The time evolution of power current in large scale digital circuits is captured in a train of the capacitor charging processes. The ANG consists of an array of a configurable capacitor to charge

(TCELL) and timing generation modules. The size of capacitor is coded as N_{Code} in the ANG.

C. On-chip monitor

An on-chip noise monitor (OCM) of Fig. 3 with multiple probing frontend (PFE) channels measures substrate noise waveforms at a variety of locations within a chip. The PFE senses the voltage variation at the point of probing by a source follower (SF), and in-place digitizes the output voltage of SF by using a latch comparator (LC) with the help of on-chip reference voltage and sample timing generators. The resolutions are typically set at mV and ps while adjustable to the characteristics of noise waveforms.

978-1-4799-5004-1/13 $31.00 © 2013 IEEE

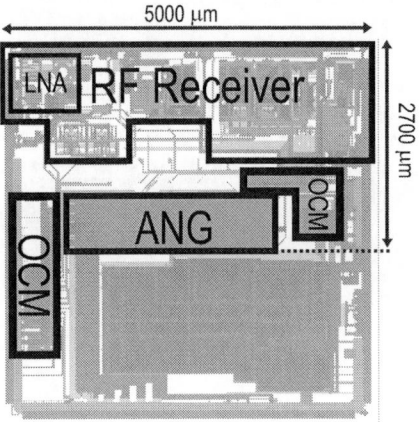

Fig. 4. Physical layout of prototype chip [5].

D. Prototype system

The prototype chip of Fig. 4 actualizes the in-system RFIC diagnosis platform. The chip with the area of 5.0 mm x 5.0 mm was fabricated with a 65 nm CMOS technology and packaged in 168-pin ball grid array (BGA). A demonstrator of the platform has a printed circuit board (PCB) mounting the BGA and providing connections to the SG and SA for the Rx chain as well as to a field programmable gate array (FPGA) device for the control of ANG and OCM. The demonstrator is tightly communicated with the ESL simulator [6] with LTE PHY models running on a work station.

The chip includes the Rx chain in the top area of the die, while locating the ANG and OCM in the middle. The signal and power I/O pins for the Rx chain are also assigned in the top side, while those for the ANG and OCM are in the bottom and both sides. The power delivery network (PDN) of Rx chain is separated from the ANG and OCM so as not to directly couple with each other both on the chip and on the board. The noise coupling between the ANG and Rx chain is considered primarily through on-chip silicon substrate coupling, while the off-chip electromagnetic coupling may also be concerned.

III. EXPERIMENTAL RESULTS

A. Substrate noise waveforms

The substrate noise waveform from the ANG is measured at the vicinity of ANG, as in Fig. 5, in the high noise configuration ($N_{\text{Code}} = 22$). Noise peaks periodically appear at every occasion of capacitor charging. The observed periodical substrate noise suggests the wide-band generation of harmonics. The frequency-domain decomposition of captured waveforms is also shown. The high order harmonics of substrate noise lies within the LTE signal bandwidth of 5 MHz centered at 2.120 GHz. The harmonics potentially appear as in-band spurs when they are contaminated with incoming RF signals and amplified by the LNA.

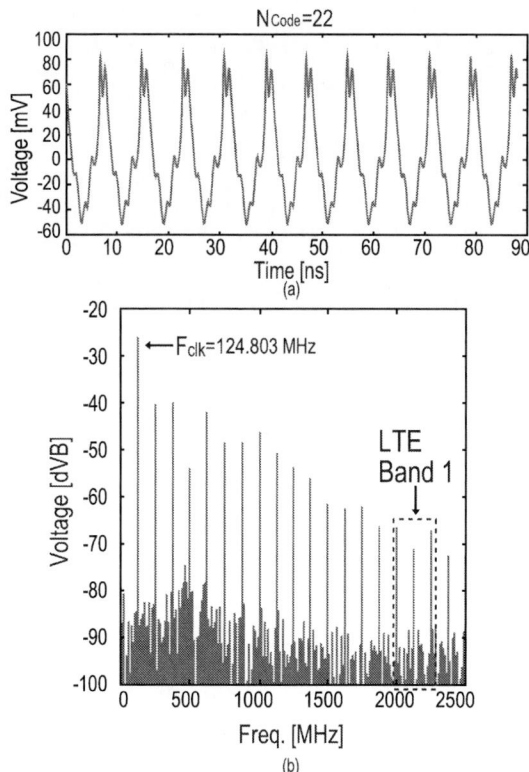

Fig. 5. On-chip substrate noise measurements. (a) Time-domain waveforms and (b) frequency components.

B. Substrate noise interference

The Rx chain receives the LTE-modulated RF signals at the input with the carrier frequency at 2.120 GHz in Band1 and outputs I/Q signals after down converting the RF signals into the baseband with the local oscillation (LO) at the same frequency. In-band frequency components are plotted in Fig. 6(a) when the Rx chain is only operating on the demonstrator. While the DC component is due to the direct down conversion, there are 300 sub frames in total with 15 kHz bandwidth each are involved in the in-band frequencies. The signal power, P_{sig}, is calculated as the integration of signal components over the bandwidth in the width of 5 MHz, as in (1).

$$P_{\text{sig}} = \int_{\text{inband}} P(f) \cdot df \tag{1}$$

When the ANG becomes active, interferers appear in-band as spurs, as observed in Fig. 6(b). The spurious power, $P_{\text{spur}}*$, is measured as the height of the largest interferer within the band. When F_{clk} of the ANG is at 124.803 MHz, the largest spur is at the 1.65 MHz. The 17th harmonics of the substrate noise by the ANG couples into the transistors of a low noise amplifier (LNA) through a silicon substrate network. The LNA is conceived as the most sensitive building block in the Rx chain. The noise components encountering the bandwidth of

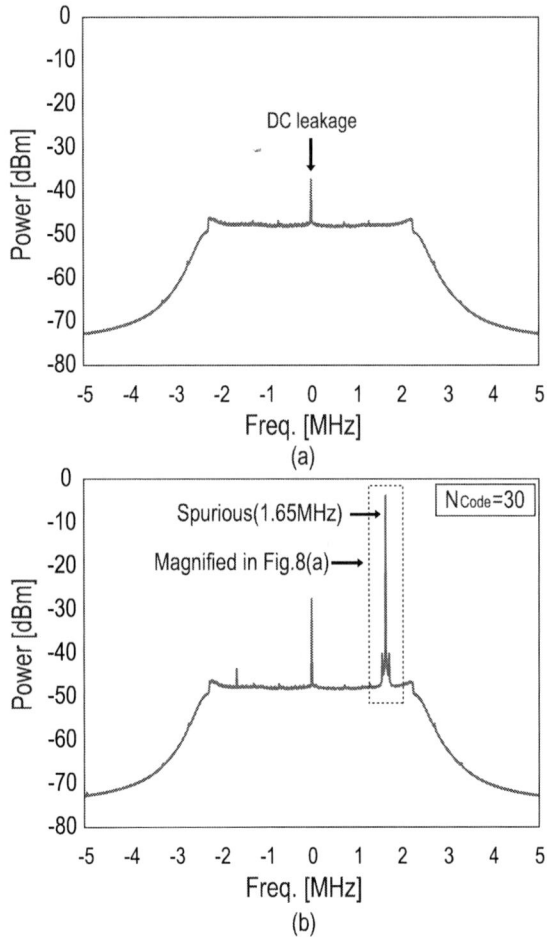

(a)

(b)

Fig. 6. I/Q frequency components in baseband. (a) LTE modulation only and (b) with ANG operation.

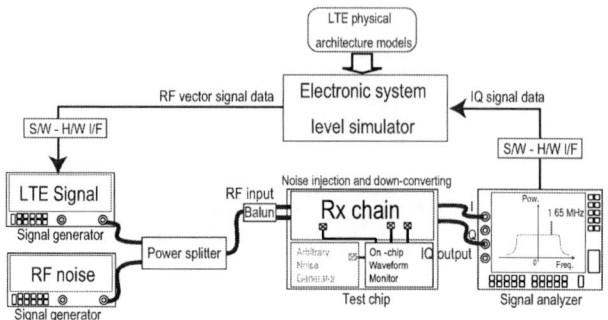

Fig. 7. Measurement setup for input referred RF noise power of substrate coupling noise.

LNA as narrow as 5 MHz in operation are amplified and transferred to the subsequent stages, and then appear as the interferers within the baseband after down conversion.

C. Impacts on wireless channels

In the criteria of LTE communication performance defined by the 3GPP, the THP of more than 95% needs to be assured with the minimum sensitivity power level of less than or equal to -100 dBmW at the input of Rx chain. The in-band spurs by the ANG degrades the THP due to some reasons like the reduction of signal-to-noise ratio within a subframe, the intermodulation among subframes, and others. As from system-level design perspectives, there is a need of noise metrics regardless of internal degradation mechanisms. We have setup the measurement system of Fig. 7 for measuring the input referred RF noise power, P_{inref}, of the RF substrate noise for producing an equivalent in-band spur with P_{spur}. This will give an idea of the strength of substrate noise coupling by the ANG, including the generation of substrate noise and

attenuation by propagation. An external signal source provides pure sinusoids at the frequency of $F_{\mathrm{inref}} = 2.12165$ GHz with the power of P_{inref}. The input referred RF sinusoidal noise is then combined with the LTE signal from another SG and given to the input of Rx chain. The measurement system of Fig. 7 constructs the HILS environment of Fig. 1 and evaluates the THP, EVM, BER of the communication channels accompanying LTE compliant input modulation and output demodulation.

The frequency components in I/Q output signals are compared for the in-band spurs due to the noise coupling from ANG in Fig. 8(a) and the input referred RF sinusoidal noise in Fig. 8(b), in a subframe. The observed insignificant difference confirms the equivalency of in-band spurs by the 17th harmonics of ANG noise components and by the input referred RF sinusoid at the same frequency.

We measured the impacts of P_{spur} by the ANG on the THP of LTE modulation in the Rx chain, as given in Fig. 9. The P_{spur} is intentionally altered by changing the N_{Code} of ANG in operation. The P_{inref} is also adjusted in response to the P_{spur}. The THP sustains higher than 98% for the in-band spurious strength of lower than -15 dBm when the input signal strength is -101 dBm at the input of Rx chain. The trends of THP versus P_{spur} and P_{inref} are consistent to each other in this condition. On the other hand, the THP significantly reduces as the P_{spur} augments with larger N_{code} and becomes higher than that point, and the trend deviates to each other. We can conclude that the impacts of substrate noise coupling by the ANG can be equivalently interpreted by the input referred large signal RF sinusoid, as long as the wireless channels are modestly within the specification. The relation of P_{spur} and P_{inref} gives an approximation of the signal gain of Rx chain but embeds the insertion losses against P_{inref} on the paths of PCB and BGA.

The HILS makes the use of the hardware emulation of substrate noise coupling for diagnosing wireless systems with versatile indicators of communication quality. The high sensitivity of THP is demonstrated, while not shown are the relatively weak sensitivity of constellation plots for calculating EVM and also that of BER in the particular wireless communication of LTE. The HILS can be generally applicable for the diagnosis of noise impacts in wireless communication standards.

978-1-4799-5004-1/13 $31.00 © 2013 IEEE

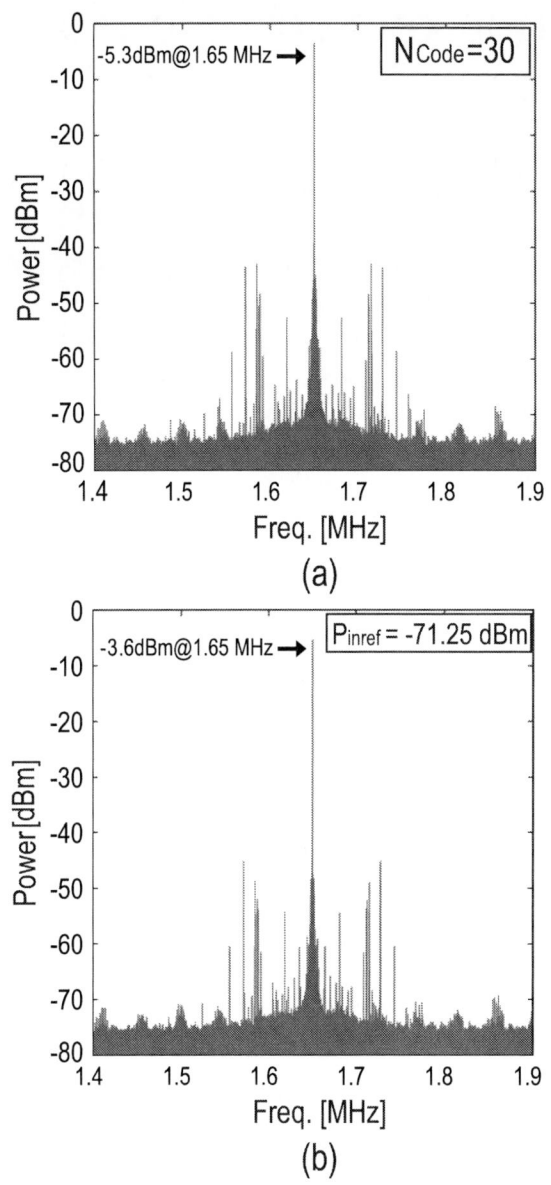

(a)

(b)

Fig. 8. Equivalency of in-band spurs (a) by ANG and (b) by RF injection.

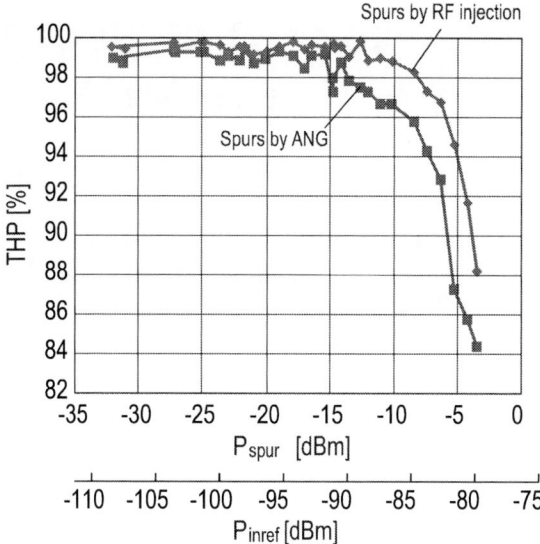

Fig. 9. THP versus P_{spur} and P_{inref}.

and provides the strong opportunities of noise awareness in the hierarchical design process of wireless communication systems.

ACKNOWLEDGEMENTS

This work was supported by the Ministry of Internal Affairs and Communications.

REFERENCES

[1] Sesia, S., Toufik, I. and Baker, "LTE - The UMTS Long Term Evolution: From Theory to Practice", John Wiley & Sons, Ltd, Chichester, UK.

[2] The 3rd Generation Partnership Project, http://www.3gpp.org/

[3] D. K. Su, M. J. Loinaz, S. Masui, B. A. Wooley, "Experimental results and modeling techniques for substrate noise in mixed-signal integrated circuits." IEEE J. Solid-State Circuits, vol. 28, no. 4, pp. 420-430, Apr. 1993.

[4] A. Afzali-Kusha, M. Nagata, N. K. Verghese, D. J. Allstot, "Substrate Noise Coupling in SoC Design: Modeling, Avoidance, and Validation," in proc. of the IEEE, vol. 94, no. 12, pp. 2109-2138, Dec. 2006.

[5] N. Azuma, T. Makita, S. Ueyama, M. Nagata, S. Takahashi, M. Murakami, K. Hori S. Tanaka, M. Yamaguchi, "In-System Diagnosis of RF ICs for Tolerance against On-Chip In-Band Interferers," in proc. of IEEE International Test Conference, pp. 12.3.1-12.3.9, Sep. 2013.

[6] Agilent Technologies, Inc., "SystemVue 2013 Documentation," http://www.home.agilent.com/en/pc-1297131/systemvue-electronic-system-level-esl-design-software?nid=-34264.0&cc=US&lc=eng

IV. CONCLUSIONS

The HILS links the quality of wireless communication with the advent of in-band interferers in RF building blocks due to interactions with environmental noises, such as substrate coupling noise from baseband digital signal processing in a single chip solution. The spurious tones by the interaction with environmental noises produce obvious impacts on wireless communication performance, and the processes are well equivalent to the input referred RF sinusoidal noise tones. The measurement based diagnosis is supported by the HILS

978-1-4799-5004-1/13 $31.00 © 2013 IEEE

Measurements and Simulation of Substrate Noise Coupling in RF ICs with CMOS Digital Noise Emulator

N. Azuma[*], S. Shimazaki[*], N. Miura[*], M. Nagata[*]

T. Kitamura[†], S. Takahashi[†], M. Murakami[‡], K. Hori[‡], A. Nakamura[‡]

K. Tsukamoto[§], M. Iwanami[§], E. Hankui[§], S. Muroga[¶], Y. Endo[¶], S. Tanaka[¶], M. Yamaguchi[¶]

[*]Graduate School of System Informatics,Kobe Univ., 1-1 Rokkodai-cho, Nada-ku, Kobe 657-8501, Japan
[†]Renesas Mobile Corp., 2-6-2 Nippon Bldg., Ote-machi, Chiyoda-ku, Tokyo, Japan
[‡]Renesas Electronics Corp., 1753 Shimonumabe, Nakahara-ku, Kawasaki, Japan
[§]NEC Corporation, 1753 Shimonumabe, Nakahara-ku, Kawasaki, Japan
[¶]Tohoku Univ., 6-6-5 Aoba, Aramaki, Aoba-ku, Sendai 980-8579, Japan
{azuma,nagata}@cs26.scitec.kobe-u.ac.jp

Abstract—Substrate noise coupling in RF receiver front end circuitry for LTE wireless communication was examined by full-chip level simulation and on-chip measurements, with a demonstrator built in a 65 nm CMOS technology. A complete simulation flow of full-chip level substrate noise coupling uses a decoupled modeling approach, where substrate noise waveforms drawn with a unified package-chip model of noise source circuits are given to mixed-level simulation of RF chains as noise sensitive circuits. The distribution of substrate noise in a chip and the attenuation with distance are simulated and compare with the measurements. The interference of substrate noise at the 17th harmonics of 124.8 MHz - the operating frequency of the CMOS noise emulator creates spurious tones in the communication bandwidth at 2.1 GHz.

Keywords—Substrate coupling, Power delivery network, Noise interference, Wireless communication

I. INTRODUCTION

A mobile wireless terminal compliant with the long-term evolution (LTE) standards uses a single chip integration of RF circuits and digital signal processing elements. The sensitivity of the device needs to be high enough for incoming signals as low as -100 dBm at the minimum in the frequency band of 2.1 GHz (Band 1) as defined in [1] [2]. A huge computation capacity is required in the digital backend processing for high-speed and wide bandwidth wireless channels where digital data are complex encoded and modulated on RF carrier signals. This directs the development of semiconductor chips toward an extremely high level of integration of RF and analog front end circuits with digital backend circuits.

On-chip power noise coupling exists between analog/RF and digital portions of a chip through the common silicon substrate. This is often called substrate noise coupling and acts as an obvious obstacle against the single chip integration of high requirements.

The origin of substrate noise coupling in CMOS digital and analog mixed signal circuits has been widely discussed

[3]–[5]. There is a strong demand of continuous efforts and breakthroughs on the study of substrate noise coupling, in response to circuits-and-systems developments toward the next-generation wireless standards.

This paper focuses on a full-chip level simulation of substrate noise coupling in an LTE compliant receiver channel.

The remaining part of this paper is organized as follows. A demonstrator of substrate noise coupling in an LTE compliant receiver channel is introduced in Sect. II. The framework of full-chip level substrate noise coupling simulation is described in Sect. III. Simulation and measurement results on the demonstrator are discussed in Sect. IV. Finally, a brief conclusion will be given in Sect. V.

II. DEMONSTRATOR OF SUBSTRATE NOISE COUPLING

A demonstrator of Fig. 1 includes an LTE compliant receiver channel in a 65 nm CMOS technology. The receiver reacts to an RF signal at 2.1 GHz (as an LTE Band 1 carrier signal) from a signal source (SG) and outputs I/Q sinusoidal signals in a baseband of a 5 MHz bandwidth to a signal analyzer (SA). The receiver channel is structured as in the simplified circuit schematic of Fig. 2. RF building circuits are in a differential topology. The incoming RF signal is amplified by a low-noise amplifier (LNA) and then down-converted to the baseband by a mixer (MIX). The baseband signals are conditioned by low frequency amplifiers and filters before received by the

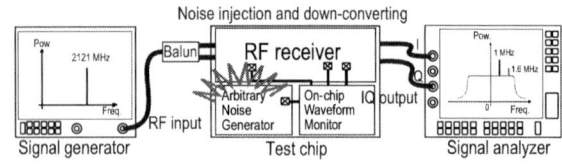

Fig. 1. Demonstrator of LTE receiver channel in substrate noise coupling.

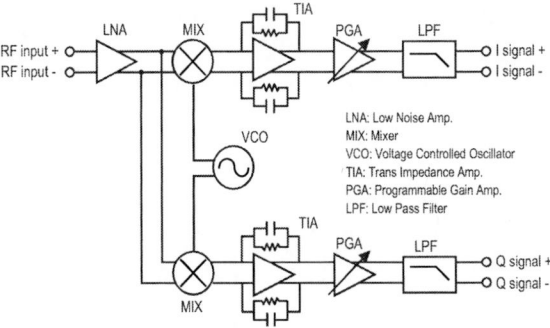

Fig. 2.   Simplified circuit schematic of LTE receiver channel.

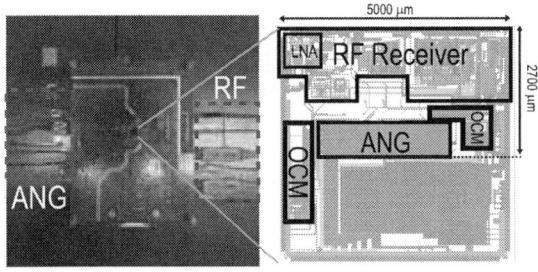

Fig. 3.   Photo of demonstrator. Test chip is also shown.

SA. The purity of incoming RF signals and the resolution of analysis for outputted sinusoids are governed by SG and SA equipment, respectively, and met the required specification of LTE standards.

An on-chip arbitrary noise generator (ANG) emulates power and substrate noise generation in digital circuits [6].

The ANG and RF receiver are co-located in a silicon chip as in wireless RF integrated circuits (IC) chips. The substrate noise from the ANG propagates to and interferes with RF circuits as substrate noise coupling, and causes the advent of in-band spurious tones (spurs). This will be captured and quantified by the SA in the frequency and the strength.

An on-chip noise monitor (OCM) with multiple probing frontend (PFE) channels is also embedded for the measurements of substrate noise waveforms at a variety of locations within a chip.

An overview of the demonstrator is given in Fig. 3. A test chip occupying the silicon area of 5.0 mm x 5.0 mm is packaged in a 168-pin ball grid array (BGA) and mounted in the center area of a printed circuit board (PCB). The PCB uses FR-4. The signals input to and output from the RF receiver are aligned in the right part of the demonstrator, while digital control signals to/from the ANG and OCM are on the left. The separation of such routing areas is carefully maintained in all levels of the demonstrator including the chip, PCB, and measurement setups. This is intended for minimizing the background noise coupling in a measurement environment.

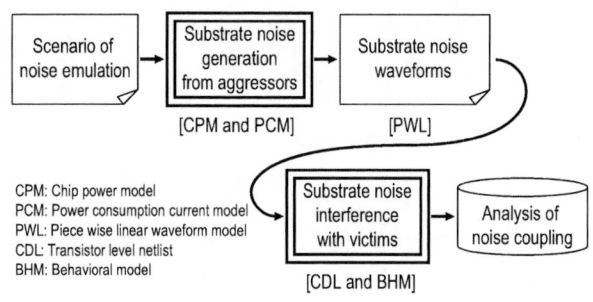

Fig. 4.   Simulation flow of full-chip level substrate noise coupling.

III. FULL-CHIP LEVEL SUBSTRATE NOISE COUPLING SIMULATION FRAMEWORK

A. Overview

The simulation flow of Fig. 4 is proposed for the analysis of full-chip level substrate noise coupling. The input to the flow is a scenario of operation in the baseband digital circuits (BB) as aggressors. This will base the derivation of power current consumption model (PCM) of the BB through hardware emulation using the ANG. A chip power model (CPM) unifies power delivery networks (PDNs) and substrate networks of an entire chip, and imports the PCM of aggressors. An integrated tool suite of power and substrate noise analysis uses the CPM for the simulation of noise generation and produces a piece wise linear model (PWL) of substrate noise waveforms at the locations of victim circuits in a chip [7].

Substrate noise interference in an RF receiver chain (Rx) is then simulated with the PWL waveforms with an RF circuit simulator. The building blocks of the Rx chain are represented in two ways, where the sensitive ones to substrate noise are in transistor-level circuit description (CDL) while the others are in behavioral models (BHM). The output of the simulation will be analyzed or post-processed to characterize the Rx's sensitivity against substrate noise.

B. Substrate noise from aggressors

The framework for the simulation of substrate noise generation is given in Fig.5 with an equivalent circuit model of aggressors. A chip often has several power domains with individual PDNs supplying various circuit cores as well as biasing a silicon substrate. The PDNs as a whole is divided into package and chip parts, and modeled in respective tool flows. The package model represents the broadband responses of power traces and their mutual couplings on the PCB as well as on an interposer. The chip model is responsible for on-chip power and substrate networks as well as internal power currents and their interaction with the networks. The package and chip models are then unified into a single netlist for the simulation of substrate noise from aggressors.

An overview of the unified package and chip model of the demonstrator is depicted in Fig. 6. The demonstrator has 6 power domains with 10 power terminals, where each of the Rx, ANG, and OCM has a pair of high and low power supply

978-1-4799-5004-1/13 $31.00 © 2013 IEEE   43

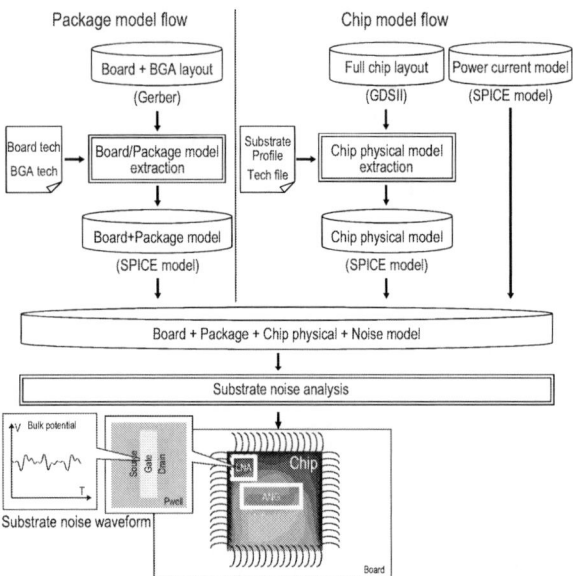

Fig. 5. Framework of substrate noise simulation in aggressors.

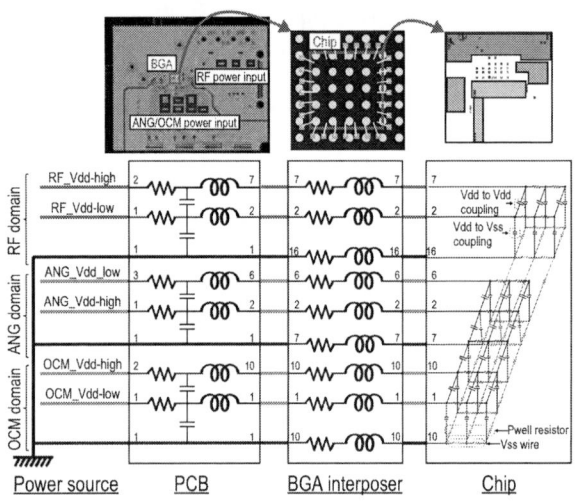

Fig. 6. Passive model of multi-domain power delivery networks and substrate networks.

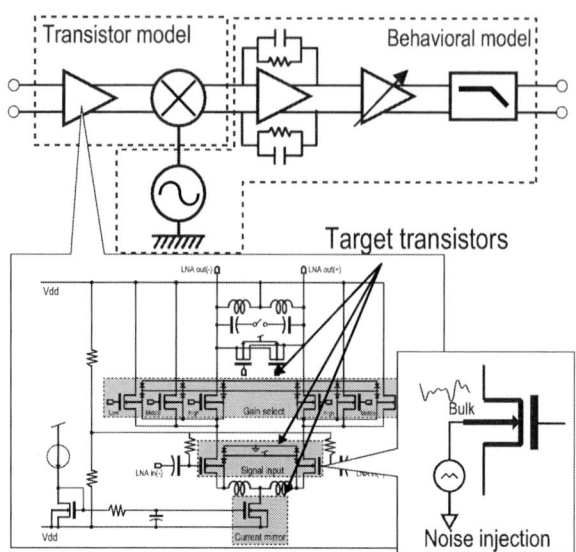

Fig. 7. Simulation model of substrate noise interference in victims.

the interposer. An equivalent LCR circuit network with the guaranteed passivity is then synthesized and matched with the S-parameter of the PCB. Since the pin-to-pin mutual coupling is negligibly small in the interposer, each trace connecting a pair of pins on the back and front sides of the interposer is represented by an LR series. The package model is then given in a single SPICE netlist.

A chip power model (CPM) has been originally developed for the integrated analysis of power/signal integrity and electromagnetic compatibility with a full-chip focus, and then extended for the inclusion of substrate noise coupling [8]. In the present paper, the authors use the CPM as the representation of on-chip high-density passive networks while detach its internal power current expressions. On the other hand, the PCM of the ANG is externally derived by circuit-level simulation and manually combined with the passive CPM. The final form of a chip model is expressed in a single SPICE netlist including the PCM as internal power current stimulus. This treatment augments the flexibility of modeling in response to a variety of the scenarios of noise emulation in the ANG.

The netlists of package and chip models are then simulated in the time-domain with a SPICE simulator. The power current consumption of the ANG interacts with the integrated on- and off-chip, unified PDN and substrate networks. This produces voltage variations as power and substrate noises and spreads them in the whole of a chip. The substrate noise waveforms are captured in the PWL form at the locations of interest within the Rx chain for the subsequent simulation of substrate noise interference.

C. Substrate noise interference in victims

The low noise amplifier (LNA) is considered the most sensitive building block of the LTE compliant Rx chain against

voltages, including the supply for in/out (I/O) pads. While ground-side pins (V_{ss}) are separated from each other in a chip, they are finally shorted to a single system ground (Gnd) defined on the PCB through off-chip PDN impedances. The total of 61 pins is in connection between the power domains on a chip and power terminals on the PCB, including V_{ss} connections to GND. Detailed numbers of pins on the package and chip models are also shown in Fig. 6.

The modeling flows of Fig. 5 are in close relation with the structure of Fig. 6. Multi-port S-parameters are individually derived by a full-wave electromagnetic solver for the PCB and

978-1-4799-5004-1/13 $31.00 © 2013 IEEE   44

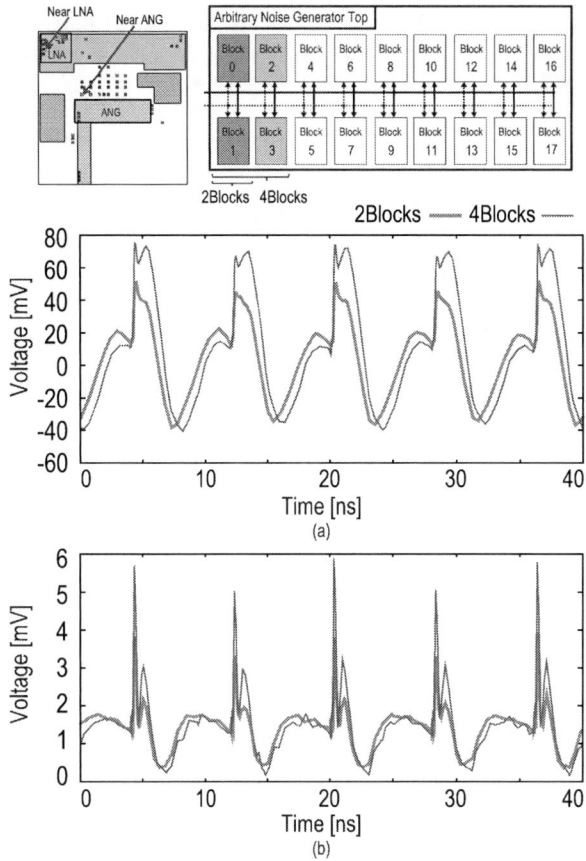

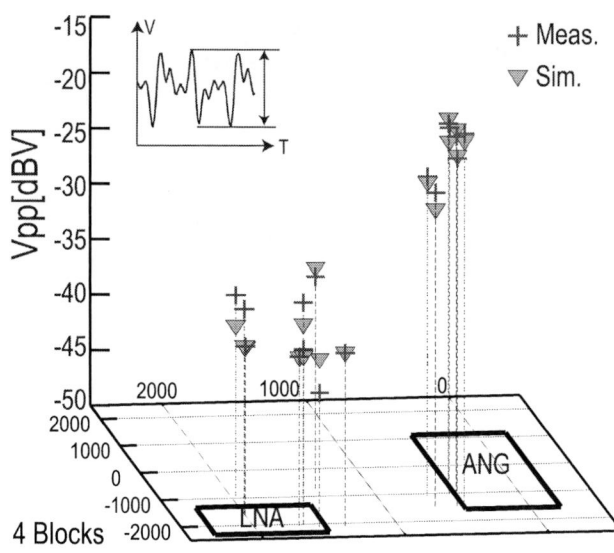

Fig. 8. Simulated substrate noise waveforms at ANG (a) and LNA (b).

Fig. 9. Simulated substrate noise distribution.

substrate noise. We assign the point of injection of substrate voltage variation (PWL models) at the body terminal of transistors in the primary gain stage of the LNA, as shown in Fig. 7. The other building blocks are behaviorally modeled. This allows us to simulate the entire Rx chain concurrently for the RF signal processing in a main channel and also for the response against substrate voltage variation in a background.

IV. EXPERIMENTAL RESULTS

A. Substrate noise generation

Substrate noise waveforms in the demonstrator are simulated with the noise generation model. The scenario of the ANG operation exemplified in this paper includes the operation frequency (F_{clk}), the size of noise generation (N_{Code}), and the distribution of active noise cells in the ANG layout. The F_{clk} was 124.8 MHz. All the noise cells in 2 and 4 blocks are activated in the waveforms, by appropriately setting N_{Code}. The noise model of the ANG is then created and simulated. The substrate noise waveforms are given in Fig. 8, captured at the vicinity of the ANG as well as around the LNA. The voltage variation is led by the capacitor charging operation of the ANG and becomes periodical with the frequency of F_{clk}.

The size of substrate noise is measured by the peak-to-peak voltage variation (V_{pp}) and given at the locations of interest within the die area, as in Fig. 9. These locations correspond to the measurable points probed by the OCM, and therefore the distribution of substrate noise is compared among the simulation and measurements. The noise amplitude immediately becomes attenuated among the positions away from the source of noise. This trend of attenuation is found to be consistent in both simulation and measurements. However, the size of noise amplitudes becomes further reduced in simulation for more distant locations, while flattened in the measurements. This can be explained by the fact that the minimum voltage variation detectable by the OCM is bounded by the background substrate noise coupling in the monitor itself and therefore irrelevant with the position of probing.

B. Substrate noise interference

The LNA is applied with the deep n-well (DNW) option in a physical layout as a well-known remedy against substrate noise coupling. The simulated PWL substrate noise waveform at the p^+ body node of the LNA transistor in DNW is plotted in Fig. 10, showing the efficient substrate noise suppression , in comparison with the noise waveforms not in DNW as given in Fig. 8 (near LNA). However the gain of the LTE compliant Rx chain needs to be programmable and capable of as high as 60 dB at the maximum for the smallest incoming signal strength with less than -100 dBm. In-band interferers by substrate noise coupling can be therefore selectively amplified to the level of significance. In-band frequency components at the baseband are obtained after down conversion, as given in Fig. 11. The spurious tones clearly present at the 0.6 MHz offset from the RF input signal at 1.0 MHz. This results from the interference led by the 17th order harmonics of the substrate noise generated by the BB with F_{clk} of 124.8 MHz

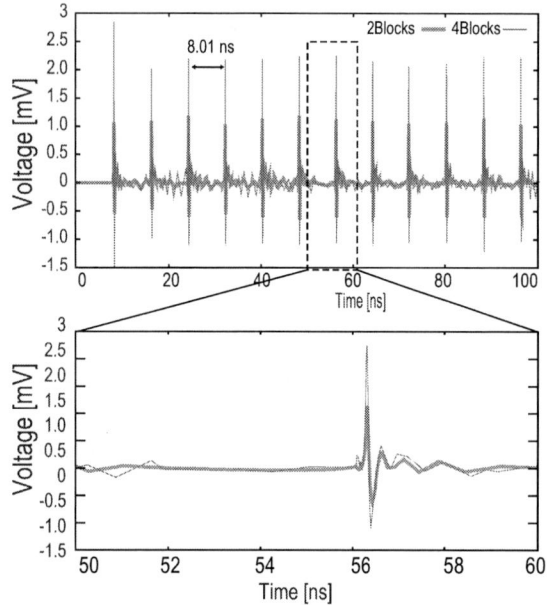

Fig. 10. Simulated substrate noise waveforms for transistors of in deep n-type well

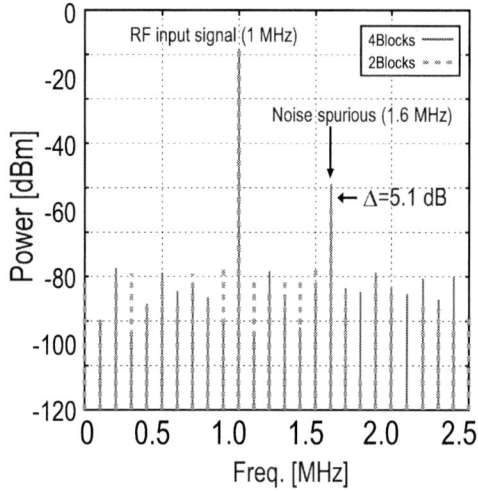

Fig. 11. Simulated in-band frequency components in substrate noise coupling.

This is the point of advancements over the past studies where substrate noise simulation is limited to the noise generation and propagation in aggressors. The accuracy of simulation needs to be pursued, and quantitative comparison of the simulation with in-depth substrate noise measurements will be made in the future study.

ACKNOWLEDGEMENTS

This work was supported by the Ministry of Internal Affairs and Communications.

REFERENCES

[1] Sesia, S., Toufik, I. and Baker, "LTE - The UMTS Long Term Evolution: From Theory to Practice", John Wiley & Sons, Ltd, Chichester, UK.

[2] The 3rd Generation Partnership Project, http://www.3gpp.org/

[3] D. K. Su, M. J. Loinaz, S. Masui, B. A. Wooley, "Experimental results and modeling techniques for substrate noise in mixed-signal integrated circuits," IEEE J. Solid-State Circuits, vol. 28, no. 4, pp. 420-430, Apr. 1993.

[4] N. K. Verghese, T. J. Schmerbeck, and D. J. Allstot, "Simulation Techniques and Solutions for Mixed-Signal Coupling in Integrated Circuits," Boston, MA: Kluwer Academic Publishers, 1995.

[5] M. Nagata, J. Nagai, T. Morie, and A. Iwata, "Measurements and Analyses of Substrate Noise Waveform in Mixed Signal IC Environment," IEEE Trans. on Computer-Aided Design of Integrated Circuits and Systems, Vol. 19, No. 6, pp. 671-678, June 2000

[6] N. Azuma, T. Makita, S. Ueyama, M. Nagata, S. Takahashi, M. Murakami, K. Hori S. Tanaka, M. Yamaguchi, "In-System Diagnosis of RF ICs for Tolerance against On-Chip In-Band Interferers," in proc. of IEEE International Test Conference, pp. 12.3.1-12.3.9, Sep. 2013.

[7] Apache Design, Inc., http://www.apache-da.com/products/totem

[8] Apache Design, Inc., "Totem User Manual Software Release 12.2," pp. 185-205, June 2013.

and down converted with the LO frequency of 2.120 GHz, even with such small voltage fluctuation of Fig. 8.

V. CONCLUSIONS

A complete simulation flow of full-chip level substrate noise coupling in an RF IC chip is established. Simulation confirms that the substrate noise from baseband digital circuits creates in-band spurious tones in RF received band, as is often the case with a real RF IC chip.

In-Band Spurious Attenuation in LTE-Class RFIC Chip using a Soft Magnetic Thin Film

S. Muroga, Y. Shimada, Y. Endo, S. Tanaka
M. Yamaguchi
Graduate School of Engineering
Tohoku University
Sendai, Japan
muroga@ecei.tohokuk.ac.jp

N. Azuma, M. Nagata
Graduate School of System Informatics
Kobe University
Kobe, Japan

M. Murakami, K. Hori
Renesas Electronics Corporation
Tokyo, Japan

S. Takahashi
Renesas Mobile Corporation
Tokyo, Japan

Abstract— A long term evolution (LTE)-class CMOS radio frequency integrated circuit (RFIC) receiver test element group (TEG) chip is developed in our project for the next generation cell phone handsets in order to clarify the on-chip-level noise coupling and demonstrate the noise attenuation using the soft magnetic thin film as an on-chip electromagnetic noise suppressor. The TEG chip equips a noise generator and a RF receiver block. The RF block amplifies and demodulates transmitted signals to IQ signals. A $Co_{85}Zr_3Nb_{12}$ soft magnetic thin film is integrated onto the TEG chip as a noise suppressor.

In this report, the noise generator is driven by a clock signal of 124.803 MHz and generates 17th harmonics of 2,165 MHz conflicts with the LTE band 1 (2,110 – 2,170 MHz). As a result, the in-band digital noise was suppressed 5-20 dB by the Co-Zr-Nb thin film as an integrated noise suppressor.

Keywords— *wireless communication, magnetic thin film, on-chip noise suppressor, inductive coupling, spurious noise*

I. INTRODUCTION

As feature sizes in the complementary metal-oxide-silicon (CMOS) technology move into the deep sub-micrometres region, the on-chip-level electromagnetic interference (EMI) from digital circuits to analog circuits in a mixed-signal chip can be serious problem.

In the case of a radio frequency integrated circuit (RFIC), the harmonic spectrum generated by the digital circuits can be transferred to the analog circuits which operate in the GHz frequency range. The noise coupling affects the analog circuit performance and causes serious desensitization [1], [2]. In this paper, we focus on the solution of the desensitization caused by in-band spurious.

To prevent the on-chip-level noise couplings, the layout of the circuit blocks and the wiring are carefully optimized based upon the experience and know-how. Furthermore, the guard rings and the triple-well shielings are used in order to reflect or bypass the conductive or displacement current. However, we need to concern the inductive noise coupling caused by the magnetic flux above the GHz range in addition to the resistive

and capacitive noise couplings caused by the conductive and displacement currents. Moreover, these currents also generate the magnetic flux and can cause the inductive noise coupling.

Therefore, we need a new countermeasure technology for inductive coupling without the increase of the footprint using magnetic materials.

As the noise countermeasure technology using the magnetic material, the noise suppression sheet using soft magnetic materials based on the ferromagnetic resonance (FMR) is major devises for those noise suppressions for printed circuit board (PCB)-level EMI problem [3].

For higher integration and performance, we proposed the magnetic thin films made of metal alloys or granular magnetic materials as the next generation of integrated type noise suppressors [4]. The soft magnetic thin film dissipates noise power only in the radio frequency (RF) range because of ferromagnetic resonance (FMR) and Joule losses [4]-[11]. For conductive loss, a representative point is that the insertion loss was only 4 dB at 1 GHz and increasing sharply to 57 dB at 6 GHz, as a 2-μm-thick $Co_{85}Zr_3Nb_{12}$ film is integrated onto a coplanar waveguide (CPW) with a 50 μm wide signal line on a soda glass substrate [6]. Owing to these outstanding noise suppressions, our group analyzes the noise suppressors using magnetic film for better design and other group also discusses in [12].

On the other hand, the chip-level noise coupling in RFIC chip is not subjected to restriction in 3GPP standard [13]. Nevertheless, it is well known that the main cause of the desensitization is spurious generated by the digital circuit.

Therefore, the clarification of the chip-level-noise coupling and its countermeasure technology is required.

In this paper, we apply the soft magnetic thin film to a long term evolution (LTE)-class CMOS RFIC receiver test element group (TEG) chip implemented in our project for the next generation cell phone handsets in order to clarify the on-chip-level noise coupling and demonstrate the noise attenuation using the soft magnetic thin film. The TEG chip equips a noise

978-1-4799-5004-1/13 $31.00 © 2013 IEEE

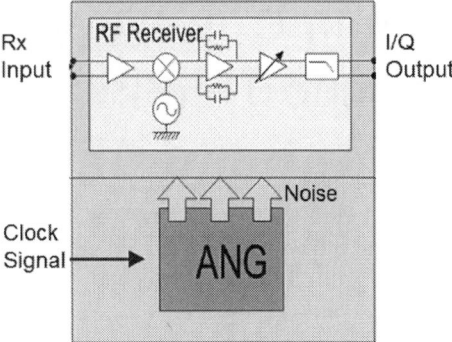

Fig. 1. The outline of the TEG chip for LTE – class cell phone hand set.

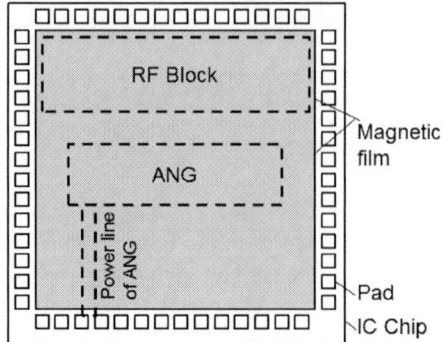

Fig. 2. Sketch of the top view of magnetic thin film integrated onto the TEG chip.

generator and a RF receiver block. The noise generator emulates the true digital noise and the RF block amplifies and implements the down conversion transmitted signals to IQ signals.

As a result, the in-band digital noise suppressed 5-20 dB by a $Co_{85}Zr_3Nb_{12}$ soft magnetic thin film as an integrated noise suppressor when the noise generator is driven by a clock signal of 124.803 MHz and generates 17th harmonics of 2,165 MHz conflicts with the LTE band 1 (2,110 – 2,170 MHz).

II. OUTLINE OF TEG CHIP

Fig. 1 shows the block diagram of the TEG chip. The TEG chip is implemented in a 65 nm CMOS technology in order to clarify the on-chip-level noise coupling and demonstrate the noise attenuation using the soft magnetic thin film in our project for the next generation cell phone handsets.

The chip equips a LTE-class CMOS RFIC receiver. The chip amplifies and implements the down conversion transmitted signals in LTE band 1 (2,110-2,170 MHz) to the IQ signals.

The arbitrary noise generator (ANG) [14] which emulates the power supply noise of the typical digital circuits in a CMOS chip is also integrated near the center of the chip. In this paper, we discuss on the solution of the desensitization caused by in-band spurious. Thus, it is enough that the in-band spurious is generated in LTE band 1 (2110-2170 MHz) regardless of the clock frequency of the digital circuit. In this case, the spurious is generated by the ANG with a clock signal of 124.803 MHz almost same as the clock signal of the analog-to-digital converter (ADC) in the RF digital circuit. As a result, the 17th harmonics of 2,165 MHz conflicts with the LTE band 1.

The chip is mounted onto the IC test board which equips the RF interfaces and DC power terminals.

III. INTEGRATED NOISE SUPPRESSOR

The frequency band and the intensity of the noise suppression are required to design the noise suppressor. Regarding the noise suppression caused by the magnetic field, the material properties of the ferromagnetic resonance frequency, the intensity and the half-bandwidth of the

imaginary part of the relative permeability is necessary to estimate the noise suppression. Additionally, the design parameter of the thickness of the film, the position and the distance between the circuit and the film is important for estimation of the demagnetizing field and the magnetic field in the film. As a result, the frequency band and the intensity of the ferromagnetic resonance loss can be estimated.

On the other hand, a soft magnetic thin film as the electromagnetic noise suppressor needs to have an isotropic permeability in the film plane since the direction of the noise current in the chip is generally random. However, the typical soft magnetic thin films, such as a permalloy (Ni-Fe) and an amorphous alloy, have in-plane uniaxial anisotropy and exhibit high permeability only in the direction of hard axis of magnetization. Meanwhile, we fabricated a composite-anisotropy multilayer (CAM) film which has isotropic permeability using the multilayering method [15] [16] by which more than 2 magnetic thin layers having different uniaxial anisotropy axis are stacked.

In this paper, a double layered film with crossing anisotropy is applied to the TEG chip as the integrated noise suppressor. Fig. 2 shows the sketch of the top view of the integrated noise suppressor integrated onto the TEG chip. The double CoZrNb layers (layer 1 and 2) are stacked having the orthogonal anisotropy.

The double layered magnetic film can have an isotropic permeability in the film plane. The permeability of the layer 1 and 2 are defined as eq. (1) and (2), respectively.

$$\mu_{i1}(\theta) = \frac{M_s^2}{2\mu_0 K_u}\sin^2\theta \tag{1}$$

$$\mu_{i2}(\theta) = \frac{M_s^2}{2\mu_0 K_u}\sin^2(\frac{\pi}{2}-\theta) \tag{2}$$

Where M_s is saturation magnetization, K_u is magnetic anisotropy energy and θ is angle between the hard axis of the layer 1 and direction of measurement in Fig. 3. Thus, the total permeability is estimate as eq. (3).

978-1-4799-5004-1/13 $31.00 © 2013 IEEE

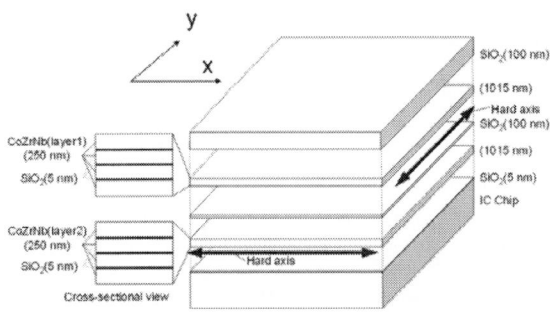

Fig. 3. Outline of the integrated noise suppressor using double layered films with crossing anisotropies.

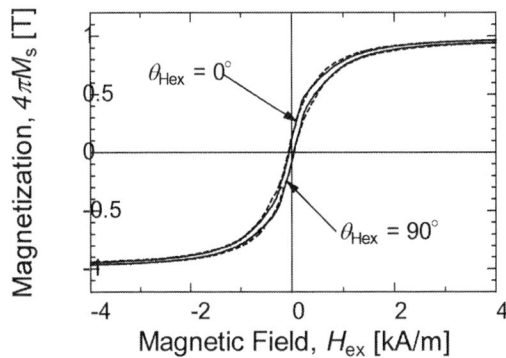

Fig. 4. Measured magnetization curves of Co-Zr-Nb double layers.

$$\mu_{i1}(\theta) + \mu_{i2}(\theta) = \frac{M_s^2 \sin^2 \theta}{2\mu_0 K_u} + \frac{M_s^2 \sin^2(\frac{\pi}{2} - \theta)}{2\mu_0 K_u}$$
$$= \frac{M_s^2}{2\mu_0 K_u}(\sin^2 \theta + \cos^2(\theta))$$
$$= \frac{M_s^2}{2\mu_0 K_u} \qquad (3)$$

As shown in eq. (3), the permeability is independent of the direction of measurement. Therefore, the double layered magnetic film has an isotropic permeability in the film plane.

The each magnetic thin film was fabricated by RF and RF magnetron sputtering system. The film pattern is formed using a photolithography and a lift-off process in order to separate the magnetic thin film above the analog and digital circuit and to avoid the contact between the magnetic thin film and pads of the chip.

Fig. 3 shows the outline and cross-sectional view of the double layered films with crossing anisotropies for use as the on-chip electromagnetic noise suppressor. The film was composed with stacked of the SiO$_2$ (100 nm) / [Co-Zr-Nb (250 nm)/ SiO$_2$ (5 nm)] × 4 / SiO$_2$ (100 nm) / [Co-Zr-Nb (250 nm)/ SiO$_2$ (5 nm)] × 4 / SiO$_2$ (50 nm) (/ Chip). Here, 5-nm-thick

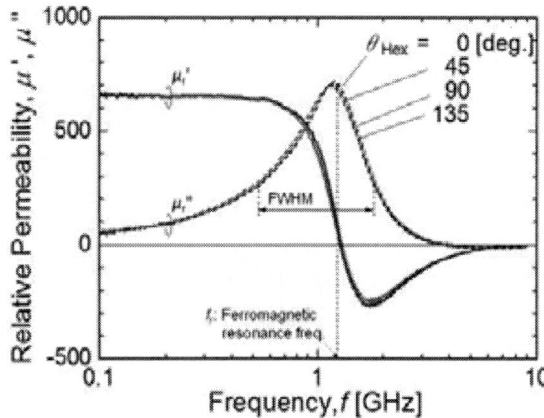

Fig. 5. Relative permeability of Co-Zr-Nb double layers with crossing anisotropies.

SiO$_2$ layers were deposited upon each 250-nm-thick Co-Zr-Nb layer for the purpose of reduction of the eddy current loss and spin wave excitation in the film-thickness direction. The hard axis of the magnetization (that is, high permeability direction) in the lower half magnetic layers is set parallel to the power lines of ANG and that in the upper half magnetic layers is orthogonal to these lines.

Fig. 4 shows a typical in-plain M-H curves of Co-Zr-Nb double layered films measured by a vibrating sample magnetometer (VSM). The θ_{Hex} shows the angle of the applied field from the easy axis of magnetization in the lower layer. The M-H curves for θ_{Hex}= 0 and 90 agree well with each other. The saturation magnetization $4\pi M_s$ and magnetic anisotropy field H_k are about 1 T and 1.2 kA/m, respectively.

Fig. 5 shows frequency dependence of relative permeability of Co-Zr-Nb double layers measured by a high frequency permeameter developed by one of the present authors (Model PMM-9G1, Ryowa Electronics Co.) [17]. The solid and dashed lines show the real and imaginary parts of relative permeability, respectively. The permeability profiles are almost identical to each other, that is, independent of the angle of the applied field. The real part of the relative permeability at low frequencies is 630 and the imaginary part of the permeability is 1,520 with an FMR frequency of 1.1 GHz.

These results demonstrate that the fabricated magnetic thin films have an isotropic feature in the film plane.

The resistivity ρ of the single layer Co-Zr-Nb film measured by the four-point method is 1.2×10^{-8} Ω· m.

In order to see how the double layer films work in the proximity of locally-designed power or signal lines, an experiment using a microstrip line was performed. Fig. 6 shows an experimental setup. The system measures near fields induced by the MSL current (supplied from the network analyzer with intensity -5dBm) with a field sensing coil (CP1S, supplied by NEC Co. Shielded loop coil with the size 20x1000 microns) and measures simultaneously MSL conduction characteristics. Co-Zr-Nb double layers with the same layer

978-1-4799-5004-1/13 $31.00 © 2013 IEEE

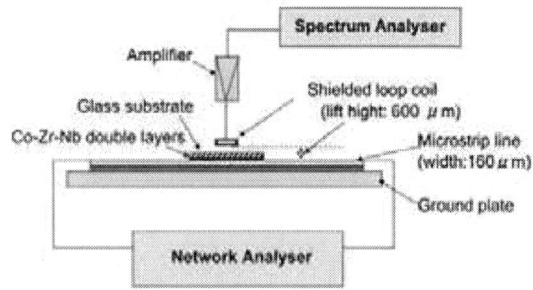

Fig. 6 Experimental setup to measure noise suppressing features for Co-Zr-Nb double layers.

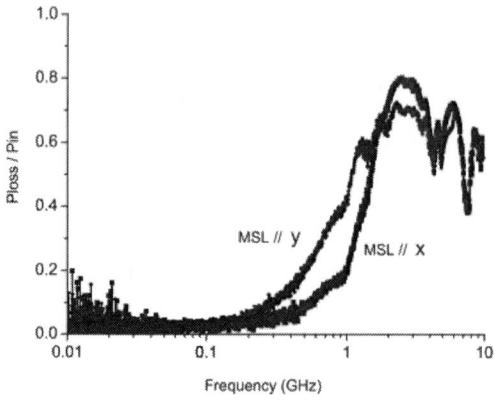

Fig. 8 Frequency dependence of conduction loss in MSL caused by the double layers. Refer to Fig.3 to identify the directions.

In this figure the loss is apparently intensified in a few GHz regions which is higher than the field reduction as seen in Fig.7. The reason why the peaks shift to higher frequencies is that the conduction loss is caused by the localized FMR under a strong demagnetizing field as mentioned previously [8].

It is clearly seen in Fig.8 that the conduction loss occurs similarly for two directions. The difference between x and y direction is due to difference of the high permeability layer position from MSL surface.

These performances of the double layers on MSL have not been fully explained theoretically and works of data accumulation and analysis are under preparation.

IV. MEASUREMENT PROCEDURE

The TEG chip equips only the RF block which amplifies and implements the down conversion from the transmitted signals to the IQ signals. There is no demodulation and IQ-imbalance calibration circuit in the TEG chip. Therefore, we developed the measurement system to evaluate the error vector magnitude (EVM), the bit error rate (BER) and the throughput using the system simulator (System Vue, Agilent Co.).

Fig. 9 shows the measurement system of the IQ signal and throughput of the TEG chip include the system simulator. This measurement system is based on the 3GPP TS 36.101 V9.10.0 [15].

Within the system simulator, the LTE - frequency division duplexing (FDD) modulated signal by quadrature phase shift keying (QPSK) or the continuous wave (CW) were transmitted to the test board using a SMA cable from the signal generator (N5182A MXG, Agilent Co.). Here, the channel band width is set to 5 MHz.

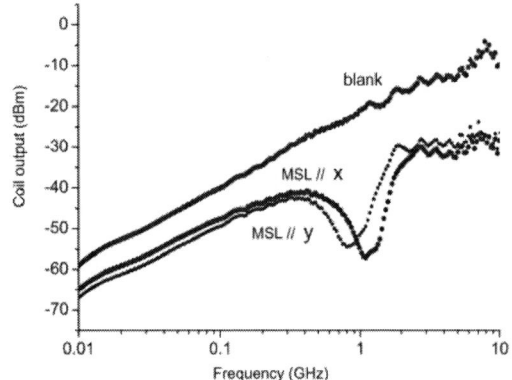

Fig. 7   Measurement of induced fields at height of 250 microns from MSL surface. The double layer sample was set in two different directions. Refer to Fig.3 to identify the directions.

structure as in Fig.3 are set on the MSL surface. They are expected to suppress induced fields and also conduction current both of which include high harmonic noise near 2.1 GHz.

Fig. 7 shows measurements of induced fields from a blank MSL and the one with Co-Zr-Nb double layers. At low frequencies from 0.01 to about 0.2 GHz the double layers work as a simple magnetic shielding layer. The field is reduced by 8-10 dB. As the frequency increases the field is reduced remarkably. This frequency region agrees with increase of imaginary part of permeability in Fig.5 suggesting a possibility that FMR is excited with a certain loss. It should be noted that the profiles are nearly the same for the two directions due to the isotropic feature as in Fig.5. However, at higher frequencies the field intensity resumes. This phenomenon was explained with a model assuming local FMR in an area very close to the MSL surface which leads to a strong demagnetizing field [5].

In Fig.8 conduction loss in MSL is shown. The loss was calculated using the S parameters measured with the Network analayzer [5][8].

$$P_{loss1} / P_{in} = 1 - \left(s_{21}^2 + s_{11}^2 \right) \qquad (4)$$

The signals are implemented the down conversion to the IQ signals by the TEG chip and the output IQ signals are evaluated by the spectrum analyzer (N9020A MXA, Agilent Co.) through the active differential probe (1330A, Agilent Co.). The output IQ signals are analyzed by the system simulator. Finally, we compared the in-band spurious level of the TEG chip with and without the soft magnetic thin film.

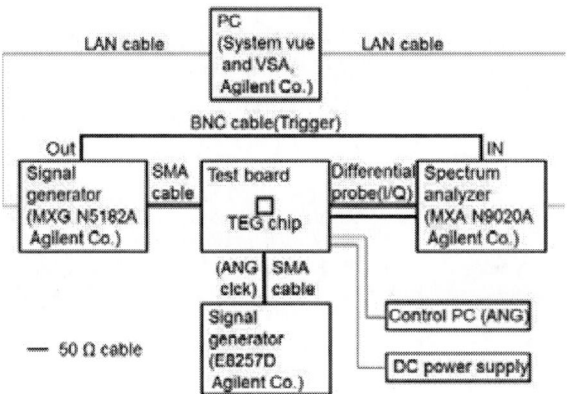

Fig. 9. Measurement system of the IQ signal and throughput of the TEG chip.

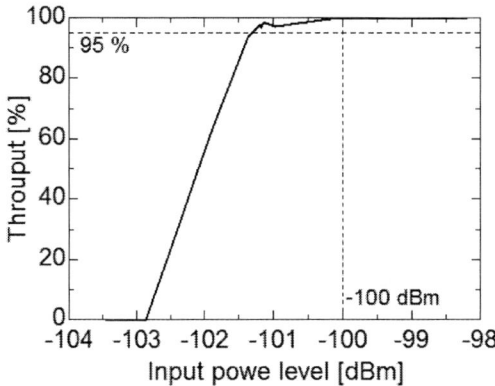

Fig. 10. Throughput dependence on the input power revel of RF signal.

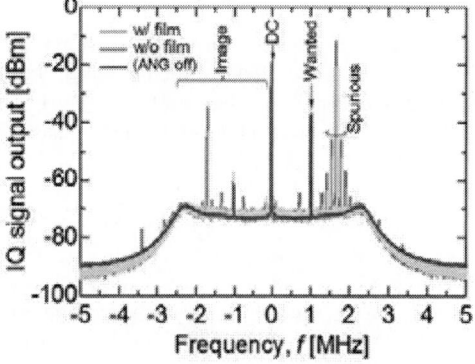

Fig. 11. Spectrum of the IQ signal in the TEG chip with the Co-Zr-Nb film.

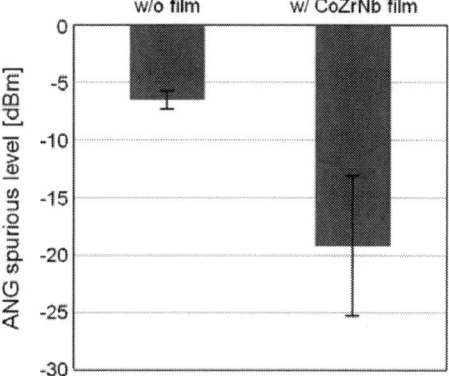

Fig. 12. The spurious power level in the IQ signal at 1.65 MHz.

V. RESULTS AND DISCUSSION

Fig. 10 shows the throughput dependence on the input power level of RF signal where the LTE - FDD modulated signal with the carrier frequency of 2,120 MHz is transmitted. In the case of the input power below -100 dBm, the throughput decreases as the input power decreases and takes zero around -103 dBm. The input power is -101.4 dBm when the throughput is 95%. From this result, the performance of the TEG chip can be satisfied with a 3GPP standard specification of ref. [9] which regulates that the throughput shall be ≥ 95% when the input power is above the -100 dBm in the case of the bandwidth of the channel set to 5 MHz.

Fig. 11 shows the spectrum of the IQ signal in the TEG chip with and without the Co-Zr-Nb film. In this case, the CW of 2,121 MHz is input to the TEG chip as the RF signal and the carrier frequency is set to 2,120 MHz. As can be seen in this figure, the wanted signal and the spurious generated by the ANG are successfully obtained at 1 MHz and 1.65 MHz, respectively. Here, the spectrum at the negative frequency is

image signals and the spectrum at 0 Hz is DC offset. In the case of the TEG chip with the Co-Zr-Nb film, the IQ signal output was about 5 dB lower than that without the film at 1.65 MHz. On the other hand, the wanted signal of the TEG chip with and without the Co-Zr-Nb film at 1 MHz is almost same level.

Fig. 12 shows individual difference of the spurious power level in the IQ signal at 1.65 MHz. The in-band spurious noise is suppressed 5-20 dB by a Co-Zr-Nb film as an integrated noise suppressor. This suppression can be caused by the integration of the Co-Zr-Nb film.

The factor of this large suppression can be discussed as follows. The shielding effect of the magnetic film integrated onto the IC chip is evaluated as about 5 dB with 1-μm-thick Co-Zr-Nb film in [8]. This evaluation is lower than the spurious suppression of 5-20 dB in this report. Oh the other hand, the conductive noise suppression was demonstrated as 18-58 dB in [4], [7] and [10]. According to these references, this noise suppression can be caused by the conductive noise attenuation from the FMR loss in the magnetic thin film.

As a consequence, these results successfully demonstrate the on-chip noise suppression of the integrated noise suppressor using the soft magnetic thin film applied to the LSI chip.

VI. CONCLUSIONS

An arbitrary noise generator (ANG) which emulates a power supply noise of digital circuits in a CMOS chip is located near the center of chip, and operates by the clock signal of 124.803 MHz to obtain a harmonic noise spectrum at 2,165 MHz in the range of LTE band 1 (2,110-2,170 MHz). As a result, the in-band digital noise was suppressed 5-20 dB by a Co-Zr-Nb soft magnetic thin film as an integrated noise suppressor.

This result indicates that the Co-Zr-Nb film well suppresses the on-chip noise coupling. Hence, these results successfully demonstrate the on-chip noise countermeasure with the soft magnetic thin film as an integrated noise suppressor and its application to a LTE-class RFIC chip.

ACKNOWLEDGMENT

The authors thank Mr. E. Hankui, Mr. J. Sakai, Mr. M. Iwanami, Mr. K. Tsukamoto (NEC Co.) for thoughtful discussions. The authors thank Prof. O. Kitakami, Associate prof. S. Okamoto and Assistant prof. N. Kikuchi (Tohoku Univ.) for providing facilities for the use of the photolithography.

This work was supported in part by Development of Technical Examination Services Concerning Frequency Crowding from the Ministry of Internal Affairs and Communications of Japan.

REFERENCES

[1] S. Rodriguez, A. Rusu, N. M. Ismail, "WiMAX/LTE receiver front-end in 90nm CMOS," *IEEE International Symposium on Circuits and Systems 2009 (ISCAS 2009)*, pp.1036-1039, 2009.

[2] H. Hwang, H. Yoo, M. Kim, Y. Na, "A Design of 700 MHz frequency Band LTE Receiver Front-end with 65 nm CMOS Process," *Asia Pacific Microwave Conference, 2009 (APMC 2009)*, pp. 720-723, 2009.

[3] S. Yoshida, M. Sato, and Y. Sato, "Permeability and electromagnetic-interference characteristics of Fe–Si–Al alloy flakes–polymer composite," *J. Appl. Phys.*, vol. 85, pp. 4636-4637, 1999.

[4] M. Yamaguchi, K. H. Kim, T. Kuribara, "Thin film RF noise suppressor integrated in a transmission line," *IEEE Trans. on Magn.*, 3183-3185, 2002.

[5] K. H. Kim, S. Ohnuma, M. Yamaguchi, "RF integrated noise suppressor using soft magnetic films," *IEEE Trans. Magn.*, vol. 40, pp. 3031-3033, 2004.

[6] T. Fukushima, S. Koya, H. Ono, N. Masuda and M. Yamaguchi, "Evaluation of RF Magnetic Thin film Noise Suppressor Integrated onto an Operating LSI Chip," *J. Magn. Soc. Jpn.*, vol. 30, pp. 531-534, 2006.

[7] J. H. Kim, S. M. Seo, H. H. Lee, M. Yamaguchi, S. H. Lim "Si-based Electromagnetic Noise Suppressors Integrated with a Magnetic Thin Film," *Appl. Phys. Lett.*, vol. 90, 2007, 143520

[8] S. Muroga, Y. Endo, Y. Mitsuzaka, Y. Shimada, M. Yamaguchi, "Estimation of Peak Frequency of Loss in Noise Suppressor Using Demagnetizing Factor," *IEEE Trans. Magn.*, vol. 47, pp. 300-303, 2010.

[9] S. Muroga, Y. Endo, W. Kodate, Y. Sasaki, K. Yoshikawa, Y. Sasaki, M. Nagata, M. Yamaguchi, "Evaluation of Thin Film Noise Suppressor Applied to Noise Emulator Chip Implemented in 65nm CMOS Technology," *IEEE Trans. Magn.*, Vol. 47, pp. 4485-4488 (2011).

[10] S. Muroga, Y. Endo, W. Kodate, K. Yoshikawa, Y. Sasaki, M. Nagata, M. Yamaguchi, "Performance of integrated magnetic thin film noise suppressor applied to CMOS noise test chips," *Microwave Conference (EuMC) 2011*, pp. 49 – 52, Oct. 2011.

[11] S. Muroga, Y. Endo, Y. Shimada, M. Yamaguchi, "Analysis of Magnetic Flux Through Magnetic Film With Negative Permeability," *IEEE Trans. on Magn*, Vol. 48, pp. 4320-4323, 2012.

[12] J. Sohn, S. H. Han, M. Yamaguchi, and S. H. Lim, "Tunable electromagnetic noise suppressor integrated with a magnetic thin film," Appl. Phys. Lett. 89, 103501, 2006.

[13] 3GPP: TS 36.101 V9.10.0 (2011-12) Annex A Table A.3.2-1.

[14] T. Matsuno, D. Fujimoto, D. Kosaka, N. Hamanishi, K. Tanabe, M. Shiochi, M. Nagata "An Arbitrary Digital Power Noise Generator Using 65 nm CMOS Technology," *IEICE Trans. Electronics*, vol. E93-C, pp.820-826, 2010.

[15] E. Sugawara, F. Matsuomoto, H. Fujimori and T. Masumoto, "Magnetic properties of composite anisotropy CoNbZr/ceramics multi-layers," *IEEE Trans. on Magn*, vol. 7, pp. 969-974, 1992.

[16] Y. Shimada, E. Sugawara and H. Fujimori, "Initial permeability of composite ‑ anisotropy multilayer films," *J. App. Phys.*, vol. 76, pp.2395-2398, 1994.

[17] M. Yamaguchi, Y. Miyazawa, K. Kaminishi, K-I. Arai, "A New 1MHz-9GHz Thin-Film Permeameter Using a Side-Open TEM Cell and a Planar Shielded-Loop Coil," *J. Magn. Soc. Jpn.*, vol. 3, pp. 137-140, 2003.

Conducted immunity of three Op-Amps using the DPI measurement technique and VHDL-AMS modeling

Siham HAIROUD, Tristan DUBOIS, Angelique TETELIN, Geneviève DUCHAMP
Univ. Bordeaux, IMS, UMR 5218, F-33400 Talence, France

siham.hairoud-airieau@ims-bordeaux.fr

Abstract— **This paper presents an application of the ICIM-CI model to the prediction of the susceptibility of ICs (Integrated Circuits) to environmental disturbances in avionic boards. The method is illustrated by the obsolescence study of three commercial operational amplifiers (Op-Amps) showing quasi-identical electrical characteristics and pin-to-pin compatibility, through the comparison of their respective conducted immunities. The model is developed in VHDL-AMS language, and the simulation results are validated through comparison with Direct Power Injection measurements.**

Keywords— *Integrated-circuit, EMC, obsolescence, immunity measurement and modeling, ICIM-CI, VHDL-AMS, Op-Amps*

I. INTRODUCTION

The continuous technological evolutions of integrated circuits (such as complexity and miniaturization), and integration processes (lithography) lead designers to be seeking for new modeling methods to anticipate EMC (ElectroMagnetic Compatibility) problems, and, to avoid additional post-production costs, while providing both functional and successful circuits. Numerous techniques have thus emerged to increase the immunity of circuits against electromagnetic disturbances, as shown in [1]-[2], to name a few. In this context, two recent models have been developed to simulate the emissivity, and the immunity, of integrated circuits, respectively: ICEM (standard IEC [3]), and ICIM (proposed to the IEC standard [4]). Both the models are sub-modeled according to the disturbance nature, that is, conducted or radiated. In this paper we have focused on the ICIM-CI model (Integrated Circuit Immunity Model-Conducted Immunity), and we have considered a black box modeling approach, in which electromagnetic disturbances are described by pure mathematical models (strong abstraction level), as it is the equipment manufacturer mostly preferred method. The ICIM-CI model, suitable for analog, digital and mixed-signal circuits [5], is based on four standards: (i) IBIS model (IEC 62014-1), (ii) immunity measurement techniques (IEC62132-1 to 6), (iii) supply system modeling (IEC62433-1), and, (iv) the standard ICEM-CE (Integrated Circuit Emissivity Model-Conducted Emissivity) (IEC62433-2).

This work falls within SEISME-CEM [6] project that handles the obsolescence problems encountered in the microelectronics industry. In this paper, the conducted immunity of three Op-Amps from various manufacturers, pin-to-pin compatible and showing similar electrical characteristics have been compared. A generic Op-Amp ICIM-CI model, fed by Op-Amp-design-dependable external parameters is developed in VHDL-AMS, and is validated thanks to DPI (Direct Power Injection) measurements. The simulated and the measured immunities are then compared and analyzed.

After a brief description of the ICIM-CI model in section II, section III presents the electrical characteristics of the considered Op-Amps. Section IV then introduces the measurement methods used to develop the model and the resulting ICIM-CI model implementation is detailed in section V. The experimental and simulation results are finally compared in section VI.

II. ICIM-CI MODEL DESCRIPTION

The ICIM-CI model is a new standard proposal [4] developed within the UTE (Union Technique de l'Electricité, French organization for normalization in electronics, IEC member). The standardization main objective is to provide macro-models to simulate the conducted immunity of electronic components during the design phase in order to predict and enhance immunity levels before manufacturing. The IC under test is modeled with two blocks: the PDN (Passive Distribution Network), and the IB (Internal Behavioral). The model output is then compared to an application-dependent user-definable threshold (Fig. 1).

A. PDN (Passive Distribution Network)

The PDN block describes the impedance network of one or more IC ports with passive elements (resistors, capacitors, inductors). The equivalent impedance (Z_{eq}) measurements should be carried out in the standard operating conditions [4]. The experimental set-up dedicated to the conducted immunity measurements of each Op-Amp considered as a standalone component (that is, without accounting for the possible disturbances caused by surrounding components) is presented in section IV.

978-1-4799-5004-1/13 $31.00 © 2013 IEEE

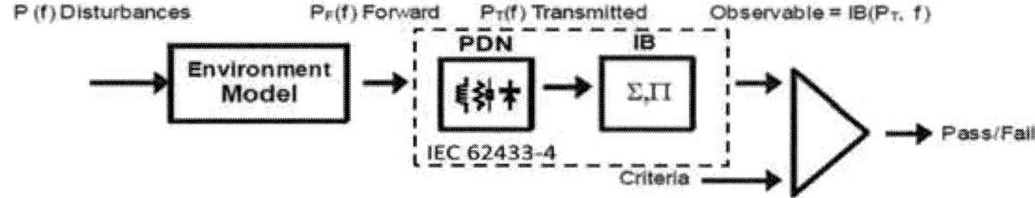

Figure 1-ICIM-CI model [4]

Note that the PDN is defined in the frequency domain and can have various descriptions such as S- or Z- or Y-parameters.

B. IB (Internal Behavioral)

The IB block represents the active part of a device and characterizes the device dysfunction due to electromagnetic aggression. The IB observable output describes the IC behavioral response to a disturbing signal applied to the IB block input, in other words, how the IC reacts to internal perturbations [4]

C. Immunity criterion

The Immunity criterion may be applied to the observable output of the IB block and it is currently fixed by the IC model user. The choice of the immunity criterion should be done in accordance with the application in which the IC is used [4].

TABLE I.   THREE OP-AMPS' ELECTRICAL CHARACTERISTICS

	LT1498	OPA2277U	AD8622
Input Resistance, Differential Mode	Not informed	100 MΩ	1 GΩ
Input Resistance, Common Mode	Not informed	250 GΩ	1 TΩ
Input Capacitance, Differential Mode	Not informed	3 pF	5.5 pF
Input Capacitance, Common Mode	Not informed	3 pF	3.3 pF
Slew Rate(V/µs)	6	0.8	0.48
Bandwidth Product	10.5 MHz	1 MHz	560 kHz
Open-Loop Gain Min (dB)	120	126	125
Output Voltage Swing	No load V_{Low}= 18 mV V_{High}= 2.5 mV	Not informed	Rail-to-rail output swing
Common Mode Input Range	Not informed	$(V^-) + 2V$ $(V^+) - 2V$	$(V^-) + 1.2V$ $(V^+) - 1.2V$

III.   COMPARISON OF THE THREE OP-AMPS' CHARACTERISTICS

In this paper, we compare the conducted immunity of three different manufacturer Op-Amps: the LT1498 from Linear Technology [7], the OPA2277U from Texas Instruments [8], and the AD8622 from Analog Devices [9]. Therefore, S-parameters and DPI measurements were carried out for each of the aforementioned Op-Amps. Those components are very similar in their electrical behavior and are pin-to-pin compatible. Table I gathers some of their electrical characteristics (given for an operating temperature of 25°C and for a supply voltage in the range [±2.5V, ±15V]), such as the open-loop gain, the bandwidth product, the slew rate, the input resistance and capacitance (in common, and in differential mode), the output voltage swing and the common mode input range.

IV.   MESUREMENT TECHNIQUES

For this study case, an experimental setup (Fig. 2) dedicated to DPI measurements as well as S-parameters measurements, and developed within SEISME project [6] allows repeatable testing of the DUT (Device Under Test) under identical environmental and measurement conditions without welding it.

Figure 2-Experimental setup used for measurements (S-parameters and DPI) considering each Op-Amp solely

A. PDN electrical schematic extraction

Among the many methods used for the extraction of the PDN structure and its component values, the most common one consists in measuring the S_{11} parameter of the DUT pin in which the disturbance is injected [4]. For this measurement, each Op-Amp is connected as a follower circuit, and fed by a supply voltage of ± 3 V. The V⁺ input, which is used as the entry point for interferences, is linked to a DC voltage of 0 V through a bias tee, as presented in Fig. 3. The bias tee is obviously taken into account in the calibration procedure of the VNA (Vector Network Analyzer). Note that each Op-Amp supply pin is correctly decoupled by a capacitor of 0.1 µF, as shown in Fig. 3.

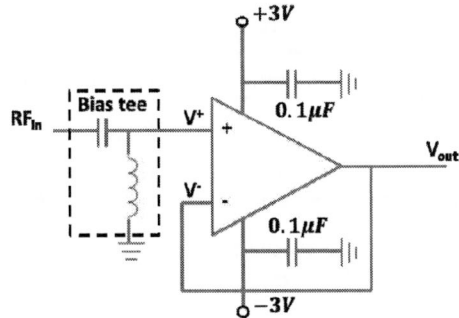

Figure 3-Studied Op-Amp follower circuit

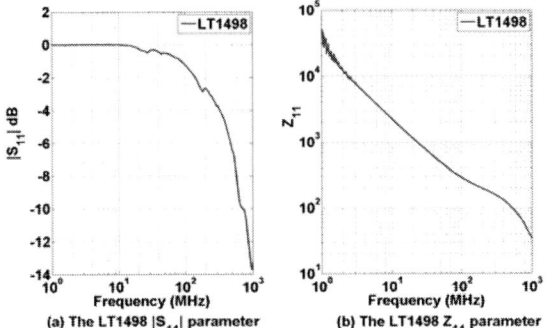

(a) The LT1498 |S$_{11}$| parameter   (b) The LT1498 Z$_{11}$ parameter

Figure 4-S$_{11}$ and Z$_{11}$ parameters of the LT1498

The S$_{11}$ parameter measurements enable to build the PDN block for each Op-Amp. Fig. 4.a displays the S$_{11}$ magnitude measurement for the V$^+$ pin of the LT1498 and Fig. 4.b displays its corresponding Z$_{11}$ parameter. The latter is deduced from the S$_{11}$ measurements by using *Eq. 1*.

$$Z_{11} = Z_0 \times (1+S_{11})/(1-S_{11}) \qquad (1)$$

In Eq. 1, Z_0 corresponds to the characteristic impedance (50 Ω).

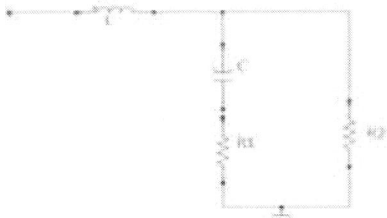

Figure 5-PDN electrical schematic

TABLE II.   PDN ELECTRICAL COMPONENT VALUES FOR EACH OP-AMP

	LT1498	OPA2277U	AD8622
Capacity (pF)	6.5	1.4	5.53
Inductor (nH)	4.996	6.533	5.006
Resistor 1 (Ω)	32	115	15
Resistor 2 (Ω)	354875	146450	235775

The PDN electrical schematic (Fig. 5) is extracted from Z_{11}. Note that this characteristic impedance is actually equivalent to a high value resistor for very low frequencies and it behaves as a RLC circuit when the frequency increases. The same PDN electrical structure was used for the three studied Op-Amps because they show similar frequency responses. The PDN passive element values, displayed in Table II, differ for each Op-Amp, and were given by ADS (Advanced Design System from Agilent). The relationship between the PDN component values and the actual electrical characteristics given in Table I has not been investigated so far.

B. DPI (Direct Power Injection) measurement description

DPI measurements were carried out by injecting an interference signal on the DUT (on V$^+$ pin or input) through a bias tee, whose frequency bandwidth is 100 kHz – 4 GHz. As in the ICIM-CI standard proposal [4], the absorbed power is assumed to be the power entering the device and the absorbed power threshold is the intrinsic power that induces a device dysfunction. The absorbed power is considered as a relevant parameter to characterize the signal variation at the IB block observable output. DPI measurements were carried out considering the forward power. The absorbed power (*Pabs*) is deduced from the forward and reverse powers (P_F and P_R, respectively) (*Eq.2*). Note that P_F and P_R are measured from the bidirectional coupler with a power meter and, as for the S$_{11}$ measurement, each Op-Amp is used as a follower circuit and is fed by a supply voltage of ± 3 V. The V$^+$ pin is set to 0 V DC. The interference signal is injected via the bias tee on this pin (Fig. 6).

$$Pabs = P_F - P_R \qquad (2)$$

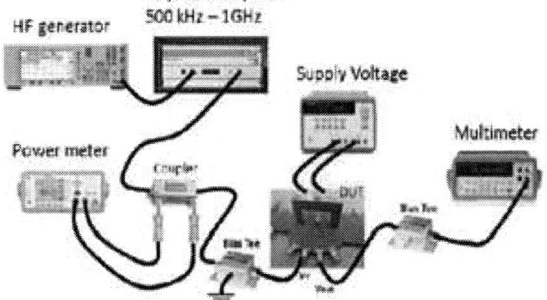

Figure 6-DPI test bench

V.   ICIM-CI MODEL IMPLEMENTATION IN VHDL-AMS

VHDL-AMS is a powerful language that simulates mixed analog-digital functions possibly combined with behavioral descriptions of physical phenomena [10]. This language allows both structural and behavioral modeling. For this study, we used ADVance-MS [11] tool associated with Questa-ADMS simulator [12]. The ICIM-CI model was developed in VHDL-AMS with the modeling approach presented in Fig. 7.

978-1-4799-5004-1/13 $31.00 © 2013 IEEE

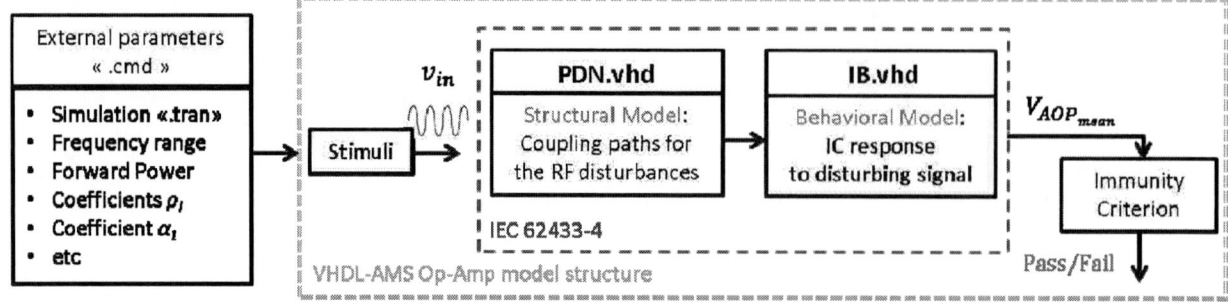

Figure 7-ICIM-CI model in VHDL-AMS

The PDN and the IB blocks were developed as a structural (mapping of sub-models) model, and as a behavioral (algorithm-based) model, respectively. A block was added to test the immunity criterion on the IB block output, that is, on the Op-Amp output mean voltage ($V_{AOPmean}$). A command file '.cmd' is run to configure simulation conditions (ie. transient simulation, frequency domain to be tested, etc.)

More details on VHDL-AMS are given in [10] and [15-20], which provide an overview of the hardware description language for analog and mixed-signals, and propose a large set of code examples to generate accurate behavioral models.

Note that our objective is to develop a generic model, in which only external parameters are to be changed by end-users. The algorithm is thus identical for the three Op-Amps with only the external parameters changed according to Op-Amp specific features (Fig. 7).

A. PDN : Structural Model

A structural model has been developed based on the electrical schematic extracted in section IV.A. The S_{11} parameter is then deduced from the characteristic impedance Z_{eq} of the PDN thanks to Eq. 1. The absorbed power is computed from both the forward power P_F and the S_{11} parameter:

$$Pabs_{simu} = (1-|S_{11}|^2) \times P_F \qquad (3)$$

The structural block is stimulated by a sine function to create an alternative voltage *vin* dependent on P_F:

$$Vin = sqrt(P_F \times Z_0) \times sin(2\Pi ft) \qquad (4)$$

The observed V_{PDNrms} (PDN output voltage) can then be modeled by the following equation:

$$V_{PDNrms} = \beta_0(f) \times sqrt(Pabs_{simu}) \times \beta_1(f) \qquad (5)$$

where β_0 and β_1 are constants extracted from linear curve-fitting with experimental data. Note that these two parameters are frequency-dependent. Thus, a VHDL-AMS block has been added to the Op-Amp model (Fig. 7) to compute β_0 and β_1 for each considered frequency.

B. IB : Behavioral Model

The IB block is developed in two steps. At first, we set an analytical model called "model 1", which successfully fits the experimental curves of $V_{AOPmean}$ as functions of *Pabs* for the considered frequency range (*Eqs.6-7*). In the analytical model $V_{AOPmean}$ stands for the average value of the measured output signal.

$$V_{AOPmean} = \alpha_0(f) \times sqrt(Pabs_{meas}) + \alpha_1(f) \qquad (6)$$

$$\begin{cases} \alpha_0(f) = \rho_3 \times f^3 + \rho_2 \times f^2 + \rho_1 \times f + \rho_0 \\ \alpha_1(f) = cste \end{cases} \qquad (7)$$

As a first approximation, we have considered α_1 as a constant coefficient, as its frequency variation is actually not significant on $V_{AOPmean}$ value. As a second step, the relationship between V_{PDNrms} (Op-Amp Input) and $V_{AOPmean}$ (Op-Amp Output) is established by solving the system of linear equations (*Eqs.8*) composed of *Eq.5* and *Eq.6*. We assumed here that the simulated and the measured *Pabs* should be equal.

$$\begin{cases} V_{PDNrms} = \beta_0(f) \times sqrt(Pabs_{simu}) \times \beta_1(f) \\ V_{AOPmean} = \alpha_0(f) \times sqrt(Pabs_{meas}) + \alpha_1(f) \end{cases} \qquad (8)$$

The analytical model (called "model ICIM-CI") that connects V_{PDNrms} to $V_{AOPmean}$, is finally given by *Eq.9* and *Eq.10*.

$$V_{AOPmean} = \delta_0(f) \times V_{PDNrms} + \delta_1(f) \qquad (9)$$

$$\begin{cases} \delta_0(f) = \alpha_0(f)/\beta_0(f) \\ \delta_1(f) = -(\delta_0(f) \times \beta_1(f)) + \alpha_1 \end{cases} \qquad (10)$$

C. Immunity Criterion

In this work, we consider that the output voltage $V_{AOPmean}$ (ideally equal to 0 V) is allowed to vary of ± 10% of the DC supply voltage without causing a failure of the system, which corresponds to a variation of ± 300 mV. The immunity curves present the frequency-dependence of the absorbed power for which the Op-Amp is in fail operating conditions.

978-1-4799-5004-1/13 $31.00 © 2013 IEEE

VI. COMPARISON BETWEEN MEASUREMENTS AND SIMULATIONS

In this section, we compare results obtained from DPI measurement techniques with the ICIM-CI model implemented in VHDL-AMS. We first compare the S_{11} magnitude curves, then, the output voltages of each Op-Amp as functions of the absorbed power, and, finally, the immunity curves of each Op-Amp.

A. S_{11} Parameter

Fig. 8 highlights that the comparison between the simulated and the measured S_{11} show quite a good agreement for the AD8622 component, and are close to one another for the LT1498 and the OPA2277U. A difference between experimental data and simulation of about 1 dB is measured around 100 MHz (for the LT1498 and the OPA2277U). Although satisfactory, the accuracy of the model could be improved by optimizing the electrical schematic describing the PDN.

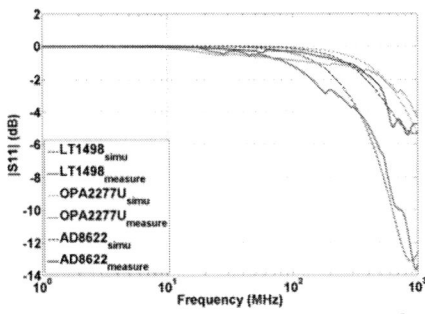

Figure 8- S_{11} parameters obtained by measurements and simulations for every Op-Amp

B. Mean Output Voltage

Figs. 9-11 show that Op-Amp output voltages obtained through the DPI measurements and the ICIM-CI model have similar values when the $V_{AOPmean}$ absolute value is below 300 mV. Regarding measurements, we observe at first that the output DC voltages increase for the OPA2277U and the AD8622, and decrease for the LT1498 when the absorbed power level increases. This behavior is actually due to a rectification phenomenon appearing on one of the transistor stages of the Op-Amp [13]-[14]. Secondly, we notice that for all of the considered Op-Amps, the higher the frequency, the smaller the variation of the output voltage as a function of the absorbed power. Consequently, it probably implies that the higher the frequency, the more robust the component. Finally, simulations and measurements show a good agreement as long as the magnitude of $|V_{AOPmean}|$ is less than 300 mV.

C. Immunity Curves

In this part, we discuss the immunity curves derived from the measurements and the ICIM-CI model. We observe for all the considered Op-Amps that their immunity curves increase with frequency as expected from the results of Figs. 9-11. Figure 12 shows that the OPA2277U is less

sensitive than the others. For example, at 100 MHz, the OPA2277U requires approximately 10 dBm to be in fail mode while the LT1498 and the AD8622 fail as low as approximately 0 dBm.

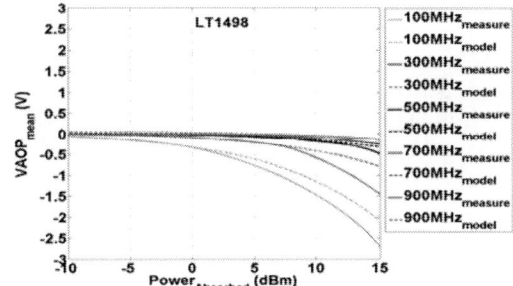

Figure 9-LT1498 output voltage vs. absorbed power- DPI measurements (solid lines) and VHDL-AMS simulations (dashed lines)

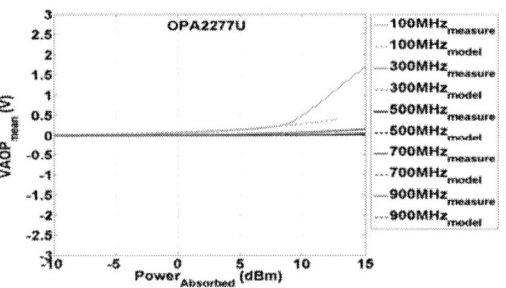

Figure 10-OPA2277U output voltage vs. absorbed power- DPI measurements (solid lines) and VHDL-AMS simulations (dashed lines)

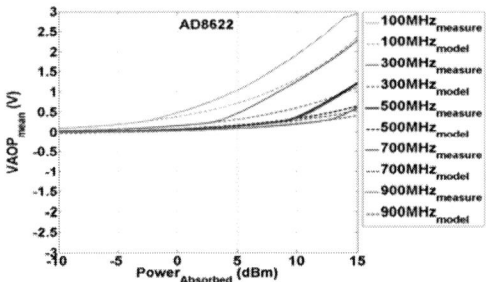

Figure 11-AD8622 output voltage versus absorbed power- DPI measurements (solid lines) and VHDL-AMS simulations (dashed lines)

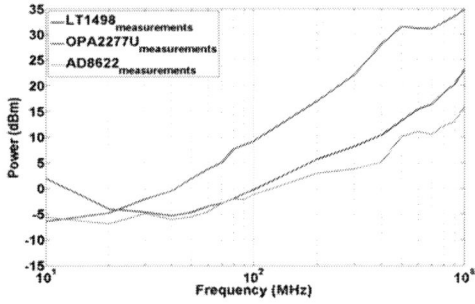

Figure 12-Measured Immunity Curves of the three Op-Amps

978-1-4799-5004-1/13 $31.00 © 2013 IEEE

Actually, even though the AD8622 and the OPA2277U have close S_{11} values, their immunity curves are totally different. The comparison between simulations and measurements show a good agreement for the LT1498 and the AD8622 as presented in Fig. 13. However, regarding the OPA2277U, Fig. 13 shows close results between measurements and simulations only from 100 MHz. For lower frequencies the ICIM-CI model overestimated the power value putting the OPA2277U in fail mode. An improvement of the model for the low frequency range should be performed, although it is satisfactory for frequencies higher than 100 MHz.

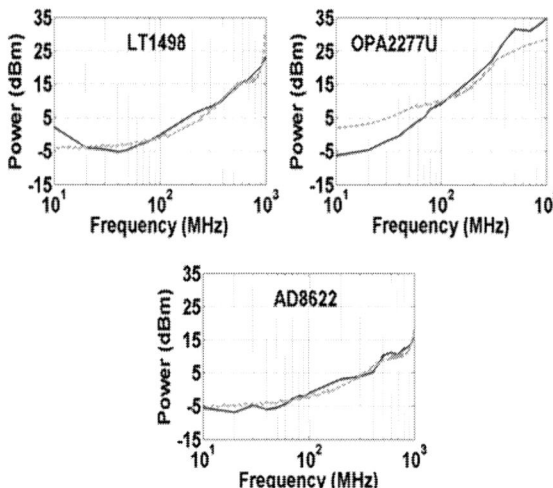

Figure 13-Immunity Curves of the three Op-Amps – measured (solid lines) and simulated (dotted lines)

VII. CONCLUSIONS AND PERSPECTIVES

In this paper, we presented an implementation of the ICIM-CI model for three pin-to-pin compatible Op-Amps, the LT1498 from Linear Technology, the OPA2277U from Texas Instruments, and the AD8622 from Analog Devices. The PDN and the IB structures of the model are identical for all the Op-Amps but with different user-definable input parameters. The simulated immunity curves are in a good agreement with those obtained through DPI measurements, except for the OPA2277U. Actually, although the model is satisfactory for frequencies higher than 100 MHz, it should be improved for lower frequencies, the first step of which would be to determine a more comprehensive PDN structure that would simulate the frequency behaviors of the components with a better accuracy. This study shows that for different pin-to-pin compatible Op-Amps, the immunity curves can vary up to 15 dBm from one another. Therefore, the choice of a suitable component has a crucial influence on the susceptibility of the system it is included in.

The presented ICIM-CI model has been developed, and will be improved, to become one of the main parts of a comprehensive immunity model of an avionic board that includes commercial Op-Amps, called "IBIM-CI" (Integrated Board Immunity Model-Conducted Immunity).

REFERENCES

[1] Pelliconi, R., Speciale, N., "Criteria to reduce failures induced by EMI conducted on the power supply rails of CMOS operational amplifiers," *Electromagnetic Compatibility, 2001. EMC. 2001 IEEE International Symposium on* , vol. 2, pp. 1102,1105, 2001.

[2] Fiori, F., "Operational amplifier input stage robust to EMI," *Electronics Letters* , vol. 37, no. 15, pp. 930-931, 2001.

[3] "IEC62433-2: EMC IC modelling - Part 2: Models of integrated circuits for EMI behavioural simulation - Conducted emissions modelling (ICEM-CE)," IEC, 2008.

[4] Marot, C., Levant, J-L., "Future IEC62433-4: integrated circuit - EMC IC modelling – Part 4: Models of Integrated Circuits for EMI behavioural simulation, Conducted Immunity modelling (ICIM-CI), " Now work Item proposal, 2008.

[5] Ben Dhia, S., Boyer, A., Vrignon, B., Deobarro, M., "IC immunity modeling process validation using on-chip measurements," *Test Workshop (LATW), 2011 12th Latin American*, pp. 1-6, 2011.

[6] Marot, C., Sicard, E., "EMC standards at IC level - status of IEC and technical goals of the SEISME project," *Electromagnetic Compatibility (APEMC), 2012 Asia-Pacific Symposium on* , pp. 9-12, 2012.

[7] Linear Technology, LT1498 datasheet [Online]. Available: http://pdf.datasheetcatalog.com/datasheet/lineartechnology/14989fa.pdf

[8] Texas Instruments, OPA2277U datasheet [Online]. Available: http://www.datasheetcatalog.com/info_redirect /datasheet2/4/0908gdjy3f2tig1lppao3hz2l8ky.pdf.shtml

[9] Analog Devices, AD8622 datasheet [Online].Available: http://www.analog.com/static/imported-files/data_sheets/AD8622_8624.pdf

[10] Sabiro, S-G., "Mixed-mode system design: VHDL-AMS", Microelectronic Engineering, vol. 54, pp. 171-180, 2000.

[11] Mentor Graphics corporation, ADVance MS presentation [Online]. Available:http://www.mentor.com/products/ic_nanometer_design/analog-mixed-signal-verification/

[12] Mentor Graphics corporation, Questa-ADMS presentation [Online].,Available:http://s3.mentor.com/public_documents/datasheet /products/ic_nanometer_design/analog-mixed-signal-verification/advance-ms/ADMS_Datasheet.pdf

[13] Dubois, T., Laurin, J-J, Raoult, J., Jarrix, S., "On the effect of amplitude modulated EMI injected on a PLL active filter," *Electromagnetic Compatibility of Integrated Circuits (EMC Compo), 2011 8th Workshop on* , pp. 170-175, 2011.

[14] Fiori, F., "A new nonlinear model of EMI-induced distortion phenomena in feedback CMOS operational amplifiers," *Electromagnetic Compatibility, IEEE Transactions on* , vol. 44, no. 4, pp. 495-502, 2002.

[15] Huss, S-A., "Model engineering in mixed-signal circuit design: A guide to generating accurate behavioral models in VHDL-AMS"- *The springer international series in engineering and computer science v.649*

[16] Ashenden, P-J., Peterson , G-D., Teegarden, D-A., "The System Designer's Guide to VHDL-AMS", *2003.*

[17] Christen, E., Bakalar , K., Dewey, A., Moser, E., "Analog and mixed signal modeling using the VHDL-AMS language", *IEEE Design Automation Conference Tutorial, 1999.*

[18] Doboli, A., Vemuri, R., "Behavioral modeling for high-level synthesis of analog and mixed-signal systems from VHDL-AMS", *IEEE Trans. Comput. Aided Des. Integrated Circuits Syst. 22 (1) (2003) 1504–1520.*

[19] Nikitin, P-V., Richard Shi, C-J., "VHDL-AMS based modeling and simulation of mixed-technology microsystems: a tutorial", *VLSI Integration Journal vol. 40, pp: 261-273, 2007.*

[20] Christen, E., Bakalar, K., "VHDL-AMS—a hardware description language for analog and mixed-signal applications", *IEEE Trans. Circuits Syst. II Analog Digital Signal Process. 46 (10) (1999) 1263–1272.*

Improvement of Reproducibility of DPI Method to Quantify RF Conducted Immunity of LDO Regulator

Tohlu Matsushima, Nobuaki Ikehara, Takashi Hisakado, Osami Wada

Department of Electrical Engineering,
Kyoto University
Nishikyo-ku Kyoto Daigaku Katsura, Kyoto, Japan
Email: {matsushima, hisakado, wada}@kuee.kyoto-u.ac.jp

I. INTRODUCTION

Intrinsic performance capability of a logic technology exceeds the frequency requirements and power consumption of low voltage processers. The internal supply voltages of the core logic circuits typically become lower to improve the energy efficiency. However, the low-voltage devices have thin relative noise margins, and internal and external disturbances limit amount of the supply voltage reduction. If the external disturbance from the voltage supply cable reaches this voltage regulator and it cannot output desired voltage, the LSI which is connected with the voltage regulator causes circuit failure. Therefore, the immunity of the voltage regulator is important for the LSI to operate normally [1].

In this paper, we focus on the Direct RF Power Injection (DPI) method described in IEC 62132 part 4[2]. In this method, the disturbance generated by a RF signal generator propagates 50 Ω transmission line such as coaxial cables and microstrip line on a PCB, and arrive in a pin of the DUT. Because of 50 Ω transmission line, the DPI method is regarded as having independency of the measurement setup. However, several differences of the measurement setup, for example output impedance of the RF amplifier and placement of DC supply cable, causes degradation of the immunity reproducibility. In this paper, the reproducible measurement setup is described.

II. DIRECT RF POWER INJECTION METHOD

Direct RF Power Injection method (DPI method) [2] is used for evaluation of immunity against conducted RF disturbances of an integrated circuit. The measurement setup of DPI method

in Fig. 1 consists of an RF generator, a RF amplifier, a directional coupler, and DC block with decoupling circuits. The disturbance as a sinusoidal wave generated by the RF generator is injected in IC pin of a device under test (DUT) through the 50-Ω transmission line after it be amplified by the RF amplifier. The directional coupler inserted between the RF amplifier and the DUT separately samples the forward wave power P_f and the reflected wave power P_r in the transmission line. The immunity level of the DUT is evaluated by the transmitted power P_t which is described as $P_f - P_r$ when the DUT operates incorrectly due to the disturbance.

A. DUT and Evaluation Criteria of Malfunction

Low dropout (LDO) voltage regulator, L4949 (ON Semiconductor) as shown in Fig. 2, is adopted as a DUT because immunity characteristics is published in the reference work[3]. The device has 8 pins and monolithic integrated 5V voltage regulator with low dropout and additional functions such as reset circuit and uncommitted voltage sense comparator. In this paper, only the voltage regulator is focused for the sake of ease. Therefore, the VCC (= 12 V), VSS and Vout are connected with the lines or plane on the PCB. The value of load resistor is 200 Ω, and the other pins are opened.

As the immunity criteria of false operation of the DUT, output voltage fluctuation of ± 2% is selected in this report. Namely when the output voltage becomes higher than 5.1 V or lower than 4.9 V due to the disturbance, we judge that the DUT performs incorrectly and we record the incident power. After that, we must distinguish the false operation and breakdown. Therefore, we need to turn down the output level of the RF generator and check that the DUT operate correctly.

If the frequency of disturbance is in the operation frequency band of the amplifier in the DUT, the output voltage is maintained essentially constant value owing to the high-gain negative feedback circuit. However, out-of-band disturbance propagates to the output pin of DUT through parasitic couplings in the DUT. Therefore, the output voltage fluctuates represents by a black line in Fig. 3. However, this fluctuation of the output voltage can be reduced by the stabilization capacitor placed on the output terminal. The false operation of the voltage regulator is defined as voltage shift from the target DC voltage. To obtain the variation of DC voltage, the output voltage is measured through the low pass filter (LPF) in the oscilloscope (the cut off frequency is 20 MHz).

Figure 3 shows the waveform of output voltage when the

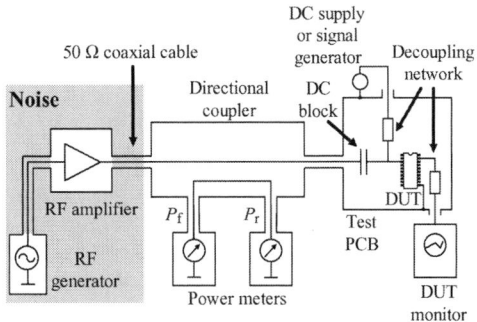

Fig. 1. Direct RF power injection method.

978-1-4799-5004-1/13 $31.00 © 2013 IEEE   59

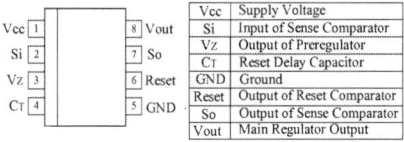

Vcc	Supply Voltage
Si	Input of Sense Comparator
Vz	Output of Preregulator
CT	Reset Delay Capacitor
GND	Ground
Reset	Output of Reset Comparator
So	Output of Sense Comparator
Vout	Main Regulator Output

Fig. 2. Pin assign of L4949.

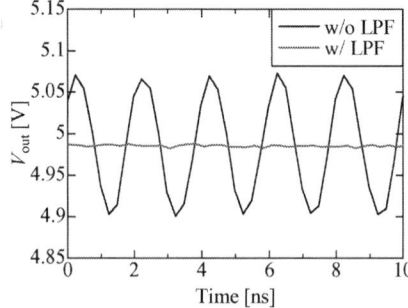

Fig. 3. Comparison between output voltage with or without LPF (20 MHz) when frequency of disturbance is 500 MHz.

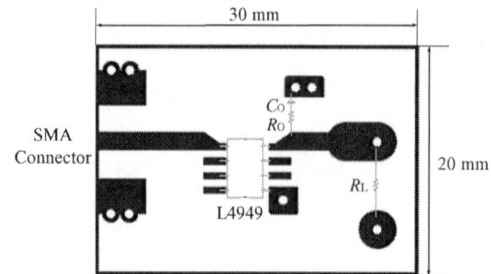

Fig. 4. Top layer of test printed circuit board.

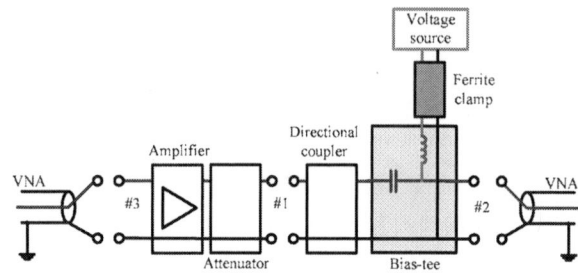

(a) Port definition.

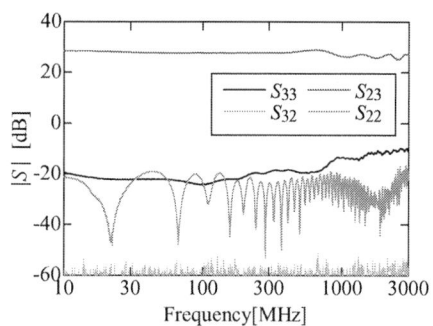

(b) With RF amplifier.

Fig. 5. Measurement of transmission and reflection characteristics of noise path (from RF amplifier to DUT.

frequency of the disturbance is 500 MHz and the incident power is 6.1 dBm. The fluctuation of the output voltage without the low pass filter reaches 4.9V. On the other hand, the output voltage measured through the low pass filter differs slightly from 5 V. Because we focus DC offset of output pin, the output voltage with LPF is estimated in this paper [1].

B. Factors Affecting Reproducibility of DPI Measurement Setup

The reproducibility of DPI method can be affected by the following 4 factors:

1) Electrical characteristics of a test printed circuit board on which DUT is mounted.
2) Transmission and reflection characteristics of measurement setup.
3) Characteristics of decoupling circuit of DC supply.
4) Common-mode current flowing on DC power supply cable.

In this paper, we discuss the influence of these factors to the reproducibility of the DPI to construct a reproducible measurement setup.

1) Electrical characteristics of PCB: The LDO regulator is mounted at the center of two-layer PCB illustrated in Fig. 4. The VCC pin is connected with SMA connector through 50 Ω the microstrip line. The VSS pin is linked to the solid ground plane on the bottom layer of the PCB. The load resistance R_L is placed between the Vout pin and the ground. Additionally, the stabilizing capacitor C_O with the resistor R_O for phase compensation is placed in parallel with the load resistor.

A DC biasing network is connected to the RF power injection path through the decoupling network. In IEC 62132-4, the decoupling network must be placed near the DUT to

[1] When we consider a generic device such as an opamp, the high frequency coupling from an input pin to an output pin must be evaluated

minimize the effect of the DC cables. However, it is difficult to design a wide band decoupling network with SMD capacitors and inductors on the PCB. Therefore, a commercial bias-tee (60 kHz - 20 GHz) is used for decoupling of DC path.

2) Transmission and Reflection Characteristics: The characteristic impedance of the transmission line of RF disturbance is designed as 50 Ω. However, usually input impedance of a DUT is not 50 Ω, and a part of RF disturbance reflects and reaches to the RF amplifier or signal generator. If VSWR of output terminal of the RF amplifier is high, the multiple reflection is generated and it affects the reproducibility of the immunity evaluation depending on the length of cable.

It is necessary to improve the termination of the output

of the RF amplifier for reduction of the efficiency of the multiple reflection. In order to improve matching characteristic of the RF amplifier, an attenuator (3 dB, 6dB or 10 dB) is connected at the output terminal of RF amplifier. As a result, 10 dB attenuator can reduce the reflection coefficient at the output terminal of RF amplifier to 0.11 as will hereinafter be described in detail.

C. Characteristics of Decoupling Network of DC Supply

When the immunity of DUT is evaluated, the DUT device is supplied power from the DC power source to operate normally. However it is difficult to control the high-frequency characteristics of the DC supply cable and the DC source. Therefore the input impedance of the power supply system from the noise injection path must be higher than 400 Ω to decouple the DC supply system. In reference[2], the decoupling network should be designed near the DUT on the test PCB. Although interested range of the frequency is from 10 MHz to 3 GHz, the decoupling network consisting of surface mount devices on the PCB has narrow band. Therefore bias tee outside the test PCB is used for decoupling of the DC power supply system.

The bias tee (Picosecound Pulse Labs, 5545-107) which is used in this paper, contains 340 μH ± 20% inductance for the decoupling DC supply system. Therefore, at the lowest frequency, 10 MHz of the immunity evaluation, the impedance, Z_L, is expressed as

$$Z_\mathrm{L} = \mathrm{j}\omega L \approx \mathrm{j}6400\pi, \qquad (1)$$

and is sufficiently larger than 400 Ω.

D. Common-mode Current Flowing on DC supply

Whole DPI setup, injection path of the RF disturbance is 50 Ω devices and coaxial cables. The 50 Ω microstrip line is designed on the test PCB. When this microstrip line has enough wide ground plane, there is no mode conversion from the injected RF disturbance to the common mode because the difference of imbalance factors is vanishingly small [4].

However, the common mode can be generated at the bias tee because usually a balanced two wire cable such as a twisted pair cable is used for DC supply. Therefore, the imbalance difference at the bias tee generates mode conversion, and common-mode current flows on the DC supply cable. The placement of the DC supply cable can affect the reproducibility of the immunity evaluation. In IEC 62132-4, the regulation of decoupling circuit is described about only the normal mode, the regulation of common-mode decoupling must be establish.

In this DPI setup, the twisted cable for DC supply places on the reference ground with keeping a constant height from the reference ground, and the ferrite clamp (EMC Kashima, FC22, 30MHz - 3000MHz, insertion loss $\succeq$ 15 dB) as the common mode absorption device, which is shown in Fig 6, is placed near the bias tee for reduction of the common-mode current flowing on the DC supply cable.

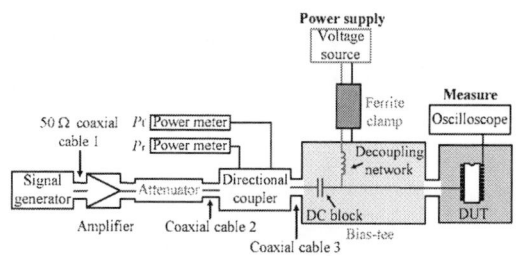

Fig. 6.   Measurement setup after improvement.

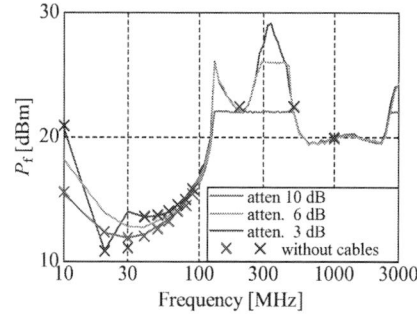

Fig. 7.   Comparison of immunity between several lengths of cable.

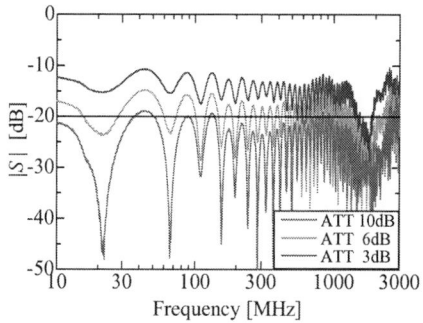

Fig. 8.   Reflection of measurement system from DUT.

III.   REPRODUCIBILITY

A. Suppression of Multiple Reflections

In order to measure reproducibility, we measure the immunity of the DUT changing the following condition;

1)   The value of the attenuator is changed from 3 dB to 10dB.
2)   The length of the cables is changed.

Figure 7 shows the measurement results of the immunity of the DUT, and compare the difference among the attenuator (3 dB, 6 dB or 10 dB) at the output terminal of RF amplifier. The maximum input power to the DUT by the measurement setup with the attenuators (3 dB, 6 dB and 10 dB) are 23 dBm, 27 dBm and 30 dBm, respectively, because the nominal maximum power of these attenuators are 33 dBm. Above 100 MHz, there is no difference among the measurement results using different attenuators. On the other hand, the results are significantly different below 100 MHz. From the point of an intention that the attenuator is inserted in the measurement setup, we can understand that the multiple reflections due to the reflection characteristics of the output terminal of the RF amplifier affect the reproducibility of immunity measurement.

In order to change the resonant frequency due to the multiple reflections, the coaxial cables 2 and 3 in Fig. 6 are removed and the RF amplifier, the directional coupler and the bias tee connects directly. The results of the immunity are shown as × marks in Fig. 7. The reproducible measurement result is obtained by using 10 dB above 100 MHz because the multiple reflections are reduced. On the other hand, the 2.9 dB difference is obtained at 30 MHz when the 3 dB attenuator is inserted in the measurement setup.

Next, the reflection coefficients of noise injection path from the DUT are measured and shown in Fig. 8. The resonances are generated at the 45 MHz interval. We suggest that the resonances caused by the length of the cable. When the 10 dB attenuator is inserted, the reflection coefficient is less than -20 dB (VSWR: 1.2) and the resonance is suppressed. As a result, the reproducibility of the immunity measurement is improved.

B. Reduction of Common-mode Current

The coaxial cable and the microstrip line as the path of the disturbance are completely asymmetric transmission line. On the other hand, general power supply cable is symmetric transmission line consisted of the ground and power lines. Because of the difference of symmetric property [4], the common-mode current is generated and flows on the power supply cable.

In order to estimate the common-mode generation, the common-mode characteristics of the power supply cable are fixed by placing the large ground plane below the power supply cable. The distance between the power supply cable and the ground plane is 100 mm or 0 mm (vinyl-covered cable is placed on the ground conductor) as shown in Fig. 9(a). Using current probe, we observe the common-mode current flowing on the power supply cable and shows the result in Fig. 9(b). The common-mode current becomes large because the cable is placed near the ground plane. In order to reduce the common-mode current, the distance between the cable and the ground plane is fixed more than 300 mm or a common-mode absorption device is placed near the DUT.

Figure 9(c) shows the comparison between immunity measurements with or without the common-mode absorption device. At the frequency range of more than 500 MHz, there is no reproducibility of the immunity measurement because the large common-mode current flows on the power supply cable as shown in Fig. 9(b).

IV. CONCLUSION

In this paper, we discussed the reproducibility of the direct RF power injection method. Transmission and reflection characteristics of the measurement setup affected the evaluation of immunity. Also, the common-mode generation at the connection of the DC supply cable should be reduced using the common-mode absorption devices.

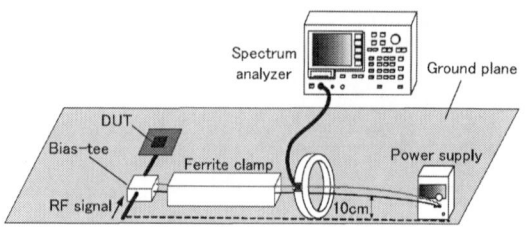

(a) Measurement setup for common-mode current.

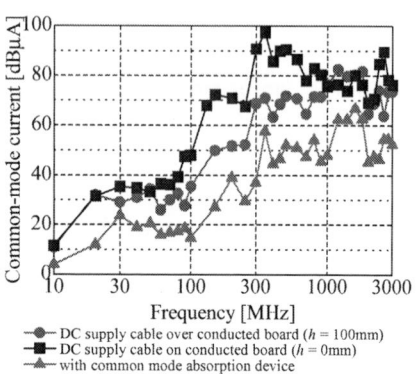

(b) Common-mode current flowing on power supply cable.

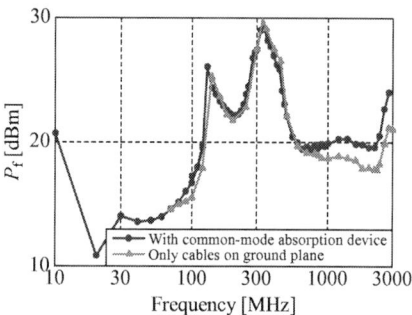

(c) Reproducibility of immunity measurement.

Fig. 9. Evaluation of common-mode current flowing on power supply cable.

REFERENCES

[1] P.S. Crovetti and F.Fiori, "A Linear Voltage Regulator Model for EMC Analysis," IEEE Trans. Power Electron. Vol. 22, No.6, pp.2282-2292, Dec. 2008.

[2] IEC 62132-4: "Integrated circuits - Measurement of electromagnetic immunity, 150 kHz to 1 GHz - Part 4: Direct RF power injection method," 2006.

[3] W. Jian-fei, E. Sicard, A.C. Ndoye, F. Lafon, L. Jian-cheng, and S. Rong-jun, "Investigation on DPI effects in a low dropout voltage regulator," 8th Workshop on Electromagnetic Compatibility of Integrated Circuits (EMC Compo)., Nov. 2011, pp. 153–158.

[4] T. Watanabe, H. Fujihara, O. Wada, R. Koga, and Y. Kami, "A Prediction Method of Common-Mode Excitation on Printed Circuit Board Having a Signal Trace near the Ground Edge," IEICE transactions on communications, Vol.E87-B, No.8, pp.2327–2334, Aug. 2003.

A Generalized Accurate Modelling Method for Automotive Bulk Current Injection (BCI) Test Setups up to 1 GHz

Sergey Miropolsky, Alexander Sapadinsky, Stephan Frei
Technische Universität Dortmund, Germany
sergey.miropolsky@tu-dortmund.de

Abstract — Development of accurate system models of immunity test setups might be extremely time consuming or even impossible. Here a new generalized approach to develop accurate component-based models of different system-level EMC test setups is proposed on the example of a BCI test setup. An equivalent circuit modelling of the components in LF range is combined with measurement-based macromodelling in HF range. The developed models show high accuracy up to 1 GHz. The issues of floating PCB configurations and incorporation of low frequency behaviour could be solved. Both frequency and time-domain simulations are possible. Arbitrary system configurations can be assembled quickly using the proposed component models. Any kind of system simulation like parametric variation and worst-case analysis can be performed with high accuracy.

Keywords — *Automotive EMC, IC EMC, Virtual EMC Tests, Bulk Current Injection (BCI), Vectfit Macromodelling*

I. INTRODUCTION

The EMC failures detected at the test-and-measurement stage may lead to expensive device redesigns and cause serious delays of a product launch. The success of a chip-level EMC test (e.g. Direct Power Injection, DPI, [1]) does not necessarily imply the success of a following system-level EMC tests (e.g. Bulk Current Injection, BCI, [2]). To evaluate the RF immunity at early design stages it is helpful to perform virtual EMC tests using accurate models. The variation of test setup parameters, e.g. cable harness length, is necessary for a worst-case analysis.

Models for system-level EMC setups have been developed by multiple groups e.g. [4-8]. In this work a generalised approach to develop an accurate component-based model of a system-level immunity test setup is described on the example of the bulk current injection (BCI) method. Equivalent circuit modelling in LF range is combined with measurement-based macromodelling in HF range. Any setup configurations can be assembled using the component models, thus parametric variation and worst-case analysis become possible.

According to [2] the BCI test is performed in the frequency range up to 400 MHz. The internal requirements often extend the range up to 1 GHz. The developed model must also support transient simulations involving RF and LF signals. Therefore the model must be valid from LF (DC) up to at least 1 GHz.

II. BULK CURRENT INJECTION (BCI) TEST SETUP

The bulk current injection for system-level applications (ISO 11452-4, [2]) is widely used for RF immunity testing of electronic components, especially in automotive industry. The sketch of such test setup is shown in fig. 1.

The equipment under test (EUT) consisting of a test PCB with one or more ICs to be tested is connected to the peripheral equipment with a cable harness of a specified length. The EUT is usually floating, i.e. it is not grounded locally, but only connected to the peripheral devices with a long cable harness. The PCB is coupled to ground due to stray fields. An artificial network (AN, LISN) is used to supply DC or LF signals or to measure the HF signal levels. Additional devices might be used to provide the LF functionality.

In the BCI test a common mode RF current of a specified amplitude or simply a specified forward power is injected into the cable harness using an injection clamp, and the EUT / DUT functionality is observed under this RF disturbance.

The setup modelling can be split into test-case dependent and independent parts. Many components involved in an RF immunity test, namely the LISN, the cable harness and the BCI injection clamp, should be modelled only once and can be re-used for further test cases. The supplementary equipment, PCB and DUT impedances are test-case dependent, therefore the models should be developed for each case individually. In some application cases the supplementary equipment is not used, the DC signals are supplied to the cable harness with the LISN and the DUT monitoring is performed with optically-decoupled probes. This configuration is considered in current work for simplicity purposes.

A cable harness of any type can be used in real BCI tests. A twisted pair cable harness will be considered in this work. A similar approach can be applied for any other cable type including homogenous cable bundles.

As it was proposed and confirmed by multiple authors [7,8,12], the test setup can be described as a complex multiple-port system. The goal of modelling is to reproduce the setup behaviour in a simulation environment.

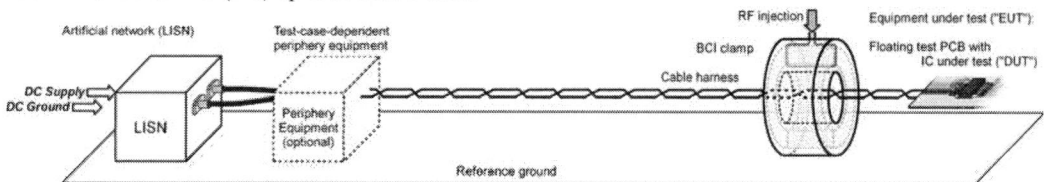

Fig. 1 System-level bulk current injection setup overview

978-1-4799-5004-1/13 $31.00 © 2013 IEEE

III. MODELLING METHODS FOR PASSIVE COMPONENTS

A. Equivalent circuit setup modelling

For some components the equivalent circuit models can be easily found in LF range. The model extension with parasitic couplings and precise parameters fitting extends the model validity to higher frequencies. Due to necessary simplifications, these models are usually valid up to 100-200 MHz.

B. Measurement-based macromodelling

Passive structures can be modelled with a measurement-based macromodelling method. The device network parameters are measured with a network analyser (VNA), approximated with rational functions [9], converted to a state-space model, and implemented as a circuit [11]. The macromodels reproduce the passive electrical behaviour as it was captured at the original measured object. Commonly a high accuracy can be reached, whereby it depends on the quality of the measurement dataset, data approximation order, and circuit implementation.

This method shows high efficiency and accuracy, but has several drawbacks. The frequency range and dynamic impedance range covered by the model do not exceed those of the VNA (commonly from 300 kHz up to 1 GHz and from 10 mΩ up to 1 MΩ). The lower frequency limit is important, since many properties necessary for transient simulation can only be captured in LF range. The measurement-based data artefacts might be approximated within the original dataset. Another significant method limitation is the necessity of a 50 Ω measurement access to all involved nodes. This is especially critical for EMC setups with floating PCBs due to the missing common ground connection. Finally, the measurement-based macromodels can only reproduce the transfer function of an existing physical setup. For any parameter variation of e.g. cable harness length or injection clamp position, the entire modelling procedure must be repeated completely.

C. Proposed combined modelling procedure

The advantages of both methods can be used to develop the combined simulation models [11]. The component must be characterized with VNA measurement up to the highest involved frequency, e.g. 1 GHz. Precise deembedding might be used to exclude the influence of test fixtures or other irrelevant components. An equivalent circuit model must be created to be valid from DC up to at least the lowest frequency covered with the measurement data, e.g. 100 MHz. The circuit parameters should be optimized so that a smooth transition of simulation model to VNA measurement data is observed in the boundary frequency range for each S-, Y- and Z-parameter curve.

The scattering network parameter datasets of LF simulation (DC to e.g. 100 MHz) and the deembedded HF measurement (e.g. 100 MHz to 1 GHz) must be concatenated. The data smoothness at the edge frequency can be enforced by

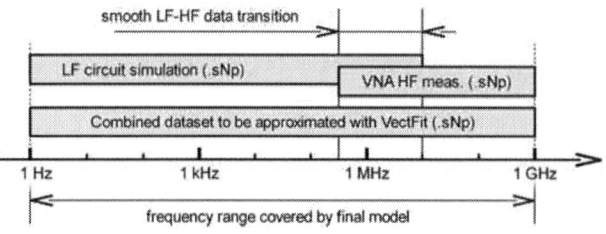

Fig. 2 S-parameter dataset concatenation for combined macromodelling

using a linear or low order polynomial transition from LF to HF data in some frequency window around the boundary frequency. A Vectfit approximation [9] can now be applied to this concatenated dataset [11] and a macromodel can be generated. The macromodels based on such combined datasets show correct results in the range from DC to 1 GHz and can be used in both frequency and time domain simulations.

IV. BCI COMPONENT-BASED SETUP MODELLING

The test setup is modelled with an electrical circuit (fig. 11) consisting of parameterised sub-circuits of cable harness, BCI coupling, LISN and floating PCB with DUT.

A. Cable harness: multiconductor TL model

The cable harness is one of the most important components of the test setup. In most existing models for BCI setups, e.g. [5-6], a quasi-TEM mode signal propagation along the cable harness is assumed. The cable harness is modelled with a multiconductor transmission line (MTL). Circuit simulators, e.g. Synopsys HSPICE, provide good support for MTL devices, so the model of this type is used for simulation. The MTL devices in HSPICE can be described with per-unit-length RLCG parameters. The frequency dependencies are either listed in tabular form, or approximated using R_0, L_0, G_0, C_0, the skin effect R_S and the dielectric loss G_D [17] matrices.

Multiple methods for measurement-based characterisation of MTLs are available, e.g. [14]. A significant issue of all types of measurement-based MTL analysis is to separate the self-properties of a homogenous MTL from the test fixtures (fig. 3). The cable properties close to the fixtures are also different from those of the homogenous cable over ground due to the stray couplings to the fixtures. A high parameter extraction accuracy for both homogenous cable harness and stray fields is critical for further deembedding procedures. Therefore a very high attention should be paid to this seemingly-simple step.

B. BCI coupling modelling

Multiple injection clamp models have been developed, e.g. [3-6]. Here the circuit model for the BCI clamp (FCC F140) was developed similar to [6]. The parameters were found by fitting the circuit model to measurement data. The coupling to the secondary and tertiary windings was considered to be concentrated at a single cable point and implemented with an ideal three-port transformer as shown in [4]. The clamped cable (7 cm) was initially included as an MTL with the same RLCG values as for the cable over ground.

A precise 5-port dataset of the BCI coupling to the cable was obtained by measurement of the setup shown in fig. 4 and a deembedding procedure. The test fixture RF ports were deembedded as lossless 60 ps port extensions. The cable with stray effects at the fixtures and the remaining cable harness were deembedded up to the side plane of the BCI clamp, so that the ports of the dataset were connected between the cable pins and the ground (ports 2-5 in fig. 4). The procedure was repeated for three cable lengths (15, 20 and 25 cm). The same dataset (up to numerical noise and smaller measurement artefacts) was obtained after deembedding. Thereby the deembedding method validity and accuracy were assured.

The smaller differences in the RLCG properties of the clamped cable to those of the main harness were extracted and implemented into the MTL model. The same setup was assembled and simulated in HSPICE and the S-parameters

978-1-4799-5004-1/13 $31.00 © 2013 IEEE

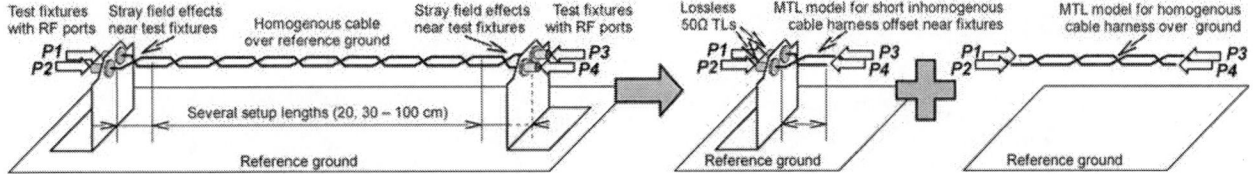

Fig. 3 Twisted cable harness characterization: measurement setup (left) and goal models for separate setup components (right)
The procedure of the parameter extraction method for homogenous cable harness and test fixture offsets will be shown in details in the coming publications.

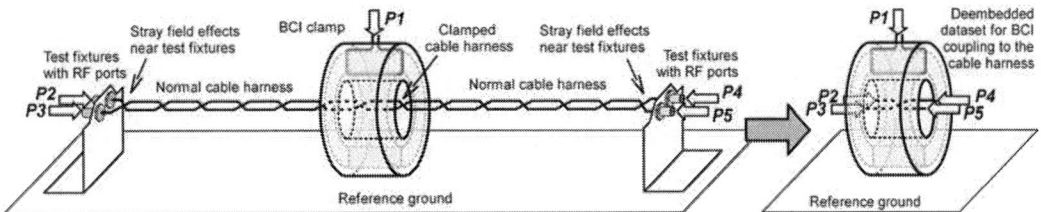

Fig. 4 BCI coupling to a twisted cable characterisation – measurement setup (left) and deembedded 5-port dataset (right)

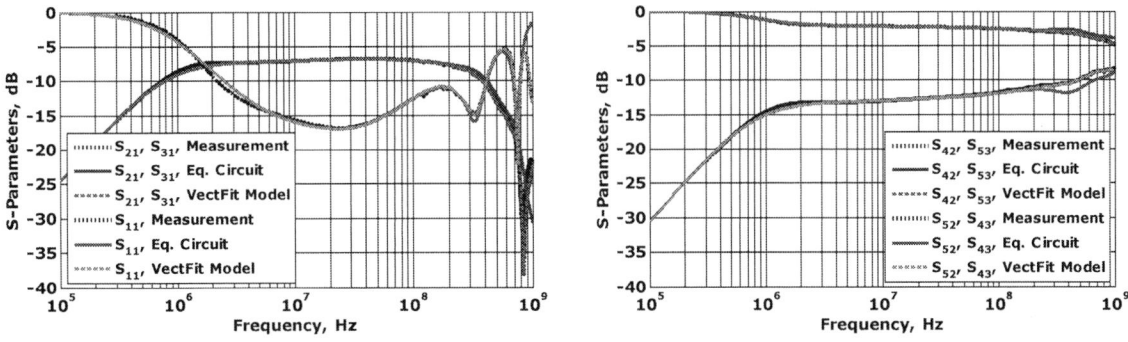

Fig. 5 BCI clamp model vs. deembedded dataset a) BCI port reflection and coupling to cable b) signal transfer through the cable within BCI clamp

were compared with measurements (fig. 5). A good correlation could be observed up to at least 400 MHz, while the internal BCI clamp resonance at 800 MHz was missing in the model.

To extend the model validity up to 1 GHz, the simulated dataset (1 Hz to 400 MHz) was concatenated with the deembedded measurement dataset (100 MHz to 1 GHz). A smooth linear transition was enforced in the range of 100 to 400 MHz. The concatenated dataset was approximated with vector fitting [9] to a macromodel. The S-parameters for the BCI coupling were simulated with the model and compared to the deembedded data (fig. 5). High model accuracy in the entire frequency range up to 1 GHz could be reached.

C. HF artificial network (AN / LISN)

A common LISN device specified in [2] must have an input impedance of $50\,\Omega$ at the cable harness port up to 108 MHz. The properties in higher frequency range are unspecified. It is possible to model an existing LISN device in the HF range using the same combined macromodelling approach. However, clean RF measurements are not possible without additional measures because of the specific non-RF connectors at available LISNs. Also by reproducing the HF issues of a LISN in a model, the known measurement problems are simply transferred into the virtual EMC tests.

A LISN with a $50\,\Omega$ input impedance and smooth transfer functions up to 1 GHz was designed (fig. 6). The SMD devices reproduce the LF behaviour. The $50\,\Omega$ PCB traces handle the signal transfer in HF range. All LISN ports are designed as RF connectors to allow device characterization with VNA. A circuit model for such device is easily developed up to 1 GHz.

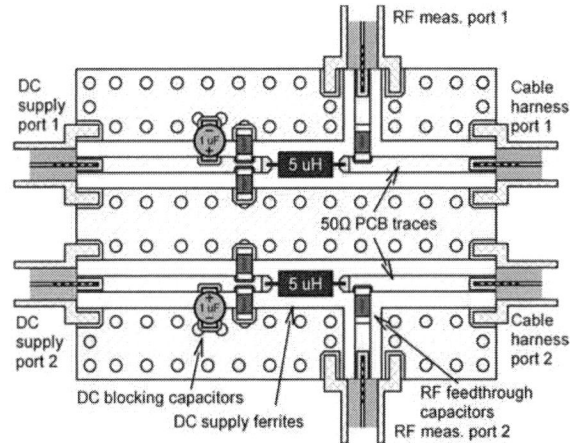

Fig. 6 HF artificial network (LISN) internal structure

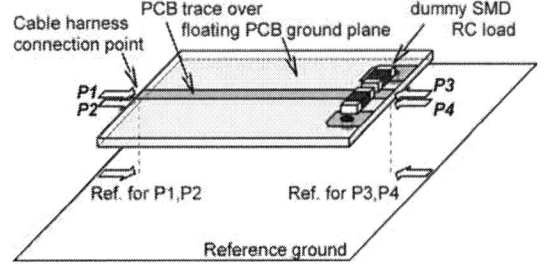

Fig. 7 Dummy EUT: floating test PCB with a single trace and dummy DUT (here: SMD RC load of 100 nF ∥ 1 kΩ)

978-1-4799-5004-1/13 $31.00 © 2013 IEEE   65

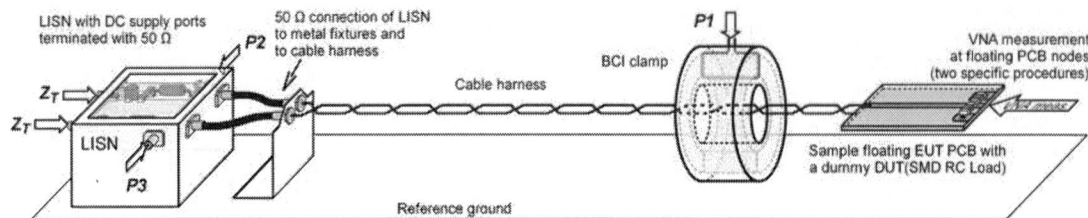

Fig. 8 Test setup for verification of developed models, BCI injection into cable harness with LISN and floating PCB with dummy RC load
Note, that signal transfer in the setup depends not on the setup only, but also on the measurement ports assignment.

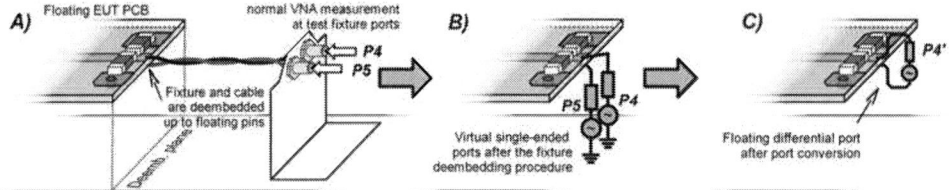

Fig. 9 Measurement of small-signal transfer function to the floating PCB nodes by deembedding the pre-characterized metal fixtures and short cable harness
A) Measurement structure with test fixtures B) Single-ended ports to reference ground after deembedding C) Floating differential port after port conversion

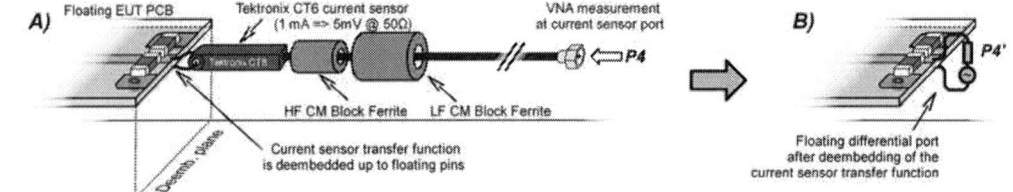

Fig. 10 Measurement of small-signal transfer function to the floating PCB nodes by deembedding the pre-characterized small-size current sensor
A) Measurement structure with current sensor B) Floating differential port after current sensor deembedding
The technical details and possible issues of the mentioned deembedding and port conversion procedures will be shown the coming publications.

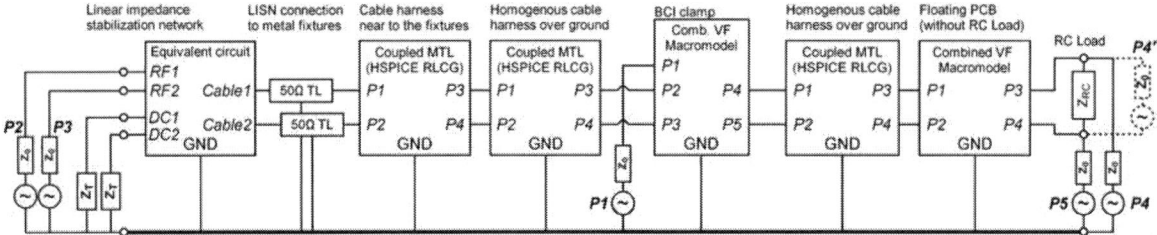

Fig. 11 Simulation model with two possible port assignments: a 5-port setup with single-ended ports P_4 and P_5 and a 4-port setup with a differential port P_4'

D. Floating test PCB modelling

One of the significant problems in modelling the system-level BCI setups is a floating PCB. According to the specification, the EUT is located in 50 mm height above the ground and is connected to the cable harness. Plastic EUT packages do not affect the couplings and therefore can be neglected for modelling. Hence the EUT can be reduced to a test PCB with an IC under test and, optionally, some SMD periphery necessary for LF functionality or RF protection.

The common-mode RF current injected is converted into a differential-mode signal due to the asymmetric impedances at the floating PCB. This CM-DM conversion was analysed in details in [13]. Starting with approx. 100 MHz a distributed conversion should be considered. In real test cases the 3D simulation of the PCB can be performed with an EM field solver. The virtual ports should be defined between the floating nodes, e.g. at the trace nodes, at the IC pins, and at the relevant points of the PCB ground plane. The S-parameter dataset can be simulated, approximated and used in the further simulation.

A simplified test PCB with a single signal trace over PCB ground is used in current work (fig. 7). The complex model of the CM-DM signal conversion at floating conductors can be simplified to an asymmetric MTL model with an approach similar to [13], where the PCB ground is considered to be just one more conductor over main reference ground.

The floating PCB structure without SMD components is measured in test fixtures in the same way as in section IV.B. The fixtures and the cable are precisely deembedded and a 4-port dataset of the floating PCB is obtained. The virtual ports are connected between the floating nodes and ground (fig. 7). The combined Vectfit macromodel is developed for the structure in the same way as discussed in section III-C.

E. DUT and dummy RC load modelling

To exactly reproduce a real RF immunity test, a physical transistor-level IC model has to be combined with a package model and must be attached to the ports of the floating PCB model. Since IC modelling for RF immunity testing is a separate complex topic which should not be covered within this work, a simple linear load of 1 kΩ ‖ 100 nF is used here. The load is represented by two SMD components soldered between signal trace and PCB ground as shown in fig 7. The load impedance is measured with a VNA, modelled with a simple passive equivalent circuit, and connected to the corresponding pins of a floating PCB model.

978-1-4799-5004-1/13 $31.00 © 2013 IEEE

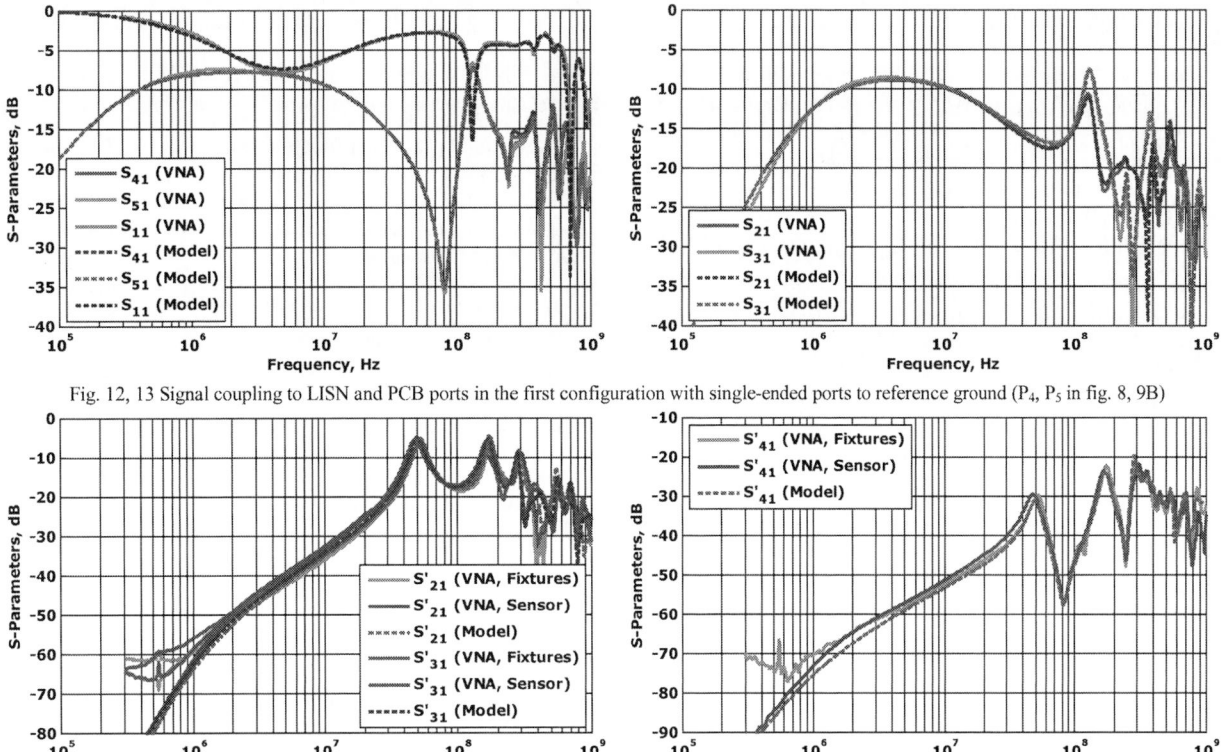

Fig. 12, 13 Signal coupling to LISN and PCB ports in the first configuration with single-ended ports to reference ground (P₄, P₅ in fig. 8, 9B)

Fig. 14,15 Signal coupling to LISN and PCB ports in the second configuration with a floating differential port (P₄' in fig. 8, 9C, 10B) obtained by current sensor deembedding (fig. 10A-B) and by fixture deembedding and port conversion (fig. 9A-C)

V. MODEL VERIFICATION

The modelling approach is verified with measurements. The verification setup, consisting of 1m cable harness, LISN and dummy floating EUT with passive RC load, is assembled as shown in fig. 8. Two short 50 Ω cables are used to connect the LISN to the test fixture ports. The BCI clamp is located in 15 cm distance from the edge of the floating EUT PCB.

A significant issue to be handled during model verification is to access the floating PCB nodes. A physical connection of a VNA port is not possible due to obvious reasons. Two following solutions were used.

A. Deembedding cable harness up to floating PCB nodes

The floating pins can be accessed by precise deembedding. The metal fixtures (section IV.A) are connected to the PCB pins with a 5 cm piece of the cable harness. The measurement is performed up to fixture ports (fig. 9A). The fixtures and the cable are then deembedded as in section IV.B. The resulting single-ended 50 Ω ports are attached between the floating nodes and the ground (fig. 9B). By applying a port conversion procedure to two single-ended ports, the signal transfer to the *floating differential port* P4' can be obtained (fig. 9B,C).

B. Floating differential measurement using current sensor

In the second method the usage of the metal fixtures is avoided. A current sensor (Tektronix CT6) is used to transform the pin-to-pin current into the signal at the VNA port (P₄' in fig. 10A).The transfer function of the sensor is deembedded [15,16]. The resulting *floating differential port* is attached between the cable pins (P4' in fig. 10B). The method presumes that the floating nodes are not disturbed by the sensor. Still

some additional CM impedance to ground ("to infinity") is introduced into the system, and cannot be deembedded in this configuration. This CM impedance can still be characterized by measuring the current sensor in a known setup and can be appended to the simulation model for the comparison purposes.

The results for both methods are compared to each other to assure the methods validity, and to the simulation results.

C. Results and discussions

The simulation and measurement results for the test setup in the first configuration with two single-ended ports (P₄ and P₅ in fig. 8, 9B) are shown in fig. 12-13. A closed common mode (CM) current loop is present in the system due to two 50 Ω ports to the main reference ground at the floating PCB nodes (P₄, P₅ in fig. 9B). The BCI magnetic coupling induces significant CM current in the cable harness. Therefore a high signal level is observed at all setup ports already in LF range.

This configuration with virtual single-ended ports doesn't correspond to a typical BCI application with floating EUT, but can be efficiently used to validate the model. The simulated BCI coupling to the virtual ports between the floating pins and the ground (S₄₁, S₅₁) shows very good correlation to the measurement even in HF range. The same accuracy can be observed for the coupling to LISN ports (S₂₁, S₃₁). The signal transfer between other setup nodes (e.g. from floating PCB to LISN ports) also show sufficient accuracy. The rest curves are not shown here due to brevity purposes.

The results for the second setup configuration with floating port (P'₄ in fig. 9C and 10B) are shown in fig. 14-15. Here the common mode current loop in the system is open, and the

978-1-4799-5004-1/13 $31.00 © 2013 IEEE   67

magnetic coupling induces a high CM voltage (but not the CM current) over the cable harness. Therefore a weak LF coupling is observed in LF range (S'_{21}, S'_{31}, S'_{41} in fig. 14-15). With rising frequency the capacitive coupling of the floating PCB to main reference ground increases, and the CM current rises. This leads to the increasing signal at LISN ports (S'_{21}, S'_{31}). The impedance asymmetry leads to the appearance of the differential signal at the cable harness in general and locally at the floating port (S'_{41}). Starting with 100 MHz the signal transfer is determined by cable resonances.

Both measurement methods show the same results (up to two insignificant deviations) in the entire frequency range. The measurement with fixtures shows an artefact in LF range (below 1 MHz). This is caused by the applied port conversion, which is rather sensitive to the VNA measurement noise, especially for the analysis of weak differential signals.

The current sensor measurement in its turn shows a smaller offset in signal levels in LF range from 5 to 50 MHz. This signal offset is caused by the current sensor CM impedance (approx. $0.5 - 0.75$ pF with some HF losses). By characterising this impedance with a separate measurement and deembedding procedure, modelling it with a simple equivalent circuit and attaching it to the single-ended ports before the port conversion an even better model correlation to the measurements can be reached for verification purposes.

A very good fitting can be observed for both measurement methods and simulation models. Even for the signal transfer to the floating differential port, which is normally very complex to both measure and model in such configurations, a good correlation of model to measurement data is observed. This confirms the accuracy of both measurement methods and the simulation model.

VI. MODEL APPLICATION

The developed test setup model can be used in virtual EMC test during IC design. Here a floating PCB model has to be created for each test case e.g. with EM field solver. The IC models (either transistor-level or behavioural) wrapped in package models have to be attached to the floating PCB nodes, and the circuit simulations of any kind can be performed. The test setup model can also be efficiently used for substitution methods mentioned in [7,8,12].

The main advantage of a component-based EMC test setup model is the possibility to simulate any test configuration, including different cable harness lengths and clamp positions, passive protection at PCB level, complex cable networks, etc., purely virtually, i.e. without running a real EMC test. A worst-case analysis can be performed. The EMC failures can be detected and necessary built-in IC solutions or an external PCB-level EMC protection can be developed in advance.

CONCLUSIONS

A generalized method to develop an accurate component-based model of a system-level BCI test setup is proposed. The components are characterized with VNA measurements and precise deembedding. An equivalent circuit modelling in LF range is combined with measurement-based macromodelling for HF range. Such models are valid in the up to 1 GHz for both frequency and time domain simulations.

The setup model is verified with measurements. Two independent measurement procedures are used to capture the signal transfer to a dummy DUT (RC load) at the floating PCB. Both methods confirm the accuracy of the simulated results.

The models can be used for advanced EMC simulations involving complex transistor-level or behavioural DUT models. The EMC failures can be detected and necessary built-in IC solutions or an external PCB-level EMC protection can be developed in advance.

ACKNOWLEDGEMENT

The reported R+D work was carried out within the CATRENE project CA310 EM4EM (Electromagnetic Reliability and Electronic Systems for Electro Mobility). This particular research is supported by the BMBF (Bundesministerium fuer Bildung und Forschung) of the Federal Republic of Germany under grant 16 M3092 I. The responsibility for this publication is held by the authors only.

REFERENCES

[1] IEC 62132-4: Integrated circuits – Measurement of electromagnetic immunity, part 4: Direct Power injection (DPI) method.

[2] ISO 11452-4: Road vehicles – Component test methods for electrical disturbances from narrowband radiated electromagnetic energy, part 4: Bulk current injection (BCI)

[3] M. F. Sultan, "Modeling of a bulk current injection setup for susceptibility threshold measurements," IEEE Int. Symp. on EMC Proceedings, San Diego, CA, 1986, pp. 188-195.

[4] F. Grassi, F. Marliani, S. A. Pignari, "Circuit Modeling of Injection Probes for Bulk Current Injection," IEEE Tran., EMC 49, no. 3, Aug. 2007, pp. 563-576

[5] F. Lafon, Y. Belakhouy, and F. De Daran, "Injection probe modeling for bulk current injection test on multiconductor transmission lines," IEEE Symp. on Embedded EMC Proceedings, Rouen, France, 2007.

[6] Miropolsky, S., Frei, S., and Frensch, J.: Modeling of Bulk Current Injection Setups for Virtual Automotive IC Tests, EMC Europe, 2011, Wroclaw, Poland, 2011.

[7] Durier, A., Pues, H., Vande Ginste, D., Chernobryvko, M., Gazda,C., and Rogier, H.: Novel Modeling Strategy for a BCI setup applied in an Automotive Application, EMC Compo 2011, Dubrovnik, Croatia, 2011.

[8] Oguri, Y., Ichikawa, K.: Simulation Method for Automotive Electronic Equipment Immunity Testing, EMC Europe 2012, Rome, Italy, 2012.

[9] Gustavsen, B., Semlyen, A.: Rational Approximation of Frequency Domain Responses by Vector Fitting, IEEE Tran. On Power Delivery, 14, 1052–1061, 1999.

[10] The Vector Fitting Website, http://www.sintef.no/projectweb/vectfit

[11] Miropolsky, S., Frei, S., Optimierung der Makromodellierung von Übertragungsstrecken mit Vector-Fitting-Methoden durch Anpassung der Eingangsdaten, EMV Düsseldorf, 2014 (accepted for publication)

[12] Miropolsky, S. and Frei, S.: Comparability of RF Immunity, Test Methods for IC Design Purposes, EMC Compo 2011, Dubrovnik, Croatia, Nov. 2011.

[13] Crovetti, P. S., Fiori, F.: Distributed Conversion of Common-Mode Into Differential-Mode Interference, IEEE Tran. on Microwave Theory, Vol. 59, No. 8, Aug. 2011.

[14] Degerstrom, M.J., B.K. Gilbert, and E.S. Daniel. Accurate resistance, inductance, capacitance, and conductance from uniform transmission line measurements, IEEE-EPEP, 18th Conf., Oct. 2008, pp. 77–80.

[15] zur Nieden, F, Frei, S., Pommerenke, D., "A combined impedance measurement method for ESD generator modeling, EMC Europe 2011 York , vol., no., pp.476-481, Sept. 2011

[16] zur Nieden, F., Scheier, S., Frei, S., "Circuit models for ESD-generator-cable field coupling configurations based on measurement data, EMC Europe 2012, 17-21 Sept. 2012

[17] Synopsys HSPICE User Guide on Signal Integrity Modelling and Analysis, version F-2011.09, September 2011

IC-Stripline Design Optimization Using Response Surface Methodology

Tvrtko Mandic*, Renaud Gillon[†] and Adrijan Baric*

*University of Zagreb, Faculty of Electrical Engineering and Computing, Croatia
[†]ON Semiconductors BVBA, Belgium

E-mail: tvrtko.mandic@fer.hr

Abstract—This paper presents design optimization of closed version of an IC-Stripline. The geometrical parameters of the IC-Stripline are optimized in order to improve VSWR performance in the frequency range up to 6 GHz. Since the optimization performed directly in 3D EM simulator is computationally expensive, the response surface methodology (RSM) based on design of experiment (DOE) is used to develop surrogate models and to accelerate IC-Stripline design optimization. The design of experiment approach is used to systematically vary geometrical parameters of the IC-Stripline structure which are then simulated in a 3D electromagnetic simulator. The equivalent model circuit parameters are extracted according to the 3D EM simulation results. The response surface formed by the extracted parameter values is modelled using a higher order polynomial. The VSWR optimization of the IC-Stripline is performed by optimizing the equivalent circuit model having the parameters defined by the response surface models. The proposed design of the IC-Stripline shows the improvement in VSWR performance over the frequency range up to 6 GHz.

Index Terms—IC-Stripline, circuit model, 3D EM simulations, response surface methodology.

I. INTRODUCTION

Electromagnetic compatibility (EMC) analysis of integrated circuits (IC) is an important factor of IC performance testing. Due to the rapid increase in operating frequencies and decrease in overall dimensions of electronic circuits, the EMC performance of ICs can have a great impact on the system reliability. This is especially critical for high density printed circuit boards (PCB) or system-on-chip (SOC) where interactions between integrated circuits/chip modules can not be neglected.

In recent years much attention is paid to the EMC analysis of ICs [1]. The emerging standards IEC 61967-8 [2] and IEC 62132-8 [3] define procedures for measuring IC electromagnetic emission and immunity, respectively. The proposed structure for performing the emission and immunity measurements is an IC-Stripline. The IC-Stripline method presents the improvement over the transversal electromagnetic mode (TEM) cell method in terms of sensitivity and efficiency. For the same level of the input power compared to the TEM cell, the IC-Stripline generates a higher level of the electric field. This makes the radiated immunity and emission tests efficient more than 20 dB. Furthermore, the TEM mode of propagation in the IC-Stripline is ensured up to several GHz.

The construction of the IC-Stripline and maximum frequency of 3 GHz for testing the ICs is proposed in [4]. In [5] the impact of the different height, differences between the open and closed version and the maximum dimension of the IC have been analysed. In [6] the design and modelling of an open version of the IC-Stripline having improved VSWR is presented. Although the open version of the IC-Stripline presents a cheaper and simpler set-up comparing to the closed version, interference with surrounding equipment is possible.

This paper presents the design and optimization of closed version of an IC-Stripline with respect to the VSWR performance. The low VSWR value ensures smaller variation of the vertical electric field component in the longitudinal direction of the transmission line, i.e. along the central section of the IC-Stripline [7], [8]. The optimization procedure is based on a response surface methodology (RSM). This methodology is used to construct computationally cheaper models which can be used for the optimization of system performance [9]–[11]. The response surface models are obtained by fitting a 2nd order polynomial to the extracted parameter values of the equivalent circuit model. The equivalent circuit model parameter extraction is performed according to the results of 3D EM simulations. The 3D EM simulation results are obtained according to the design of experiment (DOE) which systematically vary geometrical parameters in a predefined design space. The presented workflow for design and optimization is semi-automated and requires minimal user intervention after the definition of the set-up, i.e. the definition of the EM model and circuit model.

The paper is structured as follows. Section II explains step by step the RSM used for the IC-Stripline design optimization and the result of optimization is presented. Section III presents analysis of variations of E-field in longitudinal direction, while conclusion is given in Section IV.

II. RESPONSE SURFACE METHODOLOGY

The RSM flow diagram is shown in Fig. 1. This flow diagram presents the adaptation of the RSM methodology for the modelling of the IC-Stripline. Each step of the diagram is explained in the following sections.

978-1-4799-5004-1/13 $31.00 © 2013 IEEE

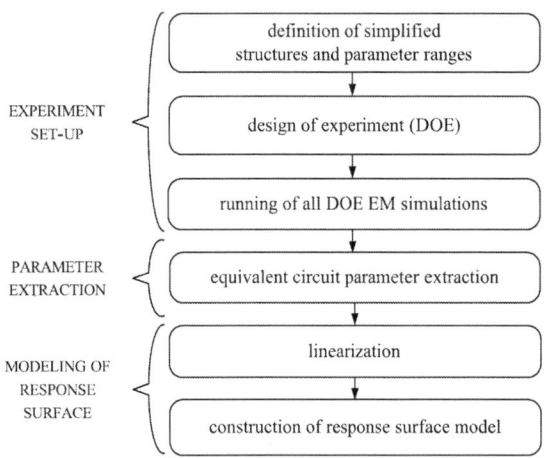

Fig. 1. Response surface methodology flow diagram for the modelling of IC packages [10].

A. Experiment Set-up

The IC-Stripline simulated in a 3D EM simulator is presented in Fig. 2. The 3D EM model is constructed using perfect electric conductors. The simulations are performed using the waveguide ports (WP). The cavity where the septum of the IC-Stripline is located is embedded in the metal enclosure which makes the IC-Stripline completely shielded. The coaxial lines used for the connection of the waveguide ports to the IC-Stripline tapers have fixed parameters $d_1 = 8.02$ mm and $d_2 = 3.48$ mm. These parameters are chosen to model the cross-sectional dimensions of a 50 Ω N-connector. The overall size of the IC-Stripline is fixed and equals $w_{IC} = l_{IC} = 100$ mm, which is a typical size of a PCB used in EMC analysis set-ups. The thickness of the IC-Stripline septum is equal to $t = 0.5$ mm.

The design factors of the performed experiment are presented in Table I. The variation of the septum length l_S with respect to the length of the cavity l_C affects the length of the tapers. The shape of the tapers has a significant impact on

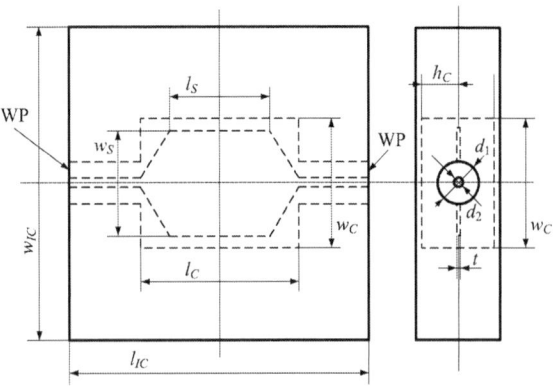

Fig. 2. Closed IC-Stripline simulated in 3D EM simulator.

the VSWR performance [12]. The height h_C of the cavity as well as the width of the cavity w_C and septum w_S have a direct impact on the IC-Stripline characteristic impedance and consequently on the VSWR. Each design factor is assigned a coded level in the range $[-1, 1]$ and, instead of the original symbols (w_S, l_S, ...) capital letters are used (A, B, ...).

A D-optimal design of experiment is performed according to the values of the design factors presented in Table I. The design is created to deliver 40 experiment runs for the 2nd order response surface model. The number of 40 experiment runs is chosen as two times the value of the minimum number of the experiment runs needed for the generation of D-optimal design for the 2nd order response surface model. The design is generated by the coordinate exchange algorithm implemented in MATLAB. The D-optimal design is found to be suitable for fitting of the 2nd order polynomials [13], [14]. The execution of experiment takes approximately $\approx$3 hours on a personal computer (Intel® i7 930, 2.8 GHz, 12 GB RAM).

B. Parameter Extraction

The extraction is performed for the parameters of the equivalent circuit model and it can be implemented using the circuit simulators with integrated optimizer, e.g. ADS [15], or using open-source circuit simulator, e.g. Ngspice [16], and passing the simulation results to optimizer implemented in e.g. MATLAB. The parameter extraction takes place after the completion of the experiment runs, i.e. EM simulations.

The gradient ADS optimizer is used and the goal function is defined as follows [10]

$$g(i,j) = \frac{1}{N_f} \sum_f \left| S_{i,j}^{EM}(f) - S_{i,j}^{CIR}(f) \right|^2 \quad (1)$$

where $g(i,j)$ is the goal function to be minimized at the frequency point f, N_f is the total number of the frequency points, $S_{i,j}^{EM}$ and $S_{i,j}^{CIR}$ are the complex S-parameters of the ports i and j obtained from the EM simulations and the equivalent circuit simulations, respectively. The final value of the error function EF_k is calculated for each experiment run k as

$$EF_k = \sum_{i,j} g(i,j). \quad (2)$$

The parameter extraction procedure, implemented using ADS for the experiment having 40 runs of the 21 frequency point 2-port S-parameter simulations, takes $\approx$10 minutes to complete on a personal computer. The values of the error function EF_k for the parameter extraction performed for the experiment

TABLE I
DESIGN FACTOR VALUES USED IN EXPERIMENT FOR IC-STRIPLINE MODELLING.

	w_S (A) [mm]	l_S (B) [mm]	w_C (C) [mm]	h_C (D) [mm]	l_C (E) [mm]
max	25	40	38	10	55
min	20	30	28	5	45

presented in Table I are shown in Fig. 3. The presented values of EF_k for all experiment runs are small and relatively uniform which leads to the conclusion that the equivalent circuit model accurately models the simulated structure in 3D EM simulator.

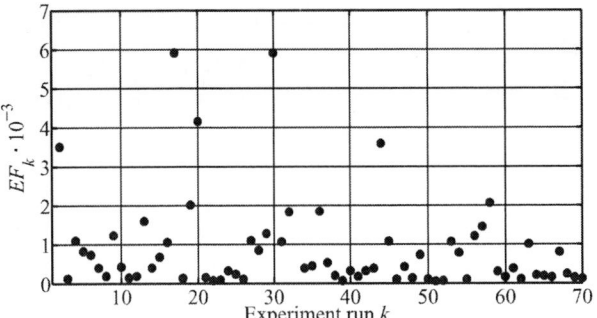

Fig. 3. Values of error function EF_k of the parameter extraction performed for the experiment defined in Table I.

The IC-Stripline equivalent circuit model which is used for the parameter extraction, is presented in Fig. 4. The IC-Stripline taper is segmented into several parts in order to model the change in the characteristic impedance more accurately. The two transmission line sections used for the taper modelling also enable more accurate modelling of the transition between the coaxial line and the septum. In order to keep the number of the extracted parameters to a minimum, the length of the taper sections are equal, i.e. $E_1 = E_2$. Furthermore, the characteristic impedance of the coaxial part of the structure Z_0 is fixed to 50 Ω. The length of the coaxial part E_0 is calculated as $((100\text{mm} - l_C)/2) * f/c_0$, while the electrical length of the central section is calculated as $l_C \cdot f/c_0$. The frequency at which the electrical lengths are calculated is $f = 10$ GHz.

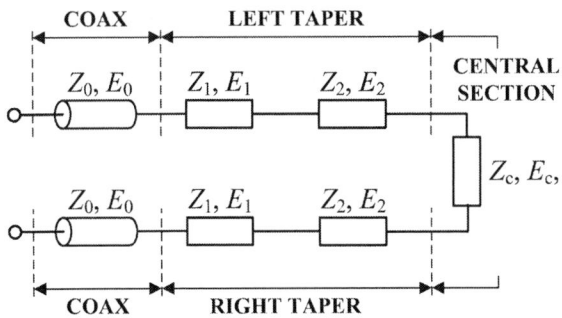

Fig. 4. IC-Stripline equivalent circuit model.

C. Modelling of Response Surface

The ordinary least squares algorithm is used to calculate the response surface model coefficients as the final step of the flow diagram presented in Fig. 1. The resulting response

surface model for 5 design factors can be written as

$$
\begin{aligned}
\mathbf{y} = {} & \beta_0 + \beta_1 \mathbf{A} + \beta_2 \mathbf{B} + \cdots + \beta_5 \mathbf{E} \\
& + \beta_6 \mathbf{AB} + \beta_7 \mathbf{AC} + \cdots + \beta_{15} \mathbf{DE} \\
& + \beta_{16} \mathbf{A^2} + \beta_{17} \mathbf{B^2} + \cdots + \beta_{20} \mathbf{E^2} \quad (3)
\end{aligned}
$$

where $\mathbf{y}$ is the vector of the extracted values, $\mathbf{A} \cdots \mathbf{E}$ are the vectors of the coded design factor values and $\beta_0 \cdots \beta_{20}$ are the unknown response surface model coefficients. Each element of the design factor vectors is assigned a coded level in the range $[-1, 1]$ according to the values presented in Table I. The number of elements of the coded design factor vectors is equal to the number of experiment runs, and in this work equals 40. The 2nd order response surface model is suitable for modelling the response surfaces which can be non-linear. Consequently, the usage of the 2nd order models makes the workflow presented in Fig. 1 more automated.

D. Optimization of VSWR Performance

The equivalent circuit model of the IC-Stripline (Fig. 4) is optimized using the Ngspice as a circuit simulator and the gradient constrained optimizer available in MATLAB. The optimization parameters are the same as those listed in Table I and their initial values are set in the middle of the presented ranges. The gradient-based (central finite difference) constrained optimizer takes ≈ 1 minute to complete, during which the circuit model simulations are started ≈ 500 times. The result of the optimization is presented in Table II. The presented results are introduced back to the 3D EM simulator (tolerance of ± 0.2 mm is allowed since the mapping between these two models is not perfect) and the optimization goal is verified. The commonly used requirement that the VSWR should be around 1.2 is met by the proposed design.

TABLE II
OPTIMIZED DESIGN PARAMETERS OF THE CLOSED VERSION OF IC-STRIPLINE.

	w_S (A) [mm]	l_S (B) [mm]	w_C (C) [mm]	h_C (D) [mm]	l_C (E) [mm]
opt	22.6	39.9	33.7	9.3	47.1

III. ANALYSIS OF LONGITUDINAL E-FIELD VARIATION

The variation of the electric field component E_z along the IC-Stripline length is verified in an EM simulator. In Fig. 6 three surfaces are shown for the calculation of the area-weighted average of the E_z electric field component. The area-weighted average on each of the defined surfaces is calculated as

$$
E_{zx} = \frac{1}{A} \int |E_z| \, \mathrm{d}A, \quad (4)
$$

where A is the area of the surface and E_z is the electric field component normal to the surface. Each of the presented surfaces has a dimension of 7 mm $\times$ 7 mm, spacing between them is 5 mm (i.e. the pitch is 12 mm) and they are located

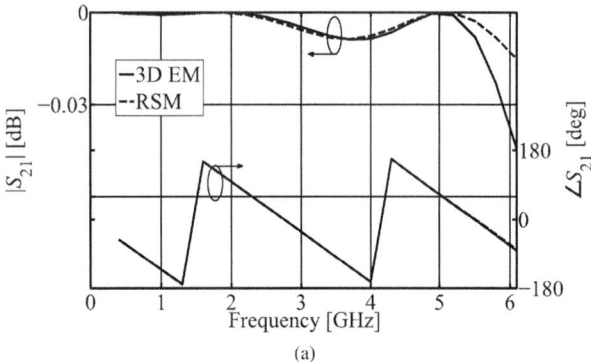

(a)

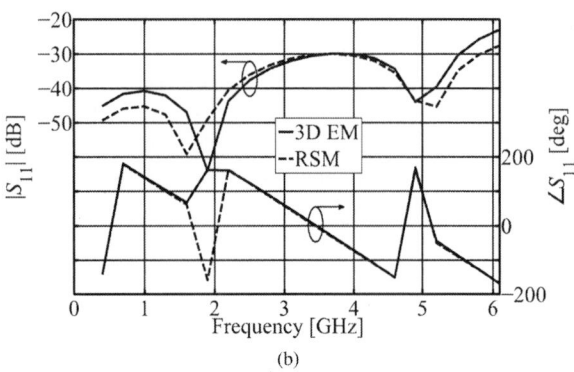

(b)

Fig. 5. Comparison between the simulated S-parameters of the optimized IC-Stripline by circuit simulator and 3D EM sover: (a) S_{21} parameter and (b) S_{11} parameter.

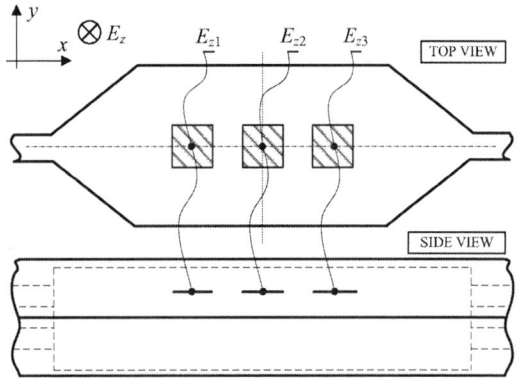

Fig. 6. Position of the surfaces for calculating area-weighted average of the electric field component E_z.

in the middle with respect to the distance between the IC-Stripline septum and upper wall of the enclosure. The area-weighted average values are calculated for three different frequencies. Since the area-weighted average is used, the calculated field magnitudes do not depend on the surface area as long as the field component E_z is uniform. The same results can be obtained using e.g. 5 mm × 5 mm surfaces. The simulation results are presented in Table III.

TABLE III
SIMULATED VALUES OF THE E_z COMPONENT ON THREE DIFFERENT
POSITIONS ALONG THE LINE LENGTH

	initial design [kV/m]			optimized design [kV/m]		
	E_{z1}	E_{z2}	E_{z3}	E_{z1}	E_{z2}	E_{z3}
1 GHz	1.39	1.43	1.15	1.14	1.14	1.13
3.5 GHz	1.39	1.13	1.10	1.09	1.13	1.12
6 GHz	1.37	1.08	1.16	1.01	1.16	1.00

The calculated values of the electric field E_z are larger for the initial design since the parameter h_C is smaller. The maximum relative difference between the calculated values of E_{z1}, E_{z2} and E_{z3} is larger for the initial design of the IC-Stripline. Even at the 1 GHz which is lowest considered frequency in this analysis the maximum relative difference for the initial design is ≈21%, while for the optimized design is ≈1%. Therefore, the longitudinal variation of the calculated E_z values along the IC-Stripline length is smaller for the optimized design of the IC-Stripline. The results presented in Table III also show a decrease in the magnitude of the average value of the electric field components E_{z1}, E_{z2} and E_{z3} with respect to the frequency. This observation should be taken into account when performing EMC tests, and 1-2 dB of tolerance in immunity and emission test results should be taken into account.

IV. CONCLUSION

This work presents design optimization of the closed version of an IC-Stripline. The design of experiment approach is used to systematically vary geometrical parameters of the IC-Stripline structure in a 3D EM simulator. The extraction of the equivalent circuit model parameters is performed from the 3D EM simulations. The response surface methodology is applied to the extracted parameters and the 2nd order models are calculated. The calculated models relate the IC-Stripline geometry to the parameters of the equivalent circuit. The fast optimization of the equivalent circuit model having the parameters defined by the 2nd order models is performed and improvement in the VSWR performance is obtained in the frequency range up to 6 GHz. The analysis of the longitudinal variations of the E-field is presented which confirms improvement of the longitudinal E-field variations with respect to the improvement in VSWR performance.

The future work will focus on realization of the presented IC-Stripline set-up which will enable comparison with the measurements. The impact of a device-under-test on the IC-Stripline performance will be assessed. Furthermore, the presented workflow will be verified on several other test cases to further improve its flexibility and usability.

REFERENCES

[1] M. Ramdani, E. Sicard, A. Boyer, S. Ben Dhia, J. Whalen, T. H. Hubing, M. Coenen, and O. Wada, "The electromagnetic compatibility of integrated circuits - past, present, and future," *Electromagnetic Compatibility, IEEE Transactions on*, vol. 51, no. 1, pp. 78–100, Feb. 2009.

[2] *Integrated Circuits Measurement of Electromagnetic Emissions Part 8: Measurement of Radiated Emissions IC Stripline Method*, IEC 61967-8, Pre-release of the official standard, 2011.

[3] *Integrated Circuits - Measurement of Electromagnetic Immunity - Part 8: Measurement of Radiated Immunity - IC Stripline Method*, IEC 62132-8, Pre-release of the official standard, 2011.

[4] B. Koerber, M. Trebeck, N. Mueller, and F. Klotz, "IC-stripline: A new proposal for susceptibility and emission testing of ICs," *Electromagnetic Compatibility of Integrated Circuits (EMC Compo), 6th International Workshop on*, pp. 1–5, Nov. 2007.

[5] B. Koerber, M. Trebeck, N. Mueller, F. Klotz, and V. Muellerwiebus, "IC-stripline for susceptibility and emission testing of ICs," *Electromagnetic Compatibility of Integrated Circuits (EMC Compo), 7th International Workshop on*, Nov. 2009.

[6] T. Mandic, R. Gillon, B. Nauwelaers, and A. Baric, "Design and modelling of IC-Stripline having improved VSWR performance," *Electromagnetic Compatibility of Integrated Circuits (EMC Compo), 8th International Workshop on*, pp. 82–87, Nov. 2011.

[7] X.-D. Cai and G. Costache, "Theoretical modeling of longitudinal variations of electric field and line impedance in TEM cells," *Electromagnetic Compatibility, IEEE Transactions on*, vol. 35, no. 3, pp. 398–401, Aug. 1993.

[8] L. Huadong and I. Peoria, "Field distribution in a stripline and its influence on immunity testing," *IEEE EMC Society Newsletter*, vol. 231, pp. 40–43, fall 2011.

[9] A. Sathanur, V. Jandhyala, and H. Braunisch, "A hierarchical simulation flow for return-loss optimization of microprocessor package vertical interconnects," *Advanced Packaging, IEEE Transactions on*, vol. 33, no. 4, pp. 1021–1033, Nov. 2010.

[10] T. Mandic, B. Nauwelaers, and A. Baric, "Simple and scalable methodology for equivalent circuit modeling of IC packages," *Components, Packaging and Manufacturing Technology, IEEE Transactions on*, pp. 1–13, July 2013 accepted for publication.

[11] S. Cove, M. Ordonez, F. Luchino, and J. Quaicoe, "Applying response surface methodology to small planar transformer winding design," *Industrial Electronics, IEEE Transactions on*, vol. 60, no. 2, pp. 483–493, Feb. 2013.

[12] T. Mandic, R. Gillon, B. Nauwelaers, and A. Baric, "Characterizing the TEM cell electric and magnetic field coupling to PCB transmission lines," *Electromagnetic Compatibility, IEEE Transactions on*, vol. 54, no. 5, pp. 976–985, Oct. 2012.

[13] D. Montgomery, *Design and Analysis of Experiments*, 7th ed. New York: Wiley, 2009.

[14] R. Johnson, D. Montgomery, B. Jones, and J. Fowler, "Comparing designs for computer simulation experiments," *Proceedings of the 2008 Winter Simulation Conference*, pp. 463–470, Dec. 2008.

[15] *Advanced Design System 2011 Documentation Set*, Agilent Technologies, Inc., Santa Clara, US, 2011.

[16] *Ngspice User Manual Version 25*, Paolo Nenzi, Holger Vogt, http://ngspice.sourceforge.net/, 2013.

Study of radiated immunity of an electronic system in a reverberating chamber

L. Guibert, P. Millot, X. Ferrières
ElectroMagnetic and Radar Department
Onera-The French Aerospace Lab
F-31055, Toulouse, France
laurent.guibert@onera.fr

E. Sicard
GEI (Génie Electrique et Informatique)
INSA (Institut National des Sciences Appliquées)
F31077, Toulouse, France

Abstract—In this paper, we are interested in measuring the level of immunity of the radiated mode of an electronic system. The behavior of this system has been studied in a Mode Stirrer Reverberating Chamber (MSRC). Nowadays, many equipment manufacturers use the MSRC as a measurement facility to describe their electronic systems in the field of EMC. In the first part, we present the use of the MSRC for measuring EMC susceptibility. Then we present the electronic system under test (DUT) and the method allows us to characterize the functional electronic behavior. In a second part, we present a novel method that allows on the one hand measurement the level of immunity and on the other hand derivation of a model of the level of immunity of the electronic system studied.

Keywords—reverberating chamber, radiated immunity, EM susceptibility, nonlinear effect, harmonic frequency level

I. INTRODUCTION

For studies in the field of EMC, and more particularly on the radiated immunity of electronic systems, the EMC team of ONERA's ElectroMagnetism Radar Department uses MSRCs (Mode Stirrer Reverberating Chambers). In the fields of avionics or automotive, from an EMC point of view of, the MSRC is a widely used facility to qualify electronic equipment. The first part of this paper is devoted to the experimental setup; we describe on the one hand the MSRC used, and on the other hand we present the system under test (DUT).

In a second part, we present a novel method that allows measurement of the level of immunity of the DUT by observing the levels of radiated harmonic frequencies. The third part is dedicated to the presentation of results of measurements. We also associate simulation results that allow us to develop a model representing the level of immunity of an electronic card.

II. EXPERIMENTAL SETUP

A. Presentation of the MSRC used

In order to be able to illuminate our system with an EM field, we use a Mode Stirred Reverberating Chamber (MSRC). This EMC test facility enables us to reproduce a homogeneous and isotropic EM environment on the entire surface of the systems under tests. This homogeneous condition on the DUT must be understood statistically in terms of average value of EM field. It is therefore well adapted for testing embedded electronic systems. Figure 1 presents the test facility with the DUT.

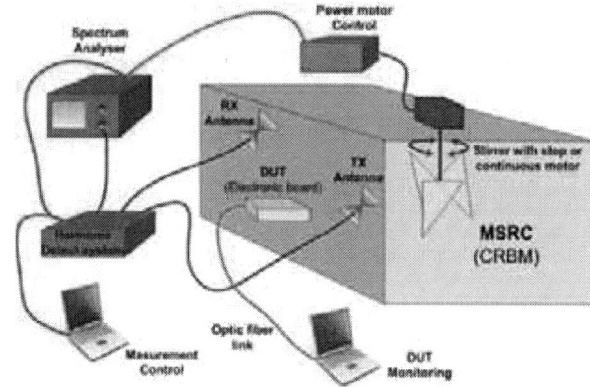

Figure 1 : Schematic view of the MSRC with the DUT installed inside

Another advantage of the MSRC is to be able to generate a high level incident electromagnetic field on the system with a reasonable power injected with any type of wideband antenna. This phenomenon comes from the reverberation properties of the metal walls of the chamber. One of the important parameters is the quality factor, noted Q, expressed by the following relationship

$$Q = 2\pi f . \frac{W}{P_t} \tag{1}$$

W is the mean energy stored in the chamber, in Joule (J), P_t is the mean power available in Watt (W). We suppose that the whole power injected in the antenna is absorbed by the walls of the chamber.

Subsequently, one can express the maximum mean power absorbed by the system, P_r, (in W) when this one is introduced in the chamber:

$$P_r = \frac{\lambda^3}{16\pi^2} . \frac{Q}{V} . P_t \tag{2}$$

V being the volume of the chamber in cubic meter (m³). Relation (2) is valid for an empty MSRC but the presence of the DUT in the chamber modifies the field structure. However we consider that its Q factor is very small compared to the one of the cavity alone.

With a few Watts injected (1 to 10 Watts) into the antenna via a Continuous Wave (CW) generator and an amplifier, we can obtain fields of a few hundreds of Volts per meter (100V/m to 500V/m) at the level of our system under test. The homogeneity of this field on our system is ensured via the stirrer. When rotating, it produces a variation of the modes of the cavity. The homogeneity is observed from a statistical point of view on a complete rotation of the stirrer. To estimate the amplitude of the field on the level of our system, we use a field sensor placed near the DUT [1] [2].

B. DUT presentation

In a research project named SEISME (Simulated Emission and Immunity Systems and Electronic Modules), we developed an electronic board containing mainly digital components. This board, representative of an electronic system, has been subjected to a variety of EMC tests. Figure 2 presents the global architecture of the global system under test which involves several manufacturers, including ONERA.

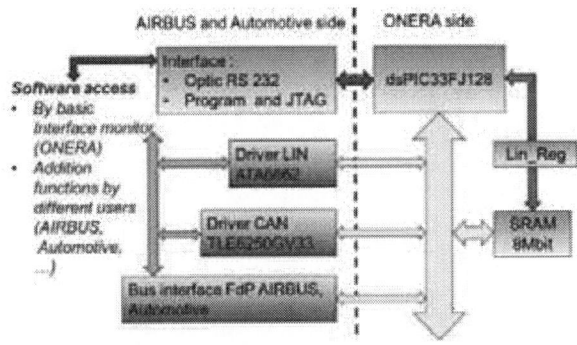

Figure 2 : Block diagram of the board system architecture

The part of the electronic board that interests us is made up of a SRAM memory and a dsPIC33 Microchip microcontroller. We conducted immunity measurements on two types of cards with 8Mb a SRAM coming from two different manufacturers: Alliance and Brillance Semi-Conductors.

Before installing the DUT in the chamber, the board is fixed on the top of a metal box that contains the power supply battery and all peripheral components to perform functional testing in real-time. This metal box allows measurement of the radiation emitted by the top face of board alone. Figure 3 shows the stages of electronic preparation of the board before to performing the measurements in a reverberating chamber.

The dsPIC33 microcontroller contains a program that is stored in the program memory flash. This program has been designed to generate and monitor in real time all communications between the SRAM and the microcontroller [3].

Figure 3 shows the electronic board (DUT) placed inside the reverberating chamber. The communication between the board and the monitoring laptop is performed by an optic fiber link.

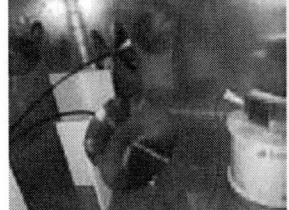 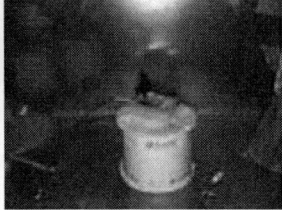

Figure 3 : Preparation of the DUT and installation in the MSRC

III. MEASUREMENT SYSTEM AND RADIATED IMMUNITY

In our situation, the MSRC is used to measure EM susceptibility but also to measure the levels of harmonic frequencies. For this, we have developed a specific detection system adapted to MSRC. Figure 4 presents the general block diagram of the device.

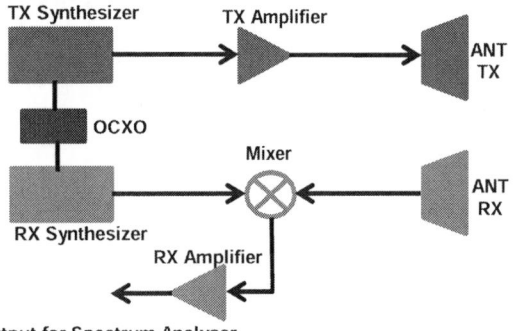

Figure 4 : General block diagram of the harmonic frequency measurement system

This device contains a chain of transmission and a chain of reception. The transmission chain includes a frequency synthesizer, an amplifier and a transmitting antenna placed in the MSRC. The receiver chain contains an antenna placed also in the MSRC, a mixer and a low noise amplifier connected at the input of a spectrum analyzer. Both frequency synthesizers are synchronized via an ultra-stable oscillator type (OCXO). The mixer allows performing a heterodyne detection in order to obtain a measurement signal at a lower frequency. The receiving amplifier ensures a good dynamic range [4].

To use this device in the MSRC, we had to make improvements. The energy budget of the chamber of course increases the dynamic of the detection, but in return, the multipath introduced by the metal walls of the chamber greatly complicate the detection. Figure 5 shows the improvements made on the transmission chain by the introduction of low pas filters and insulators.

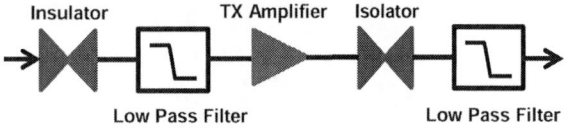

Figure 5 : TX chain

The transmission system generates the fundamental frequency that can illuminate the DUT in the chamber. However the frequency synthesizer and the amplifier emission chain generate unwanted harmonic frequencies. In order to significantly reduce them, we use low pass drum cascaded filters. In addition ferrite isolators provide good impedance matching at the output of the amplifier and the frequency synthesizer.

This technique allows measuring the value of incident frequency that we want to inject into the chamber as well as to removing most of the parasitic frequencies introduced by the injection system.

To improve the quality of our measurement, we also make changes on the receiving chain (Figure 6).

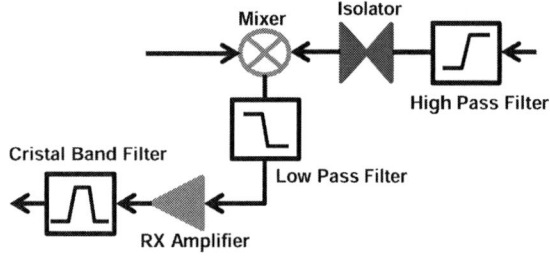

Figure 6 : RX Chain

This receiving chain performs measurement of the levels of harmonic frequencies of orders 2 and 3. For this, high-pass filters connected in cascade can remove most of the signal of the fundamental frequency generated by the transmission chain. The isolator provides impedance matching between the antenna and the RF input of the mixer.

The measurements of harmonic levels are down converted in an Intermediate Frequency (IF) by using the mixer. A low-pass filter inserted between the IF mixer output and the input of the receiving amplifier eliminates all frequencies above the desired IF. Finally, a narrow-band-pass filter at the exit of the reception amplifier enhances the selectivity of the choice of the IF. The signal is recorded using a spectrum analyzer.

IV. EXPERIMENTAL RESULTS

We performed measurements of immunity on the two identical boards with the two different types of SRAM memory components. One board was equipped with a SRAM from the Alliance (ALL) manufacturer and the other SRAM

from the Brilliance (BSI) manufacturer. For immunity tests, we measured real-time functional electronic behaviors that correspond to singular events of "Watchdog" type. For this, we connected an optical fiber interface and we recorded these reset-of-DUT events with the appropriate software.

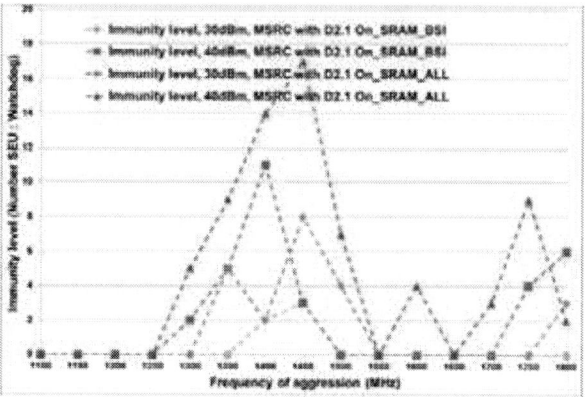

Figure 7 : Watchdog event (immunity level)

Figure 7 shows the results of immunity levels for injected frequencies ranging from 1.1 GHz to 1.8GHz with two power levels of 1 and 10 Watts (30 and 40dBm). The injected waveform is CW. It is interesting to note that we obtain different levels of immunity for both types of SRAM mounted on cards and despite the fact that these memories have the same types of packages (44-pin TSOP-II).

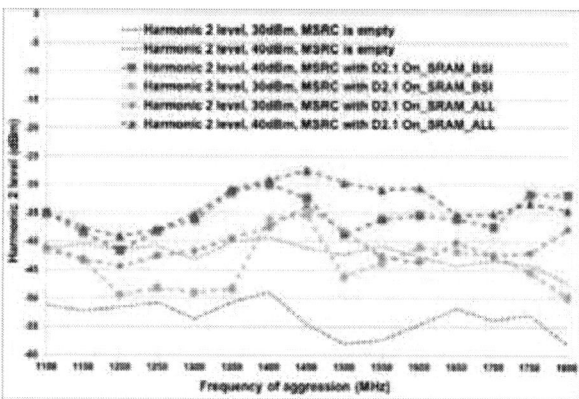

Figure 8 : Measured harmonic levels (nonlinear effect) in presence of the two types of SRAM mounted on electronic cards

Figure 8 shows the levels of harmonic frequencies radiated by the electronic board under test (DUT) when stressed in the MSRC. These levels are measured at the same time that the immunity tests; they come from the non-linear effects of the electronic components on the board which result in the emission of harmonic frequencies of rank 2 from the injected fundamental frequency.

The result in Figure 9 allows comparing the levels of immunity with the levels of harmonic frequencies. The plot shows that a significant functional malfunction of the DUT corresponds simultaneously to a low level of immunity and significant nonlinear effects corresponding to high levels of harmonic frequencies.

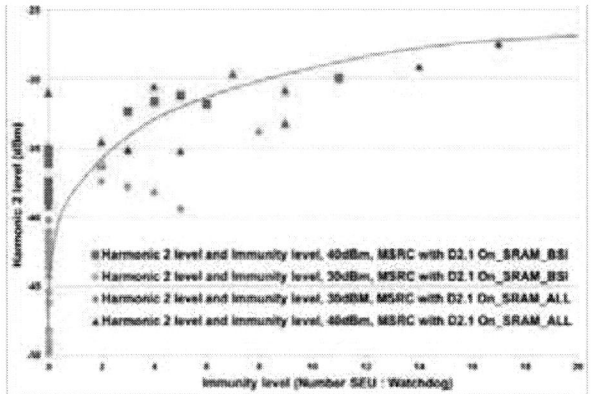

Figure 9 : Comparison of harmonic levels and Watchdog type of functional event

The level of harmonic frequency radiated by an electronic system could therefore be correlated to the radiated immunity levels.

V. SIMULATION RESULTS

In the SEISME research project, we also tried to conduct and update a kind of roadmap that presents a set of models used to characterize the embedded electronic systems in the field of EMC.

In Figure 10 we propose a model named EBIM-RI (Electronic Board Immunity Model - Radiated Immunity). This schematic model on the one hand describes the DUT installed in the CRBM and on the other hand simulates the EM emission level of harmonic frequencies.

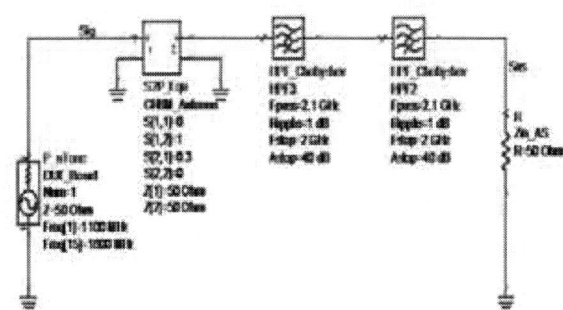

Figure 10 : Radiated immunity Model (DUT placed in MSRC)

Figure 11 shows the level of radiated immunity of the electronic board equipped with a SRAM manufacturer Alliance. Figure 12 is the simulation result obtained for the

same configuration of the DUT. The measurement result is very similar to the envelope of the simulation curve.

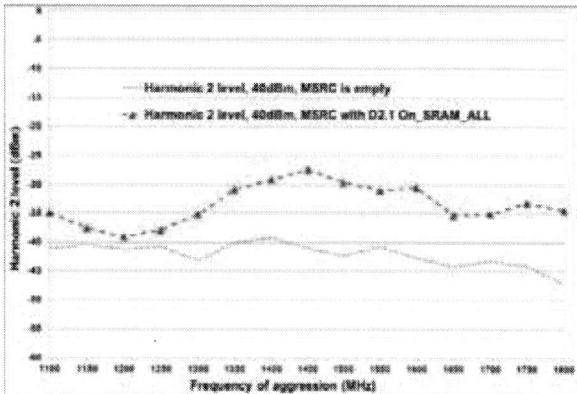

Figure 11 : Immunity level measurement (harmonic level) with Alliance SRAM card

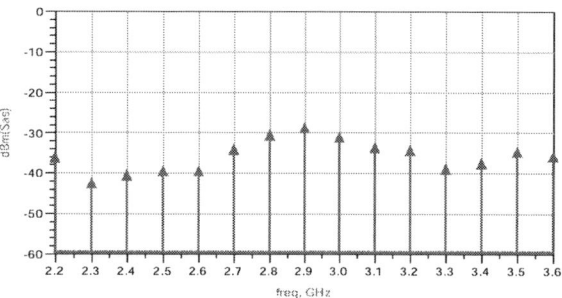

Figure 12 : Immunity level simulation (EBIM-RI model)

VI. CONCLUSION

In this paper, we first demonstrated the use of the MSRC for measuring radiated immunity. Then, we have presented a novel method of measurement which, in a non-intrusive manner, could allow qualification of the immunity frequencies. Levels of harmonic frequencies radiated by the DUT seem to represent the level of immunity of the DUT and led to the development of an EBIM-RI model. The challenge for future work is now to confirm the correlation of those frequencies with the level of radiated immunity of the DUT.

[1] C. Fiachetti, "Modèles du champ EM pour le calcul du couplage sur un équipement électronique en chambre réverbérante à brassage de modes et validation expérimentale," Ph.D. thesis , Université des sciences de Limoges, 2002. In French.

[2] P.E. Huc, Modélisation probabiliste du couplage d'un champ electromagnétique sur un equipement électronique, Thèse de l'université Toulouse III, Octobre 2004. In French.

[3] L. Guibert, J-P Parmantier "EMC study of an embedded electronic system placed in a hostile electromagnetic environment" EMC Compo 2009.

[4] A Statistical Approach to Radiated Immunity Testing of Digital Hardware in a Reverberation Chamber (EMC Europe 2010) J.Chen, A. C. Marvin, I. D. Flintoft and J. F. Dawson, Department of Electronics, University of York.

Transient Analysis of EM Radiation Associated with Information Leakage from Cryptographic ICs

Yu-ichi Hayashi[#1], Naofumi Homma[#2], Takafumi Aoki[#3], Yuichiro Okugawa[*1], and Yoshiharu Akiyama[*2]

[#]*Tohoku University,*

6-6-05 Aramaki Aza Aoba, Aoba, Sendai 980-8579, Japan

{[1]yu-ichi@m, [2]homma@aoki.ecei, [3]aoki@eceig}.tohoku.ac.jp

[*]*NTT Energy and Environment System Laboratories, Nippon Telegraph and Telephone Corporation,*

3-9-11 Midori, Musashino, Tokyo 180-8585, Japan

{[1]okugawa.yuichiro, [2]akiyama.yoshiharu}@lab.ntt.co.jp

Abstract This paper presents a time-domain visualization method for tracing electromagnetic (EM) radiation associated with information leakage from cryptographic ICs on the printed circuit board (PCB) surface. In recent years, security threats based on EM analysis attacks on cryptographic devices are attracting considerable attention due to their relative simplicity in practice. Some of the most cost-effective countermeasures against such attacks can be implemented at the PCB level. In order to implement such countermeasures effectively, critical parts (i.e., information sources and information propagation paths) on the board should be identified in advance. The key idea behind this identification is to calculate a correlation between measured EM traces and EM intensity values estimated from correct information (secret key) in the time domain. Transient analysis can reveal information propagation paths even if the EM signal carrying information is weak in comparison with noise generated from other components. Through an experiment, we confirm that EM radiation associated with information leakage can be traced even in situations where the information signal is obscured by background noise.

Keywords—Side-chennel attack, Cryptographic IC, Electromagnetic radiation, Information leakage, Transient analysis

I. INTRODUCTION

The exploitation of information leakage from cryptographic modules (software or hardware implementations of cryptographic algorithms) by side-channel attacks is of major concern for designers and evaluators of such modules. When a cryptographic module performs encryption or decryption, secret parameters related to the intermediate data being processed can be revealed from side-channel information, such as operation timing and power dissipation. A typical type of side-channel attack is the power analysis attack [1][2]. Electromagnetic (EM) analysis using the EM field generated by a cryptographic module has also been proposed as an extension of the power analysis attack [3].

Conventional EM analysis attacks usually require direct physical access to the target module in order to acquire side-channel information. However, several studies [4]-[6] have demonstrated the possibility of obtaining secret keys from EM radiation measured at a distance from the cryptographic module. In particular, successful EM analysis of an SSL accelerator was carried out in [4] by measuring EM radiation emitted from the accelerator. In [6], it was reported that side-channel information was successfully acquired from power and communication lines connected to a device equipped with a cryptographic module. Thus, EM analysis at the printed circuit board (PCB) level could pose a nontrivial threat to a vast group of cryptographic devices (i.e., electronic devices containing cryptographic modules), even if the modules are equipped with conventional countermeasures against direct access and EM radiation measurement, such as EM shielding packaging.

To date, a number of countermeasures against side-channel attacks have been considered at the algorithm and logic-gate level in cryptographic modules [2], [7], and their effectiveness has been demonstrated. However, emerging EM analysis techniques can be applied to a wider variety of legacy devices which have not been equipped with conventional countermeasures. Here, we assume that an attacker attempts to perform an emerging type of EM analysis attack from a distance (e.g., above the device or via power/communication cables), thus rendering the requirement for direct access to the cryptographic module unnecessary. The target design is in accordance with the EM compatibility (EMC) standard but without any specific countermeasures against attacks. Under such an attacker model, electrical-level countermeasures (e.g., noise suppression and reduction techniques) derived from the viewpoint of EMC can provide a sufficiently robust and cost-effective solution for existing devices equipped with unprotected cryptographic modules.

To take advantage of conventional electrical-level techniques (and corresponding countermeasures) at the PCB level, critical paths (i.e., signal propagation paths) on the board should be properly identified in advance. A straightforward method for identifying such critical paths is to map the intensity of EM radiation onto the board. However, it is difficult to understand the mechanism of information leakage

978-1-4799-5004-1/13 $31.00 © 2013 IEEE

from such intensity maps alone since a PCB usually contains many noise sources radiating similar EM waves, and information-carrying signals are sometimes weaker than signals from other components. In addition, it is fairly difficult to predict the propagation of EM waves carrying information on the PCB surface or along cables connected to the device, which can change depending on the wiring pattern and capacitive and/or inductive coupling between the patterns. Currently conventional noise suppression techniques (e.g., filter circuits consisting of capacitors and inductors, EM shielding, and installation of magnetic cores) are usually applied at the terminals of devices for reducing the total noise intensity in compliance with the existing public standards. However, these are only supportive measures from the viewpoint of information leakage prevention.

Addressing the above issue, this paper presents a transient analysis method for EM radiation associated with information leakage from cryptographic devices. First, we measure EM intensity values from cryptographic modules during encryption processing and subsequently calculate a correlation between the obtained EM traces and the estimated EM intensity values in the time domain. The proposed method makes it possible to trace the propagation of EM signals carrying secret information on the PCB even if there are complex wiring patterns and other active components on the board. From the results of transient analysis, we can understand the mechanism of propagation of EM radiation associated with information leakage and then develop effective electrical-level countermeasures. We confirm the validity of the proposed method with a simple experiment.

II. PROPOSED METHOD FOR TRACING SIGNAL PROPAGATION

Figure 1 shows the flow of the proposed method. First, we perform a set of encryption processes from the target cryptographic module with controlled timing. At the same time, we obtain the corresponding outputs. Next, we calculate hypothetical EM values from the outputs and subsequently the correct key information. In the case of hardware implementation, we can estimate the intensity of EM waves from the number of switched bits. For example, there are 128-bit registers simultaneously storing intermediate/final results in a common Advanced Encryption Standard (AES) circuit. Therefore, hypothetical power values range from 0 to 127. We measure transient EM waves at the points of interest during the above encryption processes. Note that the measurements are exactly aligned with the execution of the encryption processes. We calculate the correlation between measured and hypothetical EM values at each time step. The trace shows a high peak value where the EM wave carrying secret information from the encryption process reaches the measurement point. Due to the orthogonal nature of modern cryptographic algorithms, such a high peak will appear at a single time step. Therefore, we can obtain the propagation time directly from the time step. After repeating such measurements for all the points of interest, we can eventually visualize the propagation of information leakage via EM waves in the target device.

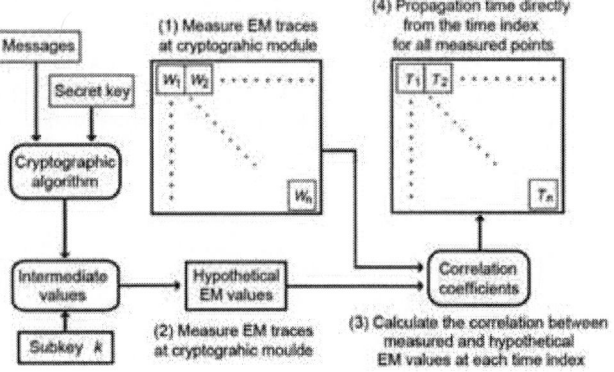

Fig 1. Analysis flow.

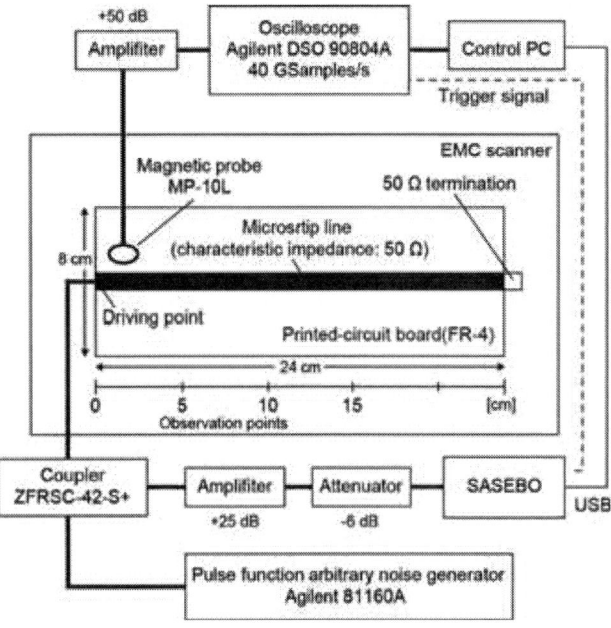

Fig. 2. Experimental setup.

Fig. 3. Side-channel Attack Standard Evaluation Board.

III. EXPERIMENT

A. Experimental setup

This section demonstrates the validity of the proposed method through two experiments using a microstrip line (MSL). Figure 2 shows a block diagram of the experimental setup. We employed a simple board with only an MSL mounted on the surface.

The source signal to the MSL was extracted from a Side-channel Attack Standard Evaluation Board (SASEBO) [8], which is a standard platform for evaluation of side-channel attacks. Figure 3 shows an overview of SASEBO used in this experiment. It is equipped with two field-programmable gate array (FPGA) chips denoted as FPGA1 and FPGA2. An AES circuit supporting a 128-bit key length [9] was implemented on FPGA1, and a control and communication circuit was implemented on FPGA2. The AES circuit was based on a loop architecture capable of processing one round in 1 clock cycle, requiring 11 cycles (including 1 I/O cycle) to perform a single encryption operation. The secret key used in the AES circuit was a reference value (0x2b7e151628aed2a6abf7158809cf4f3c) given in the algorithm specification [10]. We induced a transient current generated from the AES circuit during encryption at the end of the MSL via an SMA port (denoted as the driving point in Fig. 2).

The EM radiation from the MSL was measured with an oscilloscope (DSO 90804A, Agilent) at 40 GSamples/s via a magnetic field probe (MP-10L, NEC) with a bandwidth between 150 kHz and 1 GHz attached to an EMC noise scanner (WM7400, Morita Tech). The probe and the oscilloscope were connected via an amplifier (LNA-1050, RF Bay). Note that the coil face of the probe was placed parallel to the MSL and the probe measured the vertical component of the EM radiation.

B. Tracing leaked EM signals on the MSL

The validity of our method was first confirmed through an experiment where the source EM waves were conducted by the MSL. We measured the EM radiation at 5-cm intervals from the driving point along the MSL. Figure 4 shows the typical source signal induced in the MSL. The induced signal had 11 distinct peaks that corresponded to the 11 cycles of the AES operation. Figure 5 shows four EM waveforms acquired at the measurement points, where the time step starts from the beginning of encryption and the distance from the driving point is 0, 5, 10, or 15 cm. We obtained 5000 waveforms from 5000 different inputs at each measurement point. Note here that the acquisition of 5000 waveforms was practical since it was possible within 30 minutes in this setup. According to the method outlined in Fig. 1, we performed correlation EM analysis (CEMA) [11] for calculating the correlation between the measured and estimated EM values at each time step.

Figure 6(a) shows the experimental results, where the hypothetical values are given by the number of switched bits at the timing of the 10th round of the AES operation. Figure 6(b) shows a magnified view of the region from 670 to 685 ns in Fig. 6(a). We confirmed from the figure that the time steps

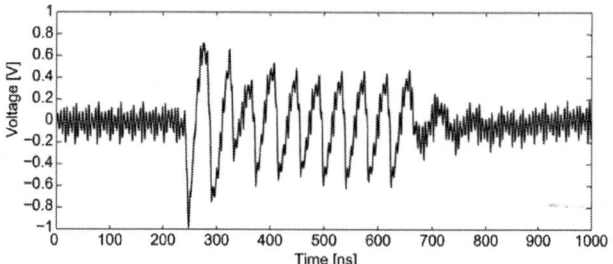

Fig. 4. Extracted source signal.

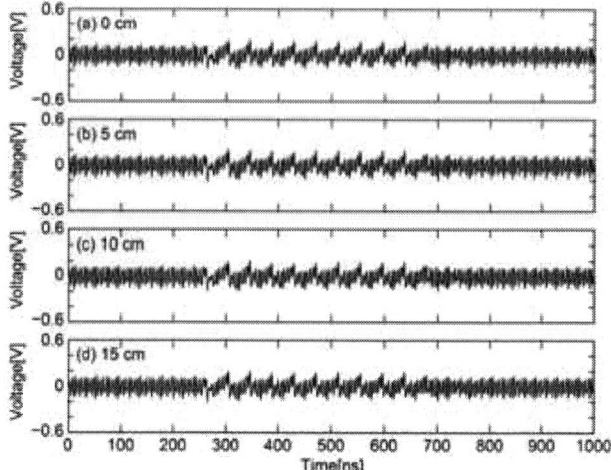

Fig. 5. Typical traces for different observation points.

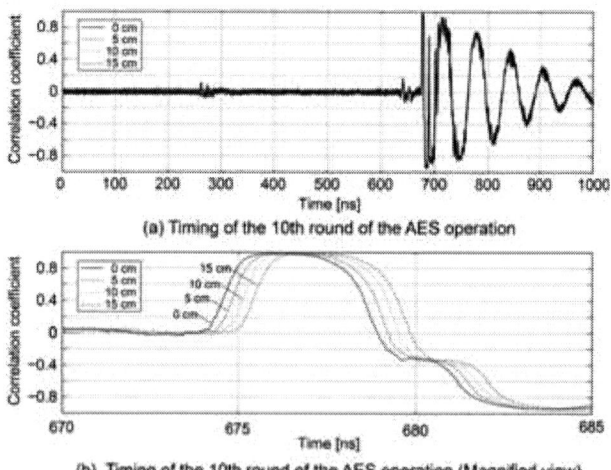

(a) Timing of the 10th round of the AES operation

(b) Timing of the 10th round of the AES operation (Magnified view)

Fig. 6. CEMA results for leaked EM signal without other noise.

with high correlation values were proportional to the distance from the driving point to the measurement point. This means that the proposed method can be used to trace the propagation of EM radiation associated with information leakage along the MSL.

C. Tracing leaked EM signals with Gaussian noise on an MSL

The validity of our method was confirmed through an experiment where the background noise level was higher than the source level that should be traced. In practice, in addition to the EM radiation emitted from the cryptographic module, EM radiation measured with a measuring device on a board also includes noise generated from other ICs and components on the same board. Thus, the signal from the cryptographic module might not be observable due to background noise from other components. In this experiment, we generated such background noise by a signal generator and coupled it with the signal from the cryptographic module. The background noise was given as white Gaussian noise with a frequency band of 1 GHz. Figure 7 shows an image of the background noise coupled to the source signal in Fig. 1.

Figure 8 shows four EM waveforms (analogous to the waveforms in Fig. 5) acquired at the measurement points on the MSL. Note that it is difficult to distinguish the 11 peaks corresponding to the AES operation due to the coupled background noise. As in the case of the above experiment, we obtained 5000 waveforms from 5000 different inputs at each measurement point, after which we calculated the correlation between the measured and estimated EM values at each time step. Figure 9(a) shows the experimental result where the hypothetical values are given by the number of switched bits at the timing of the 10th round of the AES operation. Figure 9(b) shows a magnified view of the region from 670 to 685 ns in Fig. 9(a), as in the case of Fig. 6(b). As a result, although the peak values were lower than those in Fig. 6(b), we confirmed that the time steps with high correlation values were proportional to the distance from the driving point to the measurement point, demonstrating that the proposed method can be used to trace the propagation of EM radiation associated with information leakage on the MSL even under conditions similar to real-world situations.

D. Discussion

The above result demonstrates that the proposed method can be applied even if Gaussian noise is coupled with the target signal generated from a cryptographic module. In this case, the amplitude of the background noise was assumed to be constant; however, the level of the background noise would change depending on the operation environment. In the following, we discuss the effect of background noise on the feasibility of our method.

In our method, the correlation at each time step t is given as

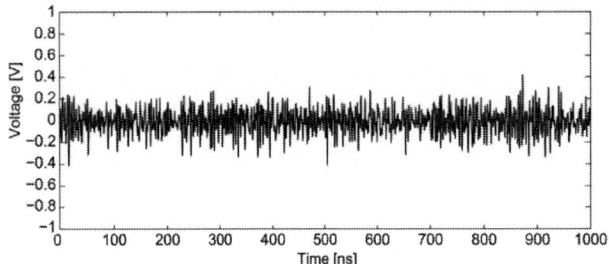

Fig. 7. Gaussian noise waveform.

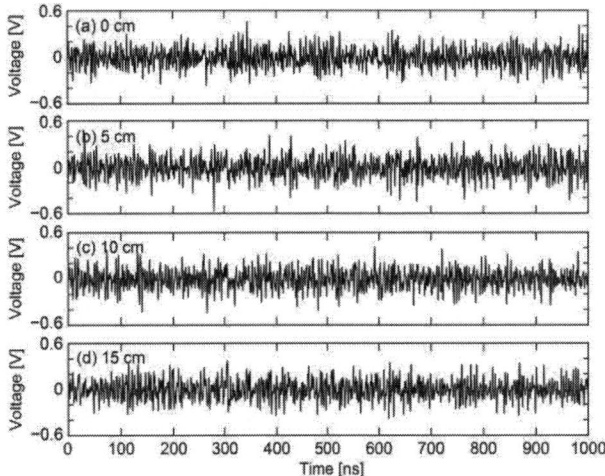

Fig. 8. Typical traces for different observation points.

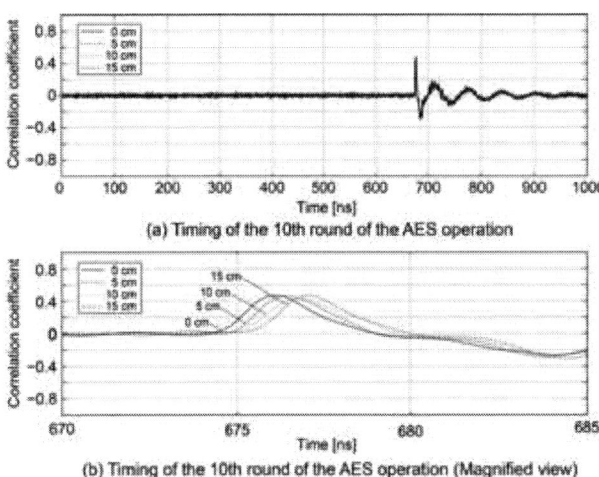

(a) Timing of the 10th round of the AES operation

(b) Timing of the 10th round of the AES operation (Magnified view)

Fig. 9. CEMA result for EM leakage signal with other noises.

$$r_t\left(E_t, W^{(total)}{}_t\right) = \frac{r_t\left(E_t, W^{(exp)}{}_t\right)}{\sqrt{1 + \frac{1}{SNR_t}}}, \qquad (1)$$

where r_t indicates the Pearson correlation coefficient, E_t denotes the hypothetical EM value, $W^{(total)}{}_t$ represents the total EM radiation from the target device, and $W^{(exp)}{}_t$ stands for the exploitable EM radiation from the cryptographic module (such as that in Fig. 4). In the proposed method, as designers or evaluators we know the value of the secret key, and thus obtain the correct EM value E_t[12]. Therefore, we can approximate the value of $r_t\left(E_t, W^{(exp)}{}_t\right)$ by

$$r_t\left(E_t, W^{(exp)}{}_t\right) \approx 1. \qquad (2)$$

As a result, we obtain the following relation:

$$r_t\left(E_t, W^{(total)}{}_t\right) = \frac{1}{\sqrt{1 + \frac{1}{SNR_t}}}. \qquad (3)$$

Equation (3) indicates that the obtained correlation coefficient is determined by SNR, which is defined as

$$SNR_t = \frac{\sigma^2_{info_t}}{\sigma^2_{noise_t}}, \qquad (4)$$

where $\sigma^2_{info_t}$ denotes the variance of the signal carrying information at time t, and $\sigma^2_{noise_t}$ indicates the variance of signals not carrying information at time t. $\sigma^2_{info_t}$ is related to secret information contained in $W^{(exp)}{}_t$, and $\sigma^2_{noise_t}$ is related to the background noise contained in $W^{(total)}{}_t$. Both $\sigma^2_{info_t}$ and $\sigma^2_{noise_t}$ are different at each time step. Equation (4) suggests that the correlation coefficient would change with $\sigma^2_{noise_t}$ under a constant $\sigma^2_{info_t}$. The correlation coefficient is dependent on the number of waveforms used in CEMA [2]. If the variance $\sigma^2_{noise_t}$ increases, CEMA requires more waveforms to obtain a distinguishable peak value, which increases the computation time of our method. In the case of the above experiment, the variance of the coupled background noise was 0.0099 while that of the information signal was 0.0021. The correlation value given from Eq. (3) was 0.4233, which was in fair agreement with the experimental result. In this regard, designers and evaluators can select a specific set of inputs in a way that would increase $\sigma^2_{info_t}$[13]. Such chosen input would reduce the computation time.

IV. CONCLUSION

This paper presented a method for tracing EM radiation associated with information leakage from cryptographic ICs on the PCB surface in the time domain. The key idea behind the method is to calculate a correlation between measured EM traces and EM values estimated in advance with correct information (i.e., the key). The proposed method makes it possible to visualize the propagation paths of EM radiation associated with information leakage in the time domain. Through two experiments, we confirmed that the proposed method is applicable even if the intensity of the leaked information-carrying signal was lower than that of background noise.

A major advantage of the proposed method is its ability to clarify the effect of the wiring pattern and component layout on the propagation of EM radiation associated with information leakage, providing useful information for the development of effective electrical-level countermeasures or guidelines against EM analysis attacks. The proposed method can also be applied to the evaluation of unintentional and intentional EM interference. In particular, it would be possible to evaluate the feasibility of fault injection attacks on cryptographic modules by using intentional EM interference [14], and the results would be useful for developing effective and efficient countermeasures for cryptographic devices.

REFERENCES

[1] P. Kocher, J. Jaffe, and B. Jun, "Differential Power Analysis," Advances in Cryptography (CRYPTO '99), Lecture Notes in Computer Science, vol. 1666, pp. 388-397, Aug. 1999.

[2] S. Mangard, E. Oswald, and T. Popp, "Power Analysis Attacks: Revealing the Secrets of Smart Cards," Springer-Verlag, 2007.

[3] K. Gandolfi, C. Mourtel, and F. Olivier, "Electromagnetic analysis: Concrete results," Workshop on Cryptographic Hardware and Embedded Systems (CHES 2001), Lecture Notes in Computer Science, vol. 2162, pp. 251-261, May 2001.

[4] D. Agrawal, B. Archambeault, R. Rao, and P. Rohatgi, "The EM sidechannel(s)," Workshop on Cryptographic Hardware and Embedded Systems (CHES 2002), Lecture Notes in Computer Science, vol. 2523, pp. 29-45, Aug. 2002.

[5] C. Kim, M. Schlaffer, and S. Moon, "Differential Side Channel Analysis Attacks on FPGA Implementations of ARIA," ETRI Journal, vol. 30, No. 2, pp. 315-325, Apr. 2008.

[6] T. Sugawara, Y. Hayashi, N. Homma, T. Mizuki, T. Aoki, H. Sone, and A. Satoh, "Mechanism behind information leakage in electromagnetic analysis of cryptographic modules," The 10th International Workshop on Information Security Applications (WISA2009), Lecture Notes in Computer Science, vol. 5932, pp. 66–78, Aug. 2009.

[7] K. Tiri and I. Verbauwhede, "A Logic Level Design Methodology for a Secure DPA Resistant ASIC or FPGA Implementation," in Design, Automation and Test in Europe (DATE 2004), pp. 246-251, Feb. 2004.

[8] " Side-channel Attack Standard Evaluation BOard(SASEBO)," http://staff.aist.go.jp/akashi.satoh/SASEBO/en/index .html.

[9] "Cryptographic Hardware Project," Computer Structures Laboratory, Graduate School of Information Sciences, http://www.aoki.ecei.tohoku.ac.jp/crypto/.

[10] NIST FIPS PUB. 197, Advanced Encryption Standard(AES), http://csrc.nist.gov/publications/fips/fips197/fips-197.pdf

[11] E. Brier, C. Clavier, and F. Olivier, "Correlation Power Analysis with a Leakage Model," Workshop on Cryptographic Hardware and Embedded Systems (CHES 2004), Lecture Notes in Computer Science, vol. 3156, pp. 16-29, Aug. 2004.

[12] T. Sugawara, Y. Hayashi, N. Homma, T. Mizuki, T. Aoki, H. Sone, and A. Satoh, "Spectrum Analysis on Cryptographic Modules to Counteract Side-Channel Attacks," EMC '09, pp. 21-24, Jul. 2009.

[13] Y. Kim, T. Sugawara, N. Homma, T. Aoki, and A. Satoh, "Biasing power traces to improve correlation power analysis attacks," COSADE 2010, pp. 77-80, Feb. 2010.

[14] Y. Hayashi, N. Homma, T. Mizuki, T. Aoki and H. Sone, "Transient IEMI Threats for Cryptographic Devices,"IEEE Trans. on Electromagnetic Compatibility, vol. 55, pp. 140-148, 2013.

Noise-immune Design of Schmitt Trigger Logic Gate using DTMOS for Sub-threshold Circuits

KyungSoo Kim, Wansoo Nah, and SoYoung Kim

IC Design and Solutions Laboratory
Department of Semiconductor Display Engineering, SungKyunKwan University
Suwon, KOREA
ksyoung@skku.edu

Abstract— **This paper presents several Schmitt trigger logic gates with enhanced noise immunity using variable threshold voltage technique for sub-threshold voltage operation. The proposed logic gates are based on buffer design using dynamic threshold voltage MOS (DTMOS) for low power operation (V_{DD}=0.4V). Our solution dramatically improves noise immunity of logic gates with much less switching power consumption and significant area reduction compared with CMOS Schmitt triggers at the expense of slight increase in delay. The proposed noise immune gate design scheme is verified with an example digital circuit.**

Keywords— Schmitt Trigger; hysteresis; DTMOS; VTMOS; noise immunity; EMI; EMC

I. INTRODUCTION

In recent years, due to the growing demands for longer battery life in mobile devices, the mobile IC designers have focused on reducing power consumption of the circuits, especially for supply voltage scaling. However, lowering the supply voltage degrades the noise immunity of the circuit at the same time. Therefore, the immunity to electromagnetic interference (EMI) became an important issue for integrated circuit (IC) designers and several solutions were proposed [1]-[7].

Among many solutions to enhance immunity of the circuits, Schmitt trigger can be one of the appropriate solutions. Unlike comparator circuits, Schmitt trigger shows different switching threshold for the direction of input signal transition called hysteresis. If there is hysteresis, the threshold voltage of Schmitt trigger is higher than that of comparators for positive transition, and lower for negative transition. If the amplitude of the input signal variation is less than the switching threshold difference, output of Schmitt trigger will not respond to input directly. For this reason, Schmitt trigger is immune to undesired noise.

There are several approaches to implement Schmitt trigger circuits suitable for low power design [8]. Since traditional CMOS Schmitt trigger circuits require too many extra transistors for implementation, dynamic threshold voltage MOS (DTMOS) design using reduced number of transistor was introduced. However, previous research on DTMOS focused on extra buffer insertion between adjacent logic gates or threshold voltage control for improvement of noise immunity [9] - [10]. Although these methods also enhance noise immunity, they still require large number of extra transistors resulting in area increase.

In this paper, we expanded the DTMOS buffer insertion method by merging the extra Schmitt trigger buffer with logic gates to improve noise immunity for various types of low power logic gates design using less number of transistors. The schematic designs of Schmitt trigger logic gates are described in section II, from basic buffer implication to multi-input logic gates with DC characteristics. Section III presents the performance and noise immunity simulation results of each gate. Finally, an example digital circuit is designed with the proposed Schmitt trigger logic gates and the noise immunity and performance benchmarks are shown in section III.

II. IMPLEMENTATION OF SCHMITT TRIGGER LOGIC GATES

A. Buffer(Inverter) Design

Figure 1 shows the basic DTMOS Schmitt trigger buffer scheme [6]. Compared to the traditional CMOS Schmitt trigger inverter shown in figure 1-(b), the VTCMOS scheme uses less

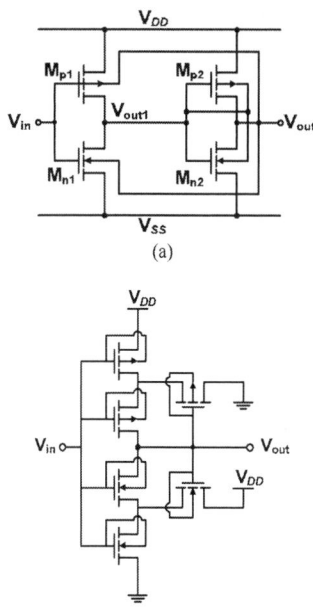

Fig. 1.   (a) VTCMOS Schmitt trigger buffer
(b) Traditional Schmitt trigger inverter

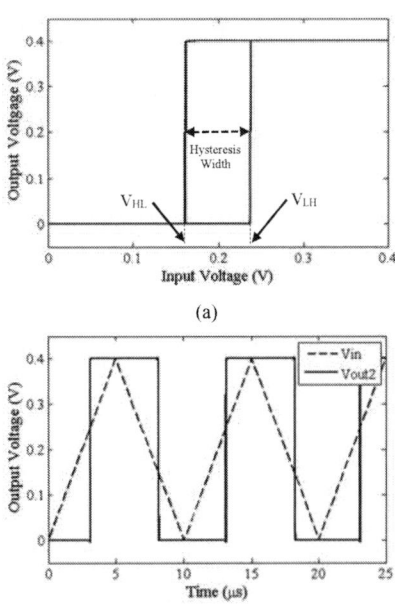

(a)

(b)

Fig. 2. Transfer characteristic of DTMOS Schmitt trigger buffer.
(a) Hysteresis characteristic of 0.4V Schmitt trigger buffer
(b) Input (V_{in}) and Output (V_{out2}) waveform.

transistor to build up a buffer with smaller device size. The buffer in the figure 1-(a) consist of two stages of CMOS inverters. Transistor M_{p1} and M_{n1} form the standard CMOS inverter with adjustable body bias. In the second stage, the configuration of M_{p2} and M_{n2} is called DTMOS inverter that ensures the inverter operation at ultra-low voltage of 0.4 V [6]. In addition, output of the second stage DTMOS inverter controls the body bias of the first stage CMOS inverter, so different V_{th} is achieved. For example, when the input signal V_{in} is low, the low output signal V_{out2} is fed back to the body of the first stage. In this case, M_{n1} is in zero substrate bias state with high threshold voltage $V_{(th0,n)}$, while M_{p1} becomes forward biased with lowered $|V_{(th,p)}|$. As a result, for low to high transition, input signal must get over logical threshold voltage V_{LH}, which is higher than $V_{DD}/2$, to turn M_{n1} on and turn M_{p1} off. In contrast, when V_{in} is high at the beginning, the high output signal V_{out2} is fed back with opposite condition. In this case, M_{n1} became forward biased with lowered threshold voltage $V_{(th,n)}$, and M_{p1} is zero biased with higher value of $|V_{(th0,p)}|$. Therefore, threshold voltage for high to low transition V_{HL} is lower than $V_{DD}/2$. This is how hysteresis characteristic is generated in a DTMOS inverter. Spectre simulation results for transfer characteristics of DTMOS Schmitt Trigger buffer with SAMSUNG 130nm process are shown in Figure 2.

The different threshold voltages, V_{HL} and V_{LH} can be calculated by following equations [6]:

$$V_{LH} = \frac{V_{DD} - |V_{th,p}| + R \times V_{th0,n}}{R+1} \qquad (1)$$

$$V_{HL} = \frac{V_{DD} + R \times V_{th,n} - |V_{th0,p}|}{R+1} \qquad (2)$$

where $R = \sqrt{(\beta_n/\beta_p)}$, β_n and β_p are transconductance parameters of NMOS and PMOS, respectively.

In this research, we expanded the concept of variable threshold voltage CMOS (VTCMOS) for AND gate and OR gates by implementing substrate bias control circuit to improve noise immunity. This is based on the previous DTMOS Schmitt trigger buffer operation. The second DTMOS inverter stage is the key idea for expanding the concept of VTCMOS to other logic gates with low power operation. The tied gate and body lowers the threshold voltage less than that of zero biased condition. Therefore, this substrate bias control unit can operate at very low voltage even though V_{th0} is around V_{DD}. In addition, the same design can be utilized for inverter itself by using V_{out1} as output, with the same hysteresis. The only difference is input to output delay of second inverter stage.

B. AND(NAND) Gate Design

Figure 3-(a) shows the proposed VTCMOS Schmitt trigger scheme for AND logic gate; the first stage is standard two input NAND gate and DTMOS inverter is attached for substrate bias control for NAND gate. In this configuration, V_{OUT1} can be used for NAND output and V_{OUT2} can be used for AND output at the same time. Compared to the traditional NAND gate shown in figure 3-(b), the proposed scheme saves four transistors.

When both input A and B is low, PMOS pulls V_{out1} up to V_{DD} and V_{OUT2} down to V_{SS}. Output does not change even if one

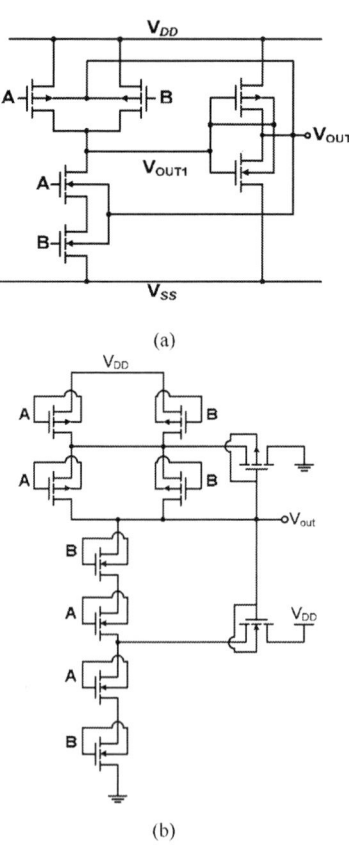

(a)

(b)

Fig. 3. (a) Proposed VTCMOS Schmitt trigger AND gate
(b) Traditional Schmitt NAND gate

978-1-4799-5004-1/13 $31.00 © 2013 IEEE

input changes from low to high because another input node is held low.

However, the situation changes if one of its inputs is set to high. For example, if input B is high, the PMOS connected to input B is turned off and the NMOS connected to input B is turned on with pulling the voltage of its drain down to V_{SS}. Therefore, the functionality of this NAND gate with only one input A becomes the same as that of the Schmitt trigger inverter presented in the previous section. This is the reason why this circuit is proposed for Schmitt trigger.

Even if the two inputs change at the same time with the same direction of transition, this circuit also behaves like a Schmitt trigger circuit. Because the value of R in equation V_{LH} and V_{HL} changes, the logical threshold voltages in (1) and (2) are adjusted. When both inputs are changing, two PMOS transistors can be replaced by one equivalent transistor with doubled width. In the same way, we can treat two NMOS transistors as a single NMOS transistor with doubled length. Therefore, we can derive effective value of R in (1) and (2) with $\beta_p{}'=2\beta_p$ and $\beta_n{}'=\beta_n/2$:

$$R' = \sqrt{\beta_p{}'/\beta_n{}'} = 2R \qquad (3)$$

C. OR(NOR) Gate Design

Figure 4 shows the schematics of proposed OR gate and traditional NOR gate. Similar to the AND gate, V_{OUT1} of proposed OR gate can be utilized for the NOR output. In addition, proposed scheme can save four transistors compared with traditional scheme.

As the behavior of AND gate described before, the proposed OR gate also have hysteresis plot for single input transition when another input is low. When input B is low, for example, the PMOS transistor with input B is turned on and NMOS transistor with input B is turned off. Therefore, the first stage with input A behaves like an inverter, so the proposed OR gate behaves just like a Schmitt trigger buffer in previous section.

When both inputs are stuck at high, output does not change even if one of its input varies from low to high because of its original functionality. Therefore, we do not have to consider this condition for noise immunity description.

Even if the two inputs changes at the same time, this circuit still operates as a Schmitt trigger with modified logical threshold voltage. That is, at the first stage of proposed schematic in figure 4-(a), the two PMOS transistors in series connection are considered as one single transistor with doubled gate length. Also, the two parallel NMOS transistors are considered as one single NMOS transistor with doubled gate width. Therefore, we can calculate the changed value of R in (1) and (2) with $\beta_p{}''=\beta_p/2$ and $\beta_n{}''=2\beta_n$ from following equation:

$$R'' = \sqrt{\beta_p{}''/\beta_n{}''} = R/2 \qquad (4)$$

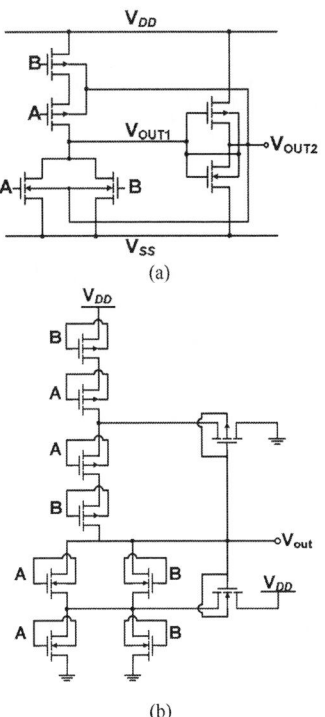

Fig. 4. (a) Proposed VTCMOS Schmitt trigger OR gate
(b) Traditional Schmitt NOR gate

III. SIMULATION RESULTS AND ANALYSIS

For the circuit implementation, we designed an unit DTMOS inverter with $w_p=0.7$ μm and $w_n=1$ μm. Based on this geometrical information of unit DTMOS inverter, all the proposed logic gates and traditional schmitt triggers were designed. For the immunity analysis, we defined the failure condition for the output of logic gate if the value varies more than 40 mV (10% of VDD) under the noise injection.

A. Schmitt Trigger Buffer (Inverter)

Figure 5 shows the DC transfer characteristic comparison results. To guarantee logical equivalence, we connected the traditional Schmitt inverter with normal DTMOS inverter to build a buffer scheme. As shown in the figure 5, hysteresis width of VTCMOS scheme is wider than that of traditional Schmitt trigger. Other parameters in detail are shown in the table I.

As we can see from the figure 1, traditional Schmitt trigger requires additional inverter to construct a buffer so eight transistors are used in total. By using VTCMOS scheme, however, the number of required transistor to construct a buffer is halved. In addition, due to the higher V_{TH} at zero-substrate-bias condition, VTCMOS scheme saved about 75% of switching current at the expense of a slight increase in delay compared to traditional Schmitt trigger. The high current consumption of the

978-1-4799-5004-1/13 $31.00 © 2013 IEEE

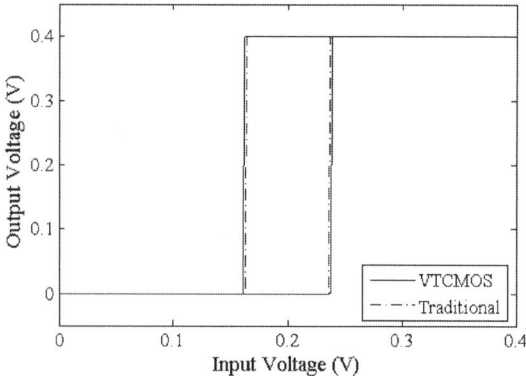

Fig. 5.   Hysteresis plot for buffer scheme using VTCMOS buffer and traditional Schmitt trigger inverter with DTMOS inverter

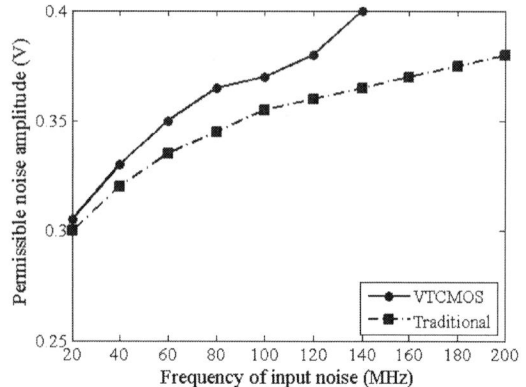

Fig. 6.   Permissible noise amplitude comparison of Schmitt trigger buffer

TABLE I. CHARACTERISTICS OF TWO SCHMITT TRIGGER BUFFER

Characteristic	Value	
	VTCMOS	*Traditional*
Number of Transistor	4	8 (6+2)
V_{LH}	237.5 mV	235.5 mV
V_{HL}	161.5 mV	163.5 mV
Hysteresis Width	76 mV	72 mV
Switching Current	310.92 μA	1.277 mA
Delay	2.88 ns	1.73 ns

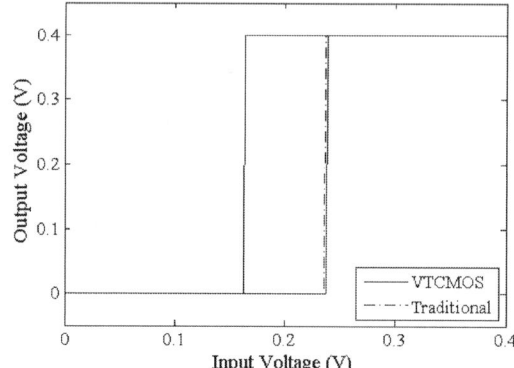

Fig. 7.   Hysteresis plot for AND scheme using VTCMOS (proposed) AND gate and traditional NAND gate combined with DTMOS inverter

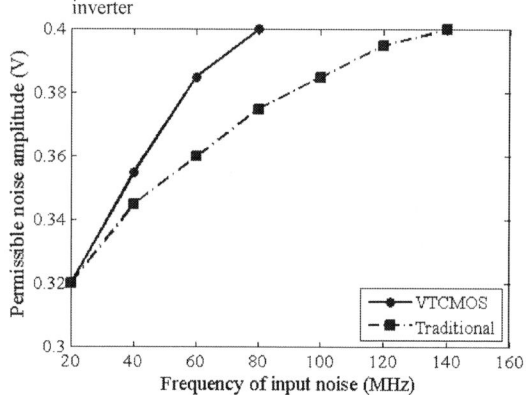

Fig. 8.   Permissible noise amplitude comparison of Schmitt trigger AND gate

TABLE II. CHARACTERISTICS OF TWO SCHMITT TRIGGER AND GATE

Characteristic	Value	
	VTCMOS(Proposed)	*Traditional*
Number of Transistor	6	12 (10+2)
V_{LH}	239.74 mV	237.41 mV
V_{HL}	164.45 mV	164.45 mV
Hysteresis Width	75.29 mV	72.96 mV
Switching Current	357.42 μA	1.106 mA
Delay	4.426 ns	2.564 ns

traditional scheme is due to additional current injection path through its feedback loop at the first stage.

Figure 6 shows the level of permissible noise amplitude as a function of frequency of input noise for two different Schmitt trigger buffer scheme. Due to the wider hysteresis width, DTMOS Schmitt trigger with VTCMOS scheme showed better noise immunity to input noise than traditional Schmitt trigger scheme.

B. Schmitt Trigger AND (NAND) Gate

Figure 7 shows the DC characteristics of Schmitt trigger AND gate implemented by proposed VTCMOS scheme and by traditional method. Detail parameters are shown in table II. As shown in the figure 7, only V_{LH} of VTCMOS scheme is

improved by 2.33 mV. This difference comes from the structural difference between proposed scheme and the traditional scheme. In the table II, the proposed VTCMOS scheme saves the area by 50% and switching power consumption by 67% compared with the traditional scheme.

In case of the proposed VTCMOS, substrate potential of the first stage MOS transistor was set to V_{SS} by V_{OUT2} for low to high transition, even though input B in figure 3-(a) is V_{DD}. For the NMOS at the bottom of traditional Schmitt trigger NAND gate in figure 3-(b), however, its drain voltage cannot be lowered to VSS even though input B is high because the tied substrate and gate makes a forward biased PN junction between substrate and drain. The channel resistance behaves like a load resistance so additional voltage drop between drain and source is generated.

978-1-4799-5004-1/13 $31.00 © 2013 IEEE

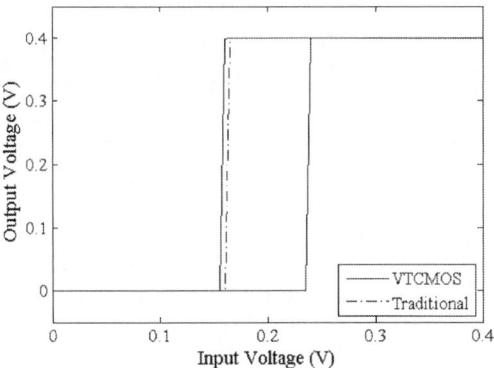

Fig. 9. Hysteresis plot for OR scheme using VTCMOS (proposed) OR gate and traditional NOR gate combined with DTMOS inverter

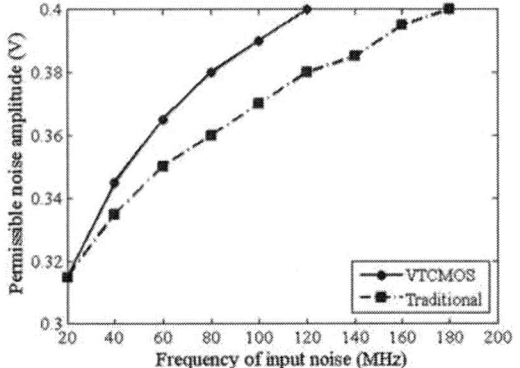

Fig. 10. Permissible noise amplitude comparison of Schmitt trigger OR gates

TABLE III. CHARACTERISTIC OF TWO SCHMITT TRIGGER OR GATE

Characteristic	Value	
	VTCMOS	Traditional
Number of Transistor	6	12 (10+2)
V_{LH}	237.5 mV	237.5 mV
V_{HL}	157.51 mV	162.5 mV
Hysteresis Width	79.99 mV	75 mV
Switching Current	123.71 μA	248.24 μA
Delay	5.576 ns	3.804 ns

Unfortunately, the drain potential lowers the V_{TH} of the adjacent NMOS with input A because its source potential is higher than V_{SS}. Therefore, for low to high transition, proposed VTCMOS Schmitt trigger AND gate with higher threshold voltage $V_{th0,n}$ requires higher V_{LH} than traditional scheme.

Figure 8 shows the noise immunity of proposed VTCMOS Schmitt trigger AND gate. Compared to traditional Schmitt trigger AND gate, VTMOS scheme showed better noise immunity, especially for the noise of which is higher than 80 MHz.

C. Schmitt Trigger OR (NOR) Gate

Figure 9 shows the DC characteristics of Schmitt trigger OR gate implemented by proposed VTCMOS scheme and by traditional Schmitt trigger NOR gate with DTMOS inverter to compare output of OR gate. Detailed performance comparison parameters are shown in table III. By using the proposed scheme, we saved 50 % of switching power consumption with halved area.

As shown in figure 9, only V_{HL} was improved by 5 mV. Similar to the AND gates, the difference comes from the structural difference of two Schmitt trigger OR gates. In this case, ground connected substrate of PMOS transistor at the top creates a forward biased PN junction and decreases V_{TH} of adjacent PMOS with input A by making the source potential less than V_{DD}.

Figure 10 shows the input noise immunity of OR gates with different implementation methods. For all the frequency range, proposed Schmitt trigger circuits with VTCMOS method shows higher noise immunity compared to traditional DTMOS Schmitt trigger for low power operation.

D. Circuit Level Noise Immunity Enhancement

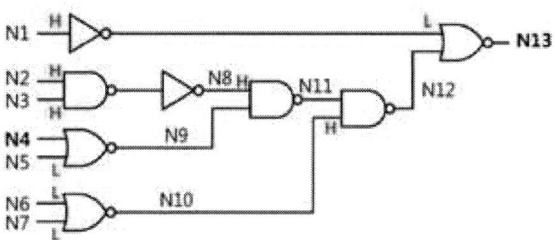

Fig. 11. Example circuit for immunity enhancement simulation

To generalize the result, we generated an example test circuit and applied the schematics of Schmitt trigger logic gates we discussed so far. Figure 11 shows the configuration of test circuit. For equivalence of circuit condition, we took the outputs of proposed Schmitt trigger for NAND and NOR.

We injected a pulse of 10 μs period with noise to node N4 and monitored the output of node N13. When input pulse is high, we injected a noise with the frequency of 20 MHz and the amplitude of 50 mV. For low input state, noise of 70 MHz frequency with 250 mV amplitude was injected.

The circuit was simulated using a supply voltage of $V_{DD}=0.4$ V. Except for node N4, other input nodes are properly controlled not to interfere the signal transmission between N4 through N13, i.e. setting N1 for V_{DD} makes variation of N12 to be transferred to N13. Figure 12 shows the simulation results for two different Schmitt trigger circuits.

As you can see in the figure 12, we injected a pulse input with high frequency noise at node N4. The signal variation of N4 was transferred through N9, N11, N12, and N13. For the small amplitude of noise injected during high input state, both configuration of circuits are remained at the value of high with

978-1-4799-5004-1/13 $31.00 © 2013 IEEE

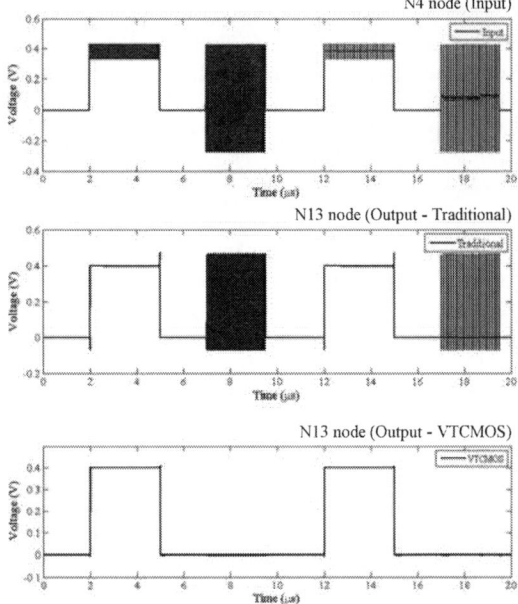

Fig. 12. Circuit level immunity test results

normal operation. For the noise with large amplitude during low input state, the voltage of N13 fluctuates for traditional Schmitt trigger circuit while proposed VTCMOS Schmitt trigger stays low.

In addition, to implement the same circuit for figure 11, traditional Schmitt trigger scheme requires 72 MOS transistors whereas proposed VTCMOS Schmitt trigger scheme requires only 44 transistors. Also, during the simulation, switching current of traditional Schmitt trigger scheme was 511 μA whereas that of proposed VTCMOS Schmitt trigger was 158 μA. As a result, the proposed scheme reduces 38.9 % of area and 69.1 % of switching current at the same time.

It has been a common practice to use Schmitt trigger to improve noise immunity. However, as we mentioned so far, the noise immunity can be further improved by using proposed VTCMOS Schmitt trigger not only for the logic gate itself but also for the circuit level. Although delay increases slightly, the proposed scheme can significantly save switching current and area.

IV. CONCLUSION

As the supply voltage of circuit scales down, noise immunity becomes more important to guarantee signal integrity. This paper presents a method of improving noise immunity applicable to sub-threshold circuits.

Traditional method for immunity enhancement was using Schmitt trigger, which requires additional current path to adjust switching threshold voltage and large area. However, by utilizing VTMOS scheme, which adjusts the threshold voltage

of MOS transistor itself to implement the hysteresis of the transfer characteristics, we can significantly save area and switching power consumption with improved noise immunity at the same time at the expense of slight increase in delay. Therefore, the proposed VTCMOS based digital logic design can be enable noise immune low power IC design.

Further research can be done on finding the trade-off between noise immunity enhancement and the circuit performance. Since the hysteresis of the DTMOS buffer can be adjusted [10], we can investigate the relation between the hysteresis width and power delay product. Therefore, we will be able to find the optimal condition for immunity enhancement considering circuit performance.

ACKNOWLEDGMENT

This research was supported by Basic Science Research Program through the National Research Foundation of Korea (NRF) funded by the Ministry of Education, Science and Technology (2012-0006847) and IDEC.

REFERENCES

[1] Golumbeanu, V., P. Svasta, and D. Leonescu. "The noise immunity for the low voltage logic circuits." *Electrotechnical Conference, 1998. MELECON 98., 9th Mediterranean.* Vol. 1. IEEE, 1998.

[2] Shepard, Kenneth L. "Design methodologies for noise in digital integrated circuits." *Proceedings of the 35th annual Design Automation Conference.* ACM, 1998.

[3] Assaderaghi, Fariborz, et al. "A dynamic threshold voltage MOSFET (DTMOS) for ultra-low voltage operation." *Electron Devices Meeting, 1994. IEDM'94. Technical Digest., International.* IEEE, 1994.

[4] Dokić, Branko L. "CMOS NAND and NOR Schmitt circuits." *Microelectronics journal* 27.8 (1996): 757-765.

[5] Soeleman, Hendrawan, Kaushik Roy, and Bipul Paul. "Robust ultra-low power sub-threshold DTMOS logic." *Proceedings of the 2000 international symposium on Low power electronics and design.* ACM, 2000,

[6] Zhang, C., A. Srivastava, and P. K. Ajmera. "Low voltage CMOS schmitt trigger circuits." *Electronics Letters* 39.24 (2003): 1696-1698.

[7] Al-Sarawi, S. F. "Low power Schmitt trigger circuit." *Electronics letters* 38.18 (2002): 1009-1010.

[8] Kim, Kyung Ki, and Yong-Bin Kim. "Ultra-low voltage high-speed Schmitt trigger circuit in SOI MOSFET technology." *IEICE Electronics Express* 4.19 (2007): 606-611.

[9] Donato, Marco, et al. "A noise-immune sub-threshold circuit design based on selective use of Schmitt-trigger logic." *Proceedings of the Great Lakes Symposium on VLSI.* ACM, 2012.

[10] Singhanath, Pratchayaporn, Varakorn Kasemsuwan, and Kittipol Chitsakul. "DTMOS Schmitt Trigger with Fully Adjustable Hysteresis." *Proceedings of the International Conference on Circuits, System and Simulation IPCSIT.* 2011.

Reliability Analysis of an On-Chip Watchdog for Embedded Systems Exposed to Radiation and EMI

C. Oliveira[1], J. Benfica[1], L. M. Bolzani Poehls[1], F. Vargas[1], J. Lipovetzky[2], A. Lutenberg[2], E. Gatti[3], F. Hernandez[4], A. Boyer[5]

[1] Catholic University of Rio Grande do Sul -PUCRS, Porto Alegre, Brazil, vargas@pucrs.br
[2] Universidad de Buenos Aires, Buenos Aires, Argentina, joselipo@gmail.com
[3] Instituto Nacional de Tecnologia Industrial - INTI, Buenos Aires, Argentina, egatti@inti.gob.ar
[4] Universidad ORT, Montevideo, Uruguay, ferhernasan@gmail.com
[5] LAAS-CNRS / Université de Toulouse, Toulouse, France, alexandre.boyer@insa-toulouse.fr

Abstract — **Due to stringent constraints such as battery-powered, high-speed, low-voltage power supply and noise-exposed operation, safety-critical real-time embedded systems are often subject to transient faults originated from a large spectrum of noisy sources; among them, conducted and radiated Electromagnetic Interference (EMI). As the major consequence, the system's reliability degrades. In this paper, we present the most recent results involving the reliability analysis of a hardware-based intellectual property (IP) core, namely Real-Time Operating System - Guardian (RTOS-G). This is an on-chip watchdog that monitors the RTOS' activity in order to detect faults that corrupt tasks' execution flow in embedded systems running preemptive RTOS. Experimental results based on the Plasma processor IP core running different test programs that exploit several RTOS resources have been developed. During test execution, the proposed system was aged by means of total ionizing dose (TID) radiation and then, exposed to radiated EMI according to the international standard IEC 62.132-2 (TEM Cell Test Method). The obtained results demonstrate the proposed approach provides higher fault coverage and reduced fault latency when compared to the native (software) fault detection mechanisms embedded in the kernel of the RTOS.**

Keywords- On-Chip Watchdog, Embedded System, Total Ionizing Dose (TID) Radiation, Electromagnetic Interference (EMI).

I. INTRODUCTION

Nowadays, several safety-critical embedded systems support real-time applications, which have to respect stringent timing constraints. In general terms, real-time systems have to provide not only logically correct results, but temporally correct results as well [1]. In this scenario, Real-Time Operating Systems (RTOSs) have been extensively adopted in order to minimize the design complexity of real-time embedded systems. Typically, these (embedded) systems exploit some important facilities associated to RTOSs' native intrinsic mechanisms to manage tasks, concurrency, memory as well as interrupts. In other words, RTOSs serve as an interface between software and hardware.

At the same time, the environment's always increasing hostility caused substantially by the ubiquitous adoption of wireless technologies represents a huge challenge for the reliability of real-time embedded systems [2]. Note that if these systems are powered by battery or this power is provided

by energy harvesting techniques, the yielded reliability is even more fragile. In detail, external conditions, such as Electromagnetic Interference (EMI), Heavy-Ion Radiation (HIR) as well as Power Supply Disturbances (PSD) may cause transient faults on electronic systems [3,4]. Currently, the consequences of transient faults represent a well-known concern in microelectronic systems. The International Technology Roadmap for Semiconductor (ITRS) predicts increasing system failure rates due to this type of fault for future generation of integrated circuits [5]. Therefore, embedded systems based on RTOS are subject to Single Event Upsets (SEUs) causing transient faults, which can affect the application running on embedded systems as well as the RTOS executing the application [6,7]. Affecting the RTOS, this kind of fault can generate scheduling dysfunctions that could lead to incorrect system behavior [1].

Up to now, several solutions have been proposed to deal with the reliability problems of real-time systems [8,9,10]. However, the vast majority of such solutions provide fault tolerance only for the application level and do not consider faults affecting the RTOS that propagate to the application tasks. Typically, these techniques are focused on detecting errors (on the application level) that corrupt data manipulated by the processor and/or induce application illegal control-flow execution [10]. Some of these approaches [8,9] deal with detecting faults degrading RTOS performance, but such approaches are very limited and cover only a few subset of possible faults affecting RTOS activity.

Regarding faults affecting the RTOS that propagate to application tasks, previous works indicate that 21% of them lead to application failure [1] and then, are liable to be detected by approaches monitoring application-level execution. Generally, these faults tend to miss their deadlines and to produce incorrect output results. Moreover, the work presented in [6] demonstrates that about 34% of the faults injected in the processor's registers led to scheduling dysfunctions. Indeed, about 44% of these dysfunctions led to system crashes, about 34% caused real-time problems and the remaining 22% generated incorrect system output results. To conclude, the fault tolerance techniques proposed up to now represent feasible solutions, but they do not guarantee detection of faults affecting the RTOS task scheduling process.

In this context, authors proposed in [11] an on-chip watchdog (RTOS-G) to monitor the RTOS's execution flow in order to detect scheduling misbehavior. RTOS-G provides detection of faults that change the tasks' execution flow in embedded systems for critical applications. In the first version, the on-chip watchdog was prototyped in systems running RTOSs based on the *Round-Robin* scheduling algorithm. As a further development, authors published [12] a more complex version of the watchdog devoted to monitor the scheduling activity of *Preemptive* RTOSs.

In the present work, we analyze the fault detection capability of such on-chip watchdog. This reliability analysis is performed by means of a case-study implementation based on the soft-core processor Plasma (opencores.org) and the RTOS-G watchdog that were prototyped into the Xilinx Virtex4 FPGA. This IC was aged under exposition to total ionizing dose (TID) radiation and the system performance analyzed under TEM-cell test method conditions [13].

The remainder of the paper is organized as follows: Section II presents a background with respect to the theory of RTOSs. Section III briefly describes the on-chip watchdog to monitor preemptive RTOSs. Section IV presents the case-study, describes the TID radiation and EMI experiments setup and discusses the obtained results. Finally, Section V draws the final conclusions of this work.

II. BACKGROUND

RTOSs represent a vital component to many embedded systems and provide a software platform upon which to build applications. An RTOS is a program that schedules execution in a timely manner, manages system resources, and provides a consistent foundation for developing application code. Basically, RTOSs can be classified in hard-RTOSs and soft-RTOSs. The main difference between the two categories is that a soft-RTOS can tolerate latencies and responds with decreased service quality while the hard-RTOS has to respect its deadlines, otherwise its tasks fail execution. In general terms, RTOSs provide four basic services to the application service: (1) time management, (2) interrupt handling, (3) memory management and (4) device management.

In computing, a *process* is an instance of a computer program that is being executed. A *computer program* is a passive collection of instructions; a *process* is the actual execution of those instructions. Several processes may be associated with the same program; for example, opening up several instances of the same program often means more than one process is being executed. Processes are often called "tasks" in embedded operating systems. The sense of "process" (or task) is "something that takes up time", as opposed to 'memory', which is "something that takes up space". In this context, a task is a set of program instructions that are loaded in memory.

A *deadline* is the time instant at which a task must finish its execution. The period of a task is the time interval between initiating two successive executions of such task. Generally, a task can be in one of the following three states: *blocked*, *ready* or *running*. Further, the transfer of CPU execution from one task to another one is called Context Switch (CS).

Every RTOS has a wide range of facilities (namely, system resources), which simplify the design of real-time applications by offering native mechanisms to manage tasks, concurrency, memory, time as well as interrupts. In comparison to other (not real-time) operating systems, the efficient use of the CPU is considered the more critical and the more important issue in real-time RTOSs. For instance, upon accessing a given blocked embedded system resource during the execution of a task, real-time RTOSs might force the task to wait for a semaphore release or some other external event before proceeding executing such task. In this context, RTOSs perform a CS to force the CPU to execute another task that is labeled *ready* to run and therefore, guarantee a more efficient usage of the CPU. If there is more than one task ready to run, the decision will be taken on the basis of task priorities.

Most RTOSs use scheduling algorithms based on the *Round-Robin* algorithm, which assigns equal Time Slices (TSs) to each task and executes them without priority in circular order. However, in a typical real-time application, there will be tasks with a shorter response time than others. Considering this situation, RTOSs usually implement a *Preemptive* algorithm with priority support. This results in a dynamic scheduling order and ensures time consistency for critical tasks.

It is important to point out that preemption will take place only if a task with higher priority than the executing one is ready to run. However, if all ready tasks have the same priority, the *Tick* signal will divide the CPU time between these tasks in equal parts (TSs) and no preemption can take place.

III. THE PROPOSED ON-CHIP WATCHDOG

This section describes the IP core RTOS-G. This IP is an on-chip watchdog that monitors tasks' execution flow according to the RTOS *preemptive* algorithm [11,12]. Figure 1 depicts the RTOS-G functional block diagram.

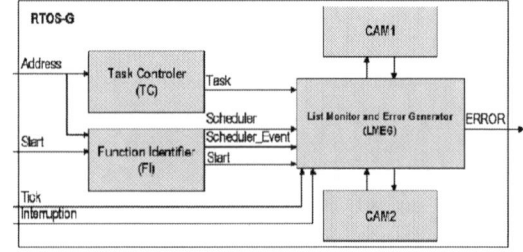

Figure 1. Functional block diagram of the RTOS-G.

The RTOS-G is connected to the embedded system's bus in order to monitor the following information: *Start*, *Tick* and

Interrupt signals as well as the RAM addresses accessed during the execution of the application code. In more detail, the RTOS-G is composed of five functional blocks. The *Task Controller* (TC) identifies the task in execution based on the address accessed by the microcontroller during the application's execution. At every clock cycle, the TC compares the address on the bus with the addresses associated to each task. If the accessed address is related to a task, the signal *Task* receives the corresponding task's number. The *Function Identifier* (FI) analyses the functions executed during the task scheduling process in order to check the scheduling process execution order. Finally, the FI identifies the event that triggered the scheduling process based on that order (e.g., occurrence of a *Tick* signal, IO request or semaphore acquisition). The block named *List Monitor and Error Generator* (LMEG) receives the *Scheduler_Event* signal and the *Task* in execution. Based on this information the LMEG classifies all tasks in two separate lists, *ready tasks* and *blocked tasks*, each one organized according to their state and priority. The LMEG implements the scheduling algorithm and indicates errors when a scheduling misbehavior is detected. As the last blocks, the two *Content-Addressable Memories* (CAM1 and CAM2) save the lists generated by the LMEG module. The tasks labeled *ready* are stored in CAM1 while the tasks labeled as *blocked* are saved in CAM2.

To implement preemption, the algorithm with priority support keeps a list of all tasks labeled *ready* (ready-list). The tasks are sorted by their priority. Therefore, every time a *CS* takes place and a scheduling event is performed, the (*ready*) task marked with the highest priority is executed. The complexity of monitoring this kind of behavior relies on keeping track of the *ready*-list: its elements must not have any pending IO requests or semaphore objects still to be acquired. In order to acomplish this task (keeping track of the *ready*-list), the RTOS-G should monitor not only the task addresses, but also the addresses related to the kernel synchronization, including: *SemaphoreLock()* and *SemaphoreUnlock()*. These functions lock and unlock a previously created synchronization object which is passed by parameter to the related functions. However, it is not possible for the RTOS-G to monitor the parameters of function calls; only the addresses of the functions are captured by the RTOS-G. Consequently, the described solution does not monitor all possible fault conditions. To counteract this limitation, an execution flow analysis is adopted as solution, since the function parameters remain unknown. In this solution, the RTOS-G observes the *order* in which the functions are being called to infer the *ready*-list constraints. To illustrate this mechanism, Figure 2 shows a situation where *Task1* is running and tries to acquire a semaphore. The system call is performed and the RTOS kernel realizes that the semaphore is already locked. In order to prevent the system from going into a deadlock as well as to increase the CPU usage, the kernel performs a *CS* calling another task into execution. The resulting execution flow for an already locked semaphore consists of: *SemaphoreLock()*

and *ReSchedule()*. When the RTOS-G detects this flow, it will infer that *Task2* is *running* and therefore is taken out from the *ready*-list. A similar analysis can be performed for other situations, always concentrating all efforts in keeping the detection algorithm generic enough for any RTOS or processor. As further positive effect, this type of analysis has rendered dispensable the *Tick* signal. In more detail, the RTOS-G detects the *Tick* by recognizing the following execution flow: *Interrupt()* and *ReSchedule()*.

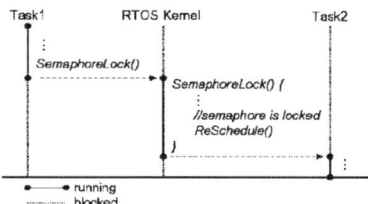

Figure 2. Typical task context switching activity executed under the control of an RTOS, which is monitored by the on-chip watchdog, RTOS-G.

IV. EXPERIMENTAL RESULTS

The fault detection capability of the RTOS-G watchdog with respect to the RTOS native fault detection mechanisms has been evaluated by: (1) aging FPGAs containing the RTOS-G and a microprocessor. Aging was performed by exposing the FPGAs to total ionizing dose (TID) radiation; and (2) in the sequence, the FPGAs were exposed to radiated EMI according to the IEC 62.132-2 standard (TEM Cell Method) [13].

A. Case Study

To evaluate the proposed approach we adopted a case study composed of a Von Neumann 32-bit RISC Plasma microprocessor running an RTOS (www.opencores.org). The Plasma microprocessor is implemented in VHDL and has, with exception of the load/store instructions, an instruction set compatible to the MIPS architecture. Moreover, the Plasma's RTOS adopts the *preemptive scheduling* algorithm with priority support composed of the following three states: *blocked*, *ready* and *running*. The Plasma's RTOS provides a basic mechanism able to monitor the task's execution flow and manage some particular situations when faults cause misbehavior of the RTOS's essential services, such as stack overflow and timing violations. This mechanism is implemented by a function named *assert()*. Generally, when the argument of the *assert()* function is false, the RTOS sends an error message through the standard output.

For the fault injection experiments, we developed two different benchmarks that exploit great part of the resources offered by the Plasma's RTOS (i.e., the use of message queues, semaphores and interrupts). Figure 3 shows the block diagram associated to the two benchmarks implementing the following tasks:

- *BM1:* 8 tasks access and update the value of a global variable, which is protected by a semaphore. Indeed, another global variable is accessed by an interrupt. The 8 tasks are assigned to the following different priorities: 1, 2, 3, 4, 1, 2, 3 and 4, respectively. The interrupt has the maximum priority.

- *BM5:* this benchmark is the most complex of the two experiments and consumes the largest amount of RTOS resources. In this benchmark, an *interrupt* communicates with a task (T1) via a message queue. Then, T1 communicates with Tasks T2 and T3 via two other message queues, which in turn send messages to Tasks T4, T5, T6, T7 T8, T9 and T10, respectively, by means of queue resources as well, as depicted in Figure 3. In this scenario, Task 1 has priority equal to 1, Tasks 2 and 3 equal to 2, Tasks 4 and 5, equal to 3, Tasks 6 to 9 equal to 4, and finally, Task 10 holds the highest priority: 5.

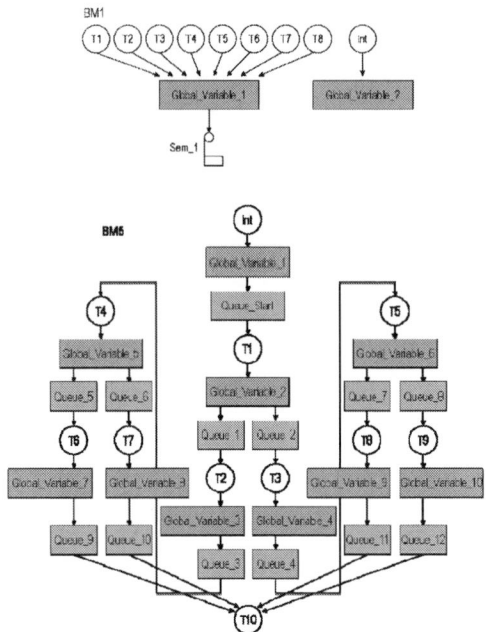

Figure 3.   Functional block diagrams of the two benchmarks.

B. Experiment Setup

The first part of the experiment was dedicated to age four Virtex-4 (XC4VFX12-10SF363) FPGAs with two different levels of ^{60}Co TID radiation: two FPGAs belonged to the same fabrication lot (Lot 1), while the other two FPGAs belonged to another lot (Lot 2). Lot 1 FPGAs received a total dose of 160 krads, whereas Lot 2 FPGAs received 336 krads.

The main effect of TID radiation on CMOS devices is the reduction of the threshold voltage (V_{th}) of pMOS transistors and increase of leakage currents due to trapped charge in insulating layers. As consequence, these transistors tend to cut-off. Therefore, the "0 to 1" response of gates along with the circuit is delayed, while the "1 to 0" remains approximately the same since nMOS devices are less sensitive to this type of radiation. Thus, it can be observed the increased occurrence of "delay faults" as long as the circuit becomes old.

The experiment was performed following the standardized 1019.8 method for TID radiation testing of the MIL-STD-883H standard. The experiment was carried out at room temperature at a 155.5 rads/s dose rate, in a PISI industrial plant. Basically, the PISI plant allows a uniform irradiation of the samples when they are physically placed on known isodose surfaces. Radiocromatic perspex amber dosimeters were employed to measure TID. The exposition time for Lot 1 was 18

minutes, while for Lot 2 it was 37 minutes. In order to have a better understanding of how much this radiation level represents in terms of aging, it is worth noting that authors published a work [12] where similar FPGAs were exposed up to 420 krads working properly. It was observed that above this radiation level the component was permanently damaged, i.e, it was no more able to properly run the application. Figure 4 gives an example of the effects of aging on the FPGA electrical parameters. In this figure, 1v2 is the core *current* (resp. *voltage*), whereas 2v5 and 3v3 are the second and third periphery *currents* (resp. *voltages*), respectively. As previously mentioned, the primary consequence of TID deposition on the IC is the increase of signal propagation delay. Thus, reducing V_{DD} to the minimum possible value for system fault-free operation increases even more the signal propagation delay. As consequence, it is more difficult for the system to respect stringent timing constraints. However, it is curious to observe that the FPGA sensitivity to power supply reduction is not proportional to the aged state of the IC. In more detail, the aged FPGAs from Lot 2 (336 krads) presented the same order of magnitude of sensitivity to V_{DD} reduction as the aged FPGAs from Lot 1 (160 krads).

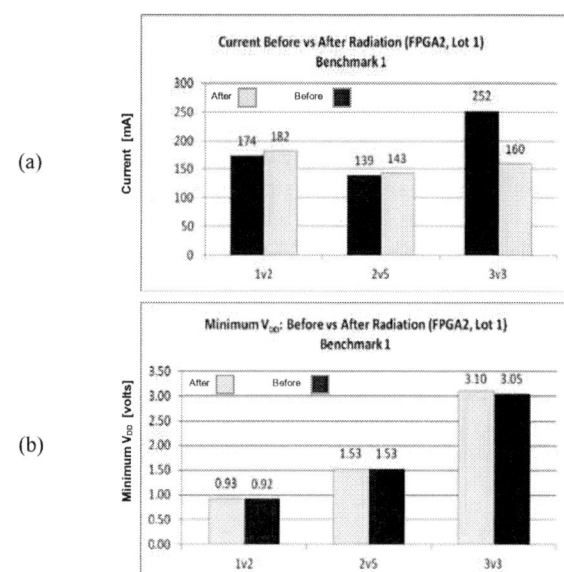

Figure 4.   Electrical parameters measurements for FPGA2 (Lot 1) before and after TID deposition (160krads): (a) Static Current (I_{DD_q}) consumption; (b) Minimum (V_{DD}) voltage for the FPGA to operate properly.

After aging the four FPGAs as described above, these components were characterized to radiated electromagnetic immunity (TEM Cell Method) according to the IEC 61.132-2 standard [13]. This method was adapted to a GTEM in order to have a larger cell volume to perform the experiment. Thus, it was possible to irradiate the whole set formed by the shielding box containing the board under test (whose test side was placed outwards the box). Figure 5 presents the test environment. The TEM cell test method conditions under which the four FPGAs were characterized to radiated electromagnetic immunity are respectively:

- EM field range*:* from 10 to 220 V/m (volts/meter);
- Radiated signal frequency range*:* [150kHz - 1GHz];
- Signal modulation format*:* AM/FM 80%.

C. Results' Discussion

Figure 6 compares the fault detection capability of the

978-1-4799-5004-1/13 $31.00 © 2013 IEEE

RTOS-G watchdog against the native fault detection mechanisms embedded in the kernel of the Plasma RTOS before and after aging the FPGAs with TID radiation. This comparison was performed for the fresh and aged FPGAs operating in a radiated electromagnetic environment according to the IEC 61.132-2 standard for a TEM-cell test method.

The terms depicted in Figure 6 mean respectively: "Only RTOS": faults detected exclusively by the RTOS native mechanisms (these faults escaped RTOS-G detection); "Both": faults simultaneously detected by the RTOS native mechanisms and by the RTOS-G watchdog; "RTOS": faults detected by the RTOS native mechanisms (note that part of these faults might eventually be detected by the RTOS-G as well); "Only RTOS-G": faults detected exclusively by the RTOS-G watchdog (these faults escaped RTOS native mechanisms detection); and "IP RTOS-G": faults detected by the RTOS-G watchdog (part of these faults might eventually be detected by the RTOS native mechanisms as well).

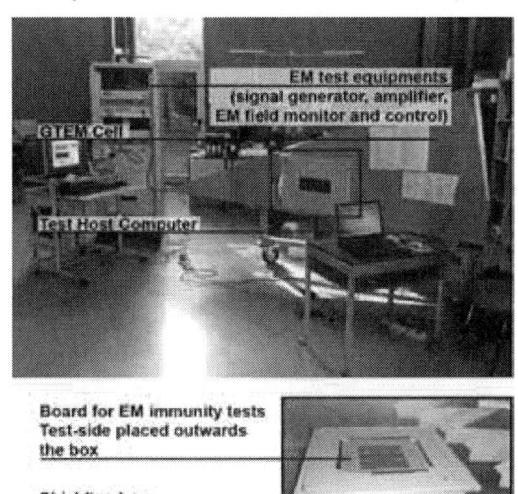

Figure 5. Environment Setup for TEM-Cell Method Measurements.

Analyzing the experimental results shown in Figure 6, one can conclude the following reasoning:

a) For any of the test cases, the fault detection capability of the RTOS-G watchdog is at least double of the one of the RTOS native mechanisms. For instance, 44.73% and 92.28% (Fig. 6a) and 17.60% and 89.28% (Fig. 6b).

b) After aging the FPGAs, this difference is even greater. For instance, 2.93% and 93.15% (Fig. 6c) and 10.98% and 97.58% (Fig. 6d).

c) After aging, it is clear the dramatic reduction of the fault detection capability of the RTOS native mechanisms: from 44.73% and 17.60% (Figs. 6a and 6b) to 2.93% and 10.98% (Figs. 6c and 6d).

d) The RTOS-G fault detection capability is independent of the aging state of the FPGAs: 92.28% and 89.28% (Figs. 6a and 6b for the fresh FPGAs) while 93.15% and 97.58% (Figs. 6c and 6d for the aged FPGAs).

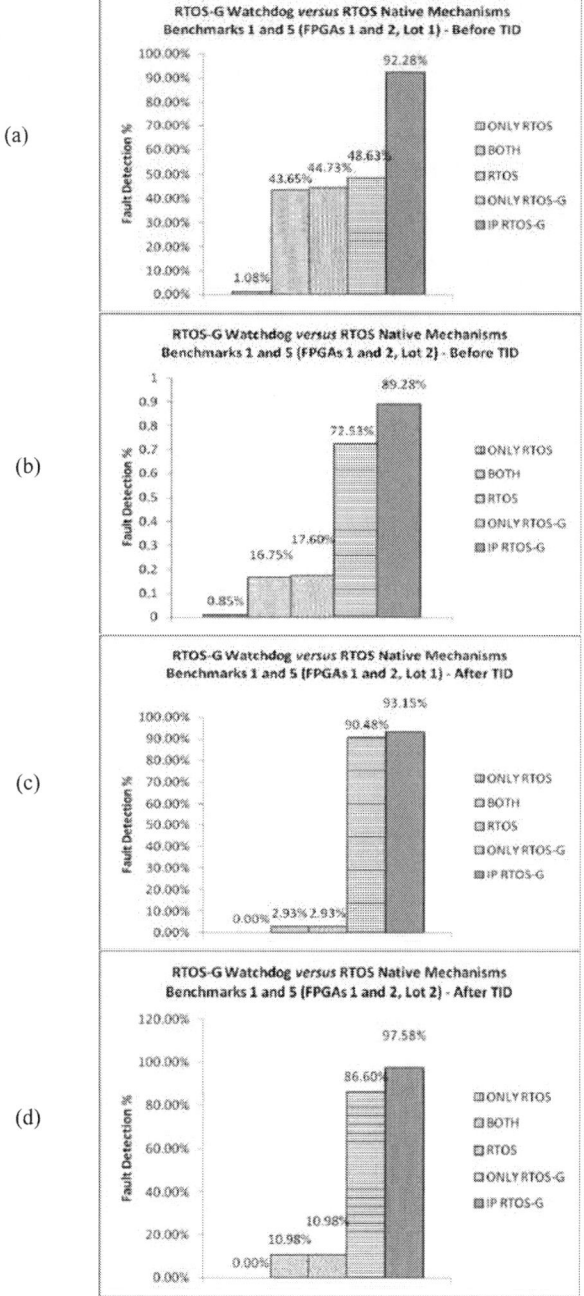

Figure 6. Comparison between the fault detection capability of the RTOS-G watchdog against the native fault detection mechanisms embedded in the kernel of the Plasma RTOS for fresh and aged FPGAs operating in an EMI-exposed environment.

We concluded that the much higher fault detection capability of the RTOS-G watchdog and its reduced sensitivity to the aging of the FPGAs is due to the following facts:

a) The watchdog is much smaller and logically less complex than the Plasma processor. Thus, rendering it intrinsically more robust to delay increase, which is the first electrical parameter to be degraded in an aged integrated circuit.

b) The processor executes applications under restrict timing control of the RTOS. Thus, the processor becomes very sensitive to delay and transient faults induced by aging the FPGAs with TID radiation.

c) As in a previous work [12], the fault detection latency of the RTOS-G watchdog was measured with the help of ChipScope (ISE-Xilinx Design Framework) and it was verified to be negligible when compared to the one of the RTOS native mechanisms. More precisely, the watchdog is capable of signaling system errors in average of 2% of the time required by the RTOS to output the same error indication. This reduced latency facilitates fault detection by the watchdog since typically, the RTOS takes several thousands of clock cycles to indicate error occurrence. Note that the processor is particularly sensitive to additional faults, if they rise up during the long time interval during which the processor is running under the control of the RTOS. If such faults occur, the system may be permanently corrupted before an error indication is yielded by the processor under the control of the RTOS.

FPGA 1 (Lot 1)	Frequencies or [frequency ranges] at which the system failed (MHz)	
	Pre-radiation	Post-radiation (160 krad)
	391	[284 – 311]
	492	[436 – 450]
	740	

FPGA 2 (Lot 1)	Frequencies or [frequency ranges] at which the system failed (MHz)	
	Pre-radiation	Post-radiation (336 krad)
	492	212
	512	[419 – 441]
	747	[492 – 507]
		522
		[538 – 543]
		560
		740

Figure 7. Frequencies at which the system failed, for fresh and aged FPGAs during the TEM-cell test method.

Figure 7 presents the frequencies where it was observed system failure during the IEC 62.132-2 TEM-cell test method. Intuitively (and as observed in this figure), the probability of coupling between an electromagnetic wave and board tracks and IC package is increased for frequencies around 600 MHz. In this case, we estimated that on-board geometries measuring 2.5 cm could behave as an efficient antenna ($\lambda/20$).

We performed this experiment by varying the electric field from 10 to 220 volts/meter. Approximately from 170 volts/meter, we started observing the occurrence of transient faults on the FPGA. As observed, the aged FPGAs failed at a larger number of frequencies when compared to the fresh FPGAs (before TID-exposure). Furthermore, the aged FPGAs failed mostly along with a whole frequency range (instead of specific frequency points). For instance, when irradiating the fresh system inside the GTEM cell, we observed that it failed precisely on the frequency of 391 MHz (we repeated several times the test execution when we observed a failure, to be sure that we were measuring the good numbers). On the other hand, when we irradiated the aged system, we observed that it failed

not only on the 391 MHz frequency, but in a large frequency range: [294 - 311] MHz (Fig. 7, FPGA 1 / Lot 1).

This scenario demonstrates experimentally that aged FPGAs are more sensitive than fresh FPGAs to transient faults induced by electromagnetic environments like the one promoted by the TEM-cell test method. Therefore, the use of the proposed approach yields even more benefits as time pass by during system lifetime.

V. CONCLUSIONS

We analyzed the fault detection capability of an on-chip watchdog (RTOS-G) designed to detect transient faults affecting the task scheduling process of real-time operating systems (RTOSs). The reliability analysis was performed by means of a case-study implementation based on the soft-core processor Plasma (opencores.org) and the RTOS-G watchdog that were prototyped into the Xilinx Virtex4 FPGA. This IC was aged under exposition to total ionizing dose (TID) radiation and the system performance analyzed under TEM-cell test method conditions. The obtained results for the fresh (pre-radiation) and aged (post-radiation) experiments indicate that the RTOS-G watchdog presents a much higher fault detection capability when compared with the conventional RTOS native fault detection mechanisms. This scenario is even more evident when the system is aged, since it is more sensitive to transient faults in a larger frequency spectrum.

REFERENCES

[1] N. Ignat, B. Nicolescu, Y. Savari, G. Nicolescu, "Soft-Error Classification and Impact Analysis on Real-Time Operating Systems", IEEE Design, Automation and Test in Europe, 2006.

[2] S. Ben Dia, R. Ramdani, E. Sicard, "Electromagnetic Compatibility of Integrated Circuits – Techniques for Low Emission and Susceptibility", Springer, 2006.

[3] J. Freijedo, L. Costas, J. Semião, J. J. Rodríguez-Andina, M. J. Moure, F. Vargas, I. C. Teixeira, and J. P. Teixeira, "Impact of power supply voltage variations on FPGA-based digital systems performance", Journal of Low Power Electronics, vol. 6, pp. 339-349, Aug. 2010.

[4] J. Semião, J. Freijedo, M. Moraes, M. Mallmann, C. Antunes, J. Benfica, F. Vargas, M. Santos, I. C. Teixeira, J. J. Rodríguez-Andina, J. P. Teixeira, D. Lupi, E. Gatti, L. Garcia, F. Hernandez, "Measuring Clock-Signal Modulation Efficiency for Systems-on-Chip in Electromagnetic Interference Environment". 10th IEEE Latin American Test Workshop (LATW'09), March 2009.

[5] http://public.itrs.net

[6] D. Mossé, R. Melhelm, S. Gosh, "A non-preemptive real-time scheduler with recovery from transient faults and its implementation", IEEE Trans. on Software Engineering, Vol. 29, N°. 8, pp. 752-767, August, 2003.

[7] B. Nicolescu, N. Ignat, Y. Savaria, G. Nicolescu, "Analysis of Real-Time Systems Sensitivity to Transient Faults Using MicroC Kernel," IEEE Transactions on Nuclear Science, Vol. 53, N° 4, August 2006.

[8] S. Gosh, R. Melhem, D. Mossé, J. Sarma, "Fault-tolerant Rate Monotonic Scheduling", Journal of Real-time Systems, Vol. 15, N° 2, Sept. 1998.

[9] P. Mejia-Alvarez, D. Mossé, "A responsiveness approach for scheduling fault-recovery in real-time systems", 5th Real-Time Technology and Applications Symposium, pp. 83-93, 1999.

[10] Ph. Shirvani, R. Saxena. E. J. McCluskey, "Software-implemented EDAC protection against SEUs", IEEE Trans. on Reliability, Vol. 49, N° 3, pp. 273-284, Sept. 2000.

[11] J. Tarrillo, L. Bolzani, F. Vargas, "A Hardware-Scheduler for Fault Detection in RTOS-Based Embedded Systems", IEEE 12th EUROMICRO Conference on Digital System Design, 2009.

[12] Benfica, J.; Bolzani Poehls, L. M.; Vargas, F.; Lipovetzky, J.; Lutenberg, A.; Garcia, S. E.; Gatti, E.; Hernandez, F. Evaluating the Effects of Combined Total Ionizing Dose Radiation and Electromagnetic Interference. IEEE Transactions on Nuclear Science, Vol. 59, Issue 4, pp. 1015-1019, Aug. 2011.

[13] "IEC 62.132-2, Ed. 1: Integrated Circuits - Measurements of Electromagnetic Immunity, 150KHz to 1GHz - Part 2: Measurement of radiated immunity - TEM-Cell Method. 2004-07". (www.iec.ch)

978-1-4799-5004-1/13 $31.00 © 2013 IEEE

An Optimizing Technique to Lower both Phase Noise and Susceptibility of a Voltage Controlled Oscillator

J. Raoult, A. Blain, S. Jarrix

Institut d'Electronique du Sud (IES)
Université de Montpellier 2
Montpellier, France
jeremy.raoult@ies.univ-montp2.fr

Abstract— **This paper deals with the electromagnetic susceptibility of a fully integrated voltage controlled oscillator. Injection locking and pulling is observed when a sinusoidal interference signal is injected on the device. To improve the immunity of the circuit we propose a procedure based on an optimization of the value of some circuit features. This work is done in relation with the optimization of the functional performances.**

Keywords—electromagnetic susceptibility, voltage controlled oscillator, lock range, phase noise.

I. INTRODUCTION

Nowadays common trend in the industry is to reduce manufacturing costs by optimizing device integration. However this trend is followed by the need to improve system reliability and as a result the electromagnetic compatibility (EMC). Electronic systems are currently getting smaller and smaller while their operating frequencies are increasing. This induces a non-negligible risk of wave coupling between emitted signals and other elements of the circuit. Furthermore the new systems generally use low power supply, which also reduces the EMC threshold. That is why EMC is becoming a major issue in current electronics [1].

In this paper our study is focused on a very common RF block circuit, a voltage controlled oscillator (VCO). This circuit is a critical block in designing wireless communications systems, and in particular PLL (Phase Locked Loop) frequency synthesizers. The circuit studied is the same as the one presented in [2] in which we have studied effects of a continuous wave (CW). Here we propose to study the impact of some main internal components on the susceptibility of the circuit. Hence in the early stage of design it would be possible to render the circuit more immune against CW interference. This optimizing step has to be done in agreement with the typical features of the circuit. Phase noise can be viewed as a random variation in the period of the oscillation signal or deviation of the zero crossing points from their ideal position along the time axis. For circuit like a VCO, to lower the phase noise is certainly the most important thing to do. The phase noise in the radio receiver's local oscillator limits immunity against nearby interference [3]. Finally we propose both an optimization of the phase noise and of the immunity of the circuit.

The structure of the paper is as follows. The first section presents the device under test and the test-bench used to inject the interference signal. The second section deals with the effects of a CW interference signal on the VCO. The third section studies the influence of some internal circuit parameters on the susceptibility of the VCO.

II. TEST-BENCH AND DEVICE UNDER TEST

A. Device Under Test

1) LC VCO

Figure 1 shows the VCO topology used for this study. It is based on a cross-coupled pair of transistors with a capacitive positive feedback.

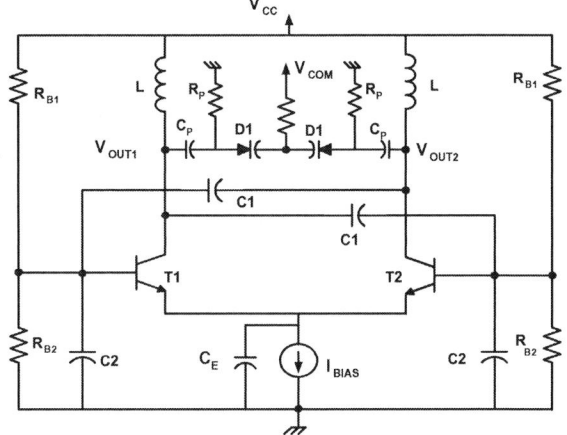

Figure 1 : Topology of the VCO

The VCO feedback is performed by a capacitive cross-coupling (C1, C2) of the collector and the base terminals of the differential pair formed by the T1 and T2 transistors. This differential pair of transistors is biased by a tail current source I_{BIAS}. The feedback ratio n is equal to:

$$n = \frac{C1+C2}{C1} = 1 + \frac{C2}{C1} \qquad (1)$$

978-1-4799-5004-1/13 $31.00 © 2013 IEEE

The oscillation frequency is determined by the LC_{eq} tank, composed of the inductor L and the equivalent capacitors of the varactor D1 in addition to the coupling and parasitic capacitors of the circuit. The frequency control is performed by the use of a P+/NWELL varactor.

The VCO is implemented in a 0.35 µm BiCMOS SiGe process. The picture of the test chip is presented in [2]. Main electrical features for NPN transistors are the breakdown collector-emitter voltage $BV_{CE0} = 3.6$ V, the peak transition frequency $F_T = 45$ GHz, and the maximal frequency $F_{MAX} = 60$ GHz. Passive and active device models have been implemented and validated by simulations with ADS software. These models will be necessary in order to simulate the effect of an interference signal on the circuit.

Figure 2 presents the VCO frequency tuning range with respect to the varactor control voltage. The VCO is biased such that $V_{CC} = 3.3$ V. A good agreement is obtained with ADS simulations results.

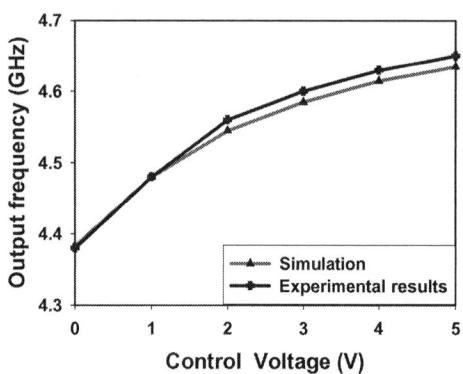

Figure 2: VCO frequency tuning range both measured and simulated.

B. CW injection Test-Bench

Susceptibility studies described in this paper are performed using a conducted mode for interference injection. The setup used for the injection depends on the track where the interference signal is injected. Bias tees are used for injection on DC tracks, such as the supply track or the tuning voltage track (Figure 3). A bidirectional coupler is used to measure the injected and reflected power. We can thus determine the interference power absorbed by the circuit.

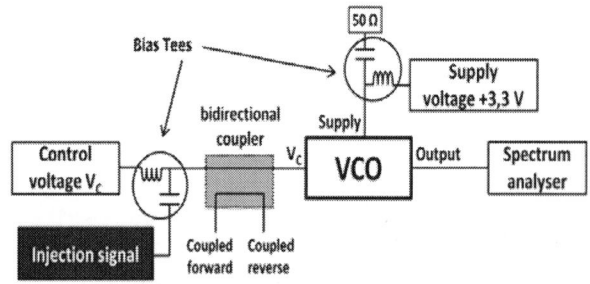

Figure 3: Test-bench for direct injection on tracks.

To inject the interference on the output of the VCO, we need to use a circulator, as shown on figure 4.

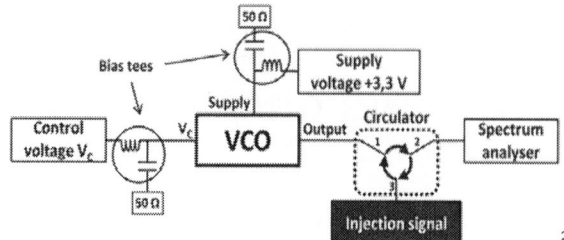

Figure 4: Test-bench for direct injection on the output of the VCO.

Bias tees are still used to prevent the interference from disturbing the power supplies. For all experiments, the output spectrum of the VCO is measured using a spectrum analyzer.

Figure 5 shows the test board on which the chip is mounted. Only the two output tracks are matched to 50 Ω.

Figure 5: Test board used. Interference is injected in a conducted mode via the connectors.

II. SUSCEPTIBILITY OF THE VCO UNDER A CW INTERFERENCE

The interference is injected on the voltage control track. Figure 6 shows the evolution of the output spectrum with respect to the injection frequency f_{INJ}. In the whole paper, the power of injection, P_{INJ}, corresponds to the power read at the synthesizer output.

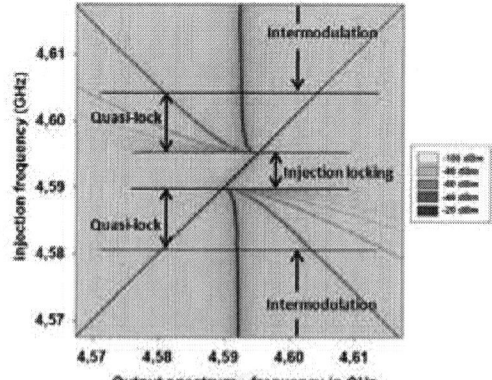

Figure 6: Evolution of the output spectrum of the VCO with injection frequency ($P_{INJ} = 0$ dBm, $V_C = 3.3$ V).

On Figure 6, each horizontal line corresponds to an output spectrum of the VCO, measured for one particular injection frequency. This operation is repeated so as to obtain the evolution of the spectrum for the complete range of interference frequency values, for both fixed control voltage and injection power. To clarify this figure an example of a spectrum of the output voltage when the VCO is under injection is presented in figure 7. The injection frequency is $f_{INJ} = 4.68$ GHz. Numerous tones are exhibited: the normal operating frequency, the injection frequency and intermodulation products between both frequencies.

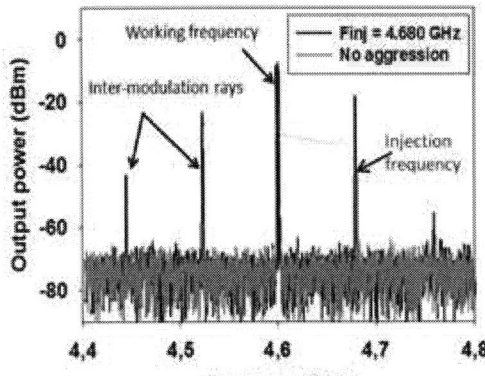

Figure 7: Example of a spectrum of the output voltage when the VCO is under injection.

Three major operating modes can be observed on figure 6. A linear oscillator would only generate two tones: one at the oscillation frequency f_0 and one at the injection frequency f_{INJ}. However we observe additional tones in the three modes; highlighting the fact that the system is not linear [4].

In the first mode numerous intermodulation products emerge. All these tones degrade the output signal of the VCO. This means thatonce the VCO implemented in a system, it can disturb the next elements of the system. The second mode corresponds to the injection locking mode. The VCO is totally controlled by the interference signal. This occurs for an injection frequency range, the lock range. In [5, 6], Adler's equation was given to account for the injection locking phenomena in a free running oscillator. This relationship, given in (2), expresses a figure-of-merit known as the lock range.

$$\delta f = \frac{f_0}{Q} \sqrt{\frac{P_{INJ}}{P_0}} \qquad (2)$$

This frequency range for which the interference signal controls the output frequency depends on the power of the interference signal P_{INJ}, the output power of the circuit P_0 and the quality factor Q.

The last mode observed corresponds to an intermediate state where the injected signal frequency value is just in the vicinity of the lock range. The VCO is then pulled or "quasi-locked" and the oscillation frequency shifts from its natural value. More details on these three major operating modes are available in references [3, 4, 6].

The goal of the next section of the paper is to present the influence of some components of the circuit on the

susceptibility of the VCO. Some design guidelines could therefore be proposed.

III. INFLUENCE OF SOME CIRCUITS PARAMETERS ON SUSCEPTIBILITY OF THE VCO

The purpose of this section is to propose some design guidelines with respect to the classical performances (phase noise, linearity) of a VCO. Two main internal parameters of this circuit play an important role on phase noise and linearity performances, the bias current of the differential pair of transistors I_{BIAS} and the feedback ratio n. Hence we choose to study the evolution of the susceptibility of the VCO to the tail bias current I_{BIAS} and the feedback ratio n. The indicator studied to evaluate the susceptibility of a VCO circuit is the lock range since it represents a frequency band in which the VCO is disrupted. We will study the lock range as a function of I_{BIAS} and n in order to obtain optimum values of these two parameters. Then we will compare these optimum values, which guarantee a minimum level of susceptibility of the VCO, to those chosen to optimize phase noise of the VCO. This work is essentially based on simulation results. Hence we have developed a simulation tool to compute the lock range.

A. Simulation tool

Circuit simulation could be an efficient tool for predicting effects induced by an interference signal. Passive and active device models have been implemented in a commercial software, ADS of Agilent. The simulation process used to compute the lock range is based on the envelope-transient method. This technique is very often used in oscillator circuits, especially to calculate the phase noise [7]. It consists in a harmonic balance analysis with time-varying envelope. Analytical study has shown that injection locking in time and frequency are well analyzed when the envelope of the VCO output is studied [4]. The envelope-transient technique is then well suited for this kind of study since it is possible to compute the time-variation of the envelope of a voltage in amplitude or in phase.

The main objective of the simulation process is to calculate the lock range for a chosen power of the interference signal P_{INJ}. The simulation process consists in sweeping the interference frequency on an interval centered on the normal oscillation frequency f_0, and in computing the envelope of the output voltage in amplitude. Figure 8 shows the variation of the envelope of the VCO output voltage with envelope-transient simulation when the interference frequency ranges within the vicinity of the initial frequency oscillation f_0.

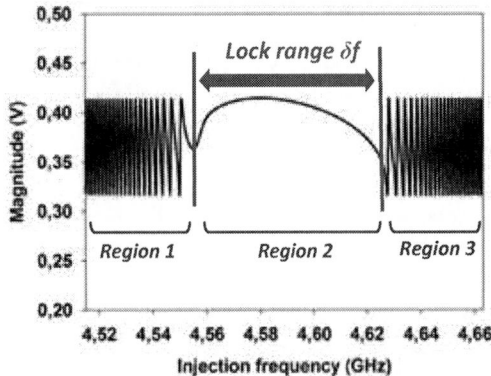

Figure 8: Variation of envelope of the VCO output voltage with envelope simulation process as a function of injected frequency. $P_{INJ} = 5$ dBm

When the envelope of the VCO output voltage is almost constant (region 2 on figure 8), the VCO is locked. The range of frequencies over which the VCO is locked is therefore easily determined. A Fast Fourier Transform of the voltage enables to describe the output spectrum of the VCO when the interference frequency is in the lock range (region 2) or above the lock range (regions 1 and 3). These results are presented in figure 9. The output spectrum in regions 1 and 3 exhibits numerous tones with a spacing of a beat frequency f_b. It corresponds to a spectrum of a quasi-locked VCO, as discussed in section II.

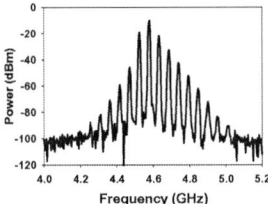 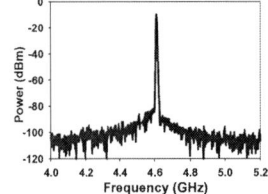

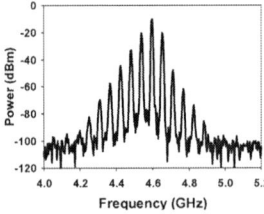

Figure 9: Output simulated spectrum of the VCO. (b) Injected frequency is in the lock range, (a) and (c) Injected frequency is respectively under and above the lock range.

To validate our simulation method we compare experimental and simulated lock ranges for injected power levels ranging from -20 dBm to 10 dBm. The interference signal is injected on the output of the VCO, using the test-bench presented figure 4. This comparison is shown in figure 10.

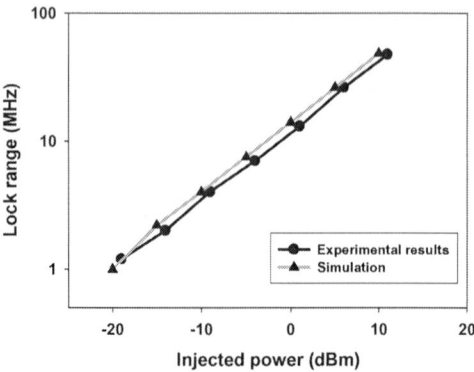

Figure 10: Simulated and experimental lock range as a function of the injected power P_{INJ}. Interference signal is injected on the VCO output.

The agreement between experimental and simulated results is quite good. This shows the ability of our simulation process to predict the behavior of a VCO when it is locked.

This simulation tool can be used to help us to predict the impact of some internal circuit parameters on the susceptibility of the VCO. This study can not be performed without considering the phase noise level of the VCO, which is the main specification of an oscillator.

B. Phase noise optimization

The phase noise qualifies the spectral purity of an oscillator [3]. When asked to visualize phase noise, we can invoke the picture of a noisy sinewave whose phase is randomly perturbed at the zero crossings. In RF applications phase noise is usually characterized in the frequency domain. For an ideal sinusoidal oscillator, the spectrum assumes the shape of an impulse, whereas for a real oscillator, the spectrum exhibits "skirts" around the carrier frequency. Phase noise mainly comes from noise in the oscillator loop which can be upconverted around the oscillation frequency by complex physical processes [8,9]. Generally speaking one of the most drastic limitations of high spectral purity oscillators in BiCMOS technologies comes from the low loaded resonator quality factor Q. Indeed integration of high Q factor passive networks is difficult because of resistive losses occurring in substrates with medium resistivity. In a 0.35μm BiCMOS SiGe process the Q factor is estimated to range between 10 and 15 at 5 GHz.

Concerning the VCO studied in this paper the main contribution to the phase noise is the low-frequency noise associated with the differential pair of transistors (T1 and T2 in the figure 1), and in particular the collector shot noise of the bipolar transistors. The phase noise level can be improved by optimizing some circuit characteristics, which are the output signal power P_0 and the operation regime of the differential pair of transistors. The current source I_{BIAS} of the transistors in the differential pair associated with the feedback ratio n set the regime in which the transistor operates. A low n implies a nonlinear regime for the transistors whereas high n sets a quasi-linear regime. Thus,

978-1-4799-5004-1/13 $31.00 © 2013 IEEE

the output voltage waveform of the differential pair and in particular the transistor conducting time depends strongly on this feedback ratio. This has an important effect on VCO phase noise performance. Indeed according to Hajimiri [8], the effective noise introduced by the transistors in the differential pair can be reduced by exploiting cyclostationary properties of the noise sources such as collector current shot noise sources of transistors in the differential pair. The differential pair injects noise into the LC resonator only over the window of time while transistors conduct. Hence the impact of the feedback ratio n and the differential pair bias current I_{BIAS} on the VCO phase noise performance has been studied, as can be seen in the figure 11.

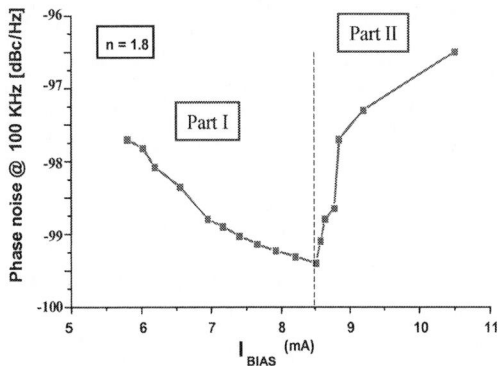

Figure 12: Impact of the bias current of the differential pair on VCO phase noise

As can be seen in figure 12, an optimum phase noise at 100 kHz of the carrier is reached for a bias current of the differential pair equal to 8.5 mA. In the first part of the curve, at a given feedback ratio, when I_{BIAS} increases, the differential pair operate in a strongly nonlinear regime. A reduced transistor conducting time is thus observed. The collector current shot noise source S_{ICE}, given its cyclostationary properties, is therefore active only over a short window of time. This seems to explain the phase noise improvement in spite of increased noise source amplitude with an increased I_{BIAS}. In the second part of the curve, when I_{BIAS} increases, the transistors in the differential pair operate in an even stronger nonlinear regime, low-frequency noise conversion into phase noise tends thus to step up. Moreover, the noise source amplitude keeps on increasing. This seems to explain the phase noise degradation.

Finally this study gives us the optimal adjustment for the bias current of the VCO and the feedback ratio for low phase noise. These optimal values have been implemented in the VCO under test used for this study.

Now we propose to investigate the influence of I_{BIAS} and n on the lock range when the VCO is under injection.

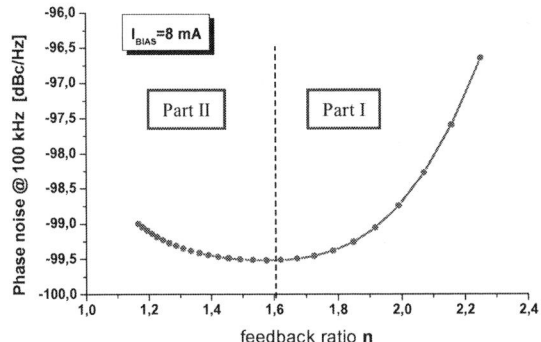

Figure 11: Impact of the feedback ratio on VCO phase noise

An optimum phase noise level at 100 kHz of the carrier is reached for a feedback ratio n equal to 1.6. In the first part of the curve, when n decreases, a reduced bipolar transistor conducting time is observed by simulation. The collector current shot noise source S_{ICE}, given its cyclostationary properties, is therefore active only over a short window of time. This seems to explain the phase noise improvement. In the second part of the curve, when n decreases, a reduced transistor conducting time is always observed but these transistors in the differential pair operate in a strong nonlinear regime. Low-frequency noise conversion into phase noise tends thus to step up. Indeed some authors have shown that a switching differential pair formed at low n acts as a mixer and the noise from the tail courant source is also up converted around the oscillation frequency. Moreover harmonics of the commutating waveform might downconvert higher noise frequencies close to the oscillation frequency band [8, 9, 10]. Hence noise conversion into phase noise is largely enhanced when the feedback ratio goes down at a fixed I_{BIAS}. This seems to explain the phase noise degradation in spite of a reduced transistor conducting time.

Now the impact of the bias current I_{BIAS} of the differential pair on the VCO phase noise performance is studied (figure 12.).

C. Lock range optimization

We have simulated the lock range when the interference signal is injected on the output of the circuit.

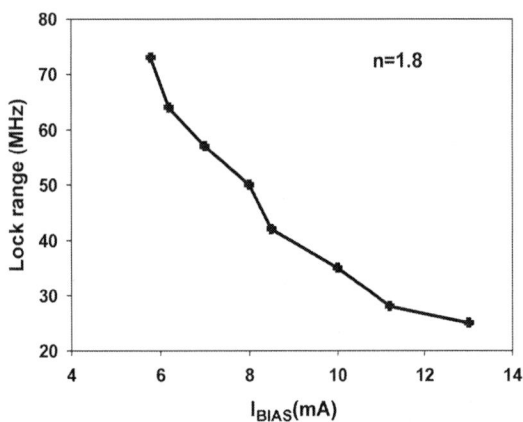

Figure 13: Lock range in function of the bias current of the circuit. The interference signal is injected on the output, and the power P_{INJ} = 5 dBm.

Figure 13 shows a decrease in the lock range when the bias current of the VCO goes up. This decrease tends to stop for the largest value of I_{BIAS} used. The voltage amplitude of oscillation is proportional to the bias courant I_{BIAS} until the "voltage-limited" regime, where output voltage is limited by the supply voltage of the circuit V_{CC}. The simplified analytical formulation of the lock range proposed in relation (2) shows that the lock range is inversely proportional to the output power P_0 of the circuit. This is confirmed by the simulation results presented in figure 13 except for the largest value of I_{BIAS} where P_0 tends towards a constant.

Finally considering the monotonic function plotted in figure 13 no optimum value of I_{BIAS} from a susceptibility point of view could be deduced. One could be tempted to therefore increase I_{BIAS}. However, as we can see in figure 12 the largest I_{BIAS} induces a large phase noise value, and at the same time an increased current consumption of the circuit.

Now let's examine the influence of the feedback ratio n on the lock range.

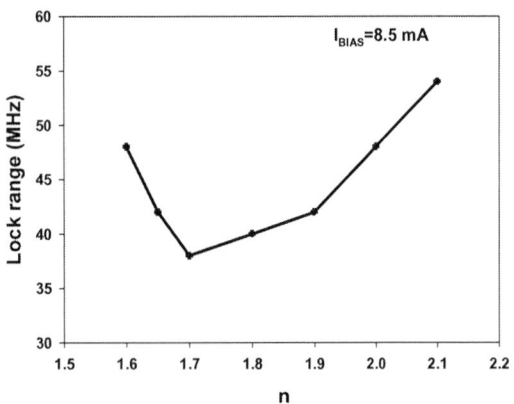

Figure 14: Lock range in function of the feedback ratio. The interference signal is injected on the output, and the power P_{INJ} = 5 dBm.

The feedback ratio implemented in the VCO studied is fixed to approximately n ≈ 1.8. According to the topology of the VCO shown in figure 1, this corresponds to the capacitance values C1 = 0.6 pF and C2 = 0.5 pF. The bias current IBIAS is fixed at 8.5 mA, which corresponds to the optimum phase noise performance. The interference signal is always injected at the output of the circuit. As can be seen in figure 14 an optimum value of the feedback ratio can be observed. This optimal value of n = 1.7 is very close to the value presented in figure 11 which improves the phase noise performance.

IV   CONCLUSION

This paper describes the different effects observed on a fully integrated VCO operating at a frequency of 5 GHz when it is subjected to a CW sinusoidal interference. We have investigated more precisely the locking phenomena which occur when the interference frequency is close to the normal oscillation frequency. The operating frequency of the VCO is therefore the interference signal frequency. The frequency range for which the interference signal can control the VCO frequency is called the lock range. From a susceptibility point of view the lock range can be regarded as a figure-of-merit, that we try to optimize in adjusting the bias current and the feedback ratio of the circuit. This optimizing process must be undertaken in considering the impact on the classical VCO performance. This process has been detailed in this paper and it was shown that it was quite possible to lower both the phase noise and the susceptibility of an oscillator.

REFERENCES

[1] M. Ramdani, E. Sicard, A. Boyer, S. Ben. Dhia, J. Whalen, T. Hubing, M. Coenen and O. Wada, "The electromagnetic compatibility of integrated circuits; past, present and future", Electromagnetic Compatibility IEEE Transactions on, vol 51, pp. 78-100, Feb 2009.

[2] A. Blain, J. Raoult, A. Doridant, S. Jarrix, T. Dubois, "Effects of CW interference on a 5 GHz monolithic VCO", *8th IEEE International Workshop on Electromagnetic Compatibility of Integrated Circuits*, Dubrovnik, Croatia, 6-9 November 2011.

[3] A.A. Abidi, "Analog circuit design-RF analog-to-digital converters; Sensor and actuator interfaces; Low-noise oscillators, PLLs and synthesizers", Kluwers Academic Publishers, Boston, Nov 1997, 428 pp., ISBN 0-7923-9968-4.

[4] B. Razavi, "A study of Injection Locking and Pulling in Oscillators", IEEE Journal of solid-state circuits, vol. 39, n°9, Sept 2004.

[5] R. Adler, "A study of Locking phenomena in Oscillators", Proceedings of the IRE, vol 34, pp 351-357, June 1946.

[6] K. Kurokawa, "Injection Locking of Microwave Solid State Oscillators", Proceedings of the IEEE, vol. 61, n° 10, October 1973.

[7] J. Dominguez, A. Suarez, S. Sancho, « Nonlinear analysis of phase noise in microwave oscillators using envelope-transient technique », Integrated nonlinear Microwave and Millimeter-Wave circuits, INMIC,2008, Workshop.

[8] A. Hajimiri, T.H. Lee, "Oscillator phase noise: a tutorial"., IEEE Journal of Solid State Circuits, vol 35, pp. 469-482, 1995.

[9] E. Hegazi, H. Sjöland, A. Abidi, « A filtering technique to lower LC oscillator phase noise », IEEE Journal of Solid-State Circuits, vol. 36, n°12, Déc 2001.

[10] C. Samori, A.L. Lacaita, A. Zanchi, S. Levantino, G. Cali, « Phase noise degradation at high amplitudes in LC-tuned VCO's, IEEE Journal of Solid-State Circuits, vol. 35, n° 1, Jan 2000.

EMC Analysis of Current Source Gate Drivers

Alexis Schindler
Robert Bosch Center for
Power Electronics
Reutlingen University, Germany
Alexis.Schindler@Reutlingen-University.de

Benno Koeppl
Infineon Technologies AG
Am Campeon 1-12
Neubiberg, Germany

Bernhard Wicht
Robert Bosch Center for
Power Electronics
Reutlingen University, Germany
Bernhard.Wicht@Reutlingen-University.de

Abstract—**The prevention of electromagnetic emissions (EME) in bridge applications is of major importance, especially in automotive applications. Current source gate drivers are a popular approach to improve the performance in electromagnetic compatibility (EMC). Despites, their EMC performance has rarely been investigated in detail. Starting with the differences in the gate drive transients versus a conventional hard switching driver this paper presents a study regarding the EMC behavior. Different application cases are considered: Switching time, direction of load current, different load current levels. Experimental evaluation boards were prepared for both driver methods, results are provided for various cases. In order to identify the potential of current source gate drivers, the influence of the dv/dt and di/dt at bridge output were studied based on simulation. The di/dt turned out to be more important, a 5dBµV improved EMC was achieved by extending the duration of the current slope from 82ns to 92ns. The simulation setup also allowed to analyze the turn-on and turn-off transitions separately regarding EMC. Mainly due to the reverse recovery the turn-on transition gives 10dBµV worse EMC noise compared to turn-off. As an overall result, EMC improvements due to current source gate drivers in general are very case dependent.**

I. INTRODUCTION

Electromagnetic emission (EME) is a major concern in many applications and especially in automotive. As more DC and brushless DC motors are introduced in cars it becomes increasingly challenging for the development engineer to suppress electromagnetic emissions. Most electric motors are driven in a pulse width modulation environment with a switching frequency in the range of 15kHz to 30kHz. Lower frequencies can be audible to the human ear and are therefore usually not acceptable. Higher frequency ranges are limited by the increasing switching losses and radiated noise. To filter the noise that is generated by the PWM, large filter elements have to be implemented in the control unit. This causes additional cost and reliability issues. On the output of the bridge snubber filter elements can be used, but they are limited to guarantee the driving characteristics of the bridge. Therefore there is a high demand for gate drive techniques to improve electromagnetic compatibility (EMC) at the bridge output. Current source gate drivers are a popular approach in this regard because they provide a more constant gate drive current, Fig. 1 [1] [2]. In contrast, conventional gate drivers consist of just switches instead of the current sources shown in Fig.1 [3]. Such a driver is depicted in Fig. 2 along with the bridge circuitry.

The outline of this paper is as follows. In Section II the general bridge drive circuit is discussed. Section III focuses on the theoretical background of EMC in a bridge configuration motor drive and the expected behavior of the two gate driver topologies. The simulation of the di/dt and dv/dt as well as the reverse recovery influences on the EMC spectrum are examined in Section IV. Section V presents EMC measurement results for different application cases.

II. BRIDGE CIRCUIT

Fig. 2 provides a system overview of the bridge circuit. The bridge consists of the power transistors T1 as the high side switch and T2 as the low side switch. L_{Motor} represents the inductance of the motor windings. To set the load current I_{Load} to the desired level the high side or low side transistor has to be switched by a PWM signal. The high side transistor is the main switching device. Once the high side is switched off and after a certain dead time, the low side transistor is turned on. While both devices are off the load current will flow through the body diode of the low side transistor. As soon as the high side is turned on again, the charges in the drift zone of the body diode have to be removed. This effect is known as reverse recovery. It causes an additional current I_{RR} that flows through the high side switch at the turn on transition.

The gate driver controls the gate to source voltage of the transistors. The output stage of the conventional gate driver in Fig. 2 is implemented as a switch and it always provides the maximum current that it was designed for. In order to slow down the switching transition, a gate resistor has to be inserted in the gate path. This results in a current progression that is mainly influenced by the characteristics of the power MOSFET.

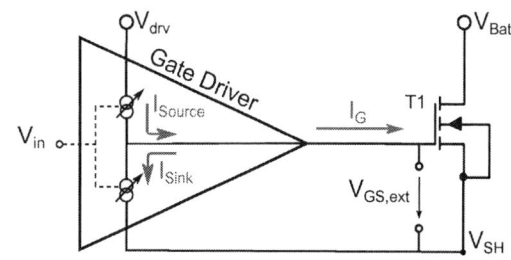

978-1-4799-5004-1/13 $31.00 © 2013 IEEE

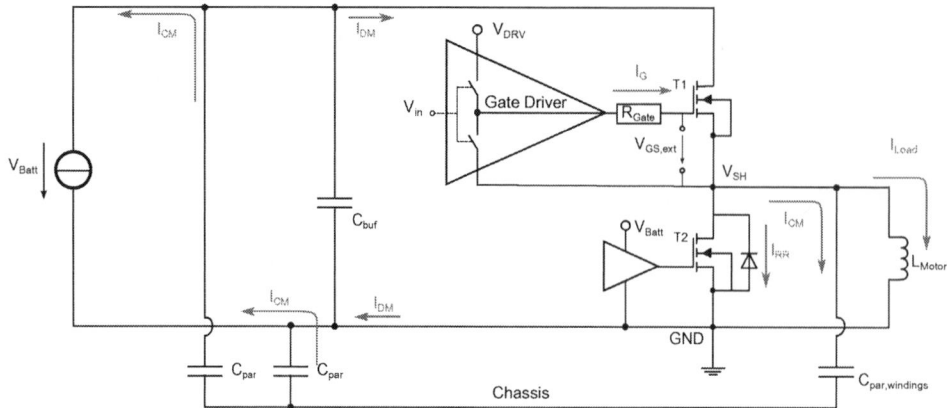

Figure 2: Half bridge configuration with gate driver and EMC coupling paths

The current source driver of Fig. 1 offers various advantages over this conventional approach. A current source gate driver does not need a gate resistor as the current is set by the current source in its output stage. The gate driver can also set the charging and discharging currents I_{Source} and I_{Sink} separately and thereby control the rising and falling edges in a detached manner. This configurable driver strength allows a change of the gate current while the application is in place. This would be much more difficult in a conventional driver with gate resistor. Additionally, with the current sources in the output stage the gate driver should be less susceptible to the impedance changes of the MOSFET.

III. THEORETICAL BACKGROUND

A. EMC in Half Bridge Circuits

To analyze the EMC of the bridge circuit it is important to understand the different coupling paths of the circuit. In EMC analyses it is usually differentiated between common mode noise and differential mode noise. In Fig. 2 both coupling paths are marked. The common mode noise is mostly generated by high frequency voltages that are transfered to the automobile chassis by the parasitic capacitance $C_{par,windings}$ between the motor windings and the case of the motor. This noise is then injected into the supply lines by the parasitic capacitance C_{par} between the supply line and the chassis. The differential mode noise is mostly accounted for by the reverse recovery of the body diode and the fast changing load current in the commutation path. They are partly filtered by the buffer capacitor C_{buf} between the supply lines.

In theory, the output signal of a half bridge circuit can be described by a trapeze (Fig. 3(a)) [4]. For simplicity it will be assumed that the rise time τ_r and the fall time τ_f are equal. With this simplification the fourier transform of the output signal can be calculated and the amplitude density in the frequency domain is gained.

$$E_{dB} = 20 \log \left(2A \frac{\tau}{T} \right) + 20 \log \left| \frac{\sin(\pi \tau f)}{\pi \tau f} \right|$$
$$+ 20 \log \left| \frac{\sin(\pi \tau_r f)}{\pi \tau_r f} \right| \tag{1}$$

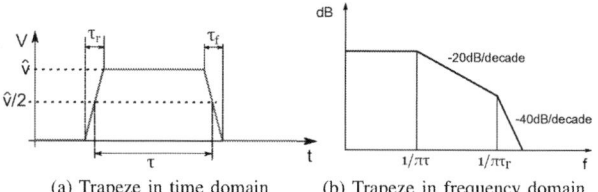

(a) Trapeze in time domain   (b) Trapeze in frequency domain

Figure 3: EMC behavior of a trapeze signal

Equation (1) can be broken down into three terms. In the first section a constant level of emission is calculated. The second and third part each give a -20dB decrease of amplitude beginning from their corner frequency. This information provides the typical amplitude spectrum of a trapeze signal which is shown in Fig. 3(b).

The real emission spectrum of the bridge circuit will differ from this ideal spectrum because the switching transition contains a di/dt, additional to the dv/dt. The di/dt influences the EMC spectrum apart from the ideal behavior in Fig. 3 by generating dv/dt contributions from:

1) The reverse recovery of the body diode
2) Parasitic inductances in the current path

The reverse recovery of the body diode causes a current through the bridge, additional to the load current. This causes a overshoot in the drain current and influences the EMC by generating a di/dt at the parasitic inductances of the commutation path. That in turn causes a voltage overshoot at turn-on.

Parasitic inductances are always present in the external transistor and the PCB traces as indicated by Fig. 4. When the load current commutates from one switch to the other it causes a high di/dt in the commutation path. This causes a voltage change $V = L_{par} \cdot di/dt$.

According to the strict automotive EMC requirements the maximum allowed electromagnetic emission is limited to specific levels. Some important measurement methods for the EMC characterization of electronic control modules and

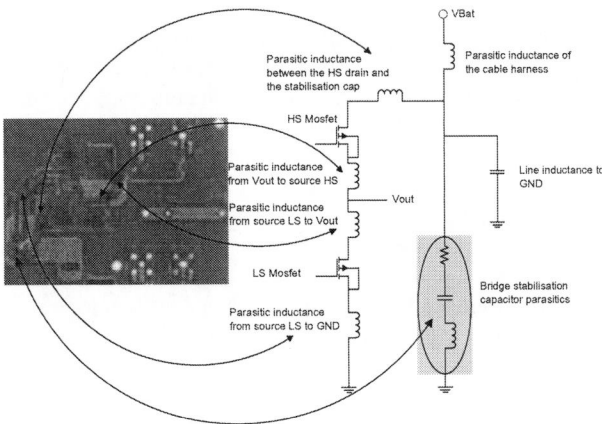

Figure 4: Parasitic elements of a half-bridge circuit on a PCB

systems are e.g. described in the CISPR25 standard [5]. One of the commonly used methods to measure the conducted emission from components/modules, the so called voltage method, is often used to characterize the emission on the global supply lines such as the battery connection. For EMC characterization of integrated circuits, i.e. the gate driver itself, the BISS (Bosch Infineon Siemens Specification) [6] is commonly used in the automotive industry. In this paper the focus is on the conducted emissions that is present at the output of the bridge and is measured according to the BISS specifications. The measurement techniques described in the BISS specification are especially useful to compare the EMC performance of different ICs e.g. up to 1GHz whereas the CISPR25 voltage method is only specified up to 108MHz and is more suitable to characterize whole modules especially regarding the influence in the differences in the layout of the PCB.

B. Expected Gate Driver Characteristics

The difference between the driving current of a current source gate driver and that of a conventional gate driver is explained in Fig. 5. By comparing the output currents over the output voltage of the gate driver, it can be observed that the driving current of the current source gate driver is constant, while the driving current of the conventional gate driver is dependent on the V_{GS} of the FET and on the external components. In the first graph of Fig. 5, the expected output characteristics of the two gate driving concepts are shown. The expression $V_{DRV} - V_{Gate,ext}$ represents the drain to source voltage of the internal driver transistor of the gate driver including the voltage across the gate resistor. The drain to source voltage of the gate driver can not be used here because the gate resistor has to be considered to get the actual gate current characteristics. According to Fig. 5 (a), the current source gate driver enters the saturation region much earlier than the conventional gate driver. Because of that the current source gate driver will provide a constant current for much longer than the conventional gate driver.

The second graph displays $V_{GS,ext}$, $V_{DS,ext}$ and $I_{Drain,ext}$

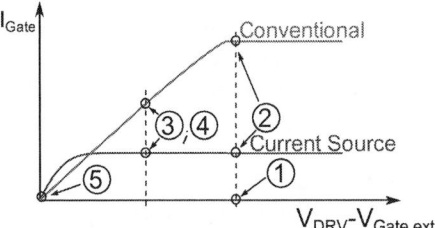

(a) Gate current characteristic of both gate driver topologies

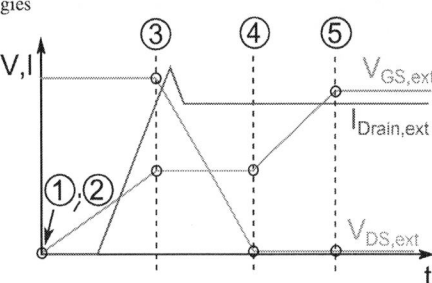

(b) Drain-source voltage, drain current and gate-source voltage of the external MOSFET while switching

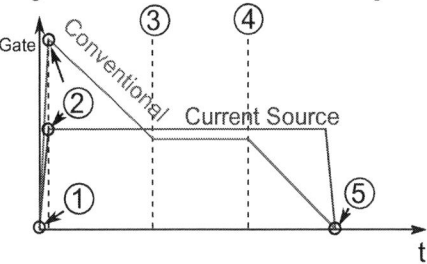

(c) Gate current progression while switching

Figure 5: Comparison of expected gate current waveforms for a current source gate driver and a conventional gate driver.

of the external MOSFET in a turn-on switching transition. At the beginning ① of the transition the external MOSFET is in the off state. As the gate driver is activated at ② both drivers jump into their respective operating point. Then V_{GS} begins to rise until V_{th} is reached and the drain current begins to flow. With the beginning of the miller plateau at ③, the maximum drain current is reached and the reverse recovery of the body diode can be observed. This current transition causes the high di/dt in the conduction path as described in Section II. In the miller plateau at ③ - ④, the voltage transition takes place.

The third graph is obtained by observing the output currents of the gate driver in Fig. 5 (a) at the five switching stages. With the much higher driving capability of the conventional gate driver, the current will peak at ② and subsequently decrease until the miller plateau is reached at ③. The current source gate driver, on the other hand, will stay in the saturation region for much longer and thus produce a constant current. While the miller capacitor is charged, the $V_{GS,ext}$ will remain constant and the driving current will also keep a constant level. As the end of the miller plateau is reached at ④, the driving current of the conventional gate driver will keep falling at a constant rate until it reaches zero. The current source driver will provide a constant current until it hits its triode region and falls off to zero very fast. At ⑤ the switching transition is finished.

By eliminating the initial peak in the gate current, the current source gate driver is expected to suppress a high current transition at the beginning of the switching event while still keeping the losses at a moderate level. Limiting the di/dt is expected to result in better EMC.

If gate resistors are implemented to reduce the EME of a conventional gate driver they have to limit the maximum gate current to the same level as the constant gate current of the current source gate driver (see Fig. 5 (c)). This will increase the switching losses because the gate current will be lower than the constant current of the current source gate driver after the initial peak.

IV. SIMULATION

In order to identify the potentials of current source gate drivers, different effects that influence the EMC of the half bridge circuit were investigated with circuit simulator spectre. A simulation model including parasitics similar to Fig. 4 and a low level MOSFET model provided by the vendor was created and verified, Fig.6.

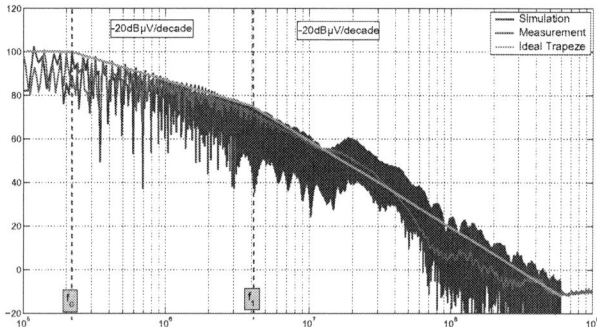

Figure 6: Verification of the EMC simulation model

To keep a strong correlation to the actual measurements, the simulation was set up to obtain the conducted emissions at the bridge output with a 150Ohm coupling network according to BISS [6] [7]. The 150 Ohm method is designed for probing

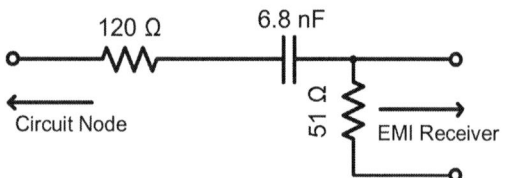

Figure 7: Matching network for the 150 Ohm measuring method according to IEC 61967-4 [7]

the conducted EMC behavior of integrated circuits at the board level. It can be applied on every pin of an IC. This

measurement method uses the matching network in Fig. 7 as impedance transformer for the measurement equipment. The spectral emission was measured by computing a fast fourier transform within Cadence on the time domain voltage signal at the output terminal of the coupling network. The EMI receiver was not emulated because these simulations are done for the purpose of comparison and not for absolute values.

First the influence of the reverse recovery was analyzed, which is only present at turn-on. The turn-on and turn-off transients were simulated separately, Fig.8. The turn-on spectrum is about 10dBμV higher than at the turn-off transition. By eliminating the reverse recovery the overall spectrum could be improved by this factor in the respective frequency range.

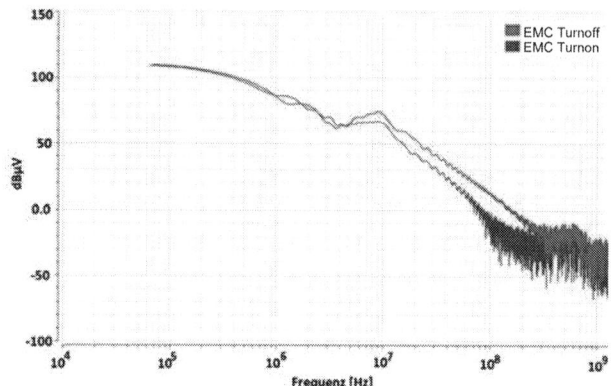

Figure 8: Comparison of the EMC at turn-on and turn-off at the V_{SH} node

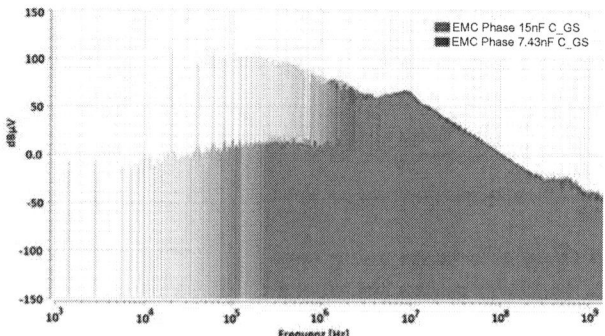

Figure 9: EMC influence of the di/dt at the V_{SH} node

The influences of the di/dt and the dv/dt were investigated by changing the capacitance of the gate-source capacitor C_{GS} and the gate-drain capacitor C_{GD}. By increasing C_{GD} the miller plateau becomes longer and the V_{DS} transition is slowed down. The same applies to the current transition if C_{GS} is increased.

The simulation in Fig. 9 shows a reduction in EME of 5dBμV in the case that the current slope is slowed down by 10ns. The slower voltage slope caused a 1-2dBμV reduction in EME. Overall the simulation promises a better EMC performance if the current slope is slowed down. The current slope is often much faster than the voltage slope and can be slowed down without much increase in the switching losses which will result in a better EMC behavior.

V. EXPERIMENTAL RESULTS

To compare both gate driver topologies, EMC measurements were performed. The gate currents were also measured and compared as these are the main difference in both topologies. In order to prevent the introduction of PCB-side influences on the measurement, both boards were layouted very carefully to obtain the same parasitic elements in the relevant traces. The layout of the half bridge with the important parasitic inductances is similar on both boards. To measure the gate currents, the gate resistor was used as a shunt and measured with an oscilloscope. To set the slopes of the conventional gate driver, different gate resistors were used. The current source gate driver can be configured with different gate currents.

A. Conventional Gate Driver

Comparing the measured gate current in Fig. 10 with the expected current in Fig. 5 good matching can be observed, especially for the turn-off transition. The small glitch at the end of the miller plateau is a result of the parasitic inductance introduced by the bond wires and the printed circuit board. The turn on transition is distorted by the fast transition of the drain to source voltage of the power MOSFET at about 1μs.

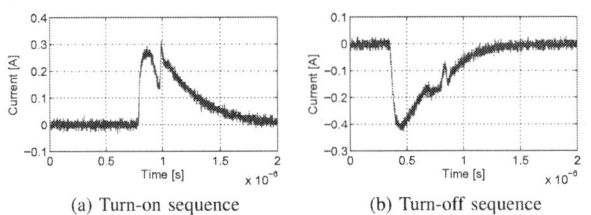

(a) Turn-on sequence   (b) Turn-off sequence

Figure 10: Measured gate currents of the conventional driver

B. Current Source Gate Driver

For the current source gate driver, a 2.2Ω gate resistor was inserted in the gate path to measure the gate current. Measured results are shown in Fig. 11. The gate current at the turn off sequence matches almost exactly the expected behavior like in Fig. 5 (c). At turn on, the gate current is more different from constant current source behavior of Fig. 5.

C. Comparison of EMC behavior

Now the EMC behavior of the bridge circuit with the different gate driver topologies is investigated. The conducted electromagnetic emission of the half bridge circuit was measured for both gate driver topologies with the parameter sets listed in Table I. This way, different application cases are covered by varying switching time, direction of load current and load current levels. The amplitude spectrum was measured with the 150Ohm measurement method (see Sec. IV Fig. 7) according to BISS [6] [7] with a peak detector.

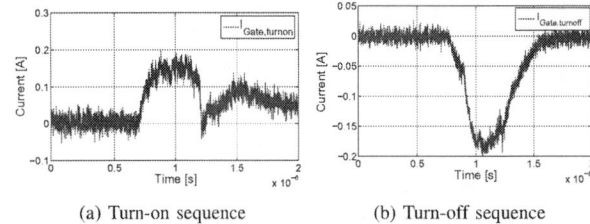

(a) Turn-on sequence   (b) Turn-off sequence

Figure 11: Current source gate driver gate currents at turn-on and turn-off transition

Transition Time	Load Current Direction	Load Current
Transition Time Variation		
100ns	VBat to GND	10A
500ns	VBat to GND	10A
Load Current Direction Variation		
500ns	GND to VBat	10A
500ns	VBat to GND	10A
Current Strength Variation		
500ns	VBat to GND	10A
500ns	VBat to GND	5A
500ns	VBat to GND	2.5A

Table I: Measurement cases for the evaluation of the EMC influences of the gate drivers

1) Transition Time Variation: The emission spectrum of both half bridge circuits is almost the same up to 100MHz at a rise/fall time of 100ns, Fig. 12. The conventional gate driver is showing a slightly better behavior in the range of 9MHz to 40MHz with up to 3dBμV less emissions. For 500ns transition time the emissions are reduced for both drivers but the overall behavior is similar to the 100ns case.

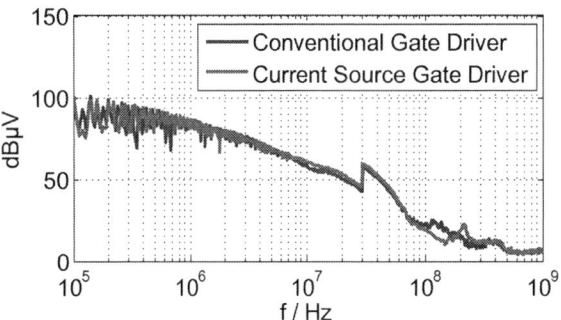

Figure 12: Comparison by EMC spectra of current source gate driver and conventional gate driver at 100ns transition time

With those measurements it is concluded that a improvement in EMC behavior is not visible. The electromagnetic emission is even slightly higher with the current source gate driver than with the conventional design. The small deviation of 3 - 4 dBμV in the range of 8MHz - 50MHz is the result of the different rising and falling slopes of $V_{DS,ext}$ with the conventional gate driver. The slopes at the external MOSFET are defined by the gate resistor R_{Gate} with this gate driver. As in most designs, the turn-off MOSFET of the gate

978-1-4799-5004-1/13 $31.00 © 2013 IEEE

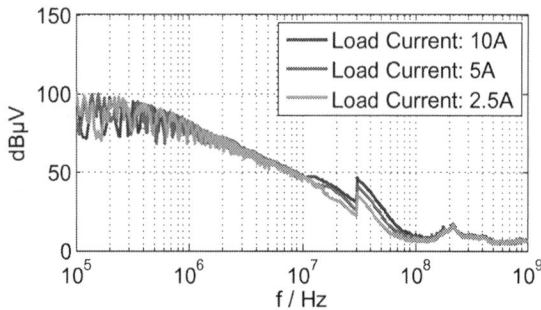

Figure 13: Comparison of EMC spectra of the current source gate driver with different load currents

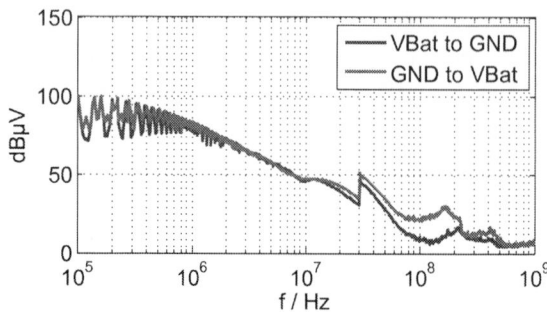

Figure 14: Comparison of the current source gate driver EMC spectrum with different load current directions

driver is stronger than the turn-on MOSFET and will provide a faster slope. Because of that, a unsymmetrical trapeze is produced and the emission spectrum changes. The slopes of the current source gate driver can be controlled separately and are configured to produce a symmetrical V_{DS} of the external MOSFET for the turn-on and turn-off slopes.

2) Load Current Variation: The variation in the load current has a similar effect on the conducted electromagnetic compatibility in both topologies. In Fig. 13 the EMC spectrum of the current source gate driver is shown as an example for both topologies. The load current affects the EMC in the area between 10MHz and 100MHz. As both chips show similar behavior it can be concluded that the gate driver topology has no influence on the EMC with respect to the magnitude of the load current.

3) Load Current Direction Variation: By changing the active switch from the high side to the low side MOSFET the commutation path of the load current is changed. In Fig. 14 the EMC behavior of the current source gate driver with different load currents is presented.

The two functions show differences mostly in the high frequency range between 15MHz and 200MHz. At 150MHz the biggest deviation of 16dBμV is measured. The main cause for these differences is the new commutation path of the load current which introduces different parasitic inductances to the circuit. From this measurement it can be deducted that the parasitic inductance has an impact on the high frequency EMC performance. It can also be used to improve the board layout symmetry. The load current direction shows almost the same impact on the EMC spectrum in the range of 10MHz to 100MHz as the load current variation measurement (Fig. 13) but extends to even higher frequencies.

VI. CONCLUSION

A current source gate driver and a conventional gate driver were investigated regarding their influence on EMC of half bridge circuits. In theory, the EMC noise at the bridge output can be approximated by the fourier transform of a trapeze signal. Additional EMC influences include body diode reverse recovery, parasitic inductances and also the fact that the switching transition consists of a current and a voltage signal transition.

The different output stages of both gate drivers provide different gate current waveforms and influence the switch in different ways. Both concepts were measured in an EMC test cabin to analyze the EMC performance. Different effects on EMC performance of the bridge circuit were investigated by simulation and measurement. Eliminating the reverse recovery of the body diode indicates a EMC improvement up to 10dBμV. In simulation the di/dt turned out to be of higher influence on the EMC than the dv/dt. The EMC of the circuit could be improved by 5dBμV with a just 10ns slower rise and 17ns slower fall time of the current. The voltage transition was slowed down to a rise time of 139ns and a fall time of 112ns and showed only an improvement of 1-2dBμV. The measurements with different load currents showed a reduction of 5dBμV between 10MHz and 100MHz. Changing the load current direction indicated the influence of the different parasitic inductances in the commutation path.

The current source gate driver and the conventional gate driver showed very similar behavior. This leads to the conclusion that simply changing the output stage of a gate driver to a current source will not necessarily improve the EMC of a half bridge. Nevertheless, the current source concept offers improvements in flexibility and is thus a very viable option for newly developed gate drivers.

REFERENCES

[1] Dinu, D.; Auer, B., *Design considerations for improved electro magnetic compatibility (EMC) behavior in high side switches design*, Semiconductor Conference, 2009. CAS 2009. International , vol.2, no., pp.417,420, 12-14 Oct. 2009

[2] Zhiliang Zhang; Eberle, W.; Yan-Fei Liu; Sen, P.C., *A new current-source gate driver for a buck voltage regulator*, Applied Power Electronics Conference and Exposition, 2008. APEC 2008.

[3] L. Balogh: *Design And Application Guide For High Speed MOSFET Gate Drive Circuits*, Texas Instruments, App. Note, 2001

[4] B. Deutschmann, R. Illing, B. Auer, *Edge Shaping to Reduce the Electromagnetic Emissions*, Proc. of the 10th Int. Symposium on Electromagnetic Compatibility (EMC Europe 2011), York, UK, September 26-30, 2011, pp. 742 - 745

[5] IEC: *Radio Disturbance Characteristics Limits and Methods of Measurement*, Publication 22, IEC International Special Committee on Radio Interference (CISPR), 1997

[6] M. Joester, F. Klotz, W. Pfaff, T. Steinecke: *Generic IC EMC Test Specification*, January 2010, ZVEI

[7] IEC 61967-4, *Integrated Circuits, Measurement of Electromagnetic Emissions, 150 kHz to 1 GHz Measurement of Conducted Emission, 1 Ohm/ISO Ohm Method*, 47A/606/CDV, 2001-04-13

EMI Resisting LDO Voltage Regulator with Integrated Current Monitor

Philipp Schröter, Stefan Jahn, Frank Klotz

Infineon Technologies AG
Automotive Power EMC Center
Am Campeon 1-12, 85579 Neubiberg
philipp.schroeter@infineon.com

Fabio Ballarin, Fabio Gini, Marco Piselli

Infineon Technologies Italia
Automotive Standard VREG
Via N. Tommaseo 65/B, 35131 Padova
fabio.ballarin@infineon.com

Abstract—**This paper introduces a low dropout (LDO) voltage regulator with an integrated current monitor, which sets a new performance benchmark in terms of RF immunity. Compared to classic topologies the proposed circuit succeeds in a very high robustness against RF disturbances which are superimposed into the regulator's battery voltage (VS) pin and output (OUT) pin. The proposed circuit was designed using a HV-BiCMOS technology for automotive applications. Circuit theory is explained in detail and the proposed approach is confirmed by RF immunity measurements.**

Keywords: analog integrated circuits, electromagnetic interferences (EMIs), RF-immunity

I. INTRODUCTION

Low dropout voltage regulators are widely used in automotive systems like in a dashboard and in a light control to provide microcontrollers, FPGAs, DSPs, sensors, active antennas with a stable and reliable voltage. A linear voltage regulator only needs few external components to generate a high precision output voltage from any voltage that enters the battery voltage pin.

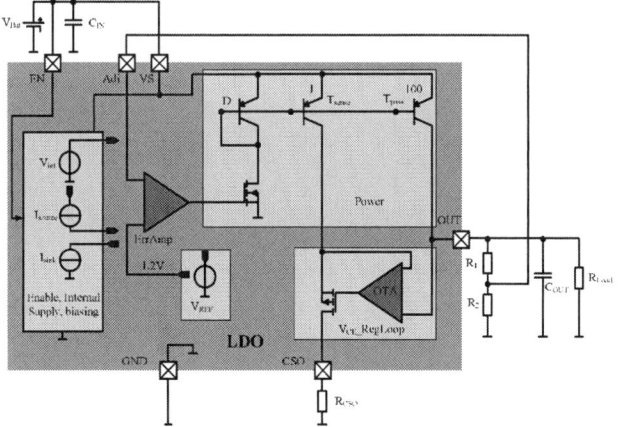

Figure 1.   LDO voltage regulator with an integrated current monitor

Figure 1. depicts the basic principle of a LDO with an integrated current monitor. The integrated circuit consists of the regulator's main loop, the current monitor, an internal supply, the biasing and the enable (EN) circuit. The PNP pass device (T_{pass}) is connected in between the VS and OUT pins

and delivers power to the load. A linear voltage regulator uses negative feedback to control the output voltage. A reference voltage (V_{REF}) and a feedback voltage coming from the feedback network (resistive divider R_1, R_2) are connected to the input of an error amplifier (ErrAmp). The error amplifier then forces its inputs to be equal and thereby drives the PNP pass device via additional buffer and driver circuits and adjusts the LDO's output voltage. The output capacitor (C_{OUT}) is part of the controlled system and is required for stability. The current flowing through the CSO pin is mirrored to be exactly 1/100th of the LDO's load current. A resistor connected to the CSO pin converts the current sourced by the CSO pin into a voltage which is proportional to the regulators output current. Internal supply and biasing provide bias signals to the regulator's basic building blocks. Switching EN to GND turns the output off and puts the LDO into a low power state.

In automotive applications global IC pins [1] of LDOs are directly connected to the automotive environment:

- VS → The battery pin is connected to the cable harness.

- OUT → The output can be considered a global pin when the voltage regulator supplies a load which is not on the application board (e.g. supplying sensors and active antennas)

Wires connected to global pins act as antennas since they receive RF disturbances from a severe electromagnetic environment. RF disturbances present on global nets of a classic linear voltage regulator cause circuit functional failures, such as generating erratic output signals and a complete breakdown of the regulator, when the level of RF disturbance magnitudes is increased. Reasons for circuit functional failures are malfunctions of basic building blocks due to induced offsets caused by nonlinear effects [2]. To make sure electromagnetic incompatibilities do not cause car accidents a high RF immunity is a key criterion of automotive LDOs in addition to conventional requirements such as low quiescent current, low noise, fast line and load response.

II. EMI RESISTING OVERALL CIRCUIT DESIGN

From an early development stage dedicated measures in the design and layout have been applied to develop a circuit with an increased RF immunity at the ICs global pins.

978-1-4799-5004-1/13 $31.00 © 2013 IEEE

A. Technology

The investigations have been done using a HV- BiCMOS technology for automotive applications. Low and high voltage bipolar and CMOS devices can be integrated on a single chip. Different devices are isolated by partial dielectric isolation. A reverse biased pn-junction ensures vertical isolation to the substrate. Deep trenches provide lateral isolation.

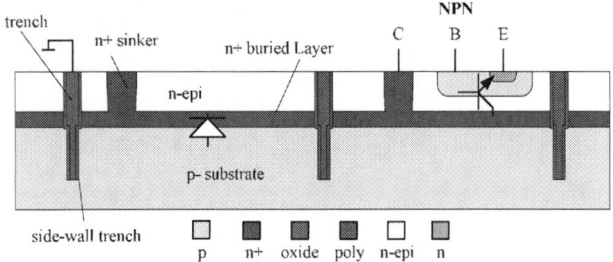

Figure 2.   Technology with partial dielectric isolation

The technology uses a highly doped n+ buried layer which can be a functional terminal of devices. These terminals are only isolated by pn-junctions to the p-substrate. Reverse biased pn-junctions are voltage dependent parasitic capacitances since they can inject RF disturbances into the substrate and sense them respectively. A proper method to minimize RF substrate couplings is to place guarding structures (guard rings) around affected devices. These guard rings will create low impedant paths to GND for RF disturbances which are present in the substrate. Guarding structures can be realized by connecting trenches of injecting and sensing devices to GND [3].

B. Design of Sub-Circuits

1) Internal Supply and Biasing

The internal supply circuit generates a pre-stabilized voltage (V_{int}) from the battery voltage which can vary over a wide range (5.5 V… 45 V). V_{int} supplies the regulators sub-circuits such as the bandgap and the error amplifier. Requirements concerning this circuit for a RF immune LDO design are:

- No DC shift of V_{int} with RF disturbances present at VS

- High PSRR between VS and V_{int}

Both requirements ensure a proper function of the regulator's sub- circuits during RF injection into VS. The proposed circuit is depicted in Figure 3. (a). A current source connected to VS creates a reference current flowing through a zener diode. The zener voltage is clamped to the gate of a source follower (M_1). V_{int} equals the zener voltage minus V_{GS} of M_1. D_1 protects the circuit against reverse polarity. This topology is quite immune against RF disturbances superimposed into VS. The mean value of V_{int} remains roughly the same as long as one of the following conditions is not fulfilled:

- I_{Ref} decreases significantly or

- M_1 leaves the saturation region due to very high RF disturbance magnitudes present at the battery voltage which cause strong nonlinear effects.

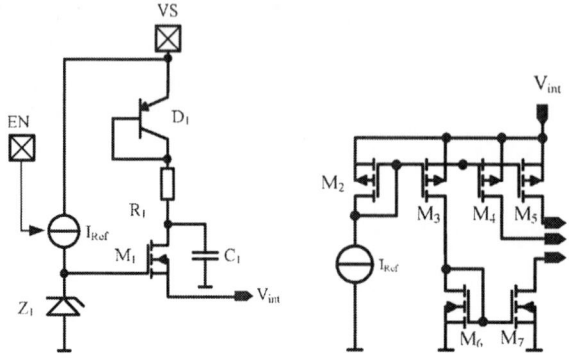

Figure 3.   Internal supply (a) and biasing (b)

A low pass filter (R_1, C_1) is added in the path between the drain of M_1 and D_1 in order to increase the PSRR between VS and V_{int}. The low current consumption of the regulator's sub-circuits which are supplied by the pre- stabilization circuit results in a voltage drop of only some millivolts across R_1. This very small voltage drop does not affect significantly the regulator's operating range. The LDO's biasing (Figure 3. (b)) is connected to the RF immune internal supply voltage V_{int}. All currents required by the regulator's sub-circuits are generated by the current sources and current sinks of the biasing.

2) Main Loop of RF Immune Topology

The schematic of the proposed main loop is depicted in Figure 4. A differential pair is used as an error amplifier. Two emitter followers are connected to the output of the error amplifier to ensure a low impedance at the input of the common source stage (gate of M_{Dr}) which drives the PNP pass device (T_{pass}). The reference voltage is generated by a standard Brokaw cell [4] which is supplied by the internal supply.

The nonlinear behavior of the error amplifier is a main cause for the EMI induced erratic output voltage of linear voltage regulators [2]. Therefore special care has been taken to decrease the amplifiers susceptibility to EMIs. Following measures have been implemented in the main loop to improve EMC performance of the LDO:

- Source degeneration resistors (R_5, R_6) linearize the differential pair [5] and reduce an EMI induced DC offset.

- The error amplifier and the bandgap are supplied internally. There are no direct connections between these sensitive sub- circuits and RF affected nets.

- A RC (C_1, R_3) low pass filter connects the adjust (Adj) pin and the input of the differential pair. Consequently RF disturbances propagating from the OUT via the adjust pin to the error amplifier's input are minimized. (C_1 is a low voltage capacitor. Consequently the additionally required area is low.)

- PMOS input devices are used for the differential pair. Since no DC gate current flows, resistors R3 and R4 do not cause a DC offset. Resistors are required for building low pass filters.

978-1-4799-5004-1/13 $31.00 © 2013 IEEE

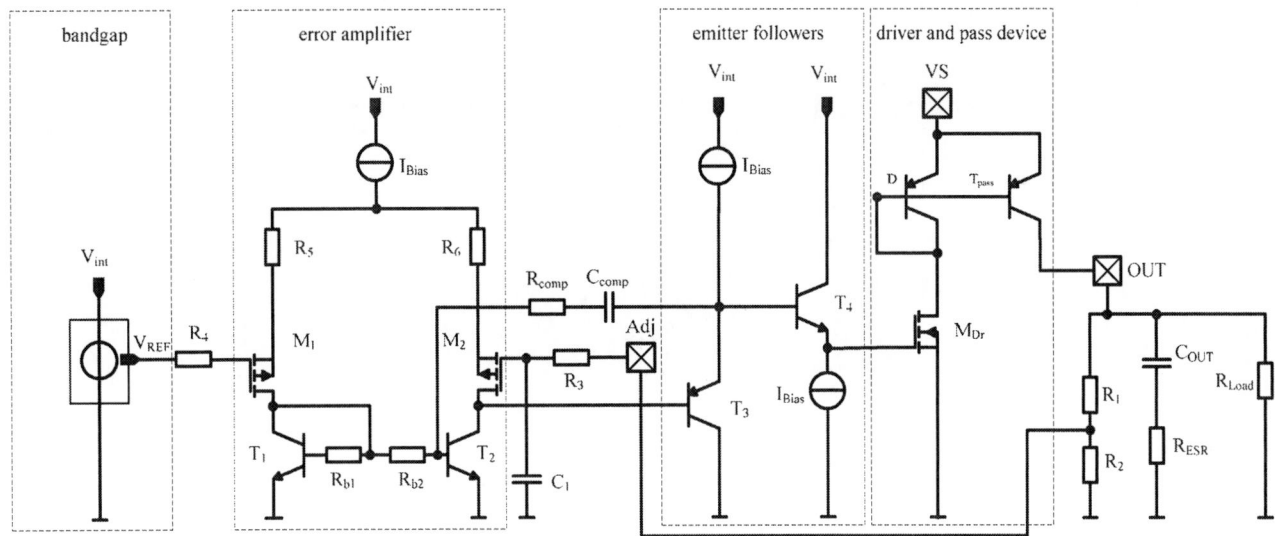

Figure 4. Main loop

The feedback signal experiences a shift in phase as it goes through the loop, since every pole shifts the phase by -90°. If not compensated by zeros, which add some phase lead, the output voltage can oscillate. The amount of shift determines whether the loop is stable or not. For stability the phase shift of the feedback signal must be less than -180° where the loop transmission (ßH(jω)) reaches unity gain (ω$_{0dB}$). H(jω) is equal to the frequency dependent open loop transfer function and ß equals the frequency independent feedback factor [6], [7], [8].

The proposed topology is compensated by the internal compensation network (R$_{comp}$, C$_{comp}$) and by the output capacitor (C$_{OUT}$). A capacitor connected to the LDO's output generates a load pole (p$_{OUT}$) and a load zero (z$_{ESR}$). Their locations depend on the load current, the capacitance value and the electro-static resistance (ESR) of C$_{OUT}$. All three parameters can vary over decades, while the regulator is supposed to function correctly. P$_{OUT}$ is a low frequency pole and z$_{ESR}$ is located in vicinity to the unity gain frequency. Two further poles and one zero occur before the loop transmission reaches zero dB. These are the pole and the zero generated by the error amplifier's output impedance and compensation network (p$_{OTA1}$, z$_{OTA}$) and the EMC pole (p$_{EMC}$). Equations (1) to (7) depict the respective calculations of poles and zeros which are relevant for the circuit's stability and the DC loop gain (A$_{VDC}$), where gm$_i$ equal small signal tranconductances and r$_{0i}$ output resistances of respective BJTs or MOSTs.

$$\omega_{p_{OTA1}} \approx \frac{1}{C_{eq} \cdot \left(R_{b2} + \frac{1}{gm_{T_1}} \right)} \tag{1}$$

$$C_{eq} \approx$$

$$C_{comp} \cdot \left(1 + gm_{T_2} r_{0T_2} || \left(r_{0M_2} \cdot (1 + gm_{M_2} R_6) \right) \right) \tag{2}$$

$$\omega_{z_{OTA}} \approx \frac{1}{C_{comp} \left(R_{comp} + R_{b2} + \frac{1}{gm_{T_1}} \right)} \tag{3}$$

$$\omega_{p_{OUT}} \approx \frac{1}{R_{Load} C_{OUT}} \tag{4}$$

$$\omega_{p_{EMC}} = \frac{1}{R_3 C_1} \tag{5}$$

$$\omega_{z_{ESR}} \approx \frac{1}{R_{ESR} C_{OUT}} \tag{6}$$

$$\beta A_{VDC} \approx \frac{r_{0T_2}}{\frac{1}{gm_{M_2}} + R_6} \cdot gm_{Dr} \frac{1}{gm_D} \cdot$$

$$gm_{T_{pass}} R_{Load} || r_{0T_{pass}} \cdot \frac{R_2}{R_2 + R_1} \tag{7}$$

Poles of the followers (p$_{followers}$), the second pole of the error amplifier (p$_{OTA2}$) and the pole of the driver (p$_{Dr}$) occur in most load/ output capacitor cases after the feedback signal reaches unity gain and then do not impact the circuit's stability. Figure 5. depicts the bode plot of the compensated loop when the output capacitor equals 1 μF and the load current is high. 1μF is in vicinity to the minimum required output capacitor, which ensures stability under all load conditions.

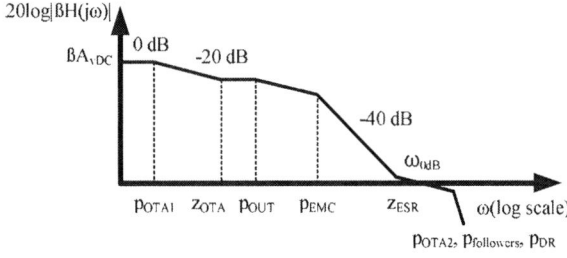

Figure 5. Frequency compensation of main loop

3) V_{CE} Regulation Loop

The current flowing through the regulator's pass device can be monitored at the CSO pin. Figure 1. illustrates the principal of the V$_{CE}$ regulation loop, Figure 6. depicts the circuit details.

978-1-4799-5004-1/13 $31.00 © 2013 IEEE

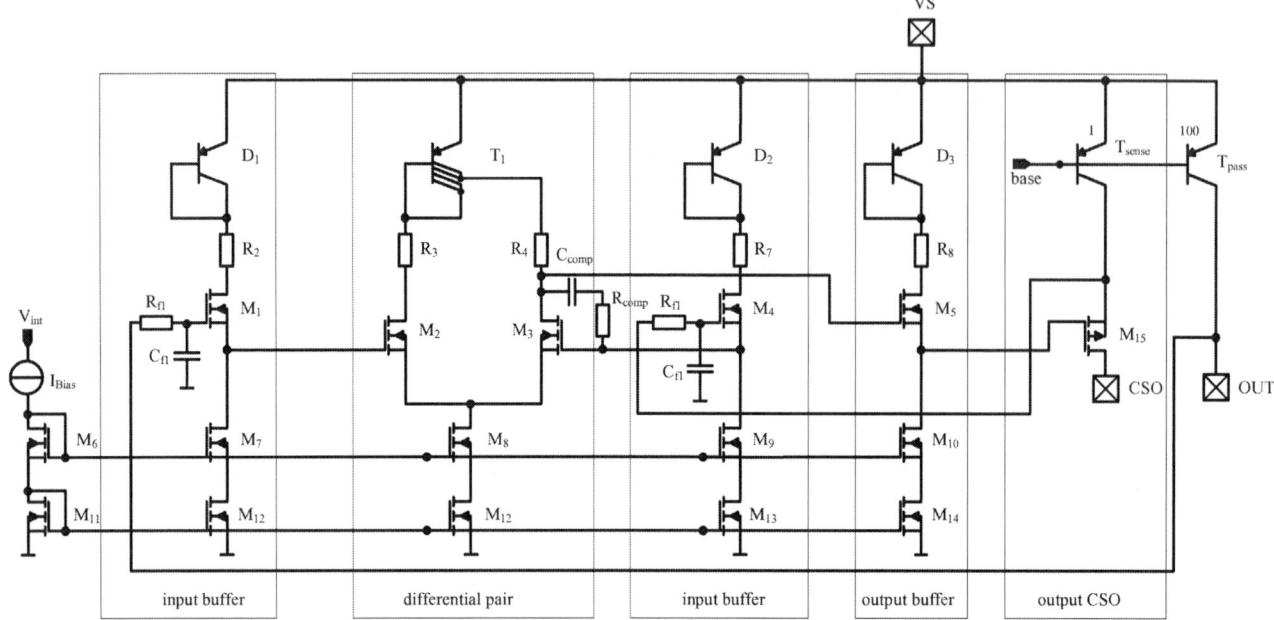

Figure 6. V_{CE} regulation loop

The ratio of the emitter areas T_{sense}/T_{pass} is 1/100. Consequently the current sourced by the collector of T_{sense} equals exactly 1/100th of the output current due to the same base-emitter and collector emitter voltages of sense and pass device.

The V_{CE} regulation loop ensures the collector voltage of T_{sense} is regulated to the collector voltage of T_{pass} in order to avoid a systematic error of the sensed current due to the Early voltage. The collector voltage of T_{sense} is monitored and applied to the input of an error amplifier which compares this sensed voltage with the collector voltage of T_{pass}. Due to the loop gain the error amplifier forces its inputs to be equal. Buffers are in front of the differential pairs' inputs to gain some voltage headroom. Additional headroom is required because of the unity gain configuration of the loop, when the output voltage is close to the input voltage. A buffer connected to the output of the differential input pair ensures a low impedance at the gate of M_{15} to speed up the loop due to the big size and the high current sourced by M_{15}.

The V_{CE} regulation loop is internally compensated using miller compensation. The circuit is unconditionally stable without placing external components. A first pole (p_{OTA1}), acting as the dominant pole, is generated from the error amplifier output impedance and the compensation network (C_{comp}, R_{comp}). A zero (z_{OTA}) located in vicinity to the unity gain frequency adds some phase margin. Equations (8) and (9) depict the calculated pole and the zero.

$$\omega_{p_{OTA1}} \approx$$

$$C_{comp} \cdot \left(R_{comp} + \left(r_{0T_1} || r_{0M_3} \right) \cdot \left(1 + \frac{gm_{M_3}}{gm_{M_4}} \right) \right) \quad (8)$$

$$\omega_{z_{OTA}} \approx \frac{1}{C_{comp} \cdot \left(\frac{1}{gm_{M_3}} - R_{comp} \right)} \quad (9)$$

The regulator's OUT pin is directly connected to the input of the V_{CE} regulation loop error amplifier, which makes the V_{CE} regulation loop especially sensitive to RF disturbances injected into the LDO's OUT pin. Low pass filtering (R_f, C_f) has been implemented in front of the inputs to increase the RF immunity. R_f and C_f add a pole in the loop:

$$\omega_{p_{EMC}} = \frac{1}{R_f C_f} \quad (10)$$

To ensure stability p_{EMC} occurs at a frequency where the loop transmission is much smaller than zero dB:

$$\omega_{p_{EMC}} > 10 \times \omega_{0dB} \quad (11)$$

Figure 7. depicts the bode plot of the compensated loop. The DC gain of the loop is given by equation (12).

$$\beta A_{v_{DC}} = gm_{M_3} \cdot r_{0M_3} || (R_4 + r_{0T_1}) \quad (12)$$

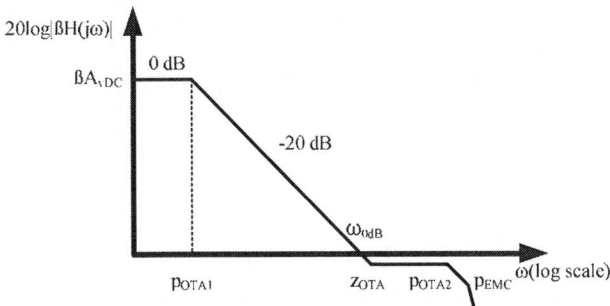

Figure 7. Frequency compensation of V_{CE} regulation loop

C. Floorplan and Layout

Floorplan and layout have a significant impact on the RF immunity of ICs, especially at frequencies higher than 200 MHz. The placement of the regulator's sub circuits has been carefully determined at an early development stage. The power stage devices such as T_{pass} (Figure 4.) can be considered as sources of interference since they inject RF disturbances into the substrate. To minimize RF substrate couplings the power devices are placed on the right side of the chip. A guarding structure "barrier" consisting of trenches is placed in between the power devices and the remaining circuitry. Analog sub circuits such as the bandgap and the error amplifier are positioned on the left side of the chip.

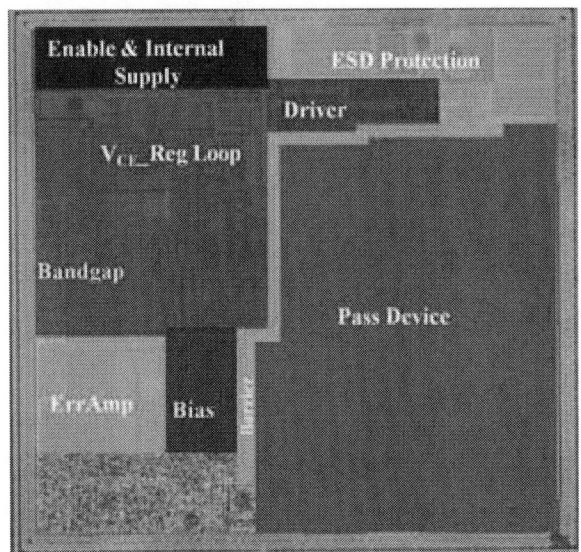

Figure 8. Microphotograph including floorplan

Distances between RF affected lines (lines connected to devices of the power stage) and lines connected to sensitive sub-circuits are determined in order to minimize coupling via lines. A microphotograph including the placement of the regulator's sub-circuits is depicted in Figure 8.

III. RF IMMUNITY MEASUREMENT RESULTS

The final circuit was packaged in a plastic package and the RF-immunity has been characterized by the direct power injection (DPI) method [9].

A. DPI Measurement Setup

Figure 9. illustrates the used DPI setup. The IC including the minimum required circuitry, artificial networks and coupling capacitors are placed on a dedicated EMC test board. RF disturbances can be injected into the VS- and into the OUT pin.

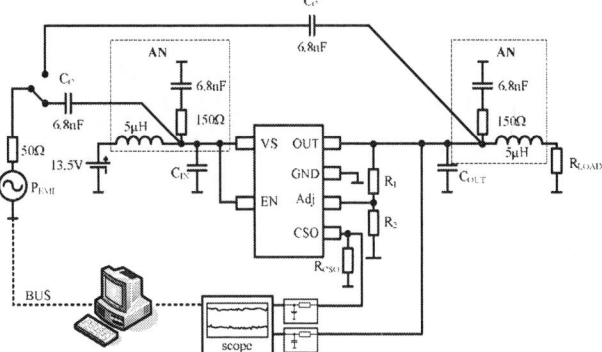

Figure 9. DPI measurement setup according to [9]

The setup consists of the following components:

- A blocking capacitor (C_{IN}) of 100 nF is connected between VS and GND. C_{IN} is part of a LDO application compensating line influences.

- An output capacitor (C_{OUT}) of 1 μF is placed at the output to ensure stability of the main loop.

- R_1 and R_2 adjust an output voltage of 5 V.

- A load resistance (R_{Load}) of 56 Ω was chosen resulting in an output current of approximately 90 mA. The chosen load current is compliant with the requirements stated in [1]: If applicable, 80% of the maximum output current is drawn, otherwise it will be ensured, that IC temperature is within normal operating ranges.

- A resistor (R_{CSO}) of 1.5 kΩ has been connected to the CSO pin. Consequently 1 V present at the CSO pin corresponds to a load current of 0.667 mA. The load current of 90 mA results in a DC voltage of 1.35 V at this pin.

- Artificial networks (AN) are placed at VS and OUT to decouple RF and DC signals and to ensure the typical common mode impedance of the cable harness.

- There are coupling capacitors (C_C) of 6.8 nF at the VS and the OUT pin to inject RF disturbances into the respective pins.

- During RF-immunity measurements the IC is supplied by 13.5 V via an artificial network, emulating the LDO is connected to the cable harness and supplied by the car battery.

978-1-4799-5004-1/13 $31.00 © 2013 IEEE   111

B. DPI Results

Continuous wave (CW) RF disturbance signals from 1 MHz up to 1 GHz are injected into VS and OUT via 6.8 nF coupling capacitors. The OUT signal is monitored within an acceptable tolerance of 5 V ± 100 mV (± 2 %) and the CSO signal within 1.35 V ± 70 mV (± 5 %). For each frequency step in a frequency range between 1 MHz and 1 GHz, the forward power is increased until one of the signals violates the defined tolerances or the maximum power of 37 dBm (5 W) is reached. A measurement result at a certain frequency equals:

- A point at 37 dBm, if no failure occurs at the maximum power level, or

- A point at the maximum power level before one of the monitored signals violates a specified tolerance.

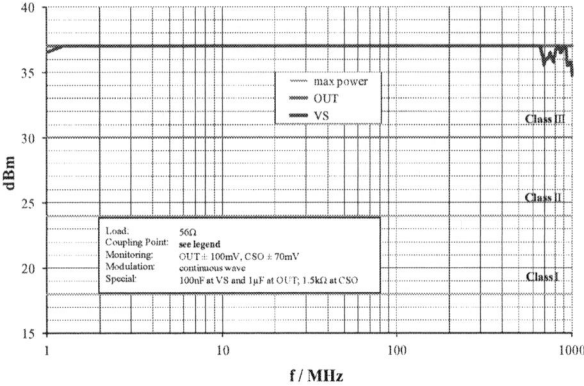

Figure 10. RF immunity results including requirements according to [1]

Figure 10. depicts the RF immunity results when injecting RF disturbances into the VS and into the OUT pins. RF disturbances up to 35 dBm at VS and 37 dBm at OUT forward power do not influence IC functionality in the checked tolerances. Failures occur when injecting RF disturbances into the VS pin at very high power levels. At 1 MHz a failure is caused by an increasing CSO signal. A decreasing output voltage violates the defined 2 % tolerance at frequencies higher than 650 MHz. In terms of RF immunity the developed IC fulfills automotive requirements with the standard circuitry used for LDOs. VS and OUT can be classified to Class III for global pins according to [1].

C_{IN} and C_{OUT} impact the IC's RF immunity, since they drain RF disturbances to ground. However, these components always belong to a LDO application. With increasing frequency the capacitors lose filter efficiency due to their inherent parasitic inductance (ESL). Maximum RF disturbance magnitudes about 8 V (peak-to-peak) are superimposed to the DC values of VS and OUT at frequencies higher than 500 MHz. These high disturbance magnitudes do not shift mean values of the LDO's OUT and CSO signals. Figure 11. shows the OUT and the CSO signals exemplarily during RF immunity measurements when a power of 37 dBm at a frequency of 250 MHz is injected into the output. Although the output signal is heavily disturbed by high RF disturbance magnitudes ($\approx$ 5 V ±

1.8V) the mean values of the OUT and CSO signals remain the same compared to undisturbed reference signals. Undisturbed reference signals are depicted in the background of Figure 11.

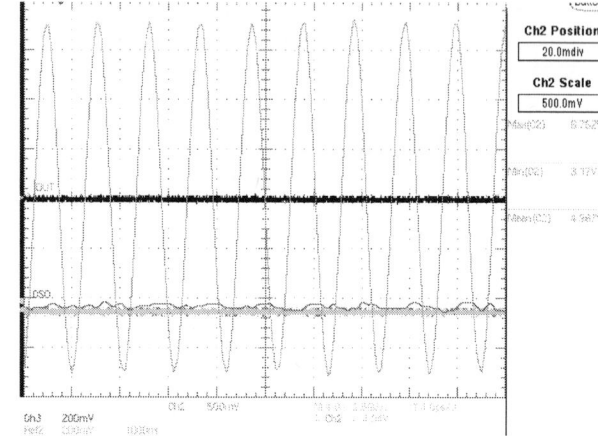

Figure 11. OUT and CSO signals during direct power injection into the OUT pin (f=250 MHz, P=37 dBm)

IV. CONCLUSION

An automotive LDO voltage regulator with integrated current monitor has been introduced fulfilling highest requirements in terms of RF immunity at the global pins with the minimum required circuitry. The EMC performance is a result of dedicated measures in the design, layout and floorplan.

REFERENCES

[1] Generic IC EMC Test Specification Version 1.2, Bosch, Infineon, Siemens VDO (BISS), 2004- 2007

[2] Fiori, P. S. Crovetti, "Linear voltage regulator susceptibility to conducted EMI", Proceedings of the IEEE International Symposium on Industrial Electronics, 2002, ISIE 2002

[3] P. Schröter, S. Jahn, F. Klotz, „Improving the Immunity of Automotive ICs by Controlling RF Substrate Coupling", EMC Compo 2011 8th International Workshop on Electromagnetic Compatibility , Dubrovnik 2011

[4] A. P. Brokaw, „A Simple Three-Terminal IC Bandgap Reference", in IEEE Journal of Solid State Circuits, S. 388-393, vol. 9, no. 6, Dezember 1974

[5] M. Corradin, G. Spiazzi, S. Buso, "Effects of Radio Frequency Interference in OPAMP Differential Input Stages", International Symposium on Electromagnetic Compatibility, EMC 2005, pp. 866-871, vol. 3, 2005

[6] B. Razavi, "Fundamentals of Microelectronic", John Wiley & Sons, 2008, ISBN 978-0-471-47846-1

[7] C. Simpson, "A users guide to compensating low-dropout regulators", National Semiconductors

[8] E. Rogers, "stability analysis of low-dropout linear regulators with a PMOS pass element", Texas Instruments Incorporated, 2005

[9] IEC 62132-4 Ed. 1 2005, "Integrated Circuits – Measurement of Electromagnetic Immunity – 150kHz to 1GHz – Part 4: Direct RF Power Injection Method"

A Study on gate voltage fluctuation of MOSFET induced by switching operation of adjacent MOSFET in high voltage power conversion circuit

Tsuyoshi Funaki (Osaka university)
Div. E. E. I. Eng.
2-1 Yamada-Oka, Suita, Osaka, 565-0871, Japan
funaki@eei.eng.osaka-u.ac.jp

Abstract—**The development of high voltage SiC power MOSFET has made the fast switching of high voltage possible. The high dv/dt caused by fast high voltage switching induces the difficulty of mal-operation of power MOSFET with the self turn-on phenomenon by the fluctuation of gate voltage. This phenomenon is recognized as intra EMC. This paper studies the mechanism of self turn-on phenomenon and discusses the suppression method.**

Keywords—self turn on; gate voltage fluctuation; power MOSFET; high voltage

I. INTRODUCTION

The bipolar power devices have been utilized in conventional high voltage power conversion circuit; such as Si PiN diode, Si IGBT, etc. The switching speed of bipolar type device is subject to the restriction of minority carrier injection, and fast switching cannot be achieved. The restrictions are known as tail current of IGBT and reverse recovery of PiN diode at the turn-off operation. The recent development of wide band gap semiconductor makes high voltage unipolar power device into reality; such as Schottky barrier diode (SBD) and MOSFET. The switching speed of unipolar type device can be made extremely fast, for it operates only with majority carrier and avoids the need for considering minority carrier behavior.

The switching loss in power device can be reduced by lessening the transition time between blocking and conducting state; for it is originated from the product of voltage and current during switching transition period. Therefore, SiC MOSFET and SBD are expected to cut down the switching loss in high voltage power conversion circuit. However, there are some incidental difficulties with the fast switching operation in the same way of Si MOSFET, for SiC power device is not an ideal switch. One is the fluctuation of gate voltage induced by switching operation in other device, which is called self (spurious) turn-on phenomenon. This phenomenon is known to occur in a bridge circuit of dc-ac inverter and dc-dc converter for bidirectional operation and/or synchronous rectification.

For example, lower arm transistor Q2 in half bridge circuit shown in Fig. 1 is conducting body diode current and blocking channel current with applying off gate voltage. The turn-on operation of transistor in higher arm Q1 abruptly changes the potential of intermediate terminal X from GND to Vcc by transition of conduction state from Q2 to Q1. The applied voltage across drain-source of Q2 flows displacement current through parasitic terminal capacitances of MOSFET. The transistor Q2 turns on, when the terminal capacitance C_{gs} is charged up to threshold gate voltage. Then, shoot through current flows in higher and lower arm, because Q1 is also in conduction state. Finally, Q2 turns back to off state. This phenomenon is caused by capacitive coupling of parasitic capacitances in the MOSFET, which is interpreted as intra EMC of power device.

Reference [1] illustrates the self turn on phenomenon and calculates the power loss associated with the phenomenon. Reference [2] shows that the power loss stemming from self turn-on increases with dv_{ds}/dt, which is applied on the transistor in synchronous rectification dc-dc converter circuit. Reference [3] analyzes the influence of parasitic inductance in the circuit and recovery current of diode to the self turn-on phenomenon. The gate voltage elevation required to be suppressed lower than threshold gate voltage especially for high voltage power conversion circuit, for power loss by self turn-on tend to be quite large. The gate voltage fluctuation is known to be suppressed with lowering the impedance of gate drive circuit, append external capacitor between gate and source terminal, and apply negative gate bias voltage at off condition [4, 5, 6]. This paper analytically evaluates the gate voltage behavior based on the simplified equivalent circuit, and discusses the circuit condition to mitigate the phenomenon.

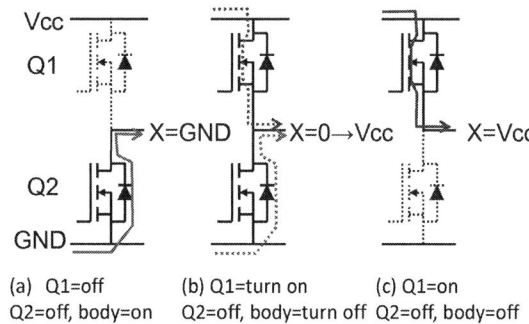

(a) Q1=off
Q2=off, body=on

(b) Q1=turn on
Q2=off, body=turn off

(c) Q1=on
Q2=off, body=off

Fig. 1. An example of half bridge circuit operation.

978-1-4799-5004-1/13 $31.00 © 2013 IEEE

II. SELF TURN-ON PHENOMENON

A. Structure of VDMOSFET and parasitic capacitance

The cross section of a vertical type planner gate n channel power MOSFET cell is shown in Fig. 2(a). The MOSFET has source and gate terminal at the top and drain terminal at the bottom. Multiple cells are connected in parallel and then constitute a power MOSFET device. The channel current flows laterally, which is same as a lateral type MOSFET, but the current in drift region flows in the vertical direction. The p+ region of the body and the n- region in the drift layer constitutes pn junction. This pn junction enables to flow current without flowing through channel, and this part is called body diode. The parasitic capacitance in the cell is superimposed on the corresponding part of cross section.

C_m, C_{oxs}, C_{oxc}, C_{oxd} respectively represents the parasitic capacitance between gate electrode and source terminal, n+ source, p+ channel, n drift region. The capacitance seen from the device terminal is shown as Fig. 2(b). The respective capacitance is obtained as following synthesis.

$$C_{gs} = C_m + C_{oxs} + \frac{1}{\frac{1}{C_{oxc}} + \frac{1}{C_c}} \quad (1)$$

$$C_{ds} = C_{dsj} \quad (2)$$

$$C_{gd} = \frac{1}{\frac{1}{C_{oxd}} + \frac{1}{C_{gdj}}} \quad (3)$$

B. Drain and gate voltage

The gate voltage v_{gs} of MOSFET fluctuates, when applying drain voltage abruptly, and turns on when v_{gs} exceeds the threshold gate voltage. The mechanism for gate voltage change in static condition is discussed based on the simplified equivalent circuit in Fig. 3(a). The gate drive circuit is ignored for the discussion simplicity. The capacitance C_{ds} is also excluded from the discussion, because it is directly charged and discharged with the drain power supply and is not involved in the self turn-on phenomenon. The applied v_{ds} is divided by C_{gs}

and C_{gd} as $v_{ds} = v_{gs} + v_{dg}$. The following relation is established with neglecting the voltage dependency of terminal capacitance and the gate drive circuit.

$$Q = C_{gs} v_{gs} = C_{gd} v_{dg} \quad (4)$$

v_{gs} can be obtained as following equation with utilizing v_{ds}.

$$v_{gs} = \frac{C_{gd}}{C_{gd} + C_{gs}} v_{ds} \quad (5)$$

That is, v_{ds} is divided with the capacitance ratio for C_{gs} and C_{gd} and appears as v_{gs}. When the resulting v_{gs} exceeds the threshold gate voltage, channel is established, and the MOSFET conducts. This condition is called self turn-on.

The relation between carrier concentration and electric potential in the depletion region, which is internally formed in semiconductor with applying reverse bias voltage v, is expressed with Poisson's equation (6) for 1 dimensional system. Poisson's equation is one expression of Gauss law $\nabla D = \rho$.

$$\frac{d^2 v}{dx^2} = -\frac{\rho}{\varepsilon} \quad (6)$$

Where, D, ρ, and ε respectively denote electric flux density in dielectrics, electric charge density and dielectric constant. The practical uniformly impurity doped drift layer gives uniform space charge distribution ρ in the depletion region.

The relation between depletion width w and the applied voltage v is derived by double integrating eq. (6) by space as $w = \sqrt{\frac{2\varepsilon(v - v_b)}{q\rho}}$, where q denotes unit charge and v_b denotes the junction built in voltage. Then, the relation between depleted charge Q and applied voltage v is expressed as eq. (7).

$$Q = q\rho w = \sqrt{2\varepsilon q(v - v_b)} \quad (7)$$

The voltage dependency of capacitance between gate and drain terminal C_{gd} can be obtained as the differential

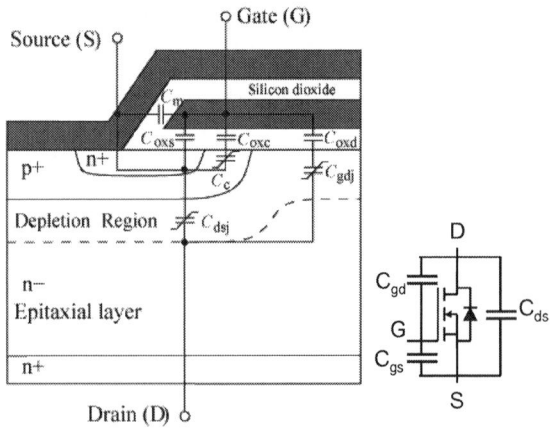

(a) cross section of MOSFET cell (b) equivalent parasitic capacitance

Fig. 2. VDMOSFET structure and parasitic capacitance

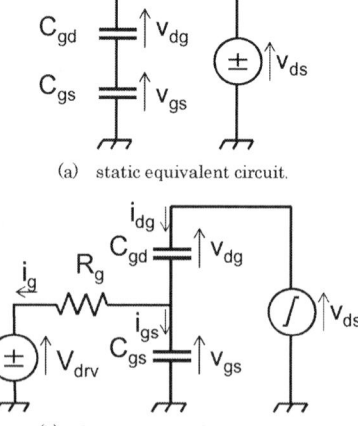

(a) static equivalent circuit.

(b) dynamic equivalent circuit

Fig. 3. Equivalent circuit to study self turn-on.

capacitance with the charge variation Q_{gd} for the drain voltage change v_{dg}.

$$C_{gd} = -\frac{dQ_{gd}}{dv_{dg}} = \sqrt{\frac{\varepsilon\rho}{2(v_{dg}-v_b)}} = (av_{dg}+b)^{-\frac{1}{2}} \quad (8)$$

Where, $a = \frac{2}{\varepsilon\rho}$ and $b = -v_b$.

The gate voltage in high voltage power MOSFET is considerably smaller than drain voltage. This enables to handle C_{gs} as constant, which results in the following relation by $Q_{gs} = Q_{gd}$.

$$C_{gs}v_{gs} = \int_0^{v_{dg}} C_{gd}dv = \frac{2}{a}\left[(av_{dg}+b)^{\frac{1}{2}} - b^{\frac{1}{2}}\right] \quad (9)$$

Then, v_{gs} can be expressed as function of v_{ds} as following eq.(10).

$$v_{gs} = \frac{-2}{aC_{gs}}\left\{ b^{\frac{1}{2}} + \frac{1}{C_{gs}} - \sqrt{\left(b^{\frac{1}{2}} + \frac{1}{C_{gs}}\right)^2 + aC_{gs}^{\ 2}v_{ds}} \right\} \quad (10)$$

This equation indicates that v_{gs} variation is small for large coefficient a in eq.(10) by mitigating the influence of v_{ds} with low impurity concentration and large C_{gs}.

The critical electric field of SiC semiconductor is 10 times higher than Si. Therefore, SiC semiconductor power device can reduce conduction resistance for high voltage device with increasing the impurity concentrations. However, high impurity concentration ρ gives small slope coefficient a in eq.(10), which leads v_{gs} to have high sensitivity for v_{ds}. This stands a potential of higher self turn on possibility of SiC MOSFET to Si MOSFET for same voltage ratings with same device structure.

C. Dynamic behavior of gate voltage

The static behavior of v_{gs} to the applied drain voltage v_{ds} was discussed in the previous section with neglecting the gate drive circuit. The dynamic behavior of v_{gs} to the transient change in build up of drain voltage v_{ds} and the influence of gate driving circuit connected to the device are discussed in this section based on the equivalent circuit shown in Fig. 3(b). The output of gate drive circuit is shunt to the ground or negative potential through gate resistance R_g to retain the off condition of MOSFET. The relation among terminal voltages in MOSFET $v_{ds} = v_{gs} + v_{dg}$ is differentiated as following.

$$\frac{dv_{ds}}{dt} = \frac{dv_{gs}}{dt} + \frac{dv_{dg}}{dt} \quad (11)$$

The voltage across the terminal capacitances C_{gd} and C_{gs} are expressed with the following differential equation.

$$i_{dg} = C_{gd}\frac{dv_{dg}}{dt} \quad (12)$$

$$i_{gs} = C_{gs}\frac{dv_{gs}}{dt} \quad (13)$$

The following circuit formula is obtained from KCL at the gate terminal $i_{dg} = i_{gs} + i_g$ and KVL for the gate drive circuit $v_{gs} - v_{drv} = R_g i_g$.

$$C_{gd}\frac{dv_{dg}}{dt} = C_{gs}\frac{dv_{gs}}{dt} + \frac{v_{gs}-V_{drv}}{R_g} \quad (14)$$

The actual drain voltage applied to the MOSFET is not step voltage and has finite slew rate. Then, the applied drain voltage is approximated by ramp function $v_{ds} = kt$ with the rate of voltage rise $k[V/s]$. The analytical solution of the circuit formula is derived for v_{gs} as following.

$$v_{gs} = R_gC_{gd}k\left[1 - e^{-\frac{t}{R_g(C_{gs}+C_{gd})}}\right] + V_{drv} \quad (15)$$

The maximum value of v_{gs} is obtained for $t \to \infty$ as eq. (16).

$$\max\ v_{gs} = R_gC_{gd}k + V_{drv} \quad (16)$$

That is, the maximum value of v_{gs} is proportional to the rate of drain voltage rise k, gate resistance R_g, and terminal capacitance between gate and drain C_{gd}, and is offset by the applied bias voltage V_{drv}. The maximum value of v_{gs} can be suppressed by increasing C_{gs}, and making the time constant of $R_g(C_{gs} + C_{gd})$ sufficiently longer than the transition period to the build up of applied drain voltage.

III. C-V CHARACTERISTICS AND GATE VOLTAGE CORRESPONDING TO DRAIN VOLTAGE

This section studies the bias voltage dependency characteristics of parasitic capacitance (C-V) in Si and SiC power MOSFET. The gate voltage v_{gs} to the applied drain voltage v_{ds} for steady state with open gate condition is also evaluated. The specification of the studied power MOSFETs are shown in Table 1. The measured capacitance between gate and source terminal C_{gs} to the applied gate voltage v_{gs} for $v_{ds} = 0V$ is shown in Fig. 4(a). The accumulation, depletion, and inversion of carrier at the channel of MOSFET appear as the change in capacitance. The boundary of v_{ds} is -2V, 5V and -6V, 5V respectively for Si and SiC MOSFET. The capacitance in accumulation and inversion condition gives 8nF and 5nF respectively for Si and SiC MOSFET with neglecting C_c in eq. (1). The depletion of channel gives 1.5nF and 2nF C_{gs}

TABLE I. RATINGS OF MOSFETs.

Type	Si MOSFET	SiC MOSFET
Model	SPW20N60S5	SCT2080KE
Manufacturer	Infineon	ROHM
Rated voltage	600V	1200V
Rated current	20A	35A
Threshold gate voltage	3.5-5.5V	1.6-4.0V
On resistance $R_{DS(on)}$	190mΩ	80 mΩ

respectively for Si and SiC MOSFET with channel capacitance C_c in eq. (1). Though, the rated current Si MOSFET is smaller than SiC MOSFET, but the device active area is larger to decrease on resistance sufficiently, which results in larger

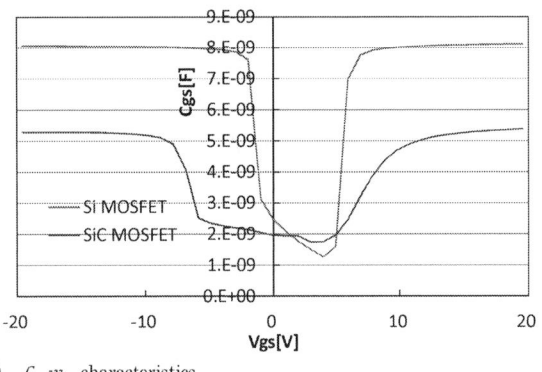

(a) C_{gs}-v_{gs} characteristics.

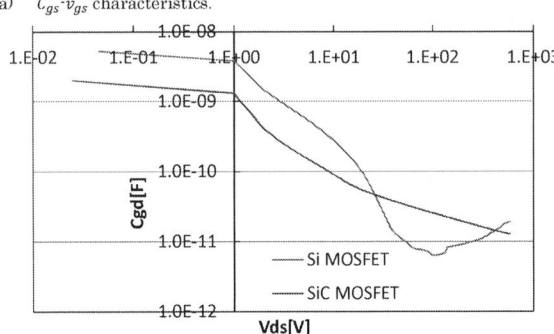

(b) C_{gd}-v_{ds} characteristics.

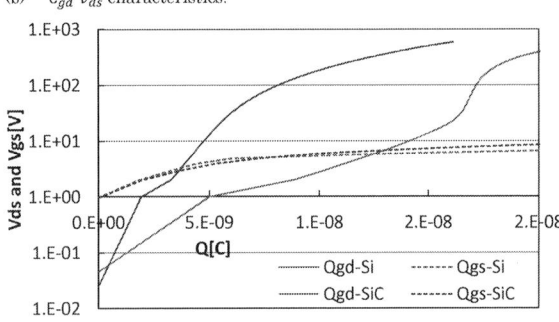

(c) Q-V characteristics.

Fig. 4. C-V characteristics of MOSFETs.

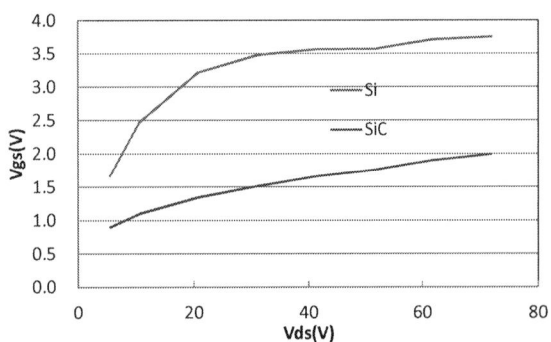

Fig. 5. v_{ds} - v_{gs} characteristics.

capacitance. These differences are originated from the difference in material characteristics of semiconductor between Si and SiC.

The negative gate bias voltage for the blocking condition alleviates the possibility of self turn-on. Furthermore, it increases C_{gs} by the formation of accumulation region and enhances the prevention of self turn-on. The drain voltage v_{ds} dependency in terminal capacitance between gate and drain C_{gd} for blocking condition with $v_{gs} = 0V$ is shown in Fig. 4(b). Si MOSFET gives larger C_{gd} than SiC MOSFET for low drain voltage, which stems from large area of super junction structure in drift region. However, C_{gd} of Si MOSFET becomes substantially small for drain voltage higher than 20V due to expansion of depletion region in vertical and lateral direction.

The steady state gate voltage to the applied drain voltage with open gate condition is estimated form these C-V characteristics. The depleted charge Q_{gd} and Q_{gs} to the applied voltage v_{ds} and v_{gs} are respectively estimated with integrating the C-V characteristics by voltage. The voltage v_{gs} and v_{ds} to the depleted charge is plotted in Fig 4.(c). The v_{ds} and v_{gs} corresponding to the same depleted charge deserves to the resulting v_{gs} to the applied v_{ds} for open gate condition. Though, the difference in v_{gs} between Si and SiC MOSFET for same depleted charge Q_{gs} is small, but the difference in v_{ds} (v_{dg}) to the same depleted charge Q_{gd} is large. That is, v_{ds} of SiC MOSFET is higher than Si MOSFET to the same depleted charge. Then, the increment of v_{gs} to the applied v_{ds} is mitigated.

The experimented steady state gate voltage v_{gs} to applied drain voltage v_{ds} in open gate condition is shown in Fig. 5. The result shows that v_{gs} of SiC MOSFET is lower than Si MOSFET to the same applied voltage. The v_{gs} variation for low drain voltage region ($v_{ds} < 20V$) is large for Si MOSFET. This result corresponds to the large variation of Q_{gd} for v_{ds} change in low voltage region of Si MOSFET.

IV. EXPERIMENTAL EVALUATION OF TRANSIENT GATE VOLTAGE BEHAVIOR

The experimental set up corresponding to the constitution shown in Fig. 3(b) is arranged with pulse generator (Agilent 81104A) and bipolar amplifier (NF corp. HSA4101). The drain voltage is applied with bipolar amplifier and the slew rate dv_{ds}/dt is regulated with the rise time of pulse generator output. The gate resistance R_g and slew rate of applying drain voltage v_{ds} is changed as a parameter. The gate bias voltage for off condition is set $V_{drv} = 0V$.

The experimental result for applying v_{ds} by $60V/\mu s$ for different parameter R_g is given in Fig. 6. The drain voltage v_{ds} builds up from 0 to 75V by $1.25\mu s$ as shown in Fig. 6(a)(c). The gate voltage response to the applied drain voltage in Fig. 6(b)(d) indicates that the increasing rate of gate voltage for Si MOSFET is higher than SiC MOSFET to same gate resistance. The gate voltage builds toward a peak and then decreases, which differs to the monotonic increase estimated from eq.

(15). This can be attributed to the nonlinear characteristics of terminal capacitance, which appear as C-V characteristics in Fig. 4. C_{gd} becomes small at high v_{ds} for both Si and SiC MOSFET. Si MOSFET gives larger C_{gd} than SiC MOSFET in low drain voltage region ($v_{ds} < 20$V). This results in larger multiplier $R_g C_{gd} k$ in eq. (15), and corresponds to the higher v_{gs}. This also leads to the difference of peak value of v_{gs} in Fig. 6(b)(d), and summarized in Table 2.

Figure 6(b)(d) shows that the large gate resistance lead to high build up rate of v_{gs}, which can also be estimated from eq. (15). Though, eq.(15) indicates monotonic increase of v_{gs} throughout v_{ds} increment, but decrement of v_{gs} is found in Fig. 6(b)(d). This phenomenon can be attributed to the drain voltage dependency of C_{gd} shown in fig. 4(b). That is, C_{gd} becomes 1/10 for v_{ds} change from 1V to 10V. Therefore, increasing rate of v_{gs} at the onset of v_{ds} application is highest due to large C_{gd} in low v_{ds}. Therefore, low v_{ds} region must be focused on to prevent self turn-on phenomenon.

Figure 7 shows the experimental result for different parameter of dv_{ds}/dt with $R_g = 51\Omega$. The high slew rate dv_{ds}/dt drain voltage application prone to cause oscillatory ringing by the interaction of parasitic inductance in device leads. The higher slew rate of v_{ds} gives higher multiplier in eq.

peak v_{gs} summarized in Table 3. The peak of v_{gs} for SiC MOSFET is suppressed lower than Si MOSFET for the same slew rate of v_{ds}.

TABLE I. PEAK v_{gs} FOR $dv_{ds}/dt = 60V/\mu s$

R_g	Si MOSFET	SiC MOSFET
10Ω	0.4V	0.2V
51Ω	1.3V	0.55V
1kΩ	3.2V	1.5V

TABLE II. PEAK v_{gs} FOR $R_g = 51\Omega$

$dv_{ds}/dt\ [V/\mu s]$	Si MOSFET	SiC MOSFET
60	1.37V	0.56V
120	1.91V	0.80V
1400	2.55V	1.47V
2300	5.26V	1.83V

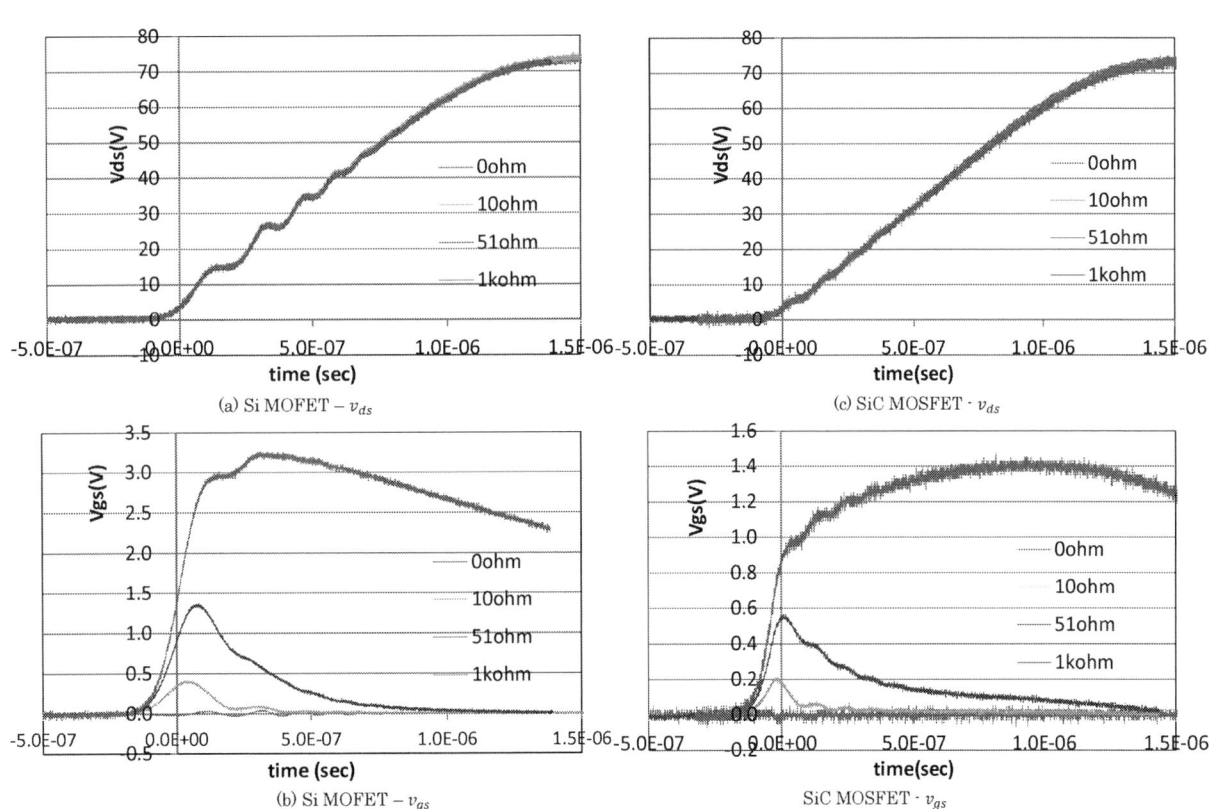

(a) Si MOFET – v_{ds}

(c) SiC MOSFET · v_{ds}

(b) Si MOFET – v_{gs}

SiC MOSFET · v_{gs}

Fig. 6. v_{gs} response to applied v_{ds} ($\frac{dv_{ds}}{dt} = 60$V/µs)..

(15), and lead to higher v_{gs}. This corresponds to the observed

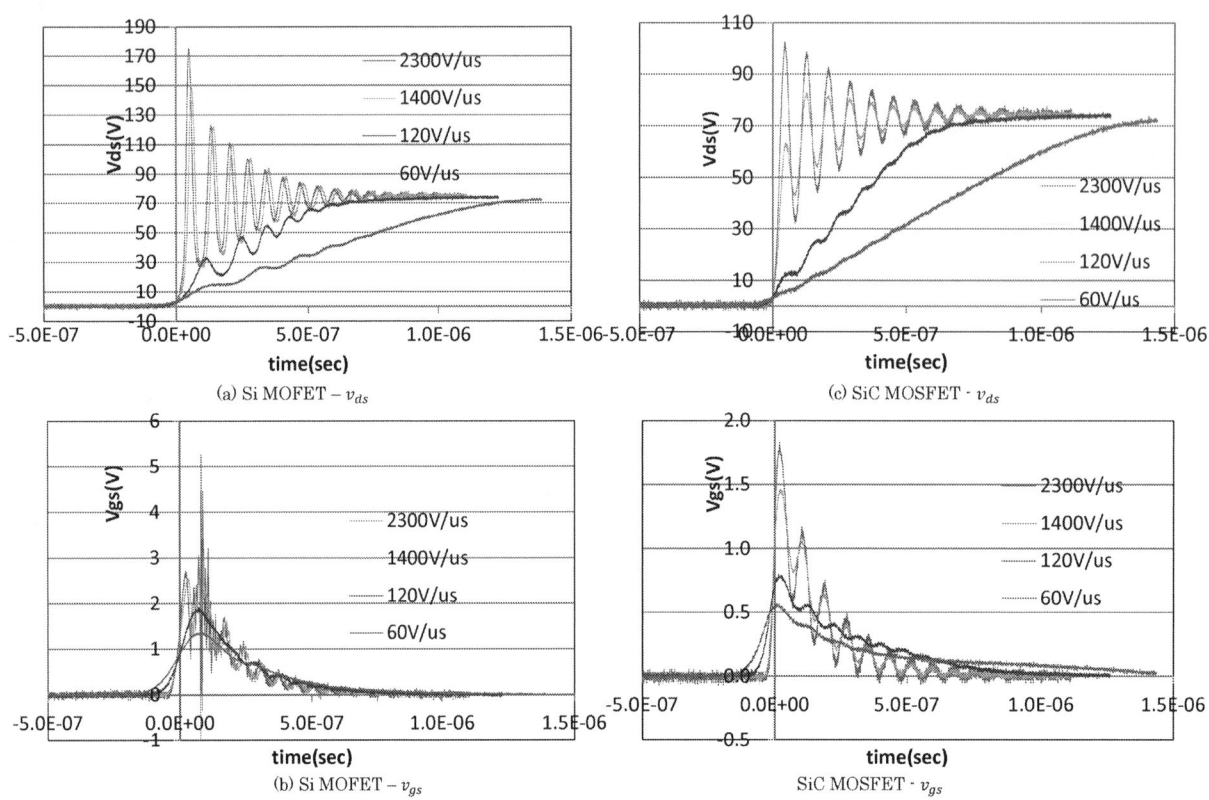

(a) Si MOFET – v_{ds}

(c) SiC MOSFET - v_{ds}

(b) Si MOFET – v_{gs}

SiC MOSFET - v_{gs}

Fig. 7. v_{gs} response to applied v_{ds} ($R_g = 51\Omega$).

V. CONCLUSION

The gate voltage of MOSFET in off state fluctuates with the switching operation of adjacent MOSFET. This phenomenon is called self turn-on and becomes severe problem especially in high voltage and fast switching power conversion circuit. This paper studied the mechanism of self turn-on phenomenon. The mathematical model of gate voltage behavior to the applied drain voltage is derived for constant terminal capacitance and ramp drain voltage application. The effect of lowering gate resistance and negative bias gate voltage application in suppressing gate voltage fluctuation is analytically presented, that the gate voltage elevation proportionate R_g, C_{gd}, and slew rate of drain voltage change $k = dv_{ds}/dt$.

The estimated gate voltage elevation is experimentally evaluated. However, the parasitic capacitance in MOSFET is non-linear and varies with applied voltage. Then, the actual gate voltage elevation becomes lower than the estimated value with assumptions of linear circuit and large C_{gd} at low drain voltage.

REFERENCES

[1] K. Murata, K, Harada,"Analysis of a self turn-on phenomenon on the synchronous rectifier in a DC-DC converter", Proc. INTELEC 2003, pp. 199-204, (2004).

[2] K. Murata, K, Harada,"A self turn-on mechanism of the synchronous rectifier in a DC-DC converter", Proc. INTELEC 2004, pp. 642-646, (2004).

[3] Y. Kawaguchi, S. Ono, K. Kinoshita and A. Nakagawa, "Study of Low side MOSFET Self Turn-on Phenomenon in Synchronous Bulk Converter by Spice Simulator," IPEC, pp.965-970, (2005).

[4] C. Licitra, S. Musumeci, A. Raciti, A. U. Galluzzo, R. Letor, and M. Melito, "A new driving circuit for IGBT devices", IEEE Trans. Power Electron., vol.10, no.3, pp.373 -378, (1995).

[5] N. McNeill, S. Kuang, B. W. Williams, and S. J. Finney, "Assessment of off-state negative gate voltage requirements for IGBTs", IEEE Trans. Power Electron., vol.13, no.3, pp.436 -440, (1998).

[6] S. Pontarollo, et al., "A new gate driver integrated circuit for IGBT devices with advanced protections", IEEE trans PELS, Vol.21, No.1, pp.38-44, (2006).

Active magnetic field canceling system

Wei-li Sun
Communication and Electromagnetic Engineering
National Taiwan University of Science and Technology
Taipei, Taiwan
essommie@hotmail.com

Feng-Chang Chuang
Department of Electrical Engineering
National Chung Hsing University
Taichung, Taiwan.
Department of Civil Engineering
National Taiwan University
Taipei, Taiwan
carsonjuang2006@gmail.com

Yu-Lin Song
High Technology Research Center
Yen Tjing Ling Industrial Research Institute
Taipei, Taiwan
d87222007@ntu.edu.tw

Chwen Yu
Taiwan Semiconductor Manufacturing Company, Ltd.
Hsinchu, Taiwan.
cyuc@tsmc.com

Tzyh-Ghuang Ma
Communication and Electromagnetic Engineering
National Taiwan University of Science and Technology
Taipei, Taiwan
tgma@ee.ntust.edu.tw

Tzong-Lin Wu
Department of Electrical Engineering
National Taiwan University
Taipei, Taiwan
wtl@cc.ee.ntu.edu.tw

Luh-Maan Chang
Department of Civil Engineering
National Taiwan University
Taipei, Taiwan
luhchang@ntu.edu.tw

Abstract—The extremely low frequency (ELF) magnetic field has significant impact on yield rate especially when the processing reaches less than 14 nanometer in next-generation nano-Fab. For sensitive equipments such as the SEMs、TEMs、STEMs、FIB writers, and E-beam writers, it suggests that the ELF magnetic field should be lower than 0.5 milli-Gauss to guarantee good yield. Therefore, mitigating the magnetic field by active/passive approaches such as the material shielding, wire permutation, and active canceling are highly demanded.

This paper focuses on the development of an active magnetic field canceling system which consists of sensors, current driven unit, and Helmholtz coils. First of all, a sensor is used to detect the magnetic field. A number of procedures such as amplifying, reversing and filtering are then involved in processing the signals. Finally, the current driven unit, i.e. the power amplifier, delivers a reversed current to the Helmholtz coil to create a canceling magnetic field which is in equal magnitude but almost out of phase with respect to the environmental one.

According to the experiments, the prototype design is able to reduce the magnetic field intensity from 10.6mG to 1mG at 60Hz.

Keywords—ELF(extremely low frequency)、sensor、Helmholtz Coil、prototype

I. Introduction

With the rapid development of modern nano-Fabs, the electromagnetic interference (EMI) has become a critical issue in setting up next-generation semiconductor foundries. It has been reported that when the CMOS progress shows a resolution less than 14 nanometer, the ELF magnetic fields show significant impacts on semiconductor equipments such as SEMs, TEMs, STEMs, FIB writers and E-beam writers. Therefore recent standard of SEMI[1] suggests that the environment magnetic field should be maintained under 0.5mG to keep such sensitive equipments from malfunction.

For achieving this objective, there are two different systems: the passive [2]-[5] and active [6] magnetic field cancelling systems. In general, passive shielding is the most straightforward way for magnetic field mitigation. However in the ELF range, shielding using ferrite composite materials significantly increases the fabrication cost of the foundry. The active magnetic field canceling system provides an alternative option to alleviate the detrimental ELF magnetic field. The active canceling system builds up reverse magnetic fields whose intensity and phase are respectively equal to and inverted with respect to the environment field for real-time canceling.

This article focuses on building up an prototype analog active magnetic field canceling system for nano-Fab. According to the experimental results, at 60 Hz the magnetic field can be significantly reduced from 10 mG to below 1 mG.

II. BUILDING BLOCKS OF THE ACTIVE CANCELING SYSTEM

Shown in Fig. 1 is the proposed active magnetic field canceling system, which consists of three major sections: the sensor, current driven unit and inductive coil. The sensor first detects the environmental magnetic field and feed into the current driven unit for processing. The output current from the driven unit is delivered to the coil and hence generates a reversed magnetic field for cancelling. In this prototype, all units are developed based on analog approaches, therefore minimizing the undesired delays in common PFGA boards with digital signal processing units. The major units are described step by step as follows.

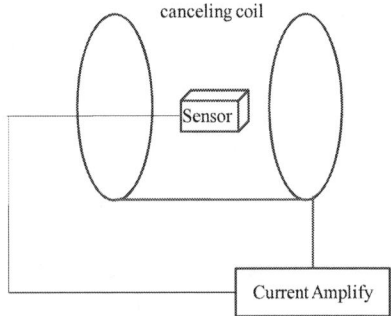

Fig. 1. Principle of the active canceling system.

A. Sensor

For detecting the environmental magnetic field, a sensor with high sensitivity is important. The common sensors include the induction coils, Hall sensor, and magneto resistive sensors. While the inductive coils are bulky in size, the Hall-effect sensor has the drawback of low sensitivity. As a result the magneto resistive sensor would be the most suitable selection for the proposed system.

B. Current Driven Unit – Power Amplifier

The common linear power amplifiers (PAs) include the class A, class B and class AB. The Class A power amplifier shows the highest linearity, but lowest power efficiency at 25%. The Class B power amplifier has the highest efficiency, but suffers from the well-known nonlinear crossover distortion effect. The power efficiency of the class AB power amplifier is somewhere in-between the class A and class B ones, and can avoid the harmful crossover distortion with good linearity. Fig 2 shows the characteristic curve of class B amplifier.

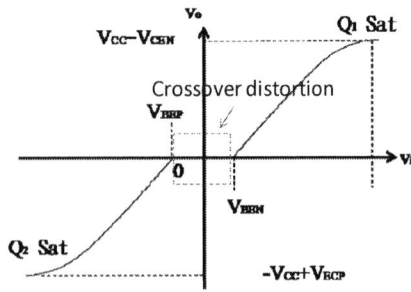

Fig. 2. Characteristic curve of a class B power amplifier.

C. Coil

For generating a magnetic field, an inductive coil is a necessity. In general, there are several types of coils which can be used such as the circular loop coil, square loop coil, and Helmholtz coils.

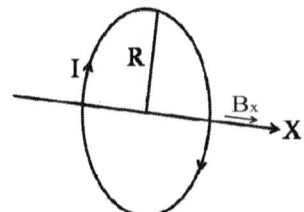

Fig. 3. A circular loop coil.

Fig. 3 shows a circular loop coil. The magnetic flux density at the center of the circular loop can be calculated by

$$\vec{B} = \frac{\mu_0 N I R^2}{2(x^2 + R^2)^{\frac{3}{2}}} \hat{x}$$

(1)

The Helmholtz coil is built by two circular loop coils; the magnetic flux density at the center of the Helmholtz coil can be calculated using equation (2), providing the two coils are equal in radius.

$$\vec{B} = \frac{8\mu_0 N I}{\sqrt{125}R} \hat{x}$$

(2)

The square loop coil is another candidate suitable for applications. It could the easiest one to build up for nano-fab. The magnetic flux density at the center of a square loop with a perimeter of $4a$ is shown in (3).

$$\vec{B} = \frac{2\sqrt{2}\mu_0 I}{\pi a} \hat{x}$$

(3)

III. PROTOTYPE AND DESIGN CONSIDERATION

The architecture of the proposed prototype canceling system is shown in Fig. 4. As mentioned above, the prototype consists of magnetic sensor, current driven unit and inductive coil.

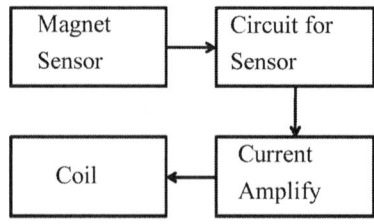

Fig. 4. The architecture of the prototype canceling system

A. Sensor

The magnetic sensor selected was HMC2003 from Honeywell [8]. It is a three-axial magnetic sensor with high sensitivity of 1V/Gauss. The dynamic range is from 40 mG up to 2 G. In addition, it has a 2.5-Volt reference output. That is, it will have a 2.5V DC output when the environment magnetic field is null. Table 1 shows the major specifications.

TABLE I SPECIFICATIONS OF HMC2003

Features	Sensor
Dimention	3-Axis
Output type	Analog
Output range	0 ~ 5 (V)
Sensitivity	1 (V/Gauss)
V Ref	2.5 (V)
Supply voltage	6~15 (V)
Field range	4 0 (μG) ~ 2 (Gauss)

B. Current Driven Unit with Intermediate Stage

The HMC2003 imposes additional issues to be dealt with. Firstly, the 2.5-V DC offset should be filtered out in order to prevent the system from emitting a constant DC magnetic field even without an input. This blocking circuit also increases the dynamic range of the power amplifier by preventing it from entering into the saturation region. In addition, the output signal from the magnetic sensor, say with a 10-mG environmental field, is still too small (only 10 mV) to drive the power amplifier. Accordingly, an intermediate stage between the sensor and class AB amplifier becomes a necessity. Here, the OPA177 is selected to fulfill the design goal. Fig. 5 shows a photo of the intermediate stage. By integrating the intermediate stage and the power amplifier, Fig 6 shows the prototype canceling system with an industrial computer as the interface.

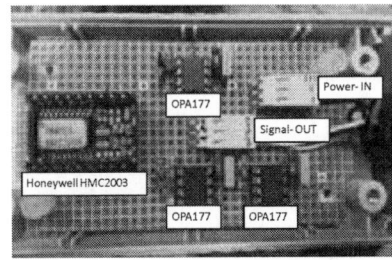

Fig. 5. The intermediate stage.

Fig. 6. Prototype active canceling system.

C. Coil

In the laboratory demonstration, a square loop coil is built up as the inductive coil. Fig 7 shows a photo of the coil. Each side has a length of 1 meter.

Fig. 7. Square inductive coil.

IV. RESULTS AND DISCUSSION

Fig. 8 shows the verification of the proposed active canceling system. A noise source is used to generate an environment magnetic field, and the prototype system is tested for its capability to alleviate the magnetic field around the center region of the cube.

The experimental results are summarized in Table II. When the noise was sourced at 60 Hz, this prototype system has effectively reduced the magnetic field from 10 mG to 1 mG. In addition, we also try to increase the operating frequency to 1 kHz. Benefitting from the all-analog system approach, the proposed system gives rise to almost identical canceling results when compared with those at 60 Hz.

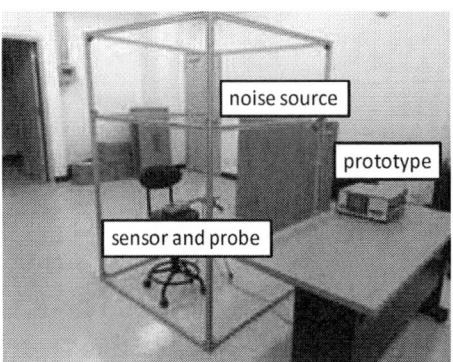

Fig. 8. Active canceling system

TABLE II   RESULT OF THE CANCELING SYSTEM

Frequency(Hz)	Field(mG)	Field after canceling(mG)
60	10.26	0.9
1K	10.2	0.93

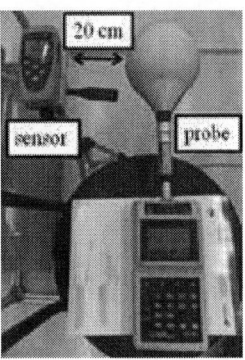

Fig. 9.   Sensor and probe

To further verify the work zone of the canceling system, an additional probe is included in this study. It is placed at a distance of 20 cm away from the magnetic sensor, as shown in Fig. 9. The field probe provides extra information for verifying the applicable range of the system. The experimental results are shown in Table III. Since the magnetic sensor is closer to the noise source, it detects a higher magnetic field than the probe does. However, the canceling capabilities at both sites are almost the same. According to the data recorded by the probe, the magnetic field drops from 5 mG to 0.77 mG at 60 Hz.

TABLE III.   RESULT OF CANCELING SYSTEM WITH DISTANCE

Frequency(Hz)	Field(mG)	Field after canceling(mG)
60	5.25	0.77
1K	5	0.8

V.   CONCLUSION

In this paper, a prototype active magnetic field canceling system has been realized and experimentally validated. With the help of the proposed all-analog canceling system, the environment magnetic field can be effectively reduced to lower than one-tenth its original strength, which almost reaches the value the standard suggests. The current version is still restricted to uni-directional field canceling along one axis. Adding extra sensors and amplifier modules with additional square loop coils to fulfill three-dimensional canceling is under development.

ACKNOWLEDGMENT

Support by the National Science Council, Taiwan, Yen Tjing Ling Industrial Research Institute, NTU, Taiwan under grants #NSC98-2622-E-002-034-CC2, Taiwan Semiconductor Manufacturing Company, Ltd. And Angeltech International Company, Ltd. are gratefully acknowledged, respectively.

REFERENCES

[1] SEMI E33-94, " Pecification For Semiconductor Manufacturing Facility Electromagnetic Compatibility " Semiconductor Equipment and Materials International, Mountanin View, CA, 1992, 1994

[2] K. Yamazaki, S. Hirosato , K. Muramatu , M. Hirayama, K. Kamata , T. Onoki , K. Kobayashi, A. Haga , K. Katada, Y. Kinomura , and Y. Kuwayama ,"Open-Type Magnetically Shielded Room Using Only Canceling Coil Without a Ferromagnetic Wall for Magnetic Resonance Imaging," *IEEE Trans. On Magnetics*, vol. 43, no. 6, pp. 2480-2482, jun 2007

[3] T. Saito, "Features of a Wall With Open-Type Magnetic Shielding Method," *IEEE Trans. On Magnetics*, vol. 44, no. 11, pp. 4191-4194, Nov 2008

[4] P. Moreno and R. G. Olsen, "A Method for Estimating Magnetic Shielding by 2-D-Thick Planar Plates for Distribution Systems Shielding," *IEEE Trans. On Power Delivery*, vol. 25, no. 4, pp 2710-2716, Oct 2010

[5] Y.-L. Song, C. Yu, F.-C. Chuang, Y.-C. Tseng, J.-Y. Zou, S.-K. Hsu, T.-G. Ma, T.-L. Wu, and L.-M. Chang, "Evaluation of magnetic field from varied permutation power transmission line at high technology nano-fab," *in Int. Conf. Power Electron. Systems Applications*, Jun. 2011

[6] F.-C. Chuang, Y.-L. Song, C. Yu, S.-K. Hsu, T.-L. Wu, and L.-M. Chang, "Active field canceling system in next generation nano-fab," *in Int. Conf. Power Electron. Systems Applications*, Jun. 2011

[7] Stefan Mayer Instruments "MR-3 Triaxial Magnetic Field Compensation System、Compensation coil design and system installation guide," January 2003

[8] Honeywell spec "3-Axis Magnetic Sensor Hybrid"

Spread Spectrum Clocking for Emission Reduction of Charge Pump Applications

Bernd Deutschmann

Infineon Technologies AG
Am Campeon 1-12
85579 Neubiberg, Germany
bernd.deutschmann@infineon.com

Abstract—Spread spectrum clock generation (SSCG) techniques are widely used in digital systems as the digital clock signals are considered as one of the major sources for conducted and radiated electromagnetic emission. Although spread spectrum techniques can nowadays be found in many clocked digital system, they are rarely found in clock signals for automotive power systems such as power switches or bridges. In this paper the ability of using SSCG techniques for automotive charge pump applications is investigated. It is shown how this technique can be used to reduce the electromagnetic emission that is caused by the high frequency switching activity of charge pumps. Additionally the assets and drawbacks when using this technique are explained. Based on conducted electromagnetic emission measurements of a charge pump EMC test chip it is shown how typical spread spectrum parameters like frequency deviation, modulation frequency and modulation signal can be optimized in order to maximize the electromagnetic emission reduction in certain frequency ranges.

Keywords-component; Spread Spectrum Clock Generation, SSCG, Electromagnetic Emission, Charge pump, Automotive.

I. INTRODUCTION

More and more electronic components are currently installed in modern vehicles. In combination with the increasing clock frequencies and signal speeds of the used integrated circuits, the amount of electromagnetic emission that is currently spread into the environment over an increasing frequency range is nowadays more dominant than before. These high electromagnetic emission leaves its signature not just in deteriorates of the reception of radio frequencies and noise in the car radio but also in an increasing amount of electromagnetic interference (EMI) problems of important safety functions, if the electromagnetic compatibility is not taken into consideration.

A main source of the electromagnetic emission is e.g. the ICs internal and external switching activity, i.e. high frequency system clock signals, data bus signals, as well as control signals such as they are used for pulse width modulation of bulbs or electric motors to mention some of them. Without any precautions the electromagnetic energy is mainly spread by the PCB and especially by attached cables harnesses into the environment. One powerful method to reduce EMI problems is based on the spread spectrum technique [1], [2]. In the beginning such spread spectrum techniques were mainly

focused on making signal transmission systems more robust and to avoid interferences by RF signals [3], [4], [5]. The specific reduction of the electromagnetic emission of an electronic system was of less focus, till the 1990ies [6], [7], [8]. Since this time spread spectrum clock generation is one of the most controversy discussed methods to reduce electromagnetic emission. Many discussions have been led since this time e.g. questioning the legality according to FCC regulations or the claim that spread spectrum is just a simple way to cheat measurements. Although many papers have been published so far that clearly prove the reduction of emission and lay down the mathematical fundamentals that prove and explain the effects of frequency modulation spread spectrum clock generation on an EMI receiver measurement result [9], [10], still there are many engineers who think that this technique is constantly shifting signals in frequency and therefore "fooling" an EMI receiver by actively "shifting" the noise frequency out of the receiver band during the measurement at one frequency position. Anyway, the spread spectrum technique is a powerful method to reduce emissions at certain frequency positions. Well, using a spread spectrum technique does not really lead to an energy reduction, but it can be used as a good way to distribute the energy that was originally concentrated only at single frequency positions (e.g. at the fundamental and harmonics of the clock signal) by modulating the switching frequency in an intelligent way. Speaking is terms of EMC measurement techniques, spreading the spectrum helps to distribute the energy over a bandwidth wider than the EMI receiver resolution bandwidth (RBW) by broadbanding a signal which was originally narrowband [7].

But as every technique has advantages so it has also disadvantages and spreading the spectrum might not always be the best choice to solve associated EMI problem. It might create interferences at frequency positions, where no problem originally existed. Also without special precautions the ability of using this technique for real time operating systems is often limited or even impossible, as attention must be given in order to minimize clock skew and to avoid that an accumulated jitter affects the correct function by causing transmission errors [11].

Therefore one should understand this technique in order to use it up to the best advantage. In this paper the ability of using spread spectrum techniques to reduce the electromagnetic emission of automotive ICs is investigated. In an experimental study based on the emission measurements of an EMC test chip, an external signal generator was used as a spread

978-1-4799-5004-1/13 $31.00 © 2013 IEEE

spectrum clock generator (SSCG) for frequency modulating the clock signal of a charge pump. It is shown that for such applications this technique can be used with minimal consideration as charge pump applications usually do not utilize a very high frequency dependency. It provides a much higher flexibility for tuning and can therefore be used up to the full swing for significant emission reductions.

II. ELECTROMAGNETIC EMISSION OF CHARGE PUMPS

Charge pumps are widely used in automotive applications, for example in gate driver circuits for power DMOS transistors [14]. E.g. if an n-channel transistor is used as a high side switch, a charge pump can be used to generate gate voltages that are higher than the battery supply voltage in order to turn on the transistor. These charge pumps are often directly integrated into the logic die. A very basic principle of a simple charge pump as an electronic circuit that uses capacitors as energy storage elements to create a higher voltage source is illustrated in figure 1. The charge pump capacitor is charged to the supply voltage Vbb when the switches S1 are both closed. S2 is closed after the switches S1 are opened and the stored charge of the capacitor is discharged via the diode to the gate capacitance of the power DMOS transistor. Integrating large capacitors directly on the die is very area consuming and therefore expensive. As a consequence only small capacitor values, usually in the range of pF's, are integrated. In order to charge up the gate capacitance of the power DMOS transistor within an adequate time, the switches are therefore operated with high frequencies (often in the MHz range).

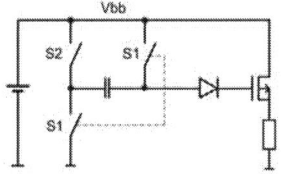

Figure 1.   Basic principle of a charge pump to create a gate source voltage higher than the supply voltage

Electromagnetic emission is created by the turning on and off of the high side power DMOS itself, especially when it is used in pulse width modulation (PWM) applications. Some additional amount of electromagnetic emission is created by the switching activity of the charge pump as it is illustrated in the conducted electromagnetic emission diagram of a high side switch in figure 2. The switching activity of the charge pump produces charge and discharge currents that spread as periodical di/dt current peaks on the supply lines and finally result in unwanted (high) emission at the fundamental and harmonics of the switching frequency.

For many applications the maximum switching frequency of the charge pump is often kept to values below 10MHz in order to be able to reduce the emission of the harmonics in the higher frequency range with simple and cheap external filtering components. Moreover, since the accuracy of the needed oscillator is not the most critical point, spread spectrum

techniques can easily be used and usually a larger frequency spread can be allowed to operate the charge pump. Especially this benefit should further be used to show the ability of reducing the emission by spreading the frequency spectrum.

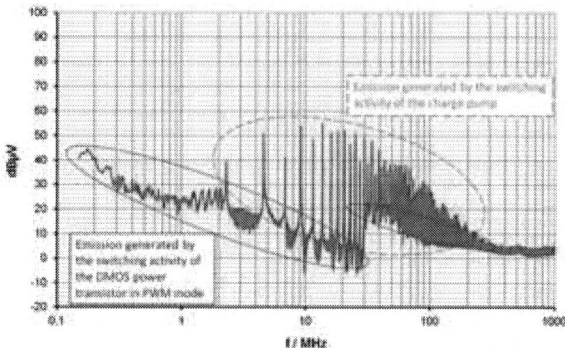

Figure 2.   Electromagnetic emission of an n-channel high side switch operated in PWM mode

For this purpose an EMC test chip was designed which incorporates several charge pumps, where one of these can be operated by an external clock signal generator. All the important spread spectrum parameters such as frequency deviation, modulation frequency, as well as the kind of modulation signal (saw tooth – down/up ramp, triangular, rectangular, sinusoidal, Hershey-Kiss[TM] function,…) can directly be controlled by the signal generator.

In figure 3 the basic schematic of the EMC test chip and the used conducted emission measurement setup based on the IEC 69167-4 is shown [15]. A 150Ohm coupling network connected to the power supply of the IC is used to characterize the conducted emission of the charge pump. A decoupling network is used to avoid that the generated emission is sorted by the DC power supply. With the external signal generator either a continuous sinusoidal frequency or a frequency modulation (FM) signal by using a single tone modulating signal is applied in order to investigate frequency spreading. The external sinusoidal clock signal is converted inside of the EMC test chip to a digital clock signal that is used to operate the corresponding charge pump.

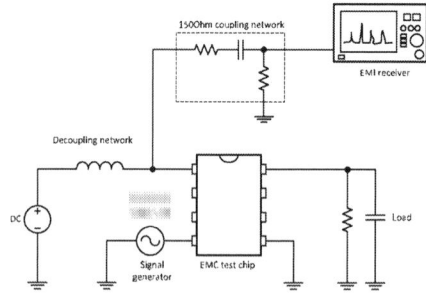

Figure 3.   Basic schematic of 150Ohm conducted emission measurement setup for the EMC charge pump test chip

An EMI receiver is used to measure the emission in the frequency range from 150kHz to 1GHz. The measurement results were evaluated by using peak, average, and quasi peak

978-1-4799-5004-1/13 $31.00 © 2013 IEEE

detectors. In table I some important EMI receiver settings such as the required resolution bandwidth (RBW) for the different frequency ranges based on CISPR16-1-1 [16] are shown. It is important to understand the relation of the spread spectrum parameters and the used resolution bandwidth and detectors in order to understand and optimize the reduction of the emission.

TABLE I.   CISPR 16-1-1 MEASURING RECEIVER BANDS AND RESOLUTION BANDWIDTH (RBW) FOR EMI TEST RECEIVERS

CISPR Band	A	B	C	D	E
Frequency range	9kHz – 150kHz	150kHz - 30MHz	30MHz - 300MHz	300MHz - 1GHz	1GHz - 18GHz
Resolution Bandwidth	200Hz	9kHz	120kHz	120kHz	120kHz

Figure 4 shows the peak, average and quasi peak detector measurement results of the conducted electromagnetic emission of the charge pump. For this measurement the charge pump was operated without spread spectrum techniques by a simple continues wave switching frequency of 2.5MHz. The clock signal can be considered as a periodic pulse train of an infinite series of time-period-shifted identical pulse functions. The corresponding spectrum in the frequency domain can be exemplified as an infinite series of Dirac pulses that are shifted by the clock frequency (i.e. each harmonic of the clock signal occupies only an infinitesimal frequency band in the frequency domain). As the clock frequency is much higher than the used resolution bandwidth of the EMI test receiver only one harmonic of the clock happens to fall within the receiver's reception channel. Therefore the measured emission can be considered as narrowband emission and the amplitudes of the fundamental and harmonics of the clock are equal for the used peak, average and quasi peak detector.

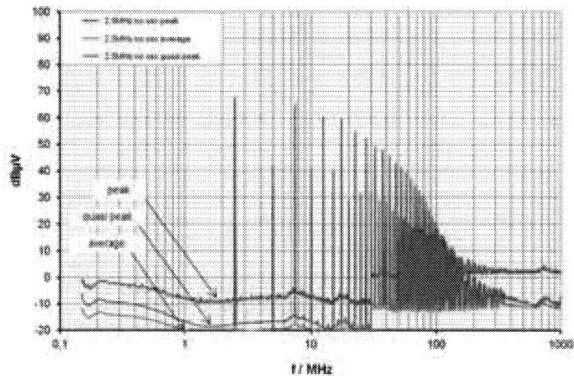

Figure 4.   Conducted electromagnetic emisson of the charge pump without using spread spectrum techniques (peak, average and quasi peak detector)

III.   SPREAD SPECTRUM BASICS

A simple and cost effective way of spreading the frequency spectrum can be achieved by using frequency modulation (FM) rather than phase or amplitude modulation. Therefore all measurement results in this paper are based on frequency modulating the clock signal of the charge pump. In order to understand the effects of frequency modulation on the emission spectrum and especially on the final EMI receiver measurement result it might be beneficial to investigate some mathematical fundamentals of frequency modulation and its corresponding frequency response.

The theoretical expression of a sinusoidal carrier frequency modulated by another sinusoidal signal is given in (1).

$$F(t) = A_C \cdot \cos(\omega_C t + m_f \sin \omega_m t) \qquad (1)$$

with: A_C … amplitude of the carrier signal
ω_C … angular frequency of the carrier
ω_m … angular frequency of the modulation signal
m_f … modulation index

The modulation index m_f gives the ratio of the frequency deviation of the carrier frequency to the frequency of the modulation signal (2).

$$m_f = \frac{\Delta \omega_C}{\omega_m} = \frac{\Delta f_C}{f_m} \qquad (2)$$

with: Δf_C … peak deviation of the carrier frequency
f_m … frequency of the modulation signal

According to Fourier's theorem any complex periodic function can be decomposed into the sum of a set of simple sine and cosine functions. In order to get a clearer picture about the different frequency spectral components of the FM signal we can also represent this periodic function (1) by its series representation. First it can be expressed by it's trigonometrically equality.

$$F(t) = A_C \cdot [\cos \omega_C t \cdot \cos(m_f \sin \omega_m t) - \sin \omega_C t \cdot \sin(m_f \sin \omega_m t)] \quad (3)$$

Next the combined sine and cosine terms can be exchanged by the corresponding Besselfunctions given in (4) and (5)

$$\cos(m_f \sin \omega_m t) = J_0(m_f) + 2 \sum_{n \, even} J_n(m_f) \cdot \cos(n \omega_m t) \qquad (4)$$

$$\sin(m_f \sin \omega_m t) = 2 \sum_{n \, odd} J_n(m_f) \cdot \sin(n \omega_m t) \qquad (5)$$

Using some further trigonometric conversion functions (1) can finally be expressed by the sum of simple cosine functions, as represented in (6).

$$
\begin{aligned}
F(t) = A_C \cdot \{ & J_0(m_f) \cdot \cos \omega_C t \\
+ & J_1(m_f) \cdot [\cos(\omega_C + \omega_m) \cdot t - \cos(\omega_C - \omega_m) \cdot t] \\
- & J_2(m_f) \cdot [\cos(\omega_C + 2\omega_m) \cdot t + \cos(\omega_C - 2\omega_m) \cdot t] \\
+ & J_3(m_f) \cdot [\cos(\omega_C + 3\omega_m) \cdot t - \cos(\omega_C - 3\omega_m) \cdot t] \\
- & J_4(m_f) \cdot [\cos(\omega_C + 4\omega_m) \cdot t + \cos(\omega_C - 4\omega_m) \cdot t] \\
+ & \dots \}
\end{aligned}
\qquad (6)
$$

Examining this equation it is easy to see that a frequency modulated signal actually contains an infinite number of frequencies and therefore occupies an infinite bandwidth. The discrete spectral components are at fixed frequency positions that are equally spaced by multiples of the modulation frequency around the carrier. As all frequencies will remain at fixed positions when using frequency modulation for spreading the spectrum no signals are constantly "shifted" in frequency out of the receiver's resolution bandwidth during the measurement. The final measurement result is mainly defined by how many of these discrete spectral components are located within the resolution bandwidth of the EMI receiver and their corresponding amplitude.

IV. INFLUENCE OF SPREADING THE SPECTRUM ON THE EMISSION OF CHARGE PUMPS

Generally there are three key parameters for reducing the emission by spreading the spectrum:

- frequency deviation Δf_C
- modulation frequency f_m
- modulation signal.

A.) Frequency deviation:

The deviation defines the peak excursion of the carrier frequency. Usually the displacement of the central frequency can be done in three different ways: Instead of shifting the carrier frequency only above and below (up-spread, down-spread), it can be also being shifted up- and down-wards symmetrically in respect to the carrier (center-spreading). Generally the higher the deviation, the higher is the reduction of the emission. The maximum frequency deviations used to implement spread spectrum techniques are usually limited to only a few % of the nominal carrier frequency as e.g. certain systems do not tolerate over clocking by using clock signals above the nominal frequency. This limitation is not that critical for charge pump applications where usually a much higher deviation can be tolerated (as it is shown in the following measurements). The comparison of the electromagnetic emission at the fundamental clock frequency of 2.5MHz of the EMC test chip with and without center spread spectrum clocking is shown in figure 5.

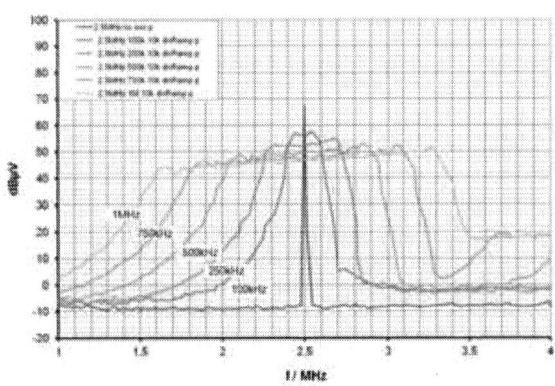

Figure 5.   Influence of the deviation on the electromagnetic emissison (f_C=2.5MHz, Δf_C=100k, 250k, 500k, 750k, 1MHz, f_m=10kHz)

A constant modulation frequency f_m=10kHz, a saw tooth (down ramp) modulation profile as well as a peak detector of the EMI receiver was used for this measurement. The frequency deviation was changed in the following steps: Δf_C=100kHz, 250kHz, 500kHz, 750kHz and 1MHz. As can be seen, by increasing the deviation Δf_C, the amount of reduction in the peak emission is also increasing. A reduction of up to ~20dB can be achieved in this case for a 1MHz displacement of the switching signal around the original central frequency.

One would now get the impression that always the highest possible deviation for spreading the spectrum should be used in order to achieve the maximum reduction of the electromagnetic emission. This is further encouraged by the fact that, as the harmonic orders increase, so does the effective reduction, because the higher order harmonics multiply the spread. Hence, one would expect that the degree of reduction is more pronounced for higher-order harmonics than for lower-order ones. But, unfortunately this effect does not continue without bound, as it is illustrated in figure 6. In this case the emission of one of the previously shown deviations of Δf_C=250kHz is plotted over the first 10 harmonics. As can be seen the higher order the harmonic, the wider the spread frequency band around that harmonic. This is because the bandwidth of the nth harmonic is n times the bandwidth of the fundamental. But as higher the order of the harmonics get, the emission of adjacent harmonic bands will sooner overlap.

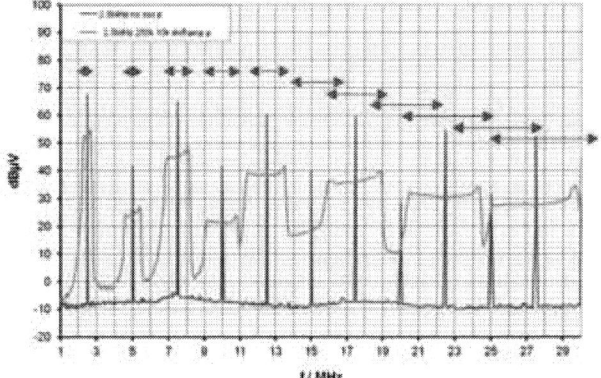

Figure 6.   Overlap of the emission at the higher order harmonics) (f_C=2.5MHz, Δf_C=250kHz, f_m=10kHz)

The order N of the harmonic where the frequency band starts to overlap with the next higher-order harmonic can be calculated by

$$N = \frac{f_C}{\Delta f_C} \cdot \left(\frac{1}{2} - \frac{f_m}{f_C} \right) - \frac{1}{2} \qquad (7)$$

For the given example in figure 6 the overlap starts between the 4th and 5th harmonic. As can be seen higher frequency deviations are to produce overlap at lower harmonic orders and tend to reduce the amount of emission reduction sooner. The influence of modulation frequency f_m in this overlap effect is significantly smaller and can be neglected in a first step.

978-1-4799-5004-1/13 $31.00 © 2013 IEEE

Another drawback that should be considered when using spread spectrum techniques is that the "empty" frequency ranges right between the discrete harmonics of the original signal without spread spectrum, are filled up with spectral components at each multiple of $f_C \pm n \cdot f_m$ when spread spectrum clocking is used. For the given example the originally narrowband emission peaks at multiples of the carrier frequency are converted to a broadband emission nearly over the whole frequency range, especially once the harmonic frequency bands start to overlap. This could lead to further problems at frequencies where originally no interference occurred. E.g. the emission at 26MHz in this example was originally around -10dBµV without spread spectrum clocking, whereas with spread spectrum clocking it is now 28dBµV.

B. Modulation frequency:

The second parameter that can be used to influence the emission is the modulation frequency. It defines how fast the carrier is varied within the deviation range. To avoid interferences it is often proposed to choose modulation frequencies higher than 20kHz in order to be beyond the audible range of the human ear. This should help to ensure that interferences coupling into the audio system are at frequencies high enough to be filtered out by the receiver, amplifier, speaker or the human ear. The influence of the modulation frequency on the peak, average and quasi peak detector results is shown in the measurement plots of the EMC test chip in figure 7. For this measurement a constant deviation of $\Delta f_C = 250$kHz and a modulation frequency $f_m = 1$kHz with a saw tooth (down ramp) modulation profile was used.

As we have seen before without spread spectrum clocking the EMI receiver gives equal readings for all detectors i.e. peak, average and quasi peak results are the same whether the RBW is 9kHz or 120kHz as the emission in this case can be considered as narrowband emission. But, with spread spectrum clocking there is a huge difference (especially between the peak and average measurement results) of the used detectors and RBW settings. Quasi peak values are not that much influenced in this example and show for all measurements just a little bit less emission compared to the peak detector values.

Knowing that an FM signal creates discrete spectral components at each multiple of $f_C \pm n \cdot f_m$ it is easy to understand that when f_m is decreased to values smaller than the RBW more than one spectral component will be present within the receiver band. Decreasing f_m leads to increasing peak measurement values and decreasing average values. In the given example in figure 7 it happens that 9 spectral components are located within the 9kHz RBW when the EMI receiver is measuring at e.g. 2.5MHz, which results in a difference between the peak and average measurement of ~19dB.

As f_m increases to values higher than the RBW, peak, average and quasi peak measurements will show nearly the same results, as again only one spectral component will fall into the receiver's RBW. This effect is illustrated in figure 8. In this case a modulation frequency of $f_m = 10$kHz which is slightly higher than the RBW of 9kHz for the measurement in the frequency range from 150kHz to 30MHz was used.

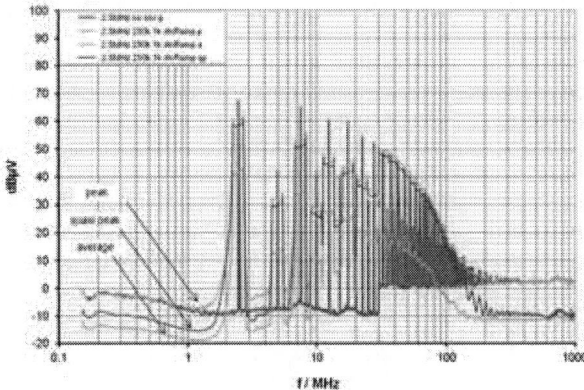

Figure 7. Influence of the modulation frequency on the electromagnetic emissison (f_C=2.5MHz, Δf_C=250kHz, f_m=1kHz)

As can be seen in this frequency range the peak, average and quasi peak values are the same. But for the higher frequency range from 30MHz to 1GHz where a RBW of 120kHz has to be used, again a huge difference between the peak and average values is present. A similar reduction for all detectors to the same emission value is achieved when the modulation frequency is slightly higher than the corresponding resolution bandwidth.

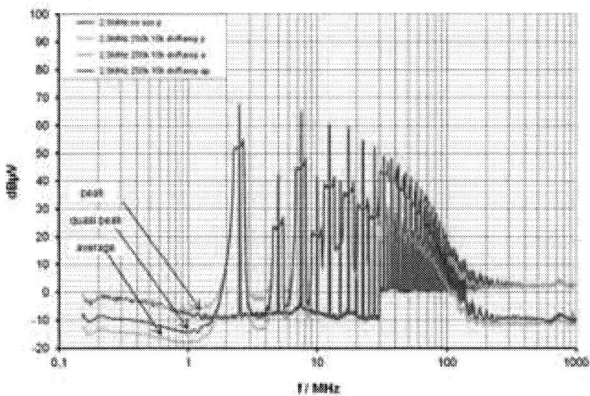

Figure 8. Influence of the modulation frequency on the electromagnetic emissison (f_C=2.5MHz, Δf_C=250kHz, f_m=10kHz)

C. Modulation signal:

The modulation signal or modulation profile defines the way how the modulation frequency is changed. It determines the shape of the spectrum outline and is another key parameter to maximize the reduction of the emission of a clock signal.

Among the variety of different modulation signals such as sinusoidal, triangular, rectangular, pos. and neg. ramp saw tooth functions, etc. there is one signal that is clearly standing out, the so-called "Hershey-Kiss[TM]" function. This function was patented in the 1990ies as it offers the best distribution of the spectral energy and optimizes the reduction by providing the flattest top of the spectrum outline [12], [13].

978-1-4799-5004-1/13 $31.00 © 2013 IEEE

In figure 9 the shape of the spectrum outline in the frequency range around the fundamental frequency of 2.5MHz for some commonly used modulation signals is shown. As can be seen the rectangular modulation provides two areas where the emission is much higher compared to the other modulation signals. Although the difference between the rectangular and the other used modulation signals is for this example ~10dB the type of modulation signal in general compared to the previously described two parameters plays a minor role.

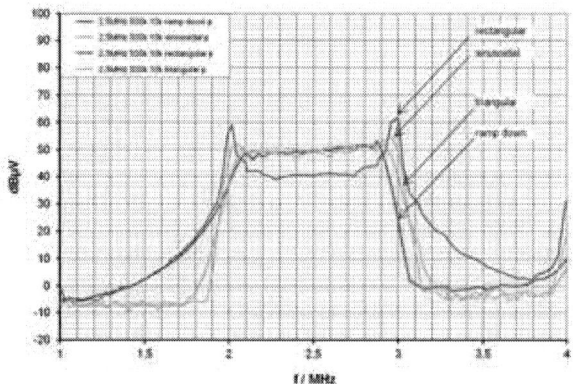

Figure 9. Influence of the modulation signal on the electromagnetic emissison (f_C=2.5MHz, Δf_C=500kHz, f_m=10kHz)

V. CONCLUSION

In this paper the ability of using spread spectrum clocking to reduce the electromagnetic emission of charge pump applications is discussed. Spread spectrum clocking is usually a very easy and cheap to implement technique that allows significant system cost savings as additional expensive external measures such as filtering capacitors, ferrite beads, or shielding that are traditionally required to pass emission regulations can be minimized or even neglected. It is shown in detail how frequency modulation used for spreading the spectrum of the clock of a charge pump implemented on an EMC test chip affect the electromagnetic emission. In general, the emission of the charge pump can significantly be reduced over a wide frequency range by using spread spectrum techniques. But it is essentially important to understand how the three described frequency modulation parameter such as frequency deviation, modulation frequency and modulation signal related to the final measurement results especially when peak, average and quasi peak detectors as well as different resolution band widths are used for the emission measurement. Otherwise spread spectrum clock generation will not always be the best choice as one might increase the likelihood of unexpected problems by filling up frequency ranges with broadband emission where originally no emission would be present without spread spectrum techniques.

ACKNOWLEDGMENT

The author would like to thank Gebhard Melcher (Infineon Technologies/Austria) for designing the EMC charge pump test chip as well as Gunter Winkler (TU Graz/Austria) for the helpful discussions in the early phase of the project and especially Arnaud Wahl (Infineon Technologies Germany) for his help in generating the numerous measurement results.

REFERENCES

[1] F. Galtie, C. Marot, "Spread spectrum clocking applied to charge pump for conducted emission improvement in automotive", Proceedings of 19th Int. Zurich Symposium on Electromagnetic Compatibility, Asia-Pacific Symposium on EMC, APEMC 2008, pp. 267 - 270

[2] D. Gonzalez, et al. "Conducted EMI Reduction in Power Converters by Means of Periodic Switching Frequency Modulation", IEEE Transactions on Power Electronics, vol. 22 , no. 6, 2007, pp. 2271 – 2281

[3] N. Tesla, "System of Signaling", U.S. Pat 725.605, filed July 16, 1900

[4] N. Tesla, "Method of Signaling", U.S. Pat 723.188, filed June 14, 1901

[5] H. Kiesler Markey, G. Antheil, "Secret communication sytem", U.S. Pat. 2.292.387, filed June 10, 1941

[6] F. Lin, D. Y. Chen, "Reduction of Power Supply EMI Emission by Switching Frequency Modulation", Virginia Power Electronics Center, Virginia Polytechnic Institute and State University, The VPEC Tenth Annual Power Electronics Seminar, Sept. 20-22, 1992, Blacksburg, Virginia

[7] K. B. Hardin, J. T. Fessler, D. R. Bush, "Spread Spectrum Clock Generation for the Reduction of Radiated Emissions," Proceedings of the 1994 IEEE International Symposium on Electromagnetic Compatibility, Chicago, IL., August 22 - 26, 1994, pp. 227 - 231

[8] K. B. Hardin, J. T. Fessler, D. R. Bush, "Digital Circuit Radiated Emission Suppression With Spread Spectrum Techniques", Interference Technology Engineers Master (ITEM) 1994

[9] H. G. Skinner, K. P. Slattery, "Why Spread Spectrum Clocking of Computing Devices is Not Cheating", IEEE International Symposium on EMC. Volume 1, 13-17 Aug. 2001, pp. 537 – 540, vol.1

[10] J. Shepherd et al., "Getting the most out of frequency spreading", Proceedings of EMC Compo 09, Nov. 17-19, Toulouse France, 2009

[11] T. Steinecke, "Low-jitter frequency-modulated PLL: Clipped-FM PLL significantly reduces the maximum time interval error for proper operation of asynchronous serial data interfaces", Proceedings of EMC Compo 11, Nov. 06-09, Dubrovnik, Croatia, 2011, pp. 176 - 181

[12] K. Hardin, et al., "Spread Spectrum Clock Generator and Associated Method", U.S. Patent 5 488 627, Jan. 30 1996

[13] K. Hardin, "Spread Spectrum Clock Generator", U.S. Patent 5 631 920, May 20 1997

[14] M. Amighini, V. Poletto, "Design Guidelines for Improvement of EME of Switching Circuit, like Charge Pump", 6th Int. Workshop on Electromagnetic Compatibility of Integrated Circuits, EMC Compo 2007, pp, 48-52, Torino, Italy

[15] IEC 61967-4, "Integrated circuits - Measurement of electromagnetic emissions, 150 kHz to 1 GHz - Part 4: Measurement of conducted emissions - 1 ohm/150 ohms direct coupling method", Ed. 1.1, 2006/07

[16] CISPR16-1-1 Ed. 3, "Specification for radio disturbance and immunity measuring apparatus and methods Part 1-1: Radio disturbance and immunity measuring apparatus – Measuring apparatus", International Electrotechnical Commission, 2010

Evaluating the Impact of Substrate Noise on Conducted EMI in Automotive Microcontrollers

Marco Cazzaniga[1,2], Patrice Joubert Doriol[1], Aurora Sanna[1],
Emmanuel Blanc[3], Valentino Liberali[2], and Davide Pandini[1]

[1]Central CAD and Design Solutions, STMicroelectronics, Agrate Brianza, Italy
[2]Dipartimento di Fisica, Università degli Studi di Milano, Milano, Italy
[3]Apache Design Inc., Grenoble, France

Abstract—Board-level I/Os signal integrity and conducted EMI have become a critical concern for high-speed circuit and package designers, and a major element of performance and reliability degradation in modern electronic systems, in particular for automotive microcontrollers, which must satisfy stringent low-EMI and noise immunity requirements. One of the most detrimental root causes of I/O signals conducted EMI is the simultaneous switching noise generated by the toggling I/Os (SSO) on the power distribution network of the I/O ring. However, this is not the only noise source that must be considered. In fact, an often overlooked contributor to SSO is the noise generated by the switching digital core that propagates to the I/Os and the noise-sensitive on-chip analog circuitry throughout the common silicon substrate. In this work, we analyze the impact of substrate noise on the I/O signals conducted EMI of an industrial automotive microcontroller, and we compare it against other noise sources. Moreover, we demonstrate the effectiveness of the technological protections against substrate noise in a leading-edge technology.

Keywords—Substrate noise; noise integrity; simultaneous switching output noise (SSO); signal integrity; I/O ring; conducted EMI; System-on-Chip; automotive microcontroller

I. INTRODUCTION

The ever-increasing performances and complexity of modern System-on-Chip (SoC) designs and the consequent increase of their operating frequencies can impair the electromagnetic compatibility (EMC) of the electronic systems (encompassing the die, package, and board), and deteriorate the electromagnetic interference (EMI) generated by these systems in the surrounding environment. Following the enforcement of strict governmental regulations and international standards [1][2] and customer specifications, mainly (but not only) in the automotive domain, EMC and signal integrity have become additional mandatory objectives for first-silicon success [3]. In particular, the I/O switching noise, or SSO, has emerged as a critical detrimental factor in high-performance SoC designs because of higher I/O count and faster signal edge rates [4][5]. Where previous generation designs could extensively exploit off-chip decoupling capacitors (i.e., decaps) for mitigating SSO effects, today the high I/O cell density makes it difficult to place PCB (printed

circuit board) decaps close to the I/O pins. Moreover, SSO also introduces a major challenge for conducted EMI on the local signals, which must be analyzed and controlled along with the EMI generated by the IC digital core power distribution network (PDN). It is worth pointing out that optimizing EMI on the I/Os introduces an additional challenge with respect to EMI on the die power/ground supply. In fact, the package choice and consequently the number of allowed I/Os are usually dictated by strict and cost-driven marketing specifications during the design conceptualization stage, and once defined cannot be changed during the other steps of the design implementation. Nowadays, the large effort spent on properly modeling and limiting the effect of the on-chip simultaneous switching noise (SSN), also known as power rail noise, allows to control the EMI generated by the power supply pads [3][6]. However, the same effectiveness does not apply to the I/O signals, because the upper limits on the number of available package supply pads, as well as the die size constraints, make it difficult to implement efficient strategies for EMI reduction on the I/O signals [7]. Therefore, to meet the design constraints, avoiding the utilization of unnecessary components, and reducing the risk of design re-spins caused by EMI failures, an accurate knowledge of all the EMI sources as well as an adequate modeling approach are mandatory.

An often overlooked noise source is the SSN that from the switching digital core propagates through the common silicon substrate to the I/Os and analog blocks. Although in modern SoC design it is common practice to keep the power supply network of the high-voltage circuitry (I/Os and analog blocks) separated from the supply network of the digital core (typically standard cells and SRAMs), the digital SSN can still propagate from the digital PDN to the noise-sensitive I/O cells and analog IPs through the propagation path of the silicon bulk. In particular, SSN can generate substrate currents through the diffusion contacts biasing the wells. Furthermore, fast voltage variations at the transistors source and drain terminals, and the impact ionization close to the drain of the n-channel transistors introduce more sources of substrate noise. The silicon wells, depending on their dopant concentration, can provide either lightly or highly resistive propagation paths for these currents, while the reverse-biased well junctions can

978-1-4799-5004-1/13 $31.00 © 2013 IEEE

be modeled by their junction capacitances. By means of this substrate RC mesh, digital SSN can reach any noise-sensitive circuitry.

In this paper we focus on the I/O cells, because we want to evaluate the impact of substrate noise on I/Os conducted EMI, since such noise will add up to the I/O ring power/ground supply fluctuations (i.e., SSO). Therefore, under the hypothesis that the two effects can be linearly superimposed, we extract the substrate noise waveforms at the I/Os supply contacts and we superimpose these waveforms with the SSO injected on a quiet victim I/O by adjacent toggling aggressor I/Os. Furthermore, we evaluate the effectiveness of a typical substrate noise guard structure such as a deep n-well (DNW) below the I/Os, which is normally used in silicon technology. This paper is organized as follows. In Section II we review some previous work in this area, while in Section III we describe the approach used to extract the substrate noise waveforms to analyze the conducted EMI. In Section IV we present the experimental results for the I/O signals conducted EMI taking into account the contribution of substrate noise on an automotive microcontroller. Finally, Section V summarizes a few conclusive remarks.

II. PREVIOUS WORK

The most important substrate noise source is the digital core SSN, which can inject noise into the substrate by means of three different mechanisms: a) power supply fluctuations and ground bounce; b) capacitive coupling of transistors source and drain; c) impact ionization current [8][9]. While mechanism c) can usually be considered negligible [8], the relative importance of the other two noise sources depends on the technology and design choices [9]. The digital core itself is not affected by substrate-related problems, and it was demonstrated that the IR-drop analysis produces more pessimistic results when the substrate parasitic network is not considered [10]. However, the same positive impact does not apply to the on-chip analog circuitry, whose performances and functionality can degrade under the effect of the noise originated by the SSN of the digital core, which propagates through the silicon substrate and reaches the PDN of these circuits. Several efforts were devoted to understand the effects of substrate noise on analog circuits such as VCOs and PLLs [11][12]. To limit its impact on sensitive circuitry, a deep knowledge of the noise propagation mechanisms and the efficiency of the guard structures is necessary. In [13] the authors presented a study on the propagation mechanisms and the shielding techniques in a lightly-doped substrate. A similar study was proposed in [14], while the efficiency of the DNW to shield the substrate noise was studied in [15], where it was demonstrated that in particular cases (e.g., at high frequencies), it can also be counterproductive.

EMC is another issue that must be considered in IC design. Therefore, different approaches were proposed to evaluate EMI on industrial microcontrollers [3][16]. In [7] SSO noise modeling techniques and design solutions were proposed, with a particular focus on I/O signals conducted EMI. In [6] the critical impact of small analog blocks on EMI was demonstrated and accurately modeled. However, in these works the substrate noise was not taken into account. It is worth noticing that the common silicon substrate is not usually considered in I/O signals EMI modeling, even if provides a coupling path from the noisy digital core to the I/O ring. The reason for this flaw is that efficient and accurate substrate models are not available, and their extraction within an industrial digital design flow is cumbersome and impractical. Macromodels for EMI estimation which considered also the substrate were presented in [16][17]. However, these approaches suffer from limitations that make them unsuitable for complex SoC designs, such as automotive microcontrollers. The substrate model proposed in [16] consists of a capacitance in parallel with a RL series, whose component values are extracted from S_{21} parameters by means of manual fit. Furthermore, this model can only connect the ground networks on the die. On the other hand, the substrate model described in [17] is derived from a highly doped substrate and is based on a resistance representing the lightly doped epitaxial layer covering the highly doped bulk, in parallel with a capacitance representing the reverse-biased junction capacitance of the wells. Since the substrate has a very low resistance it can be approximated by an equipotential node. Obviously, this simple model cannot realistically represent a process technology with lightly doped substrate, twin wells, and DNWs, like the one considered in this work, and cannot take into account the different topologies and consequently the different capacitances and resistances of the propagation paths between the digital core and the I/O cells.

III. METHODOLOGY

The approach presented in this work is based on the assumption that the different noise components on the I/O ring can be analyzed separately. Therefore, we consider the overall I/O ring power/ground rail noise as a linear superposition of the substrate noise and SSO. This assumption allows to extract the substrate noise waveforms and to represent them as independent voltage sources at the power supply/ground pins of the I/O cells. Moreover, it is also possible to evaluate the specific contribution of the substrate noise with respect to SSO, and consequently to estimate the accuracy loss which stems from neglecting the substrate noise contribution in EMI and signal integrity analysis. Although previous approaches based on simplified substrate representations (typically a single resistor, capacitance, and inductance estimated from doping profiles, process technology data and physical layout, or characterized from silicon measurements) were proposed [16][17], nevertheless, they cannot accurately consider all the different propagation paths from the PDN of the digital core to the I/O ring, especially for technologies based on a highly-resistive silicon bulk with DNW. Hence, the substrate network must be represented by a lumped-element distributed RC mesh, in order to model all the resistive propagation paths and to consider all the well junctions, from the digital PDN to the I/O ring. We used a commercial chip-level power integrity tool (Apache RedHawk [18] with the Chip Substrate Extension (CSE) [19]), which allows modeling the switching

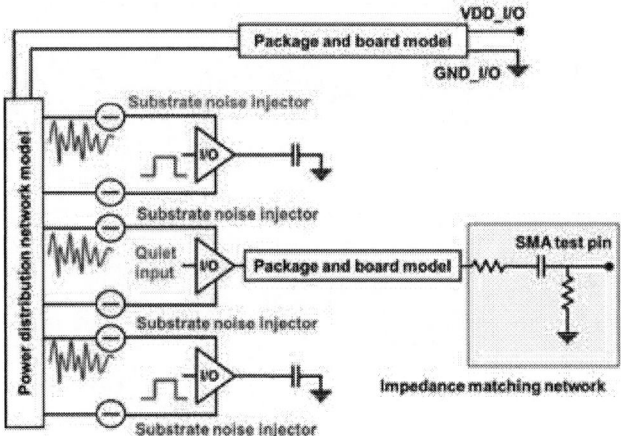

Figure 1. Test bench block diagram for conducted EMI analysis

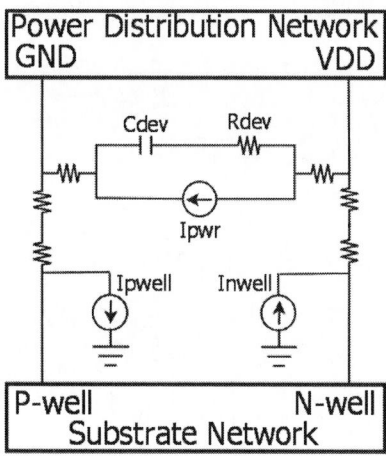

Figure 2. Standard cell model for substrate noise analysis

digital core, the substrate noise injection, and its propagation through a *RC* mesh towards the I/O cells and the analog circuitry. This tool allows extracting the time-domain waveforms of the substrate noise generated by the digital core at different physical locations on the die, in particular at the supply contacts of the I/O cells. These waveforms can be superimposed by means of independent voltage sources (the substrate noise injectors in Figure 1) connected to the SPICE netlist of the I/O ring, which includes the PDN parasitics and the transistor-level netlists of the I/Os. The macromodel shown in Figure 1 is completed with the package and PCB traces models, and the impedance matching network of the SMA (*SubMiniature version A* connector) test pin. The forthcoming subsections will describe the details of the two different steps for conducted EMI evaluation.

A. Substrate noise modeling

The substrate noise analysis was performed at gate level with an extended standard cell model. The typical cell linear model used for dynamic power integrity analysis consists of an independent current source in parallel with a resistor-capacitor series that represents the intrinsic parasitic capacitance of the standard cell. This model cannot capture the different mechanisms of substrate noise injection. Only the noise originated by SSN can be generated with such model. In order to describe the other substrate noise mechanisms, such as device parasitic capacitive coupling and hot-carrier injection, this model must be extended with two independent current sources injecting noise currents into the nodes where the cell contacts the twin wells and the DNW, as illustrated in Figure 2. These additional current sources inject in the substrate network the currents probed at the body pins of the standard cell transistors, whose device models include the aforementioned effects. Previous models which did not comprise these current sources could not capture these noise injection mechanisms. Therefore, this extended cell model is a critical component in our methodology.

The substrate noise propagation paths are described by a lumped-element distributed *RC* parasitic network. The tool builds a mesh of resistors inside the wells and the common bulk, and inserts capacitors at the reverse-biased junctions. The mesh refinement can be controlled by a set of parameters which determine the number of layers used to model the die thickness and the discretization resolution along the die surface.

B. Conducted EMI evaluation

The evaluation of I/O signals conducted EMI with the substrate noise generated by the digital core is based on linear superposition of the substrate noise waveforms propagated to the high-voltage supply contacts of the I/O cells with the SSO generated by the toggling I/Os. Once the time-domain waveforms of the substrate noise have been obtained, we build the transistor-level macromodel shown in Figure 1, and by means of SPICE simulations we obtain the noise waveform at the output of a quiet victim I/O, which propagates to the SMA test pin through the package and PCB traces. The frequency-domain spectrum obtained by FFT allows assessing the conducted EMI. It is worth noticing that the same approach can also be used for signal integrity analysis on a DDR2/DDR3 I/O bank, where all I/Os toggle simultaneously. However, where the substrate noise might have a critical impact is on conducted EMI. In fact, the noise contribution will be potentially much more important on a quiet I/O instead of a switching I/O, whose output signal amplitude is much larger than the noise amplitude.

C. Case study

The high-resistivity substrate and the presence of a DNW below the I/O cells are expected to provide an efficient protection against substrate noise propagation. To properly understand how these noise protection structures prevent conducted EMI, we replaced the substrate with a highly doped version (by reducing the substrate resistivity of three order of magnitude and adapting the junction capacitances accordingly), and we removed the DNW below the I/O cells. The flexibility

978-1-4799-5004-1/13 $31.00 © 2013 IEEE   131

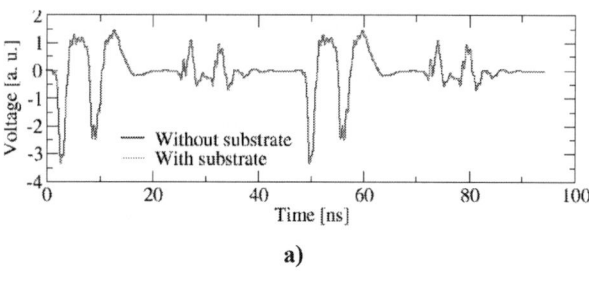

a)

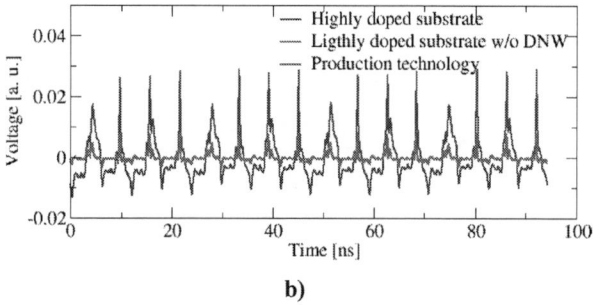

b)

Figure 3. Victim I/O time-domain power supply noise: a) SSO + substrate; b) substrate only

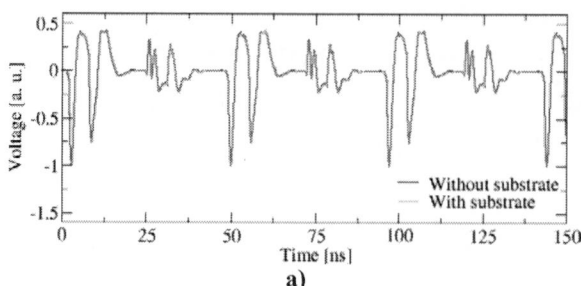

a)

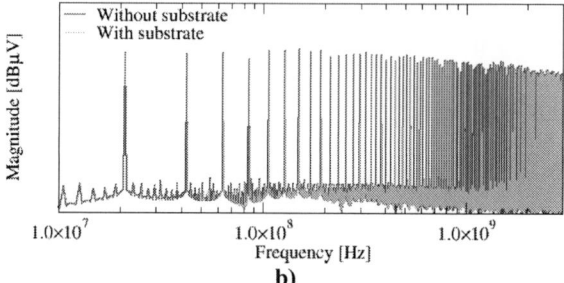

b)

Figure 4. Victim I/O output signal: a) time-domain waveforms; b) frequency spectra

of the tool allowed performing this technology exploration very efficiently, by simply modifying the technology file and without changing the design database.

IV. CONDUCTED EMI ANALYSIS WITH SUBSTRATE NOISE

We applied the approach presented in Section III to extract the substrate noise generated on the supply pins of the I/O cells of an automotive microcontroller in a leading-edge CMOS technology with embedded Non-Volatile-Memory (eNVM). The die package is a wire bonding package. This microcontroller was fabricated on a highly resistive p-substrate, which is a first shield for sensitive components against substrate noise propagation. Moreover, while the digital core is directly implanted on the p-substrate, noise sensitive analog blocks (including also the final driver stage of the I/O cells) are protected by a DNW. For confidentiality reasons, the waveforms reported in this paper were purged from their DC component and normalized, so that the peaks with larger magnitudes in the conducted EMI waveforms were bounded to one. The other waveforms were normalized accordingly. The victim I/O considered in this analysis is the same I/O where we evaluated the conducted EMI. The probing points to extract the substrate noise waveforms were placed at the diffusions biasing the wells, where the transistors driving the output pads are located. To support this approach, the tool adapts the substrate *RC* mesh in order to insert a node at the same die coordinates where the probes are located, thus giving access to the voltage fluctuations at the desired physical location.

A. Substrate noise waveforms

The substrate noise waveforms extracted at the VDD supply pin of the victim I/O are shown in Figure 3b), and were superimposed by means of an independent voltage source (i.e., the substrate noise injector in Figure 1) to the SSO waveform on the victim I/O in Figure 3a), which is generated by the toggling aggressor I/Os (including their substrate noise). As shown in Figure 3b) the substrate noise (here reported in arbitrary units) is two orders of magnitude smaller than SSO, as also confirmed in Figure 3a), where the contribution of the substrate noise is negligible.

Moreover, we studied the effectiveness of different technological options to shield noise-sensitive blocks from the detrimental effects of substrate noise. As illustrated in Figure 3b), the most effective solution is the combination of a lightly doped substrate with a DNW below the noise-sensitive blocks (i.e., our production technology). As it was expected, the major contribution to shielding stems from a highly resistive substrate. Similar conclusions can be drawn for the ground bounce noise. These results demonstrate that a technology based on a lightly doped substrate plus DNW provides an effective protection against substrate noise. Therefore, it can be used to fabricate microcontrollers with increasingly tight specifications on conducted EMI.

B. Conducted EMI analysis

Figure 4 represents the time-domain waveforms and the corresponding frequency spectra obtained with FFT at the SMA test pin for the victim I/O, when its input is set to a stable-high level. Following the results presented in Figure 3, the substrate noise contribution to the conducted EMI is negligible, as confirmed from the results summarized in Table 1, where for confidentiality only the difference with respect to

Table 1. Substrate noise impact on conducted EMI

Harmonics	1st ref.	2nd 2X ref.	10th 10X ref.	20th 20X ref.
Without substrate noise (reference)	–	–	–	–
With substrate noise (difference in dBµV)	0	0.01	-0.03	-0.14

the reference case (without substrate noise) is reported. The results presented in this work confirm that the technological solution based on highly resistive bulk and DNW is very effective at shielding the I/O cells from the noise generated by the switching digital core. Therefore, in this technology, the substrate noise is not a significant factor impacting the I/O signals conducted EMI.

V. CONCLUSIONS

In this work we discussed and analyzed the effect of substrate noise on the I/O signals conducted EMI of an automotive microcontroller. Substrate noise is often considered as a major culprit for conducted EMI degradation; however, to the best of our knowledge, a full-chip analysis on a complex SoC design of this effect has never been presented. One of the potential root causes for this gap was the lack of efficient chip-level substrate-aware tools that can be seamlessly integrated into a standard digital design flow.

In this paper, we proposed a flow and methodology to effectively evaluate the impact of the noise generated by the switching digital core that propagates to the noise-sensitive circuitry like the I/O cells, through the common silicon bulk. Our approach was exploited during the implementation of an automotive microcontroller, to analyze the impact of substrate noise on the conducted EMI of the I/O signals. We demonstrated that for this kind of SoCs in this technology (a leading-edge eNVM technology) the impact of substrate noise on conducted EMI is negligible. However, it is important to point out that this positive outcome is mainly due to the effectiveness of technological solutions like the combination of a lightly doped substrate and a deep-n-well, which proved to be very efficient at shielding the I/O cells from the digital noise. Therefore, we believe that for these process technologies, in order to limit I/O signals conducted EMI, designers should focus their efforts mainly on damping the I/O ring power/ground fluctuations generated by the simultaneous toggling of several aggressor I/O cells.

ACKNOWLEDGEMENTS

This work was partially supported by the Dote di Ricerca Applicata of Region Lombardy.

REFERENCES

[1] IEC 61967, 2001, *Integrated circuits – Measurements of electromagnetic emissions, 150 kHz to 1 GHz,* IEC standard; www.iec.ch.

[2] IEC 62132, 2003, *Characterization of integrated circuits electromagnetic immunity,* IEC standard; www.iec.ch.

[3] P. Joubert Doriol, Y. Villavicencio, C. Forzan, M. Rotigni, G. Graziosi, and D. Pandini, "EMC-aware design on a microcontroller for automotive applications," in *Proc. DATE,* Apr. 2009, pp. 1208-1213.

[4] JEDEC Standard, JESD8-22A, Oct. 2012, *HSUL_12 LPDDR2 and LPDDR3 I/O with optionl ODT,* www.jedec.org.

[5] JEDEC Standard, JESD79-3F, Jul. 2012, *DDR3 SDRAM Standard,* www.jedec.org.

[6] C. Forzan, P. Valente, A. Kumar, and D. Pandini, "Understanding and modeling the impact of analog IPs on system-on-chip's EMI," in *Proc. EMC Europe,* pp. 1-5, Sept. 2012.

[7] P. Joubert Doriol, A. Sanna, A. Chandra, C. Forzan, and D. Pandini, "SSO noise and conducted EMI: modeling, analysis, and design solutions," in *Proc. Signal and Power Integrity Workshop,* pp. 107-110, May 2012.

[8] J. Briaire and K. S. Krisch, "Principles of substrate crosstalk generation in CMOS circuits," *IEEE Trans. on Computer-Aided Design of Integrated Circuits and Systems,* vol. 19, pp. 645-653, Jun. 2000.

[9] E. Salaman, E. G. Friedman, R. M. Secareanu, and O. L. Hartin, "Identification of dominant noise source and parameter sensitivity for substrate coupling," *IEEE Trans. on VLSI Systems,* vol. 17, pp. 1559-1564, Oct. 2009.

[10] R. Panda, S. Sundareswaran, and D. Blaauw, "On the interaction of power distribution network with substrate," in *Proc. Intl. Symp. on Low Power Electronics and Design,* pp. 388-393, Aug. 2001.

[11] P. Heydari, "Analysis of the PLL jitter due to power/ground and substrate noise," *IEEE Trans. on Circuits and Systems-I,* vol. 51, pp. 2404-2416, Dec. 2004.

[12] C. Soens, G. Van der Plas, M. Badaroglu, P. Wambacq, S. Donnay, Y. Rolain, and M. Kuijk, "Modeling of substrate noise generation, isolation, and impact for an LC-VCO and a digital modem on a lightly-doped substrate," *IEEE Journal of Solid-State Circuits,* vol. 41, pp. 2040-2050, Sep. 2006.

[13] R. M. Vinella, G. Van der Plas, C. Soens, M. Rizzi, and B. Castagnolo, "Substrate noise isolation experiments in a 0.18µm 1P6M triple-well CMOS process on a lightly doped substrate," in *Proc. Conf. on Instrumentation and Measurement Technology,* pp. 1-6, May 2007.

[14] G. A. Rezvani and J. Tao, "Substrate isolation in 0.18µm CMOS technology," in *Proc. Intl. Conf. on Microelectronic Test Structures,* pp. 131-136, Apr. 2005.

[15] R. Rossi, G. Torelli, and V. Liberali, "Model and verification of triple-well shielding on substrate noise in mixed-signal CMOS ICs," in *Proc. European Solid-State Circuits Conference,* pp. 643-646, Sept. 2003.

[16] C. Labussiere-Dorgan, S. Bendhia, and E. Sicard, "Modeling the electromagnetic emission of a microcontroller using a single model," *IEEE Trans. on Electromagn. Compat.,* vol. 50, pp. 22-34, Feb. 2008.

[17] Y. Villavicencio, F. Musolino, and F. Fiori, "Electrical model of microcontrollers for the prediction of electromagnetic emissions," *IEEE Trans. on VLSI Systems,* vol. 19, pp. 1205–1217, Jul. 2011.

[18] RedHawk User Manual, Release 12.2, Jan. 2013, http://www.apache-da.com/products/redhawk

[19] Chip Substrate Extension (CSE), http://www.apache-da.com/products/totem/cse

978-1-4799-5004-1/13 $31.00 © 2013 IEEE

Impedance Balance Control for Suppression of Fluctuation on Ground Voltage in LSI Package

Masaaki Maeda, Tohlu Matsushima, Osami Wada
Department of Electrical Engineering,
Kyoto University
Nishikyo-ku Kyoto Daigaku Katsura, Kyoto, Japan
Email: {matsushima, wada}@kuee.kyoto-u.ac.jp

Abstract—**Simultaneous switching current of a CMOS circuit causes power and ground bounces. The voltage fluctuation is injected into the CMOS substrate, and it degrades the performance of the circuit operation. In this report, we focus on the fact that parasitic couplings in the CMOS substrate and parasitic inductance in the power and ground connection form a bridge circuit, and we demonstrate that the voltage bounce can be suppressed by controlling variable resistances that are inserted between the substrate resistive coupling and the conductor line for the DC supply. The effectiveness of this method is verified with a scaled quad flat package (QFP) and we reduced the measured voltage bounce about 40 dBμV.**

I. Introduction

In recent years, integration of digital and RF analog circuits on a single chip LSI has been accelerated with the advance of semiconductor manufacturing technology and its application to multimedia and wireless communication such as mobile broadcasting and smartphones. Simultaneous switching current of a digital circuit causes power and ground bounces on the LSI chip, and switching regulators also generate current and voltage noise. When the voltage fluctuation is injected into the CMOS substrate, it degrades the performance of the circuit operations [1], [2], [3]. In high frequency chip substrate mounted on a package such as QFP has strong coupling to the die support [4], and the die support is also coupled to the printed circuit board (PCB) below the package via parasitic capacitance, which forms coupling paths of the RF noise. To prevent the noise coupling through substrate and die support, it is important to suppress the voltage fluctuation on the substrate and die support.

One possible way to suppress it is the impedance balancing technique [5], [6], which utilizes a bridge circuit formed by parasitic inductances on the DC power supply circuit in the package and PCB and parasitic capacitances between a chip, a package and a board. By satisfying the balancing condition of an impedance bridge, the common-mode voltage at the DC power supply terminals of an LSI or a PCB is suppressed. However, in this balancing technique, one side of the impedance bridge is formed by parasitic capacitances which are usually very small and dependent on the structure of the package and the PCB and their dimensions. As a result, sometimes it is difficult to establish the balance control, particularly in high frequency region. In this paper we extend the impedance balancing technique and propose a new balancing circuit to apply to internal circuits of an LSI. Instead of the parasitic capacitance coupling, we adopt mainly

resistive interconnection paths which have low dependency on frequency. To demonstrate the new scheme, we prepared a scaled QFP package formed by a PCB, and realize the balancing circuit with CMOS ICs and discrete components.

This technique can be compatible with existing technologies, specifically the deep n-well technology. We plan to validate the effectiveness of our method by means of deep n-well structure in the future.

II. Reduction Method of Voltage Fluctuation of Lead Frame in QFP

A. Equivalent Circuit Model of PDN

The equivalent circuit model of chip and package PDN is constructed from QFP shown in Fig.1. A unit CMOS inverter with equivalent parasitic couplings in the substrate is shown in Fig.2(a). A CMOS inverter consists of a pMOS and an nMOS. An nMOS is made on a p-substrate and connected to V_{SS} line. A pMOS is made on an n-well which is manufactured on a p-substrate and connected to V_{DD} line. A p-substrate tap is connected to V_{DD} and an n-well tap is connected to V_{DD} to prevent latchup.

When high frequency noise current generated by operation of a CMOS circuit is injected into the CMOS substrate via the body contact, it causes the substrate voltage fluctuation. The equivalent circuit model of PDN with substrate noise coupling path is shown in Fig.2(b), where V_{sub} is the substrate voltage, R_{bw} and R_{sub} are equivalent resistances in CMOS substrate, and C_{bw} is the capacitance of PN junction between the n-well and the p-substrate. The current source represents the noise current due to operation of the LSI, and Z_{core} represents the impedance between a power supply terminal and a ground terminal of CMOS circuits. The upper part of Fig.2(b) represents the parasitic coupling of the on-chip metal, bonding wire and lead frame.

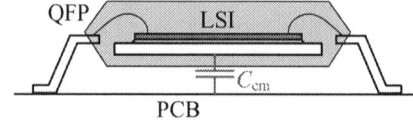

Fig. 1. Parasitic coupling between die support of a QFP package and PCB below

If the bypass capacitors are placed on the package and the input impedance of the power distribution network of the package is low enough, the equivalent circuit can be illustrated as the bridge circuit in Fig.2(c).

B. Impedance Balance Condition

The voltage fluctuation of V_{sub} depends on the ratio of impedances of the resistive couplings in the CMOS substrate. If the impedance ratio fulfills the impedance balance condition described below, the voltage fluctuation can be eliminated. The impedance balance condition of the equivalent circuit in Fig.2(c) is written as

$$\frac{R_{\mathrm{bw}} + 1/\mathrm{j}\omega C_{\mathrm{bw}}}{R_{\mathrm{sub}}} = \frac{R_{\mathrm{v}} + \mathrm{j}\omega L_{\mathrm{v}}}{R_{\mathrm{g}} + \mathrm{j}\omega L_{\mathrm{g}}}. \tag{1}$$

Separating into the real part and the imaginary part, Eq.(1) can be rewritten as

$$\frac{R_{\mathrm{bw}}}{R_{\mathrm{sub}}} = \frac{R_{\mathrm{g}}R_{\mathrm{v}} + \omega^2 L_{\mathrm{g}}L_{\mathrm{v}}}{R_{\mathrm{g}}^2 + (\omega L_{\mathrm{g}})^2} \tag{2}$$

and

$$\frac{1}{R_{\mathrm{sub}}} = \frac{\omega^2 C_{\mathrm{bw}}(L_{\mathrm{g}}R_{\mathrm{v}} - L_{\mathrm{v}}R_{\mathrm{g}})}{R_{\mathrm{g}}^2 + (\omega L_{\mathrm{g}})^2}. \tag{3}$$

Although these equations show that the impedance balance condition depends on frequency, if we can assume $1/\omega C_{\mathrm{bw}} \ll R_{\mathrm{bw}}$, Eq.(2) and Eq.(3) can be approximated as

$$\frac{R_{\mathrm{bw}}}{R_{\mathrm{sub}}} = \frac{L_{\mathrm{v}}}{L_{\mathrm{g}}} \tag{4}$$

and

$$\frac{1}{R_{\mathrm{sub}}} = \frac{C_{\mathrm{bw}}(L_{\mathrm{g}}R_{\mathrm{v}} - L_{\mathrm{v}}R_{\mathrm{g}})}{L_{\mathrm{g}}^2}. \tag{5}$$

When these conditions are satisfied, V_{sub} is equal to 0. Therefore, if the impedance ratio of the substrate couplings can be controlled, the package voltage fluctuation can be eliminated. We propose to insert variable resistances between the substrate resistive coupling and the conductor line for the power supply in order to tune the impedance ratio.

III. VALIDATION BY MEASUREMENT

A. Test Board Structure

The DUT is a two-layer scaled QFP with 48 pins formed by a printed circuit board. The circuit diagram of the DUT is shown in Fig.3. We use CMOS inverter ICs as noise sources and a ring oscillator as a clock wave generator. The fundamental frequency of the ring oscillator is 25 MHz that is tuned by a variable capacitor. The ring oscillator has two outputs with opposite phases and each of them is connected to the same number of inverters so that the same quantity of noise current flow through the power supply and ground lines.

Two chip resistances and a chip capacitance are mounted on the top of the package as the substrate noise coupling path. C_{bw} is 0.1 μF and R_{sub} is fixed to 20 Ω. R_{bw} is variable so that we can control the value of $R_{\mathrm{bw}}/R_{\mathrm{sub}}$.

L_{v} and L_{g} are calculated by inductance equipped on the package and parasitic inductance of the power supply or ground lines, and we can change the value of $L_{\mathrm{v}}/L_{\mathrm{g}}$ by

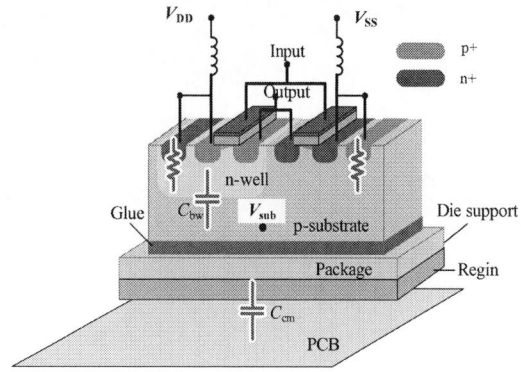

(a) Parasitic coupling in CMOS circuit.

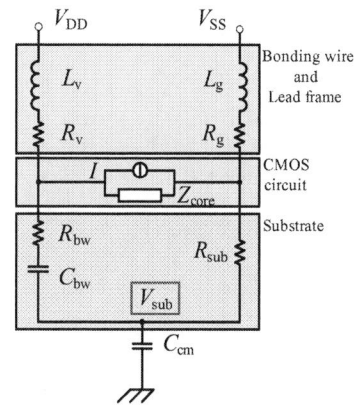

(b) Equivalent circuit model with parasitic noise coupling path.

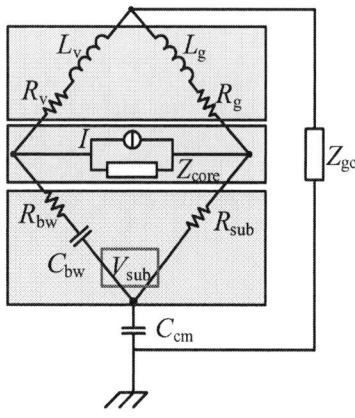

(c) Bridge circuit model.

Fig. 2. CMOS equivalent circuit model with parasitic coupling.

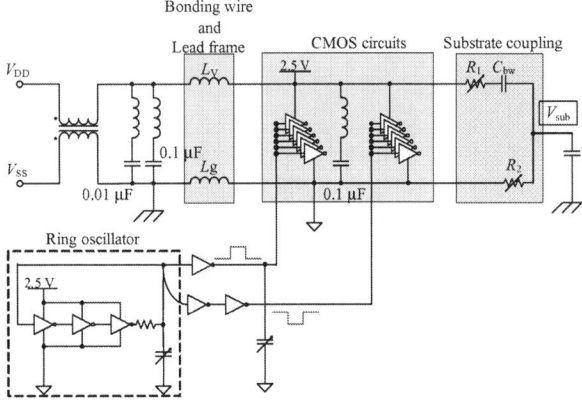

Fig. 3. Circuit schematic of test board.

changing the number of connection to the ground plane on the PCB.

The cross section of DUT is shown in Fig.4(a). The package is fixed above the PCB. The distance between the bottom of the package and the top of the PCB is 1.5 mm. The top layer of the package is shown in Fig.4(b). CMOS inverter ICs and additional components (R_{bw}, R_{sub} and C_{bw}) are mounted on this layer. The bottom layer of the package, shown in Fig.4(c), is the die support and we assume the voltage is the package ground voltage (V_{sub}).

B. Measurement Results

First, the measurement has been performed with $R_{bw} = R_{sub} = 0\ \Omega$ by a spectrum analyzer. The measurement results are shown in Fig.5. These results show that even harmonic waves are dominant in the frequency range below 200 MHz. The noise current generated by CMOS inverters flow power supply and ground lines twice in a period of output wave and it results in the dominance of even harmonic waves. Therefore, we should focus on the even harmonic waves.

The measurement has been performed again with $R_{sub} = 20\ \Omega$ and varying R_{bw}. The measurement results are shown in Fig.6. V_{sub} is minimized with $R_{bw} = R_{sub} = L_v/L_g$ in the frequency range below 200 MHz. This result shows that the effectiveness of our approach is verified at least in the frequency range below 200 MHz and 40 μVdB of suppression was achieved in maximum.

IV. CONCLUSION

In this paper, we proposed a new method for suppression of fluctuation on ground voltage in an LSI package. In this method, variable resistances are inserted between CMOS substrate and power supply/ground lines on the package. The voltage fluctuation is suppressed by controlling the variable resistances.

The equivalent circuit including the model of substrate noise coupling path is constructed. The impedance balance condition in order to eliminate the voltage fluctuation is derived from the equivalent circuit.

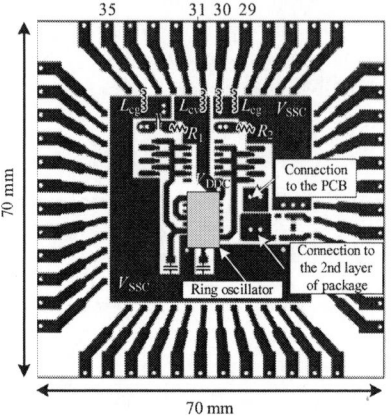

(a) Cross sectional structure.

(b) Top layer of test package.

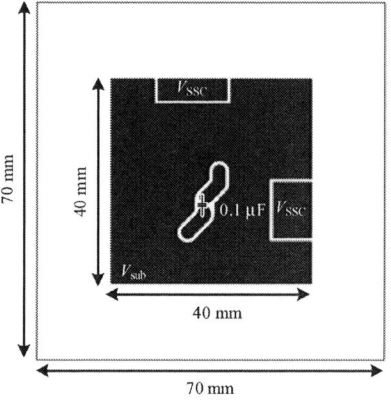

(c) Bottom layer of test package.

Fig. 4. Layout of test package.

The effectiveness of this method is proved by means of experimental test results. The test board is a scaled QFP formed by a PCB. CMOS inverter ICs as noise sources and a ring oscillator are mounted on this package, and chip resistances as substrate noise coupling path. We reduced the voltage fluctuation of the package ground by tuning the ratio of resistances and achieved 40 dBμV of suppression in maximum.

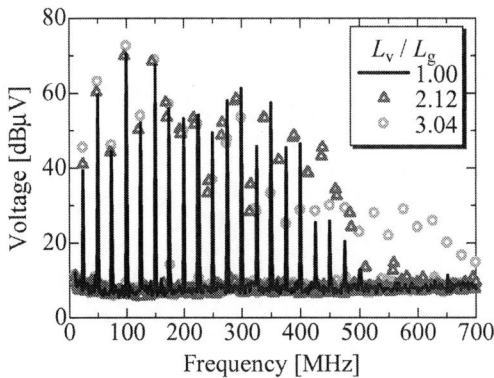

Fig. 5. Spectrum V_{sub} ($R_{\text{bw}} = R_{\text{sub}} = 0$).

ACKNOWLEDGMENT

The authors would like to thank Reresas Electronics and particularly to Mr. Atsushi Nakamura for their continuous support to our research and fruitful discussion on this work.

REFERENCES

[1] M. Jeong, et al., "A 65nm CMOS Low-Power Small-Size Multistandard, Multiband Mobile Broadcasting Receiver SoC," 2010 IEEE International Solid-State Circuits Conference (ISSCC),pp.460–461, Feb. 2010.

[2] D.K. Su, M.J. Loinaz, S.Masui, and B.A. Wooley, "Experimental results and modeling techniques for substrate noise in mixed-signal integrated circuits," IEEE J. Solid State Circuits, vol.28, no.4, pp.420–430, Apr. 1994.

[3] A. Afzali-Kusha, M. Nagata, N. K. Verghese, and D. J. Allstot,"Substrate Noise Coupling in SoC Design: Modeling, Avoidance, and Validation," Proceedings of the IEEE, vol.94, no.12, pp.1209–2138, Dec. 2006.

[4] T. Tsuda, T. Uno, K. Ichikawa, "EMI model improvement taking LSI package structure into consideration," IEIC Technical Report, EMCJ2006-84, vol.106, no.433, pp.19–24, Dec. 2006.

[5] Y. Mabuchi, A. Nakamura, A. Ohmae, T. Uno, K. Ichikawa, and H. Mizuno, "Development of a low emi micro-controller package for automobile applications," in 2009 20th International Zurich Symposium on Electromagnetic Compatibility, Zurich, Switzerland, Jan. 2009. pp.377–380.

[6] A. Nakamura, M. Maeda, T. Matsushima, O. Wada, "Substrate noise reduction based on impedance balance using tunable resistances," in 9th International Workshop on Electromagnetic Compatibility of Integrated Circuits (EMC compo 2013), Nara, Japan, Dec. 2013.

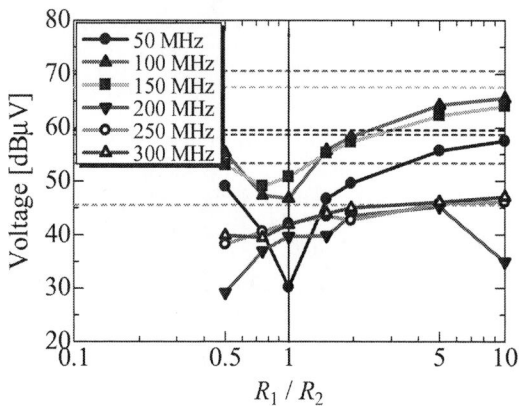

(a) $L_{\text{v}}/L_{\text{g}} = 1.00$

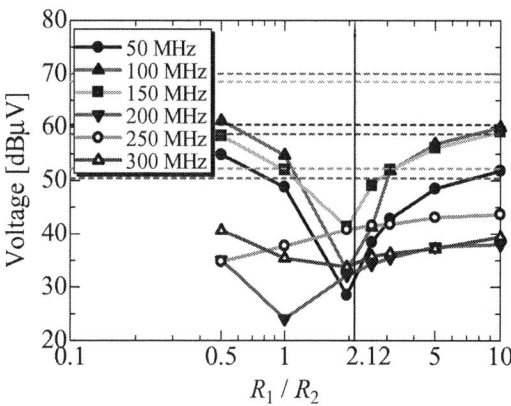

(b) $L_{\text{v}}/L_{\text{g}} = 2.12$

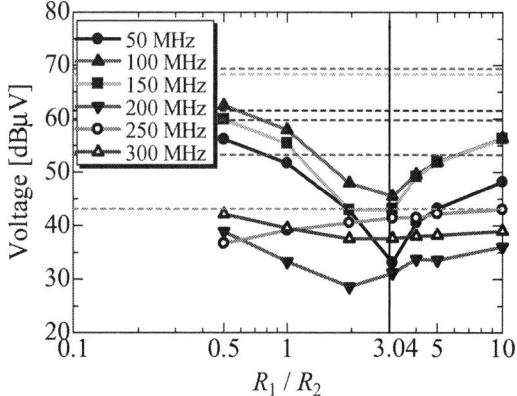

(c) $L_{\text{v}}/L_{\text{g}} = 3.04$

Fig. 6. Voltage of substrate V_{sub}.

Automatic Conducted-EMI Microcontroller Model Building

Shih-Yi Yuan
Dep. Communication Engineering,
Feng Chia University
Taichung, Taiwan, R.O.C.
syyuan@fcu.edu.tw

Shry-Sann Liao
Dep. Communication Engineering,
Feng Chia University
Taichung, Taiwan, R.O.C.
ssliao@fcu.edu.tw

Abstract—**This paper proposes an automatic algorithm for block-box electromagnetic interference (EMI) modeling of microcontroller (µC). Due to intellectual property considerations, IC design companies seldom expose internal architecture details of their µC products to EMI modelers. Since the internal module behaviors are unknown, it makes EMI modeling very difficult. This method is based on the measurement of a pre-prepared testing board(s) to build a conducted EMI (cEMI) µC model. The concept is based on block-box impulse response (BBIR) function calculation. BBIR is based on solely measurement basis and treat the target as a block-box. Through block-box type deductions and measurements, BBIR model can be built. After the model is built, the cEMI behavior of a new testing board (or module) with the same µC are estimated. A case study is given for the proposed method. In this case study, the cEMI model is firstly built and, then, followed by a real measurement of the cEMI behaviors of the new testing board. The proposed model is verified by the comparison of the estimated data and the physical measurements. From the experiment results, it shows that the proposed power model does in good accordance with the cEMI behavior of the target µC both in time-domain and frequency-domain.**

Keywords—*Power Integrity model; black-box model; Integrated Circuit modeling*

I. INTRODUCTION

Electronic devices are now essential to everyday life. These features include transportation, medical care, entertainment, communication, business, or safety ensuring. Among all these electronic devices, the central part is the microcontroller (µC). The electromagnetic interference (EMI) problem, by all means, is one of the most critical issues for µC to be used in these area. Thus, an EMI model for µC is essential for these applications.

The proposed method is based on IEC 62433 [1]. This standard is developed for the purpose of Integrated EMI (IC-EMI) modeling. The modeling standard has been successfully used to model the power and ground signal fluctuations for many µCs, ASICs, and programmable devices within the range between 1MHz–2GHz.

Although IEC-62433 is successful in some applications, there are several difficulties coming from the model building process. First, the accuracy of IEC-62433 model is greatly dependent on the internal current activity and internal structure of IC. Due to the intellectual property (IP) considerations, IC design companies seldom expose the internal architecture details of their IC products to cEMI modelers. Since the internal module behaviors are unknown, it

makes IEC-62433 modeling very difficult without the help of IC companies.

Second, conventional IEC-62433 model building process is tedious and error-prone. Without the supporting of IC designers, the estimation of gross Internal Impedance (IntZ) [1] is only based the very little public domain power property specifications available for target µC and some conjectures of a presumed internal static netlist which relies on the domain knowledge of the modeler.

Third, even a detail internal architecture of IC is supported by the IC designer, the current activity can only be modeled by several fixed-periodic linear (periodic triangle or periodic trapezoidal) current waveform. And the waveform can only estimate the first-order accuracy of the current activities which reduces the model accuracy [8].

Forth, the model cannot be applied to different modules or testing boards. Since the model building process in the original IEC-62433 is based on the fine-tuning of the model to fit the measurement data. The effects of IntZ and board-level impedance cannot be discriminated. Thus, the model can only be applied to the same module. Any other module, even with the same uC, is not applicable and a new model should be built from the beginning.

The proposed method tries to improve IEC-62433 modeling procedures by a black-box method. This method is modified from IEC-62433 and is specially targeted on µC's conducted EMI (cEMI) modeling. Through several mathematical deductions and cEMI measurement, the proposed model can be built. IntZ can be estimated by a set of proposed "black-box impulse response" (BBIR) functions. Both of the modeling efficiency and the model accuracy are increased by the proposed method.

The impulse response not only models the impedance but also models the internal activities. Since BBIR estimate the current source at frequency, the number of the equivalent time-domain current sources can be greatly increased. Actually, the number of the current sources can be as many as 2000-5000 or even higher depending on the sampling rate and accuracy of the measuring equipment. Due to the large number of current sources, the second or third order accuracy of the target can be modeled. The limit of the accuracy is dependent on the measuring equipment.

A case study is given (a Microchip$^@$ PIC microcontroller test board) for the effectiveness of the proposed method. In this case study, a cEMI model of a testing board is firstly built according to the proposed method. Its cEMI behaviors are

978-1-4799-5004-1/13 $31.00 © 2013 IEEE

estimated by the model. Then, a physical cEMI measurement of the testing board is followed. The proposed model is verified by the comparison on both of the estimated data and the physical measurements. From the experiment results, it shows that the propose model can accurately estimate the cEMI behaviors both in time-domain and frequency-domain.

The paper is organized as following: section II describes the difficulties of the original model and the proposed model concepts. Section III describes the proposed BBIR modeling method. Experimental results are described in section IV and followed by conclusions.

II. IEC-62433 DIFFICULTIES AND THE PROPOSED BLACK-BOX IMPULSE RESPONSE (BBIR) MODEL

The proposed model is for estimating the conducted time-domain power fluctuation behavior of any target μC which, in turn, can be transformed to cEMI. The IEC-62433 model is shown in Fig. 1. This model consists of a core activity current source(s) to translate IC activities to power behavior prediction parameters; other parameters include lumped passive components networks to represent IC package, contacted/non-contacted probe, and test board models.

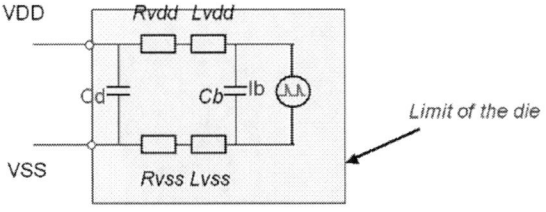

Fig. 1. ICEM (adapted from [1])

The general difficulties of IEC-62433 are described in section I. From the difficulties described above, there is a very hard problem to be solved when the modeling process is implemented.

The problem is that the conventional IEC-62433 modeling process is tedious and error-prone. Without the supporting of IC designers, the estimation of gross Internal Impedance (IntZ) [1] is only based the very little public domain power property specifications available for target μC and some conjectures of a presumed internal static netlist which relies on the domain knowledge of the modeler. Also, the estimation of the Internal Current Activity (IntCA) is a rough conjecture by experience. After the initial guessing, these two parameters (IntZ and IntCA) in the primitive model are simulated and, then, iteratively fine-tuned according to the physical measurements. As it can be expected, the trial-and-error modeling processing is both tedious and error prone.

Since the board-level impedance (Z_{PCB}) of the target board and the IntZ are modeled together, the model can only be accurate on the same board. And due to the IntZ/IntCA are dependent on each other, we can never be sure the final fine-tuned model is IntZ-accurate, IntCA-accurate, or not accurate at all. As a whole, the model cannot be applied on other boards even with the same μC. Thus, the purpose of the μC's cEMI modeling is not achieved.

The new proposed model avoids such difficulties. It is a black-box technique [2] for model building process. The concept is based on "block-box discrete impulse response" (BBIR) function calculation. BBIR is solely based on measurement basis and treat the target as a block-box. The impulse response can be easily derived from the transfer function from a circuitry theory [3].

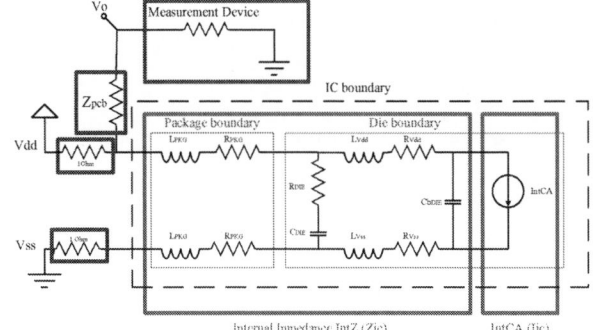

錯誤! 找不到參照來源。. (a) Netlist and its reduction for estimating IntZ ($Z_{IC}(f)$) and IntCA ($I_{IC}(f)$)

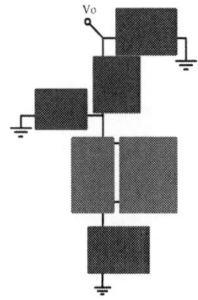

Fig. 2. (b) Reduction form of 錯誤! 找不到參照來源。. (a)

Example of a circuitry netlist and its reduction form are shown in Fig. 2. The netlist is based on IEC 61967 EMI measurement standard [4]. The blue blocks represent the controllable part of the cEMI modeler which is the board-level impedance. The red parts represent the IEC-62433 model for IC. This is only known by IC designers. It is not controllable and not known to the cEMI modeler. From Fig. 2, $V_O(f)$ can be deduced as:

$$\frac{-50 \times I_{ic} \times Z_{ic}}{(101 + 51 \times Z_{ic} + 2 \times Z_{pcb} + Z_{pcb} \times Z_{ic})} = V_O \qquad (1)$$

The unknown variables (the internal current activities I_{IC} and internal impedance Z_{IC}) are solved by the proposed method and BBIR deduction which will be detailed below.

III. BLACK-BOX IMPULSE RESPONSE DEDUCTION AND MODELING PROCEDURE

Since there are 2 variables (I_{IC} and Z_{IC}) to be solved in equation (1). Two testing boards (PCB1 and PCB2) are designed for solving the variables by different board-level impedance $Z_{PCB1}(f)$ and $Z_{PCB2}(f)$ designs. These 2 boards are embedded with the same μC. We can reasonably assume the unknown $Z_{IC}(f)$ and $I_{IC}(f)$ are generally the same. The board-

978-1-4799-5004-1/13 $31.00 © 2013 IEEE

level transfer function $Z_{PCB1}(s)$ and $Z_{PCB2}(s)$ can be easily derived from circuitry theory and easily transformed to $Z_{PCB1}(f)$ and $Z_{PCB2}(f)$. Assuming the output of PCB1 and PCB2 are $V_{O1}(f)$ and $V_{O2}(f)$. From here, we get two equations from (1):

$$\frac{-50 \times I_{ic} \times Z_{ic}}{(101 + 51 \times Z_{ic} + 2 \times Z_{pcb1} + Z_{pcb1} \times Z_{ic})} = V_{O1} \quad (2)$$

$$\frac{-50 \times I_{ic} \times Z_{ic}}{(101 + 51 \times Z_{ic} + 2 \times Z_{pcb2} + Z_{pcb2} \times Z_{ic})} = V_{O2} \quad (3)$$

By solving above equations, we get:

$$Z_{ic} = \frac{(-101 + (V_{O1} - V_{O2}) - 2 \times Z_{pcb1} \times V_{O1} + 2 \times Z_{pcb2} \times V_{O2})}{(51 \times (V_{O1} - V_{O2}) + Z_{pcb1} \times V_{O1} - Z_{pcb2} \times V_{O2})} \quad (4)$$

and

$$I_{ic} = \frac{V_{O1} \times (101 + 51 \times Z_{ic} + 2 \times Z_{pcb1} + Z_{pcb1} \times Z_{ic})}{-50 \times Z_{ic}} \quad (5)$$

Equations (4) and (5) can be solved by substituting every frequency responses of the measurements of $V_{O1}(f)$ and $V_{O2}(f)$. The only limitation is the measurement accuracy because the bandwidth of $V_{O1}(f)$ and $V_{O2}(f)$ are limited by the sampling rate and dynamic range of the measurement equipment.

From the deduction above, any groups of "Vo-pairs" (the output of the two boards) can be used to estimate the IntZ and IntCA. We consider a quasi-static condition in a short period of time that the IntZ and IntCA are assumed to remain static. Since all PBDM-style μC is digital and clock-driven, the quasi-static period can be safely estimated by the external clock-cycle or internal CPU machine-cycle (the integer multiple or fraction of the clock cycle). In the case study, the external clock-cycle is 4MHz and the machine-cycle is 1 MHz (a quarter of the external clock-cycle). The period can be safely set to 1μs.

A third board (verification board or PCBv) is designed for the verification purpose. After $V_O(f)$ of PCBv is estimated by the proposed model, the estimated $V_O(t)$ can be derived from IDFT [5]. The real PCBv responses can be measured and compared by high speed oscilloscope.

IV. EXPERIMENTAL RESULTS

From the above deductions, the proposed method estimates IntZ by BBIR method. The method is done without any prior knowledge of the netlist information. In the following verification, we try to estimate a μC's IntZ and IntCA. Three testing boards are designed (Fig. 3).

The PCB1, PCB2, and PCBv are specially designed that their board-level impedance (Z_{PCB}, or the blue part in 錯誤! 找不到參照來源。) are NOT the same. The Z_{PCB} of PCB1 and PCB2 are not following the IEC 61967 standard. The selection of Z_{PCB} can be found in [6]. The Z_{PCB} of PCBv is designed according to the standard. Table 2 shows the Z_{PCB} of the each board. The DUT to be estimated is a commercial μC PIC12F629 [7]. The oscilloscope is DPO72004 (Fig. 4).

Fig. 3. Three testing boards for physical verification (PCB1, PCB2, and PCBv)

TABLE I. Testing board ZPCB(Definition of ZPCB can be found in Fig. 錯誤! 找不到參照來源。)

	Zpcb
PCB1	1//(s×470e-6)
PCB2	1000+1/(s×6.8e-9)
PCBv	51+1/(s×6.8e-9)

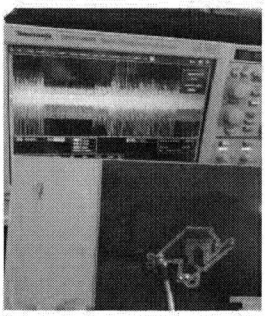

Fig. 4. Testing setups

A machine-cycle Vo-pair measurement is shown in Fig. 5.

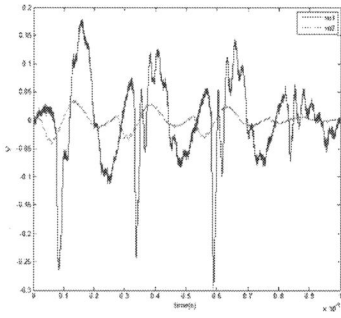

Fig. 5. Machine cycle Vo of PCB1 and PCB2

Based on the proposed method to estimate the quasi-static IntZ/IntCA, the BBIR model can be built. The PCBv output can be estimated by the model and the Z_{PCB} of PCBv board.

Fig. 6 shows the time domain and frequency domain comparisons between the measurement and the estimated result.

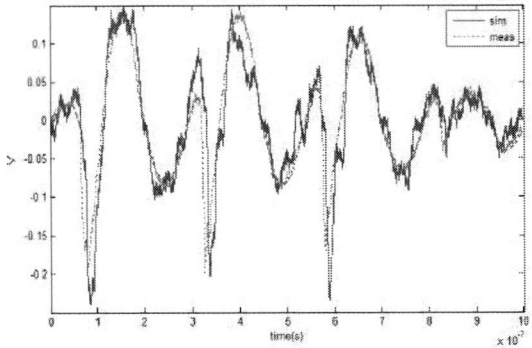

Fig. 6. (a) Time domain machine-cycle conducted-EMI measurement and estimation of PCBv

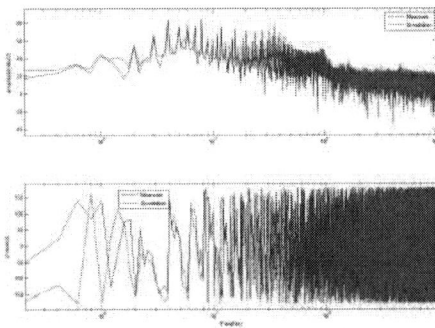

Fig. 6. (b) Frequency domain comparison of (a)

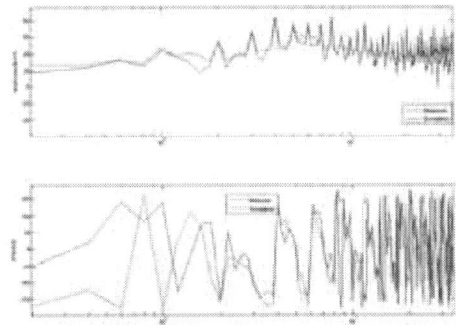

Fig. 6. (c) Zoom of Fig. 6. (b)

Fig. 6. The proposed method estimation vs. measurement

The estimated and measured waveform is very accurate within the range DC to 120 MHz.

Comparing to [8], the accuracy of the frequency domain bandwidth is increased from 80MHz to 120MHz. Time needed for the model building process is reduced. For any μC, the average model building time in [8] is 3 days. While in the new proposed method, the average building time is about 10 minutes. The most important issue is the model effectiveness. In [8], the model is only accurate in one single board. While in the new proposed method, owing to the BBIR deduction, the board-level information can be later cascaded to the IntZ model. Any new board-level power behavior can be recalculated according to the original model. Thus, the model can be applied to any board with the same target μC.

V. CONCLUSIONS

This paper proposes a method which can find block-box conducted EMI model for microprocessor (μC). The internal impedance (IntZ) and current activity (IntCA) can be analyzed and estimated by this method without any prior knowledge of μC. The IntZs/IntCAs are represented by a set of internal "block-box impulse response" (BBIR) functions. The method is verified both by theoretical deductions and a case study.

The case study is to build a cEMI model for a commercial μC. The BBIR cEMI model is build according to the proposed method. It shows that the method can be of practical use on any μC cEMI modeling. Comparing to previous researches, the frequency-domain bandwidth is increased (from 80MHz to 120MHz). The model building process is also greatly reduced (from 3 days to 10 minutes). In previous study, μC cEMI model can only be accurate in one PCB board. While in the new method, owing to the BBIR deduction, the model can be applied to any board with the same target μC.

ACKNOWLEDGMENT

This work was supported by grants from the Bureau of Standards, Metrology and Inspection (BSMI 0121136A) and the National Science Council (NSC) (NSC 101-2221-E-035-050-), Taiwan, Republic of China.

REFERENCES

[1] IEC 62433 "Models of Integrated Circuits for EMI behavioral simulation," [Online]. Available: http://www.iec.ch

[2] Foissac. M, Schanen. J. L, Vollaire. C, "Black-box EMC model for power electronics converter" in ECCE, 2009. , p. 3609.

[3] Johnny R. Johnson, and David E. Johnson, "Electric Circuit Analysis," John Wiley & Sons, 1999.

[4] IEC 61967 "Integrated circuits - Measurement of electromagnetic emissions, 150 kHz to 1 GHz," [Online]. Available: http://www.iec.ch

[5] Alan V. Oppenheim, Alan S. Willsky, and S. Hamid Nawab, Signals and Systems, 2nd ed., Prentice Hall, 1996.

[6] Shih-Yi Yuan, Jiun-Jia Huang, Chia-Yuan Hsu, Shry-Sann Liao, Chi-Chin Tang, and Haw-Yu Wu, "IC-EMC Model Extension Based on Internal Impulse Response Function," Asia-Pacific EMC-Symposium (APEMC), Singapore, May 21-24, 2012.

[7] 8-Bit CMOS Microcontroller (PIC12F629) Microchip, 2003.

[8] S. Y. Yuan, H. E. Chung, and S. S. Liao, "A Microcontroller Instruction Set Simulator for EMI Prediction," IEEE Trans on EMC, vol. 51, pp. 692-699, 2009.

978-1-4799-5004-1/13 $31.00 © 2013 IEEE

Evaluation of PDN Impedance and Power Supply Noise for Different On-Chip Decoupling Structures

Haruya Fujita, Hiroki Takatani, Yosuke Tanaka, Shohei Kawaguchi, Masaomi Sato and Toshio Sudo

Shibaura-Institute of Technology

3-7-5 Toyosu, Koto-ku, Tokyo, Japan

Ma12091{toshio}@shibaura-it.ac.jp

Abstract –Because CMOS LSIs operate at higher clock frequencies in recent years, conventional methods for obeying EMC regulations are not sufficient only at package level and board level. So chip level counter-measure is even more important to reduce EMI as an excitation source of noise. In this paper, power supply noise was evaluated by fabricating two circuit blocks in a test chip. One was with on-chip capacitance consisted of intentional MOS (metal-oxide semiconductor) capacitors and MIM (metal-insulator-metal) capacitors, and the other was without any intentional capacitors. Reduction effect of power supply noise and the impedance of PDN (power distribution network) at each circuit block were evaluated based on chip-package-board co-design. It has been found that PDN Impedance was suppressed by implementing on-chip capacitance in the high frequency region.

Keywords- power supply noise; on-chip capacitance; PDN impedance; measurement

I. INTRODUCTION

In rent years, electronic devices such as smart phones and tablet PC's are developing rapidly. They have been realized by high-speed, high-density, and low-power CMOS LSIs, but they are confronted to serious noise issues. Since CMOS LSIs flow switching current at the rising and falling edges of high speed signal, power supply noise occurs. This fluctuation spreads all over the board, and then impairs the stability of the overall system. To meet the noise margin is more difficult for the low power supply voltage devices [1][2].

Power distribution network (PDN) of the system must be designed to be as low as possible up to GHz range. Surface mount device type capacitors are normally closely placed to LSIs to reduce the switching noise. However, the impedance profile of the conventional discrete decoupling capacitor is narrow due to the equivalent series inductance (ESL) of capacitor itself [3][4].

So far, decoupling capacitors on a board were effective to suppress EMI (Electromagnetic Interference) from various electronic devices. However, both core circuits and I/O circuits operate more than several GHz in recent years, so on-board decoupling capacitor does not effectively work to reduce EMI at higher frequency range [5][6].

In this paper, two circuit blocks was fabricated in a test chip. One was with on-chip capacitance consisted MOS (metal-oxide semiconductor) and MIM (metal-insulator-metal) capacitors, and without capacitance. Power supply noises were measured and compared when core circuits were excited, and on-chip PDN Impedances were directly measured for both circuit blocks. in the high frequency region.

II. MECHANISM of NOISE GENERATOIN

Fig1 shows a simplified chip-package-board model for the PDN (power distribution network) of a typical system. This model consisted of parasitic inductances, capacitances, and resistances. And this becomes a parallel RLC resonance circuit when seen from the chip. Anti-resonance peak occurs at the cross point that inductance of package and the capacitance of the chip as shown in Fig2. As a result, power supply noise becomes larger at the anti-resonance frequency.

Anti-resonance peak impedance is described by this formula.

$$Z_{PEAK} = \frac{1}{R_{die}} \frac{L_{pkg}}{C_{die}} \qquad (1)$$

Based on this formula, increasing the value of on-chip capacitance makes lower the peak impedance and the cross point of anti-resonance frequency.

When CMOS buffers operate, charge current, discharge current and shoot-through current flow. These switching current induces large voltage fluctuation due to the parasitic inductance of the package. By implementing enough value of on-chip capacitance, switching current are supplied from the internal capacitance of CMOS LSI. Therefore, supply current from the board becomes smaller, then, power supply fluctuation becomes suppressed.

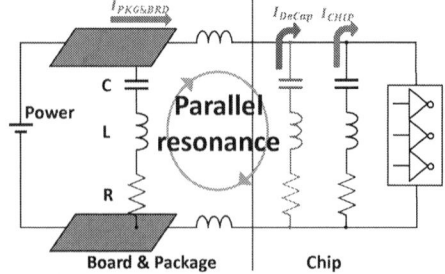

Fig.1 Total PDN model consisting of chip, package, and board

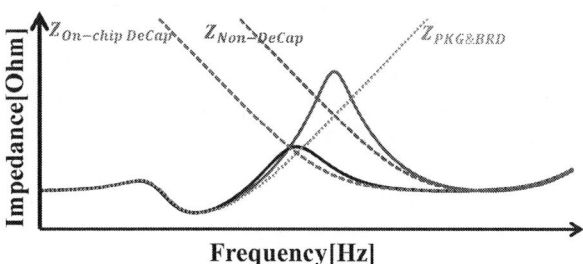

Fig.2 Anti-resonance peak in total PDN Impedance

III. CHIP-PACKAGE-BOARD CONFIGURATION

A. Chip configuration

Fig3 shows a part of the test chip that was used for this experiment. The test chip was fabricated by a 90 nm CMOS process technology with 8 metal layers. On-chip decoupling capacitances were implemented at the only right circuit block of the chip. The value of MIM (metal-insulator-metal) type on-chip decoupling capacitance was 200 pF. at the power supply network near the noise generator. And MOS (metal-oxide-semiconductor) type on-chip decoupling capacitance of 1830 pF was implemented on the whole circuit block. Consequently, total capacitance value of 2030 pF was implemented in the right side circuit block.

Conversely, floating metal was implemented instead of MIM and MOS capacitor in the left side of non-decap area.

The estimated values of on-chip capacitance ingredients for each area were shown in Table1. The value of MOS capacitance increased by approximately 3.7 times when the power supply voltage of 1.2 V was applied.

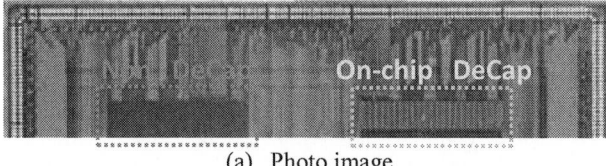

(a) Photo image

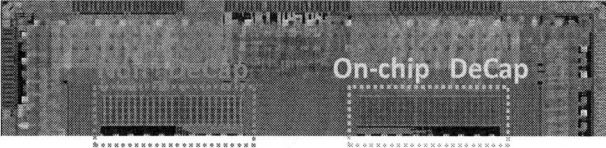

(b)GDS Image

Fig.3 Two circuit blocks with (right) and without (left) on-chip capacitance

Table 1 Estimation of On-chip capacitance ingredient

	L : Non DeCap		R : On-chip DeCap	
	0.0V	1.2V(static)	0.0V	1.2V(static)
MIM(nF)	0	0	0.2	0.2
Grid(nF)	0.24	0.24	0.24	0.24
MOS(nF)	0	0	0.5	1.83
Well&Gate(nF)	0.45	0.79	0.45	0.79
Total(nF)	0.69	1.03	1.39	3.06

B. Noise generator

Shift register circuit was used as a noise generator in each circuit block as shown in Fig.4. The shift register consisted of 1024 steps random pattern generator circuit, and it was able to drive 15channel circuits in parallel.

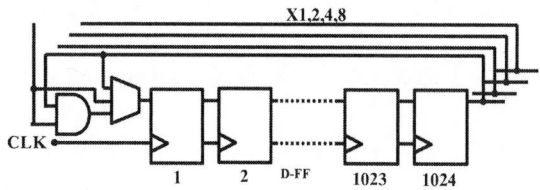

Fig.4 Shift Register Circuits

C. Package configuration

Fig.5 shows PDN traces in an organic package. The package consisted of 8 conductive layers and single SMD decoupling capacitor of 0.1 uF was placed in each trace at the bottom. The organic substrate was connected to the evaluation board using soldering balls. The power supply traces of the on-chip decap side vertically overlapped at a part of VDD trace and GND trace. On the other hand, non-decap side trace did not overlapped each other. So, non-decap side trace had bigger loop inductance than that of on-chip decap side trace.

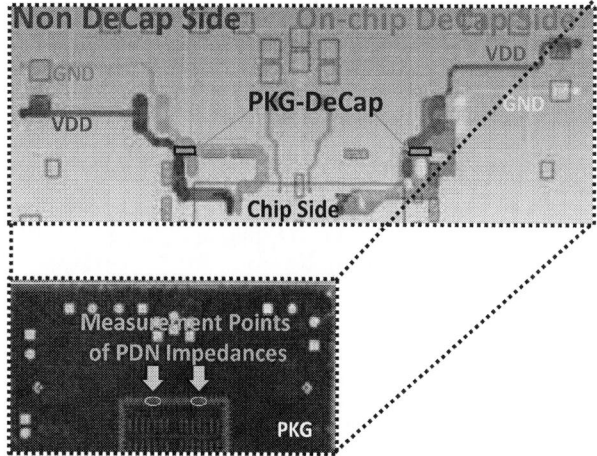

Fig.5 PDN trace in an organic package

D. Board configuration

Fig.6 shows PDN traces on an evaluation board on which a test chip was mounted. The evaluation board consisted of 6 conductive layers whose size was 210mm×250mm.

An FPGA controlled the power supply sequence to the chip. Because on-chip decap side board trace took a detour, this side seem have bigger parasitic inductance than non-decap side trace.

Fig.6 PDN traces on an evaluation board

E. On-chip noise monitoring circuit

Fig.7 shows implemented on-chip noise monitoring circuit in the chip. It consisted of source follower circuit and current mirror circuit. Once senced noise voltage at on-die PDN was converted to the current, and the extracted current by the current mirror circuit was converted to the voltage at termination resister. Measured voltages was calibrated by launching the reference voltage to the sencing circuits prior to actual measurement.

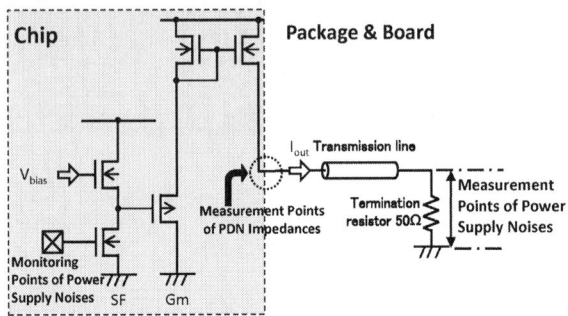

Fig.7 On-chip monitoring circuits

IV. MEASUREMENT RESULTS

A. Power Supply Noise

Figs.8-11 show power supply noises using 15 channel shift register. The voltage fluctuation of on-chip decap has been reduced by 60-75 percent compared with that for with non-decap case. With the increase of the clock frequency, peak-to-peak noise amplitude of the power supply noise became smaller.

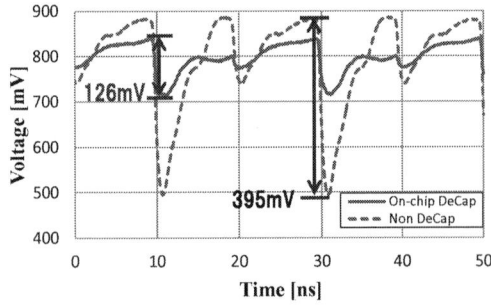

Fig.8 Measured power supply noise comparison between with and without on-chip capacitance at 50 MHz

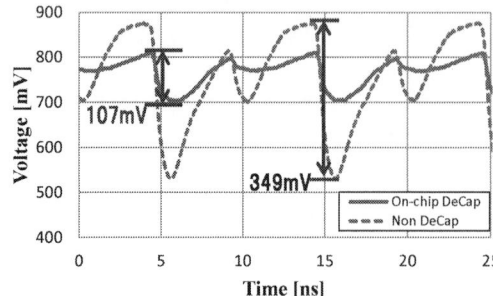

Fig.9 Measured power supply noise comparison between with and without on-chip capacitance at 100MHz

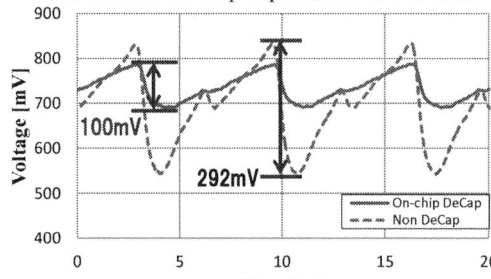

Fig.10 Measured power supply noise comparison between with and without on-chip capacitance at 150 MHz

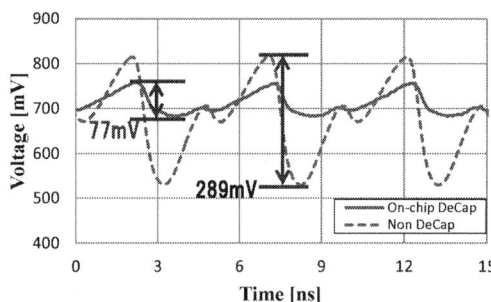

Fig.11 Measured power supply noise comparison between with and without on-chip capacitance at 200 MHz

B. PDN Impedances

The PDN Impedances for the chip, package, and board were measured by VNA (vector network analyzer). The measurement results of the on-chip PDN are shown in Figs.12-13. The on-chip capacitance for the non-decap area changed from 666 pF to 984 pF when the 1.2V was applied to the power supply terminals. In case with on-chip decap, the values of capacitance increased from 1372 pF to 3415 pF.

Fig14 shows the comparison of package and board PDN impedance between without and with on-chip decoupling capacitance. Anti-resonance peak at 1-10 MHz occurs at the cross point between the inductance of board and the capacitance of package. Inductance at 100 M-1 GHz of non decoupling capacitance area was larger than on-chip decoupling capacitance area of 1.78 nH.

Solid lines in Figs.15-16 are simulated total PDN impedances consisted of chip, package, and board by HSPICE. The anti-resonance peak occurred at the cross point of chip

978-1-4799-5004-1/13 $31.00 © 2013 IEEE

capacitance and package inductance only for the on-chip decoupling area. In case of non decoupling capacitance area, the anti-resonance peak did not occur, because the capacitance of chip was increased, and the package inductance was small.

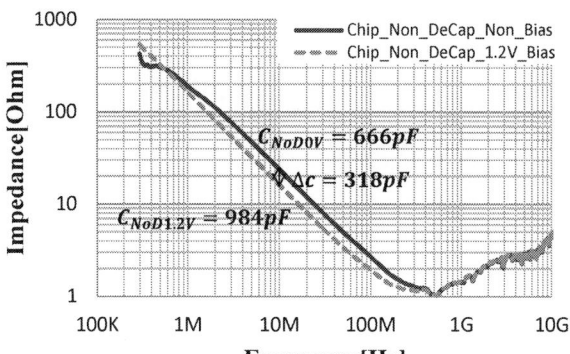

Fig.12 On-chip PDN Impedances for non-decoupling capacitance area

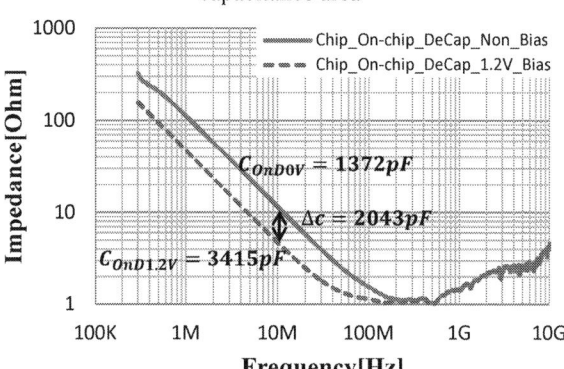

Fig.13 On-chip PDN Impedances for on-decoupling capacitance area

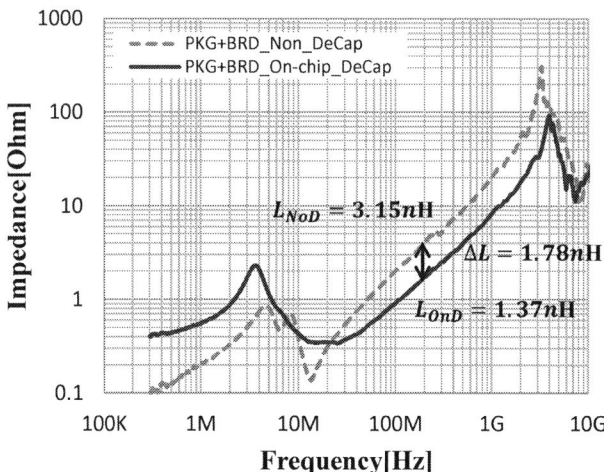

Fig.14 Comparison of package&board PDN impedance between without and with on-chip decoupling capacitance.

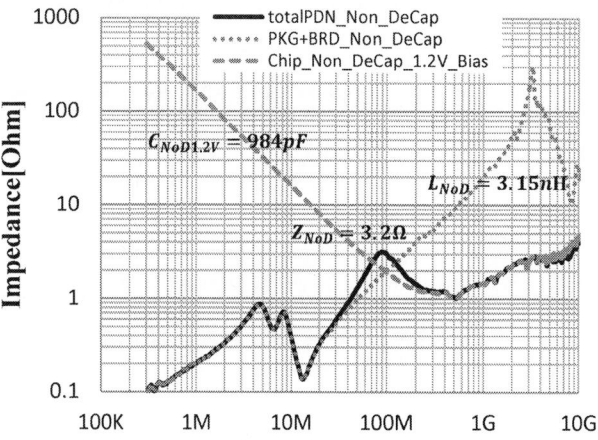

Fig.15 Anti-resonance peak generation at the cross point of Z_{die} and Z_{pkg} for the non-decap area

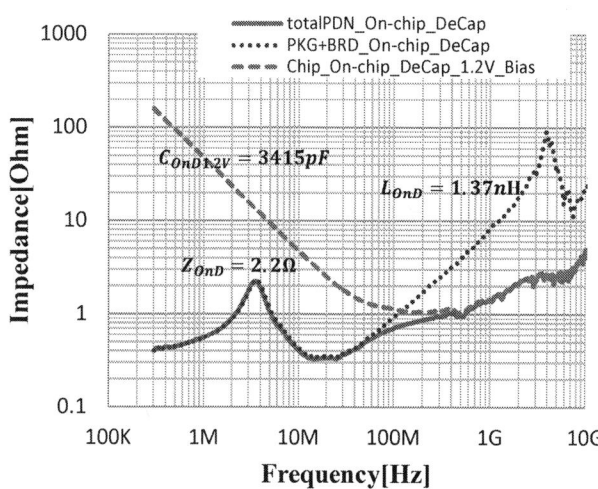

Fig.16 Anti-resonance peak generation at the cross point of Z_{die} and Z_{pkg} for the decap area

V. SUMMARY

In this paper, the effects of on-chip decoupling capacitance were characterized by comparing the two circuit blocks in a test chip by measurement. The power supply fluctuation was approximately 70 percent reduced by the effect of on-chip capacitance. The anti-resonance peak in the total PDN impedance was significantly suppressed by effect of on-chip capacitance. It has been found that on-chip capacitance is essential component to maintain power integrity in the high frequency region.

REFERENCES

[1] Eric Bogatin, "Signal and Power Integrity," Chapter 1, PRENTICE HALL, 2009.
[2] Nanju Na et al., "The Effects of On-Chip and Package Decoupling Capacitors and an Efficient ASIC Decoupling Methodology," ECTC Proceedings, 2004, pp.556-567.

[3] Madhavan Swaminathan, A.Ege Engin, "Power Integrity Modeling and Design for Semiconductors and Systems," Chapter 1, PRENTICE HALL, 2007.

[4] Y. Zhou et al., "Distributed On-chip Power Supply Noise Characterizationof the Cell Broadband Engine," EPEP Proceedings, 2007, pp.99-102.

[5] K.Hoshino, T.Sudo, et al., "Experiment and Simulation of Power Supply Switching Current Dependency on On-chip Capacitance," IEEE EDAPS 2009 Symposium, 2009, pp.69-72.

[6] Roy Leventhal, Lynne Green, "Semiconductor Modeling For simulatingSignal, Power, and Electromagnetic Integrity," Springer Science, 2006.

[7] T.Okumura,, "Power Supply Noise Evaluation with On-chip Noise Monitoring for Various Decoupling Schemes of SiP,"IEEE EDAPS 2010 Symposium.

[8] H,Fujii, and T.Sudo, "Evaluation of Power Supply Noise Reduction by Implementing On-Chip Capacitance,'' January, 2010.

Characterization of conducted emission at high frequency under different temperature

N. Berbel, R. Fernández-García and I. Gil

Department of Electronic Engineering
Universitat Politècnica de Catalunya (UPC)
Terrassa, Spain
nestor.berbel-artal@upc.edu

Abstract— **In this paper, the characterization of the EMC conducted emissions of integrated circuits under different temperature stress condition, up to 3 GHz is presented. The impact of high temperature has been measured on the input impedance of propagation paths of the electromagnetic conducted emissions, as well as on the electromagnetic noise of a clock generator.**

Keywords—Integrated circuit, EMC, switching noise, temperature impact, conducted emission.

I. INTRODUCTION

The Electromagnetic Compatibility (EMC) problem has been studied for many years, especially at printed circuit board (PCB) level [1]. Recently, due to the increasing complexity and miniaturization of microelectronic components, rising frequencies and bitrates, the demand to characterize the EMC at integrated circuits (IC) level has been enhanced. In fact, the electronics industry requires the characterization of electromagnetic emission under different environment [2].

The "Integrated Circuit Emission Model" (ICEM) was proposed by the International Electro-technical Commission (IEC) [3], and it is applicable to complex ICs such as commercial components and microcontrollers [4]. The ICEM-CE model predicts the conducted emissions of a commercial component within the frequency range of 150 kHz to 1 GHz. The ICEM-CE standard has been used in several works and applications: to solve the decoupling capacitor of an ASIC IC [5], to perform a jitter analysis of an integrated phase-locked loop [6], to model the electromagnetic emission of a microcontroller [4], to estimate the effect of Digital Signal Controllers disturbances on measurements and control systems [7] and many others.

The ICEM-CE has been revealed as an excellent model to predict the conducted emissions [8]. However, it is limited in frequency as reported above. In addition, this standard model does not take into account the environment conditions, such as temperature and humidity. According to the EMC roadmap [9], currently there is a need to extend the current emission models beyond 1 GHz due to the increasing frequency operation of many commercial electronic systems. Also, the evolution of the IC emissions due to external conditions should be evaluated [10].

This work was supported by the Spain MINECO under Project TEC2010-18550 and AGAUR 2009 SGR 1425.

The aim of the work is study the effects of temperature on the conducted emission at IC level. The integrated circuits, on real-life environment, are exposed to harsh environments, which can degrade the functionality and the electromagnetic behavior of the IC [11-13]. So, in order to predict the conducted emission at different temperatures, the standard ICEM-CE model has been extended. In this paper, a characterization of the conducted emissions of an IC up to 3 GHz under different temperature condition is presented.

This paper is organized as follows. In Section II, the experimental setup to characterize the conducted emissions is presented. The experimental results are reported and discussed in Section III. The main conclusions are summarized in Section IV.

II. EXPERIMENTAL SETUP

The conducted emissions of an IC are characterized according to the standard 61967-4 [14]. Fig. 1 depicts the switching noise basic setup according to the standard. The obtained measurements provide the electromagnetic noise including the contribution of all internal current flowing through the power supply pins. A 1 Ω resistor is added in the power bus for sensing the current, whereas a 49 Ω resistor is used to match the output impedance of the sensing network to 50 Ω, as described in the 61967-4 standard. Moreover, the supply voltage (V_{CC}) input impedance has been measured by means of a vector network analyzer. In order to reproduce different environment conditions the device under test (DUT) has been enclosed in an oven (Fig. 2).

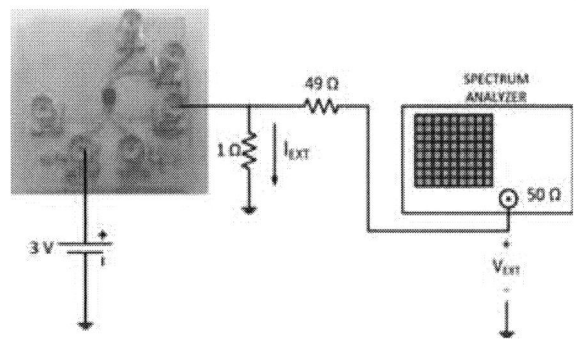

Fig. 1. Characterized DUT according the 61967-4 standard.

Fig. 2. Setup for measuring the DUT including temperature impact.

The DUT consists of a DS1088 low-cost clock generator. This device produces a square-wave output without external timing components. The DS1088 has two power supply pins and two ground pins, a power-down pin for power-sensitive applications and the square-wave output. The chip is packaged in a Micro Small Outline Package (µSOP). This IC has been chosen due to its high-frequency operation. A two-layer PCB, (area dimensions 40 mm x 40 mm and 0.55 mm thickness) has been manufactured. The PCB substrate consists of the commercial Rogers RO4350B, with relative dielectric permittivity ε_R=3.66, loss tangent *tan δ*=0.0037 and all the conducting layers made by copper with conductivity of $5.8 \cdot 10^7$ S/m. The board presents six microstrip traces (Fig. 1) with a characteristic impedance of 50 Ω. Two decoupling capacitors are added to the PCB in order to preserve the proper operation of the IC.

The DUT has been subjected to ambient temperature, 323 K, 348 K and 358 K. Each temperature has been applied to the DUT for 4 hours. The highest temperature has been limited according to the DUT specification. In order to assess the impact of the temperature condition, the input impedance and the switching noise have been measured. To achieve a significant result, ten samples have been tested and the average value has been extracted.

The characterization of the conducted emissions has two significant parts: the Passive Distribution Network (PDN) and the Internal Activity (IA). The Passive Distribution Network includes the features corresponding to the propagation paths of electromagnetic conducted emissions. The PDN takes into account the propagation paths of the Printed Circuit Board (PDN PCB) and the propagation paths of the IC (PDN IC). To extract the PDN of PCB , the impedance is measured by means of a Vector Network Analyzer (VNA) with a two-port self-impedance shunt connection measurement method [15]. Using the obtained S_{21} scattering parameter, the impedance, Z_X, is calculated according to (1).

$$Z_X = \frac{Z_0}{2} \cdot \frac{S_{21}}{1 - S_{21}} \qquad (1)$$

Where Z_0 is the characteristic impedance of the transmission line.

The Internal Activity (IA) models the electromagnetic noise and it includes the contribution of all internal current flowing through the power supply pins. The IA is measured according to the schematic presented on Fig. 1.

III. EXPERIMENTAL RESULTS

To observe the impact of temperature on the conducted emissions of a DS1088, the experimental input impedance of VCC1 and VSS1 terminals has been obtained, respectively. The input impedance of the PCB microstrip traces is measured at several temperatures, and Fig. 3 and Fig. 4 depict the results. An increase of the input impedance with the temperature in the frequency range of 1.5 GHz to 3.0 GHz can be observed and it's due to a change in the PCB substrate properties, as it has been verified with the electromagnetic simulator Agilent ADS.

The electrical model for the microstrip traces is the distributed *RLCG* lumped-element equivalent-circuit of the classical Transmission Line model [16], which has been used to model the PCB microstrip transmission lines. Assuming lossless transmission lines, the main design equations are:

$$v_P = \frac{1}{\sqrt{L \cdot C}} \qquad\qquad Z_O = \sqrt{\frac{L}{C}} \qquad (2)$$

Where v_P is the phase velocity, and L and C are the per-section inductance and capacitance of the transmission line unit cells. The phase velocity can be calculated from the measured phase of S_{21} ($\phi(S_{21})$) through (3). The parameter l is the transmission line length and f is the frequency at which the lumped model is extracted.

$$v_P = \frac{2 \cdot \pi \cdot l \cdot f}{\phi(S_{21})} \qquad (3)$$

Fig. 5 and Fig. 6 plot the electrical model of the microstrip traces, and Table I reports the extracted lumped-element inductances and capacitances per unit cell on the board for several temperatures, considering the transmission lines lossless. The inductive and capacitive part of the RLCG equivalent circuit result from the layout of the conductors. The inductive part is determined by the dimension of the conductor, permeability of the metal, and layout. On the other hand, the capacitive part is determined by the layout of the conductors, permittivity and thickness of the board material and the area of the conductor. So, decreasing the capacitive part means that the permittivity or thickness of the board material has been modified due to the temperature stress.

978-1-4799-5004-1/13 $31.00 © 2013 IEEE   148

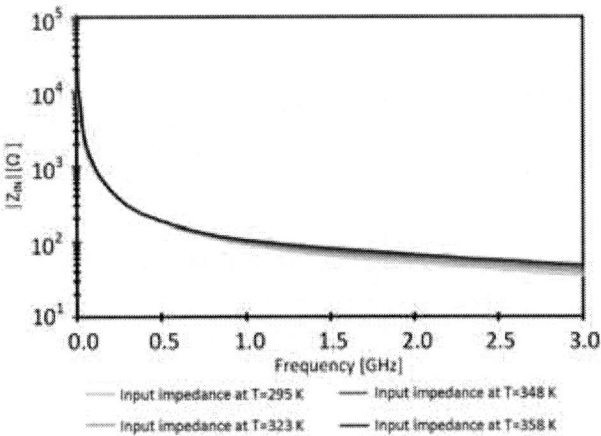

Fig. 3. Input impedance of the V_{CC} microstrip.

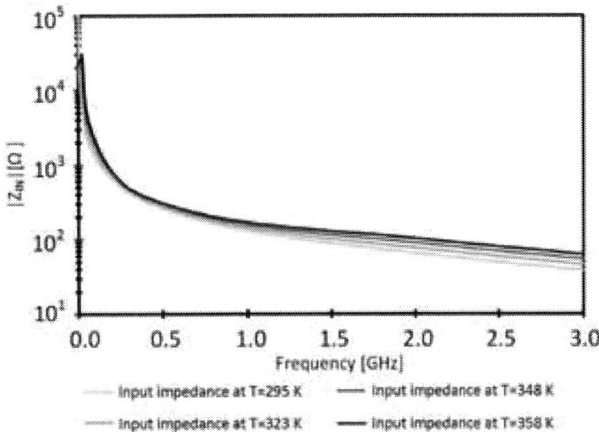

Fig. 4. Input impedance through the V_{SS} microstrip.

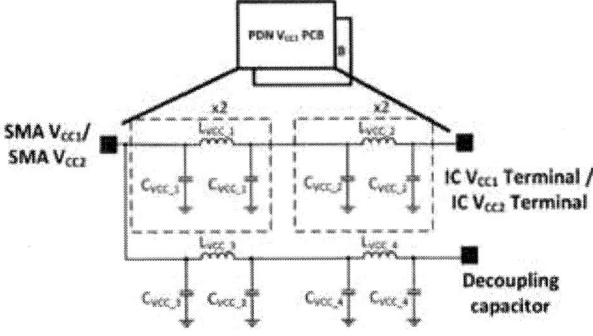

Fig. 5. Passive Distribution Network of the microstrip traces of the V_{CC1} and V_{CC2} terminals

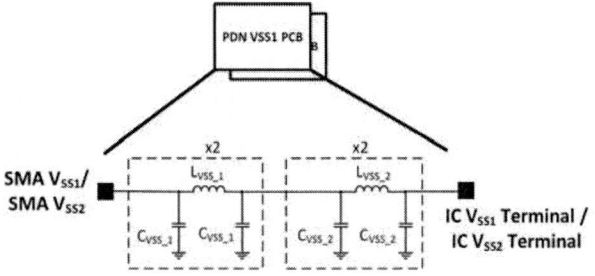

Fig. 6. Passive Distribution Network of the microstrip traces of the V_{SS1} and V_{SS2} terminals

TABLE I. Capacitance and inductance elements of PDN PCB

Component	Value			
	T=295 K	T=323K	T=348 K	T=358 K
L_{VCC_1}	1.24 nH	1.24 nH	1.24 nH	1.24 nH
L_{VCC_2}	1.30 nH	1.30 nH	1.30 nH	1.30 nH
L_{VCC_3}	1.33 nH	1.33 nH	1.33 nH	1.33 nH
L_{VCC_4}	2.99 nH	2.99 nH	2.99 nH	2.99 nH
L_{VSS_1}	1.25 nH	1.25 nH	1.25 nH	1.25 nH
L_{VSS_2}	1.25 nH	1.25 nH	1.25 nH	1.25 nH
C_{VCC_1}	172.94 fF	147.00 fF	122.35 fF	112.41 fF
C_{VCC_2}	72.10 fF	61.28 fF	51.01 fF	46.86 fF
C_{VCC_3}	188.00 fF	159.80 fF	133.01 fF	122.20 fF
C_{VCC_4}	91.45 fF	77.73 fF	64.69 fF	59.44 fF
C_{VSS_1}	207.97 fF	181.98 fF	154.24 fF	142.98 fF
C_{VSS_2}	66.17 fF	57.90 fF	49.07 fF	45.49 fF

Once the PDN of the PCB has been obtained, the PDN of the IC is obtained. Fig. 7 shows the input impedance through the V_{CC1} terminal and Fig. 8 plots the input impedance through the V_{SS1} terminal, taking into account the effects of the microstrip traces. Due to the temperature, an increase of the input impedance with the temperature in the frequency range of 1.5 GHz to 3.0 GHz has been observed, and it's due to the characteristics of the board.

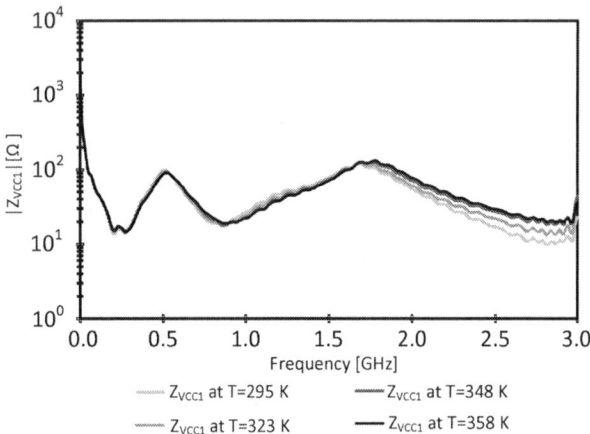

Fig. 7. Input impedance through the V_{CC1} terminal

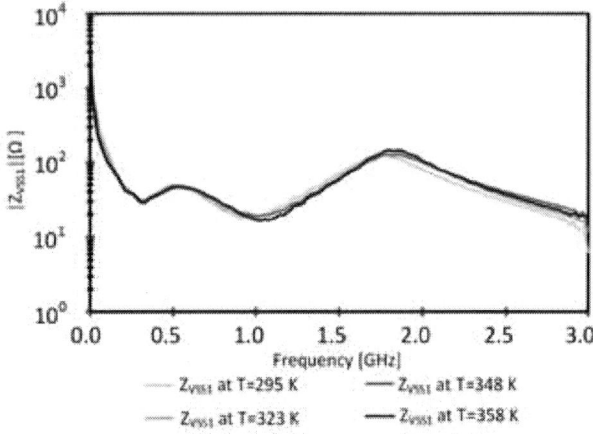

Fig. 8. Input impedance through the V_{SS1} terminal

The switching noise of the DS1088 is obtained by measuring the spectrum of the external voltage and measured in a power bus impedance of 1 Ω, and de-embedding the effects of the propagation paths of the PCB and the IC, the switching noise has been obtained. In order to quantify the impact of the temperature, the oscillation frequency of the DUT has been tested.

Fig. 9 presents the tested switching noise for a power supply of 3.0 V at different temperatures. It is observed a harmonics shift to higher frequencies when increasing the temperature, due to the oscillation frequency increase of the DUT. Fig. 10 shows the aforementioned effect from 96.6 MHz to 100 MHz. In addition, the linear behavior between oscillation frequency and temperature has been analytically obtained.

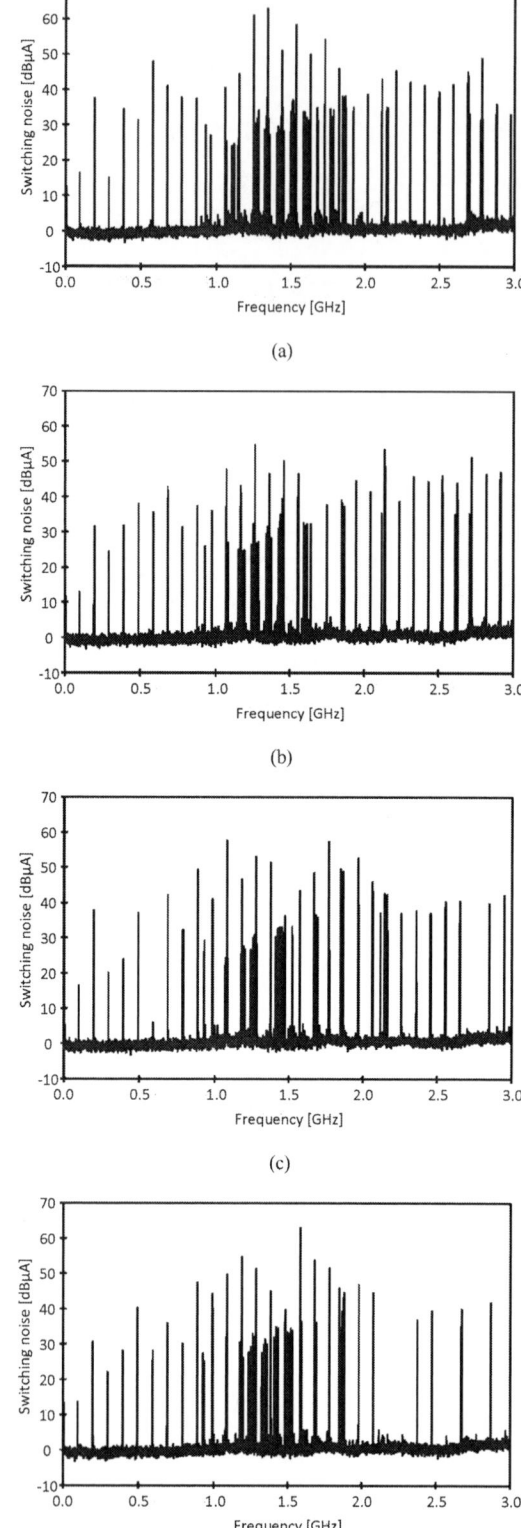

Fig. 9. Switching noise at temperatures: (a) 295 K, (b) 323 K, (c) 348 K and (d) 358 K.

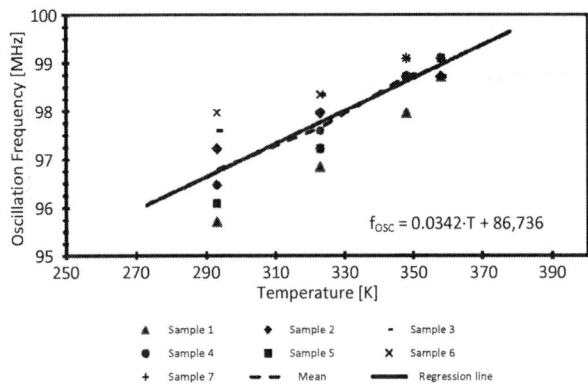

Fig. 10. Variation of the DS1088 oscillation frequency due to temperature.

IV. CONCLUSIONS

This paper presents the study of the impact of temperature on conducted emissions of a digital circuit up to 3 GHz. The propagation paths has been characterized and has been observed that the input impedance of the PCB microstrip lines increases, due to the change in the characteristics of the substrate.

On the other hand, the switching noise has been measured by means of the 1 Ω/50 Ω method, and the oscillation frequency of the DS1088 has been extracted. Due to the temperature, and increase in the oscillation frequency has been observed.

A deeply investigation should be carried out in order to determine the origin of the observed temperature drift. Further physical analyses are required to identify degradation mechanisms responsible of the conducted emission drift.

REFERENCES

[1] M. I. Montrose, *EMC and the Printed Circuit Board: Design, Theory, and Layout Made Simple*: Wiley-IEEE Press, 1998.

[2] W. Parker, W. Tustin, and T. Masone, "The case for combining EMC and environmental testing," *Proc. ITEM 2002.* pp. 54-60, 2002.

[3] I. E. C. Standard, "Models of Integrated Circuits for EMI Behavioral Simulation — Conducted Emission Modelling (ICEM-CE) IEC 62433-2," ed, 2006.

[4] C. Labussiere-Dorgan, S. Bendhia, E. Sicard, T. Junwu, H. J. Quaresma, C. Lochot, and B. Vrignon, "Modeling the Electromagnetic Emission of a Microcontroller Using a Single Model," *IEEE Transactions on Electromagnetic Compatibility,* vol. 50, pp. 22-34, 2008.

[5] J. L. Levant, C. Marot, M. Meyer, and M. Ramdani, "Solving ASIC decoupling with the ICEM-CE Model," presented at the 7th International Workshop on Electromagnetic Compatibility of Integrated Circuits (EMC Compo), Toulouse, France, 2009.

[6] L. Jean-Luc, R. Mohamed, P. Richard, and D. M'Hamed, "EMC Assessment at Chip and PCB Level: Use of the ICEM Model for Jitter Analysis in an Integrated PLL," *IEEE Transactions on Electromagnetic Compatibility,* vol. 49, pp. 182-191, 2007.

[7] C. Zhou, M. Hu, X. Lin, L. Dang, and T. Yang, "Electromagnetic compatibility analysis and design for digital signal controllers," in *Electromagnetic Compatibility (APEMC), 2010 Asia-Pacific Symposium on*, 2010, pp. 668-671.

[8] C. Zhou, J. Wang, X. Pan, F. Gao, and D. Yu, "Modelling and analysis of electromagnetic interferences for a 32-bit digital signal controller," in *Antennas, Propagation & EM Theory (ISAPE), 2012 10th International Symposium on*, 2012, pp. 1132-1135.

[9] M. Ramdani, E. Sicard, A. Boyer, S. Ben Dhia, J. J. Whalen, T. H. Hubing, M. Coenen, and O. Wada, "The Electromagnetic Compatibility of Integrated Circuits - Past, Present, and Future," *IEEE Transactions on Electromagnetic Compatibility,* vol. 51, pp. 78-100, 2009.

[10] W. Pfaff, "Industrial use of EMC: Trends in system development," presented at the EMC Compo 2005, Munich, Germany, 2005.

[11] S. Ben Dhia, E. Sicard, Y. Mequignon, A. Boyer, and J. Dienot, "Thermal Influence on 16-Bits Microcontroller Emission," in *IEEE International Symposium on Electromagnetic Compatibility, 2007. EMC 2007*, 2007, pp. 1-4.

[12] A. Boyer, A. C. Ndoye, S. Ben Dhia, L. Guillot, and B. Vrignon, "Characterization of the Evolution of IC Emissions After Accelerated Aging," *IEEE Transactions on Electromagnetic Compatibility,* vol. 51, pp. 892-900, 2009.

[13] K. Armstrong, "Specifying lifecycle electromagnetic and physical environments - to help design and test for EMC for functional safety," in *Electromagnetic Compatibility, 2005. EMC 2005. 2005 International Symposium on*, 2005, pp. 495-500 Vol. 2.

[14] I. E. C. Standard, "Integrated circuits – Measurement of electromagnetic emissions, 150 kHz to 1 GHz – Part 4-1: Measurement of conducted emissions – 1 Ω/150 Ω direct coupling method – Application guidance to IEC 61967-4 ", ed, 2005.

[15] "Impedance Measurements. Evaluating EMC Components with DC Bias Superimposed - Application Note," ed. Agilent, 2009.

[16] D. M. Pozar, *Microwave Engineering*: Wiley, 2012.

Using the EM simulation tools to predict the Conducted Emissions level of a DC/DC boost converter : introducing EBEM-CE model

André DURIER
Continental Automotive France SAS
Toulouse, France
andre.durier@continental-corporation.com

Christian MAROT
EADS France IW
Toulouse, France
christian.marot@eads.net

Olivier CREPEL
EADS France IW
Toulouse, France
olivier.crepel@eads.net

Abstract- DC/DC Boost Converters are commonly used in the electronics industry to provide a raised voltage to a specific function. These converters are constituted by a basic commutation cell (Inductor-MOS transistor-diode-capacitor) managed by an integrated circuit realizing voltage and current control typically running between 100 and 500 kHz. This control's frequency creates high conducted Electromagnetic noise which could cause troubles on the supply network. We propose to use a SPICE modeling to estimate the conducted noise on supply network during CISPR 25 CE measurements. Then, we will intend to build an EBEM-CE model of the converter from these measurements.

Keywords-component: DC/DC Boost Converter; SPICE modeling; CISPR 25 Conducted Emissions, EBEM CE model

I. INTRODUCTION

To evaluate the EMC behavior of an automotive power converter after a power component change, Continental Automotive France is deeply involved into French Aerospace Valley labeled SEISME (Simulation of Emission and Immunity of Systems and Electronics Modules) [1],[2] project which aims to study the effects of the change of one or several electronic component on EMC behavior of an electronic board.

But, prior to evaluate the effects of power component's change, a numeric model of the commutation cell must be realized and validated by a CISPR 25 Conducted Emissions measurement.

II. DC/DC BOOST CONVERTER AND BOARD DESCRIPTION

A. SEISME D31 Board

The SEISME D31 board shown figure 1 is a 65 Watts DC/DC Boost Converter as used in typical automotive engine control application. The converter is dedicated to be used on a 12V battery to deliver 1 A on a 65 Ω load.

The commutation cell is constituted by :

- a 100 V / 42 A N-MOSFET transistor (On Semi NTB6413AN),
- a 10 µH power choke (Wurth WE HCF2013),

- a 100 V / 5 A diode (STM STPS5H100B),
- 2 electrolytic aluminum 47 µF Nichicon capacitors.

The 65 Ω resistive load is built from 3 power 200 Ω / 35 W resistors (BOURNS PWR263S35-2000) placed in parallel.

We choose a Linear Technology gate driver LTC4441-1 which allows gate voltage's adjustment between 5 to 8 V and provides up to 8 A.

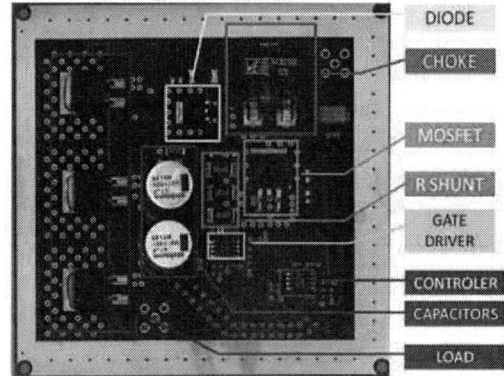

Figure 1. SEISME D31 Automotive DC/DC Boost Converter

B. Pulse width modulation principle

The DC/DC Boost converter embeds a Linear Technology LT1243 controller which is a fixed frequency, current mode, pulse width modulator. Unlike most pulse width modulators, this controller allows non-continuous operation means the current flowing into the choke could equal zero. As the non- continuous operation is not the common one used for DC/DC boost converters, we will briefly describe in this paragraph the pulse with modulation principle used by the LT 1243 controller.

The typical schematic of a DC/DC boost converter using the LT 1243 is given figure 2.

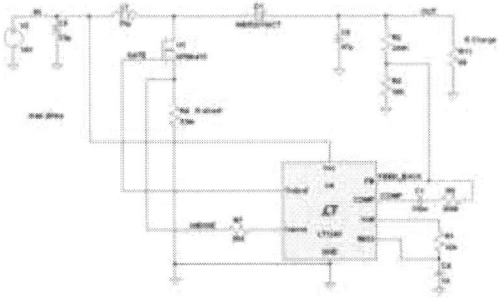

Figure 2. DC/DC Boost converter' schematic using the LT 1243

R1 resistor and C4 capacitor fix the modulation's frequency. In figure 2, the frequency is fixed to 165 kHz.

The output voltage OUT, expected to 65 V, is divided by the resistors' ratio R3 / (R2+R3) ~ 1/26 to create a feedback voltage FB which is internally compared to a 2.5V reference voltage as shown in the LT1243 block diagram illustrated by the figure 3.

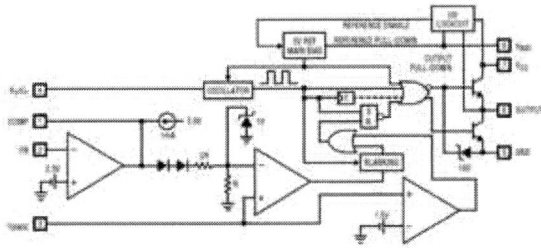

Figure 3. LT 1243 block diagram

Maximum current flowing into the MOS transistor is determined by the following formula:

$$IPeak = \frac{VCOMP - 1.4V}{3 * RSHUNT} \quad (1)$$

where VCOMP is the voltage applied to compensation pin (COMP) and RSHUNT is the shunt resistor, named R4 in the figure 2, placed in the MOS transistor source.

The current flowing into the transistor is monitored by LT 1243 ISENSE pin thru the shunt resistor R4. Otherwise, to limit the peak current flowing into the transistor, ISENSE input voltage is clamped to a voltage proportional to the voltage applied to the COMP pin. The maximum value of this clamp voltage equals 1 V. Consequently, the maximum current I Peak is clamped to $\frac{1V}{RSHUNT}$.

The figure 4 shows the MOS current (blue) and the output voltage (red) during the first 3 ms of the operation. After a start time fixed by the LT 1243 controller, we may observe 3 operating phases:

- Starting phase❶ : $V_{OUT} \ll 65V$ means $V_{FEEDBACK} < 2.5V$: consequently $V_{COMP} = 5.6V$ and the clamp voltage equals its maximum 1 V. The maximum current into the MOS is $1V / 33m\Omega = 30A$

- Transition phase ❷ : V_{OUT} is closed to 65 V, the $V_{FEEDBACK}$ will be closed to 2.5V: consequently VCOMP will approach to 2.5V and the clamp voltage decreases. The maximum current into the

MOS will decrease to reach the value defined by formula (1) $\frac{(2.5V - 1.4V)}{3 * 0.033\Omega} \sim 11$ A.

- Normal operation ❸: $V_{OUT} = 65$ V, the maximum MOS current = 11A

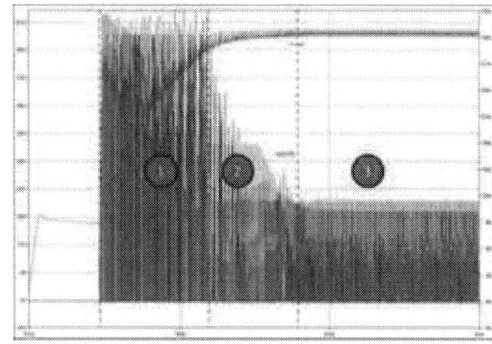

Figure 4. DC/DC Converter Operating phases

As we may observe in figure 4, the nominal operation is non continuous because the MOS current reaches 0 A.

III. CONDUCTED EMISSIONS MEASUREMENT

CE measurements have been realized by Continental according the CISPR25 2008-3 -voltage method- norm [3].

The figure 5 shows the real set–up used by Continental. As the resistive load is mounted on the D31 board, there is no harness neither external load simulator. The D31 board is placed 50mm from the ground on an isolator (Ɛr < 1.4). The length of the power supply harness up to LISN (Line Impedance Stabilization Network) is fixed to 300 mm.

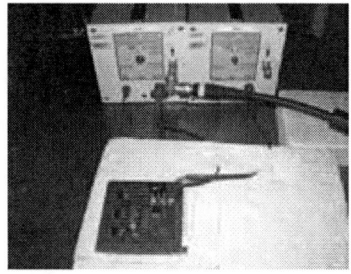

Figure 5. Real CE voltage method test set-up

The figure 6 shows the voltage versus the frequency measured on VBAT LISN (peak measurement).

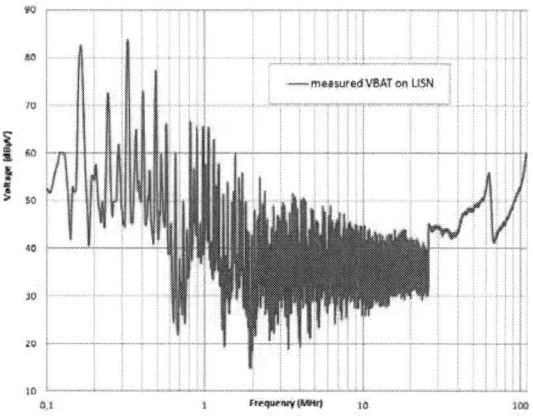

Figure 6. CE voltage measured on VBAT LISN

978-1-4799-5004-1/13 $31.00 © 2013 IEEE

IV. D31 BOARD SPICE MODELING

A. Basic Modeling

D31 Board Spice model is illustrated by the figure 7. LT1243 controller and LT 4441 gate driver SPICE models are provided by Linear Technology. NTB6413AN MOS SPICE model is provided by On Semi.

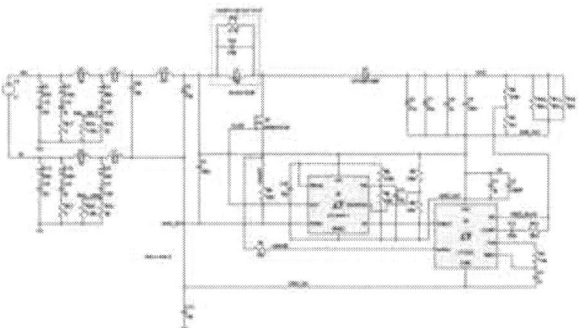

Figure 7.   D31 Board modeling using LT SPICE

The CE measurement is made on the 50 Ω of each LISN. To complete the model, power supply harness is modeled by its inductance (770 nH/m) and its capacitance (55 pF/m). The 10 μH choke is modeled with its parasitic resistor and capacitance measured with a Network Analyzer.

Calculated voltages versus time on VBAT (red) and on GND (blue) are given by the figure 8.

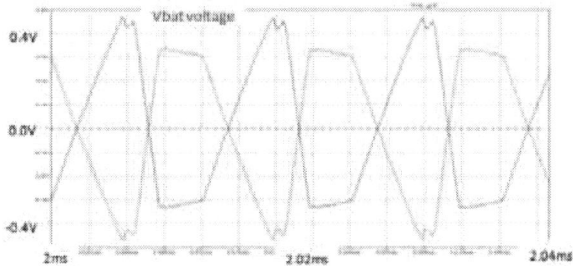

Figure 8.   Calculated voltages on VBAT and GND

Time related voltages have been measured (AC mode) during the CE measurement on VBAT (blue) and on GND (red). These measurements are given by figure 9.

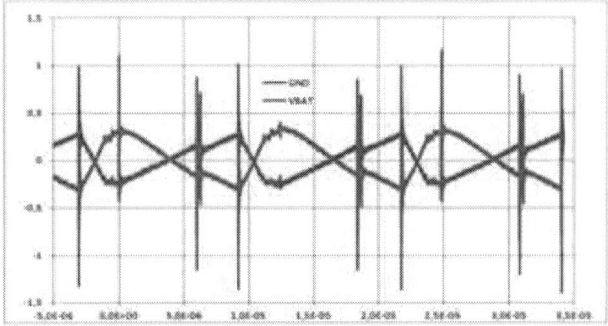

Figure 9.   Measured voltages on VBAT and GND

The comparison between calculation and measurement clearly shows that commutation's noise doesn't appear totally in SPICE transient simulation. Basic SPICE modeling is sufficient to demonstrate that Conducted Emissions for frequencies below few MHz are mainly due to differential mode noise flowing thru the choke.

B. A way to improve the modeling

Using Continental experience of DC/DC Converter design, Conducted Emissions for frequencies beyond 5 MHz are mainly due to common mode noise flowing thru the ground. According Liyu Yang [4], the noise path is mainly due to the Equivalent Series Inductance (ESL) of electrolytic capacitors and common mode capacitor of heatsinks. Parasitic elements of PCB traces have been evaluated using CST 3D solver and could be negligible. The figure 10 shows the new voltage calculation at VBAT LISN using ESL=5 nH and common mode capacitor = 3 pF.

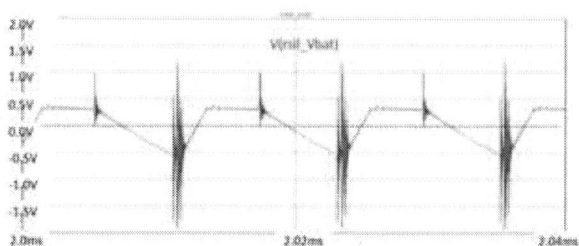

Figure 10. New calculated voltage on VBAT LISN

The frequency domain view of the conducted noise is obtained applying a FFT (Fast Fourier Transform) on SPICE transient simulation results. The figure 11 shows the comparison between calculated Conducted Emissions coming from SPICE FFT (red curve) and CE measurement (blue curve).

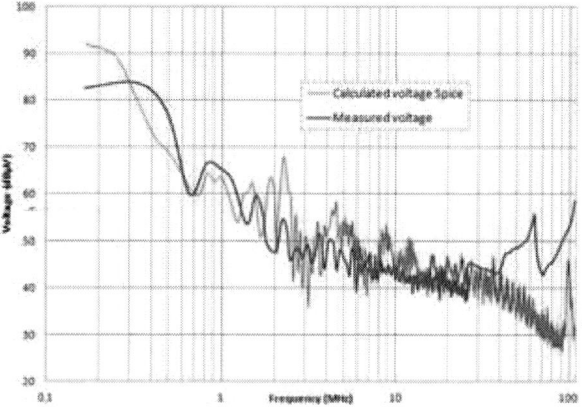

Figure 11. Comparison between measured and calculated CE on VBAT

We may observe that SPICE modeling gives limited and imprecise results even if the tendency is correct until 20 MHz. Resonances observed on measurement above 30 MHz are not reproduced with a SPICE modeling.

V. TOWARDS AN EBEM-CE MODEL

We propose now to build a conducted emission model of the whole D31 board from measurements. We propose to use the same philosophy as used to build IEC 62433-2 ICEM-CE (Integrated Circuit Emission Model – Conducted Emissions)[5]. The figure 12 intends to show the EBEM-CE (Electronic Board Emission Model–Conducted Emissions) as a white box construction.

978-1-4799-5004-1/13 $31.00 © 2013 IEEE

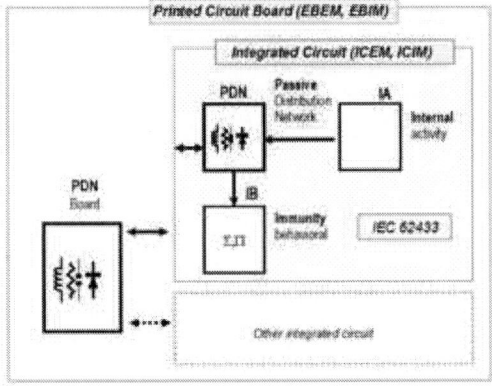

Figure 12. Proposal for IC and PCB EMC model construction

As ICEM-CE model, EBEM-CE aims to represent an electronic board as a Passive Distributed Network (PDN) associated to an Internal Activity (IA) generator as shown by figure 13.

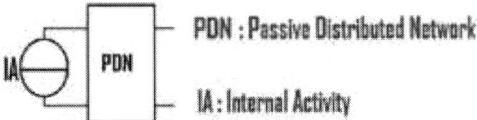

Figure 13. Basic EBEM-CE model construction

Using experience of SPICE modeling, we propose to build the EBEM-CE model of the D31 Board as described in figure 14.

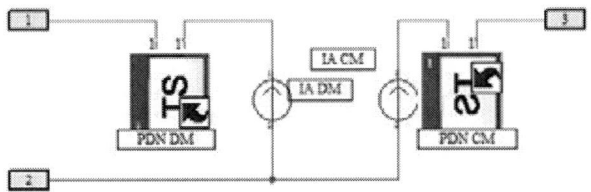

Figure 14. Improved EBEM-CE model construction

We added an Internal Activity source for Common Mode called IA CM associated to a Common Mode PDN called PDN CM.

A. PDN parameters extraction

Both PDN parameters are extracted from a S11 parameter measurement done respectively at the 12V input (S11 IN) and at the 65V output (S22 OUT).

The figure 15 shows S11 measurement for injection's power equals to 10dBm.

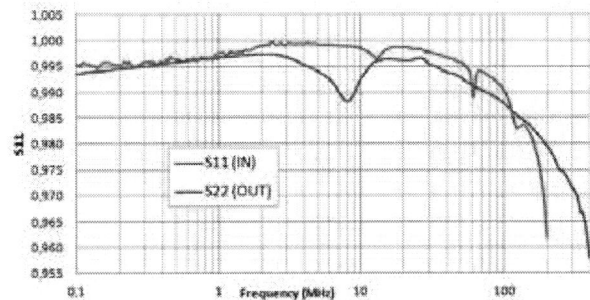

Figure 15. S11 measurement for 10dBm injection power

The figure 16 shows the PDN impedance values calculated with CST Design Studio tool. PDN DM is associated to Z11 measurment and PDN CM is associated to Z22 measurement.

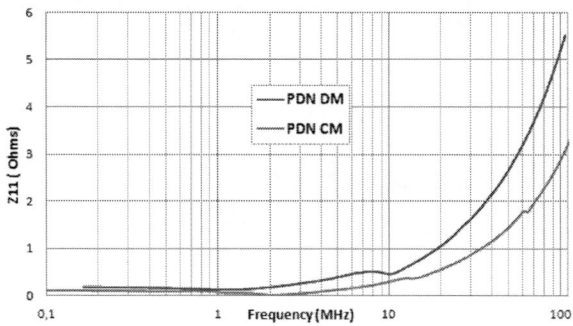

Figure 16. PDN of D31 board

B. IA DM generator extraction

The IA DM generator extraction uses a CE measurement setup using a current probe as illustrated by figure 17.

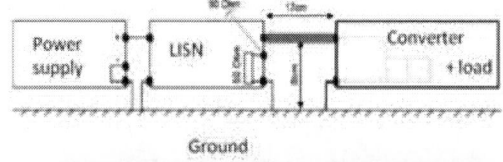

Figure 17. IA DM measurement set-up

The measurement is made on VBAT wire while the ground of the D31 board is connected to the ground plane. The board is placed vertically to reduce the common mode.

The figure 18 represents the measured IA DM current.

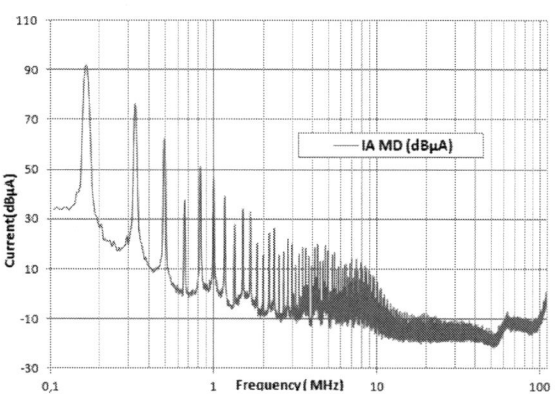

Figure 18. Measured IA current (Differential Mode)

978-1-4799-5004-1/13 $31.00 © 2013 IEEE

As IA generator will be built as a voltage noise source, we need to calculate a correction factor between this voltage source and the current measured on wire. Figure 19 shows the model used to calculate the corrective factor IA/P1. The P1 probe (red circle) represents the current probe used to measure IA. As LISN equivalent circuit can't take in account the propagation phenomena, a 30mm wire model representing the LISN's length (green circle) is added.

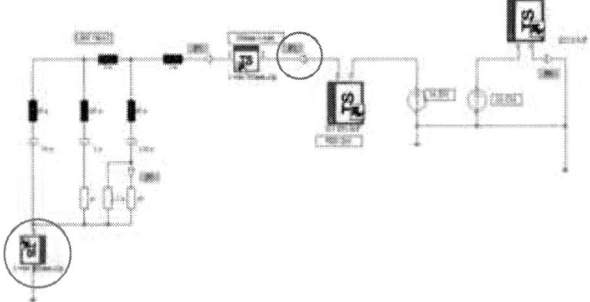

Figure 19. Set-up modeling used to calculate corrective factor IA/P1

Figure 20 shows the calculated corrective factor IA/P1 expressed in dBΩ.

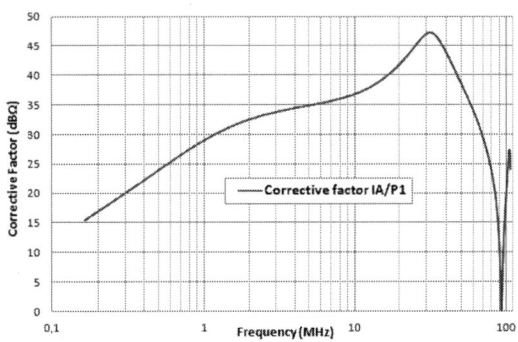

Figure 20. Corrective factor IA/P1

C. IA CM generator extraction

We use the set-up used to determine the IA DM generator as described figure 17. As the load is embedded on the board, IA CM is measured in time-domain in AC mode. The figure 21 shows the time domain measurement at D31 board output level.

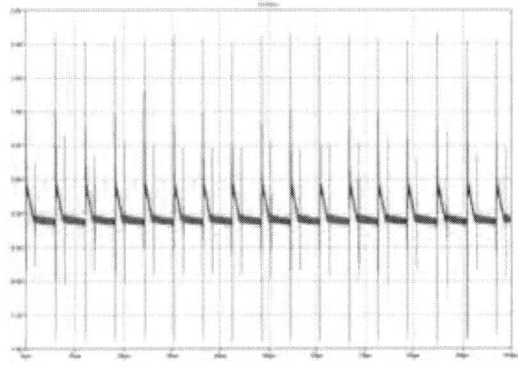

Figure 21. IA CM time domain measurement

We may now use a Fast Fourier Transform with a Hamming windowing to determine IA CM in frequency domain.

The figure 22 shows the comparison between the Differential Mode IA voltage noise source coming from current measurement on VBAT wire and Common Mode IA voltage noise source coming from time-domain voltage measurement done on Output. Note that curves have been reduced to an envelope linking together the maximum magnitude points.

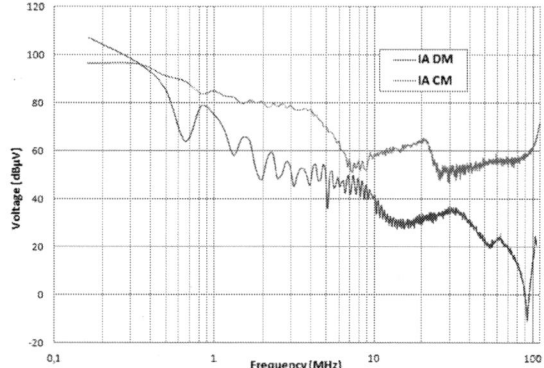

Figure 22. Comparison between DM and CM IA voltage noise sources

D. Common mode impedance measurement

We need to determine more precisely the common mode impedance between the heat sinks and reference ground. The figure 23 shows the S-parameters measurement set-up used to determine these impedances.

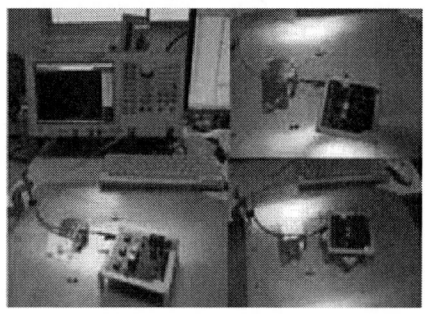

Figure 23. Common mode impedance measurement set-up

The figure 24 shows the Z-parameters obtained by measurement when the heat sinks are placed at 13 mm from reference ground (isolator=air) and when the heat sinks are placed at 50 mm from reference ground (isolator=polystyrene) .

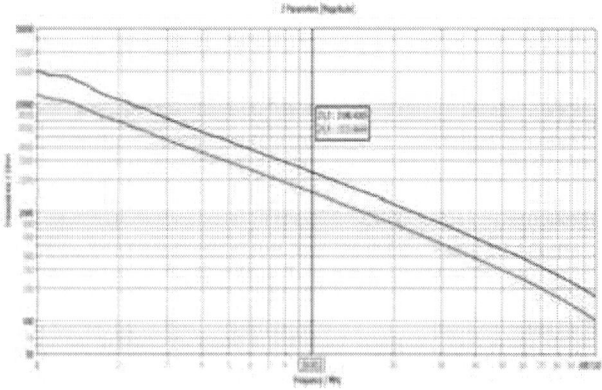

Figure 24. Common mode impedances

For each heat-sink the common mode impedance is equal to a 9.8 pF capacitor when the heat sinks are placed at 13 mm from reference ground (red curve) and 6.7 pF capacitor when the heat sinks are placed at 50 mm from reference ground (blue curve) .

E. Use of EBEM-CE model into a CISPR25 CE voltage method measurement set-up modeling

We propose now to validate the EBEM-CE of the D31 board model in a simulation of a CISPR25 CE voltage method measurement. The figure 25 shows the CISPR25 CE voltage measurement set-up modeling used to calculate the voltage at VBAT LISN. In this set-up appear all common mode impedances (red circle).

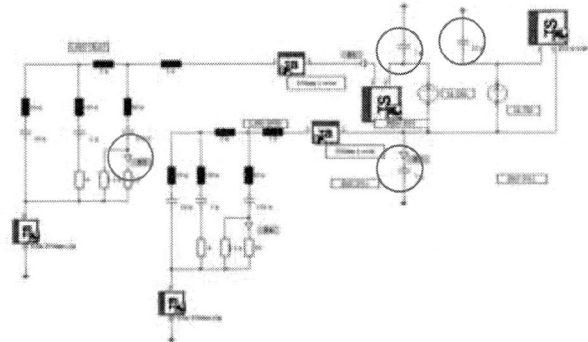

Figure 25. CISPR25 CE voltage measurement set-up modeling

This modeling allow to calculate the transfer function between IA DM to P3 and the transfer function between IA CM to P3 where P3 is a voltage probe placed at VBAT LISN 50 Ω (green circle).

The figure 26 shows the transfer functions for both differential and common modes

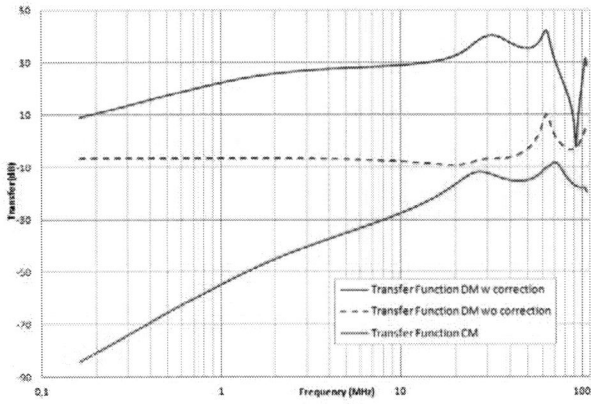

Figure 26. Transfer Function for Differential & Common Modes

Applying these transfer functions to IA DM and IA CM voltage noise sources measured figure 22, we obtained the calculated voltage at VBAT LISN (reduced to envelop curve) illustrated by figure 27.

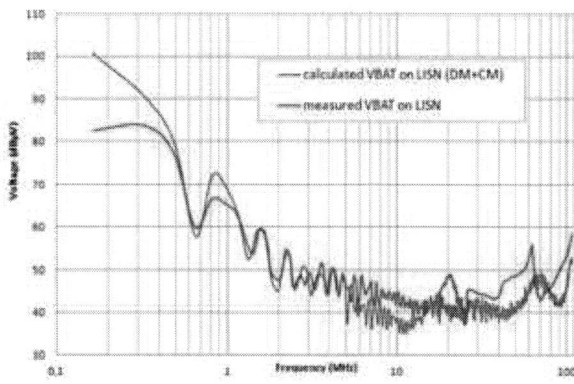

Figure 27. Measured and calculated voltages at VBAT LISN

We observe that calculated voltage fits to measured one up to 6 MHz. Beyond 6 MHz, the correlation is not so good but could be improve taking in account some corrective actions:

- Improve time-domain measurement of IA CM or replace it by a CE current method measurement using external load,
- Improve the calculation using the phase information and including receiver characteristics (RBW,…).

CONCLUSION

We have demonstrated first that a SPICE modeling taking in account main parasitic elements is sufficient to estimate roughly the CE level of a DC/DC converter up to 30 MHz.

Then, we have proposed a method to build the EBEM-CE model of a DC-DC converter using a S-parameter measurements to obtain the Differential and Common Mode PDN, a CE current measurement to obtain the Differential Mode IA and a time-domain measurement to obtain the Common Mode IA. As PDN and IA are defined from the ground of the PCB, we still may consider this model as a differential-mode one.

The advantages of EBEM-CE model are the following:

- Easy extraction of PDN and IA,
- This stand-alone modeling approach allows using the EBEM-CE model into a conducted emission common mode test set-up as described in DO160 or CISPR25 standards.

REFERENCES

[1] C. Marot "Electro-Magnetic Compatibility Embedded Electronic Units Modelisation and Simulation" APEMC 2012, Singapore, May. 2012.

[2] C.Marot « SEISME project T0+24 » Microwave &RF 2013, April 2013 PARIS

[3] CISPR25 2008-3: Radio disturbance characteristics for the protection of receivers used on board vehicles, boats and on devices :Limits and methods of measurement

[4] Liyu Yang, "Modeling and Characterization of a 1KW CCM PFC Converter for Conducted EMI Prediction", Applied Power Electronics Conference and Exposition, 2004.

[5] IEC 62433-2 -Integrated Circuit - EMC IC modeling – Part 2: ICEM-CE, Integrated Circuit Emission Model, Conducted Emissions.

978-1-4799-5004-1/13 $31.00 © 2013 IEEE   157

Design of Contactless Wafer-Level TSV Connectivity Testing Structure Using Capacitive Coupling

Jonghoon J. Kim, Heegon Kim, Sukjin Kim, Bumhee
Bae, Daniel H. Jung, Sunkyu Kong, and Joungho Kim
Terahertz Interconnection and Package Laboratory
Korea Advanced Institute of Science and Technology
Daejeon, Republic of Korea
jonghoonk@kaist.ac.kr

Junho Lee, Kunwoo Park
Advanced Design Team
SK Hynix Semiconductor Inc.
Icheon, Republic of Korea

Abstract— **Driven by the abrupt miniaturization of mobile devices and demand for 3D-IC, Through Silicon Via (TSV) has been highlighted as the key technology for compactly integrating multiple dies of various functions as a whole system. However, due to the instability in the TSV fabrication process, various types of disconnection defects can be resulted during fabrication steps, resulting in a severe decrease in the final chip yield as the number of TSVs and stacked dies increases. In this paper, we propose a novel contactless wafer-level TSV connectivity testing structure using capacitive coupling that can detect TSV disconnection defects on wafer-level. The proposed structure can detect the TSV disconnection by observing the change in the capacitance between adjacent TSVs, using only passive components such as metal pads and lines, without additional power consumption for the testing. Through time- and frequency-domain simulation results, such as transfer impedance and voltage waveforms, we verified that the proposed structure can successfully detect TSV defects, while overcoming the limitations of the conventional direct probing methods.**

Keywords—disconnection; capacitive coupling; through-silicon via (TSV); TSV test; wafer-level

I. INTRODUCTION

Driven by the abrupt miniaturization of mobile devices, the importance of realization of mixed-signal system has been extensively highlighted for multi-function yet, compact design. With the advent of 3D-IC, it became very important for the system to achieve both higher bandwidth and lower power consumption for wide I/O system. In order to achieve those requirements, Through Silicon Via (TSV) has been highlighted as the key technology, of which the number is expected to increase exponentially and allow high speed data transmission up to the speed of TB/s [1]. TSVs can provide the shortest paths between the chips; however, due to the uncertainties and instabilities in the TSV fabrication process, the final chip yield could decrease drastically with the increase in the number of TSVs and stacked dies [2]. To increase the final chip yield and ensure the reliability of the chip performance before packaging, TSVs first need to be tested on wafer-level before stacking.

Testing of TSVs is becoming of more importance as the number of TSV is expected to increase dramatically. As can be seen from Fig. 1, there are various types of disconnection

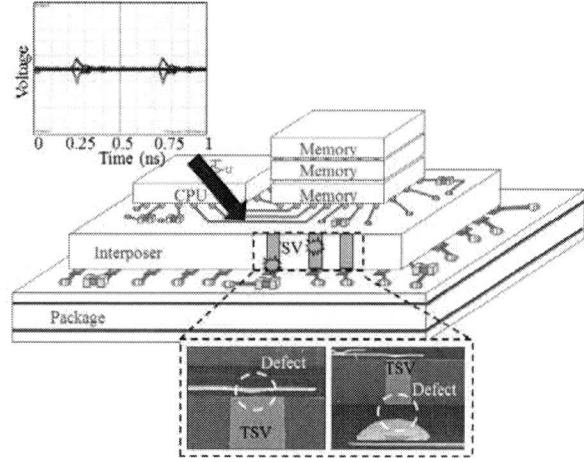

Fig. 1. Importance of TSV connectivity and possible types of disconnection defects in TSV-based 3D-IC.

defects in TSV-based 3D-IC that can severely degrade the electrical performance of the system by impeding the signal transmission. However, conventional testing methods of wafer-level testing have exhibited many limitations, such as pitch limitation of the conventional probe cards, physical damage on microbumps, and height variation of the microbumps from unstable TSV fabrication process [3].

Therefore, in order to accurately detect TSV disconnection, introduction of a new method that can surmount these problems, while maintaining accuracy is necessary, as accurate detection of TSV defects is necessary to further improve such TSV recovery methods proposed in [4-5].

In this paper, a contactless wafer-level TSV connectivity testing structure that can accurately test TSVs on wafer-level using capacitive coupling is proposed. The proposed structure can detect the TSV disconnection by observing the change in the capacitance between adjacent TSVs using only passive components such as vias and metal lines; in other words, no active components, hence extra power consumption are required. Through time- and frequency-domain simulation results, such as transfer impedance and voltage waveforms, respectively, we verify that the proposed structures can successfully detect TSV defects while overcoming the limitations of the conventional direct probing methods.

978-1-4799-5004-1/13 $31.00 © 2013 IEEE

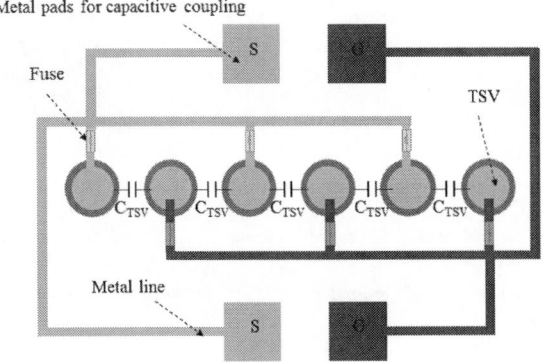

Fig. 2. Proposed wafer-level TSV connectivity testing structure using capacitive coupling. Adjacent TSVs are connected to different set of metal pads to detect series capacitance between one another.

II. STRUCTURE AND ANALYSIS OF THE PROPOSED TSV CONNECTIVITY TESTING STRUCTURES

A. Contactless TSV Connectivity Testing Structures

As can be seen from Fig. 2, adjacent TSVs are connected to different set of metal pads – each set being Signal and Ground - for the series capacitance between the adjacent TSVs to be observed. In order to test TSV connectivity in contactless manner using capacitive coupling, 4 metals pads – 2 signal pads and 2 ground pads – are required on the wafer for testing 6 TSVs, where 2 groups of 3 TSVs are connected to each set of the metal pads. All 6 TSVs are aligned in a straight line for the simplicity of analysis. Furthermore, since metal lines and metals pads connected to multiple TSVs can severely degrade the performance of TSVs with parasitics, fuses are placed on the metal lines to physically disconnect the metal lines from the TSVs by burning out the fuses with laser, once the wafer-level TSV connectivity testing is finished. Of various defects, only metal line disconnection case is analyzed in this paper, as physical disconnection can cause the most severe deterioration in the electrical characteristics and performance of the TSV channels. Although not specified in the figure, the probe card, which will be placed 1 μm vertically away from the wafer, is assumed to be a simple structure with four metal pads, two for transmitting and receiving test signals, and the other two for ground. Lastly, as is depicted in Fig.3 (a) and (b), the silicon substrate thickness is set to be larger than TSV height, since wafer testing needs to be conducted before wafer is grinded. The physical dimensions of the proposed structures designed for 3D-EM simulation are indicated in Fig 3 (a),(b), and Table I. For 3D-EM simulation, ANSYS HFSS is utilized.

B. Analysis of the Testing Structures

As indicated in Fig. 2, a series capacitance can be observed between adjacent TSVs, which is the series combination of the oxide capacitance C_{ox} and the parallel combination of

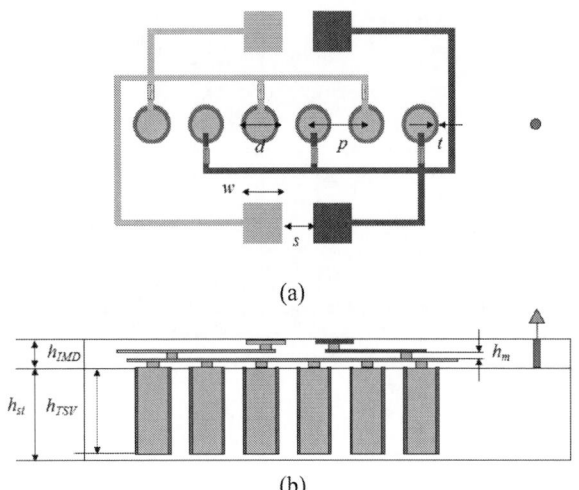

(a)

(b)

Fig. 3. (a) Top view and (b) cross-sectional view of the proposed TSV connectivity testing structure.

TABLE I.
THE PARAMETERS INDICATED IN THE TESTING STRUCTURE AND THEIR PHYSICAL DIMENSIONS.

Parameters	Parameter Description	Dimensions (μm)
h_{tsv}	TSV height	100
d_{tsv}	TSV diameter	30
t_{insul}	TSV Insulator thickness	0.3
p	TSV pitch	60
h_{via}	Metal Via height	4
t_{IMD}	IMD thickness	20
t_{si}	Silicon Substrate thickness	110
w	Pad diameter	100
s	Pad-to-Pad distance	50

substrate conductance G_{si} and substrate capacitance C_{si}, with different fitting coefficients α and β, depending on the various configurations of the structure. The reason for the use of fitting coefficients in the equation is that there are always more than 2 TSVs under test. The total capacitance value of the proposed structure can be expressed as follows :

$$C_{ox} = \frac{2\pi\varepsilon L}{\ln\dfrac{b}{a}} = \frac{2\pi\varepsilon}{\ln\dfrac{d_{TSV}}{d_{TSV} - 2t_{insul}}} . h_{TSV} \tag{1}$$

$$C_{si} = \frac{2\pi\varepsilon L}{\ln\dfrac{p}{b}} = \frac{2\pi\varepsilon}{\ln\dfrac{p}{d_{TSV}}} . h_{TSV} \tag{2}$$

$$G_{si} = \frac{\sigma \cdot C_{si}}{\varepsilon} = \frac{2\pi\sigma_{si}}{\ln\dfrac{p}{d_{TSV}}} . h_{TSV} \tag{3}$$

978-1-4799-5004-1/13 $31.00 © 2013 IEEE

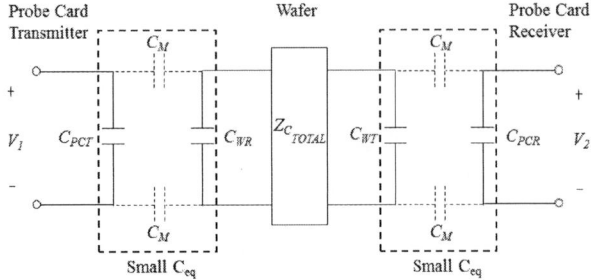

Fig. 4. Simplified circuit model of wafer-level TSV testing structure using capacitive coupling, with the structure connected in parallel to the pads.

$$\frac{1}{sC_{total}} = Z_{C_{total}} = \frac{2}{s\alpha C_{ox}} + \frac{1}{\beta(G_{si} + sC_{si})} \tag{4}$$

$$= \frac{2\beta G_{si} + s(2\beta C_{si} + \alpha C_{ox})}{s\alpha\beta C_{ox}(G_{si} + sC_{si})}.$$

$$\therefore C_{total} = \frac{\alpha\beta C_{ox}(G_{si} + sC_{si})}{2\beta G_{si} + s(2\beta C_{si} + \alpha C_{ox})}. \tag{5}$$

$$X_C = \frac{1}{2\pi f C_{total}}. \tag{6}$$

(1) to (3) are the equations used to find the oxide capacitance, silicon capacitance, and silicon conductance, respectively, based on the physical dimensions of the TSVs. It can be observed that these values can easily be calculated by considering the physical dimensions of the TSVs used in the structures. Further, as expressed in (4) to (5), total capacitance can be expressed in terms of aforementioned oxide capacitance, substrate capacitance, and oxide conductance, as well as the fitting coefficients α and β, which may vary depending on different TSV configurations. If one TSV were to be disconnected from the structure, the total capacitance would decrease, and since impedance is inversely proportional to capacitance from (6), the impedance of the structure will increase.

From Fig.4, it can be observed that the proposed structure is connected in parallel to the metal pads. The reason for this configuration is that the capacitance derived from the metal pads is relatively much smaller than the total capacitance of the proposed structure, and unlike in the case of resistor, in series configuration of two or more capacitors, smallest capacitance value is always dominant. In other words, since the capacitance of the proposed structure is much larger than the equivalent capacitance of the metal pads, for the change in the capacitance of the structure to be accurately observed and be dominant, the structure must be connected to the pads in parallel configuration.

Therefore, with a TSV disconnection, the total capacitance of the proposed structure would decrease and the impedance of the structure would increase. Accordingly, Z_{21} would increase and higher level of voltage would be transferred back

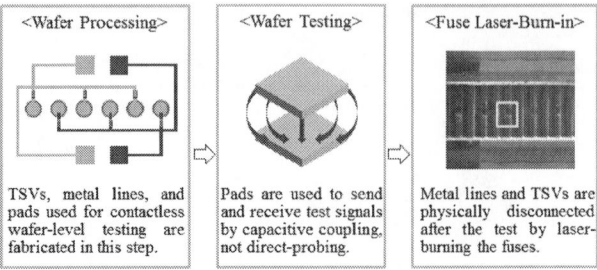

Fig. 5. Overall modified test sequence with the proposed testing structure

to the probe card. In summary, with the addition of the proposed testing structure, the test sequence is modified as depicted in Fig. 5. Firstly, TSVs, metal lines, and metal pads used for contactless wafer-level testing are fabricated. Then, the metal pads are used to send and receive test signals by capacitive coupling for determining disconnection defects. Lastly, the metal lines and the TSVs are physically disconnected once the test is complete, by laser-burning the fuses. Finally, with the fuses burnt-in, the wafer can be further processed through conventional fabrication steps, such as back-grinding, dicing, packaging, and so on.

III. SIMULATION AND ANALYSIS OF THE PROPOSED CONTACTLESS TSV TESTING STRUCTURES

As shown in Fig. 6, the transfer impedance curve found between two ports – each at probe card transmitter and probe card receiver – shows capacitive characteristics. For capacitor, impedance is very high in low frequency, which accordingly makes the Z_{21} level high near DC. As indicated in Fig. 6, the dominant factors that determine the shape of the curve are oxide capacitance, silicon capacitance, and IMD capacitance, respectively, depending on the frequency range of observation points. Further, with the increase in the number of TSV disconnections, the total capacitance of the structure decreases

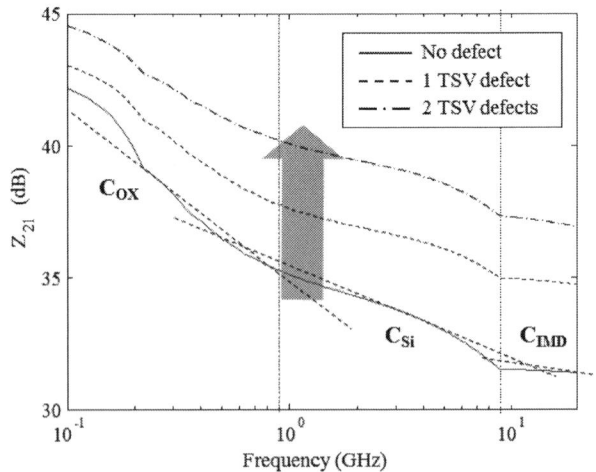

Fig. 6. Transfer impedance curve of the proposed testing structure from 3D-EM simulation in the frequency-domain.

978-1-4799-5004-1/13 $31.00 © 2013 IEEE

TABLE II.
THE FITTING COEFFICIENTS AND TOTAL CAPACITANCE FOR DIFFERENT
NUMBER OF TSV DISCONNECTIONS

Case	Fitting Coefficients		C_{total}
No defect	α	9	4.07 pF
	β	2.1	
Without one TSV	α	7	3.28 pF
	β	1.7	
Without two TSVs	α	6	2.33 pF
	β	1.2	

and accordingly increases the impedance of the testing structure. As a result, the amount of voltage transferred back to the probe card increases and the transfer impedance curve Z_{21} results in upward-shifting. By adjusting the fitting coefficients α and β from (5), we can find the total capacitance of the proposed structure, based on the transfer impedance curve from 3D-EM simulation in frequency-domain. For 3D-EM simulation, ANSYS HFSS was utilized and the probe card transmitter and receiver sides were selected to be the two ports of observation. The dimensions of the TSVs under simulation are set same as those listed in Table I. As was expected previously, it can be verified from simulation that the Z_{21} experiences upward-shifting with increasing number of TSV disconnections. Also, total capacitance of the structure decreases from 4.07 pF to 3.28 pF, and then to 2.33 pF, as the number of TSV disconnections increases from 1 to 3.

The same phenomenon can be observed in time-domain voltage waveform as depicted in Fig. 7: the peak-to-peak voltage level of the transferred signal increases by 30 mV, which is about 33% increase – from 90 mV to 120 mV, and from 120 mV to 150 mV – for every TSV disconnected from the metal line. For time-domain verification, 5V, 2 GHz sine wave is applied as input test signal to the proposed testing structure, as described in Table III. Agilent ADS was utilized for the time-domain simulation.

From the results shown in both Fig. 6 and Fig. 7, it can clearly be shown that the existence of TSV disconnection can be accurately detected using the proposed contactless wafer-level TSV connectivity testing structure. With additional disconnection of TSV, the transfer impedance curve in

TABLE III.
THE COMPONENTS IN THE TESTING STRUCTURE AND THEIR VALUES.

Components	Component Description	Values
V_S	Source Voltage	5 V
Freq.	Source frequency	2 GHz
R_S, R_L	Source and Load Resistance	50 Ohm
C_{ox}	Oxide Capacitance	3 Pf
C_{si}	Silicon Capacitance	69 fF
G_{si}	Silicon Conductance	6.7 Ms

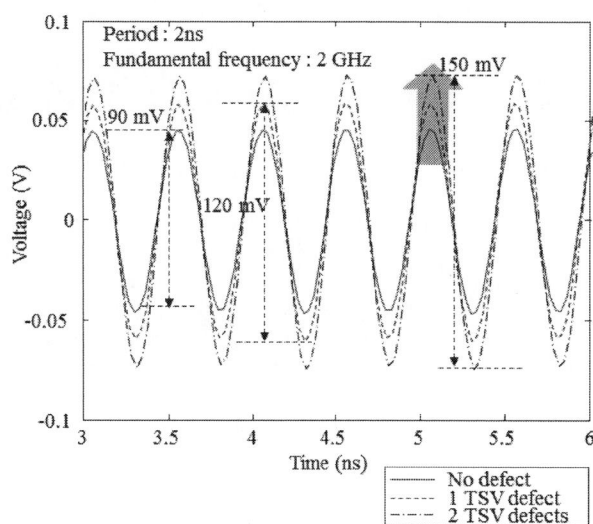

Fig. 7. Ttime-domain voltage waveform transferred back to the probe card for the proposed testing structure. The peak-to-peak voltage level increases with each TSV disconnection.

frequency domain and voltage waveform in time domain both shift upward.

Of course, with the rapidly decreasing dimensions of the TSV, the test resolution would eventually get deteriorated, as the series capacitance observed between the TSVs would also decrease. Furthermore, as the number of TSVs is expected to increase to thousands and higher, placement of a large number of metal pads on wafer for the sole purpose of testing would be burdensome and the area consumption will have to be more carefully considered. However, considering the numerous advantages that can stem from contactless testing, such as elimination of physical pressure on the wafer from direct contact of probe tips as well as substantial increase in the testing speed due to no need for fine control of the probe tips, contactless testing using capacitive coupling will bring more merits than it causes area consumption problems in the future. Furthermore, as the proposed structure consists only of passive components for the purpose of testing, it does not require any extra power consumption on wafer.

IV. CONCLUSION

In this paper, a novel contactless wafer-level TSV connectivity testing structure using capacitive coupling is proposed and successfully verified using simulation. With the increasing number of TSVs for integration of many dies of multiple functions, TSV has become the key technology for the realization of 3D-IC. However, due to the instability issues that still remain to be solved in the fabrication process of TSVs, testing of TSV connectivity has become very important, as disconnection defects can result in severe decrease in the final chip yield and deterioration of electrical characteristics. However, many conventional testing methods of wafer-level testing have exhibited various limitations, such as pitch limitation of the conventional probe cards, physical damage on microbumps, height variation of the microbumps from unstable

TSV fabrication process, and extra power consumption required for testing. Therefore, for accurate detection of TSV disconnection, contactless wafer-level TSV connectivity testing structure, that does not require power or direct contact to the wafer, is introduced.

The performance of the proposed structures was successfully validated by a series of time- and frequency-domain simulation results using HFSS and ADS, respectively. It was verified that the TSV defects could accurately be detected using the proposed structures. With further research to increase the sensitivity/resolution of the proposed testing structure while decreasing area consumption, the testing structures could later be applied to allow high yield of TSV in TSV-based 3D-IC and moreover, provide a set of guidance for improvements in TSV fabrication processes.

ACKNOWLEDGMENT

This work was supported by the Smart IT Convergence System Research Center funded by the Ministry of Education, Science and Technology as Global Frontier Project"(STRC-2011-0031863) and by the R&D program of ISTK

[Development of an image-based, real-time inspection and isolation system for hyperfine faults].

REFERENCES

[1] S. Lim, "Physical design for 3D system on package," Design & Test of Computers, IEEE , vol.22, no.6, pp.532,539, Nov.-Dec. 2005

[2] A. Hsieh, T. Hwang, "TSV Redundancy: Architecture and Design Issues in 3-D IC," Very Large Scale Integration (VLSI) Systems, IEEE Transactions on , vol.20, no.4, pp.711-722, April 2012

[3] J.J. Kim, H. Kim, S.Kim, C.Cho, D.H. Jung, J.Kim, J.S. Pak Heegon Kim ; Sukjin Kim ; Changhyun Cho ;Jung, D.H. ; Joungho Kim ; Jun So Pak, "Contactless wafer-level TSV connectivity testing method using magnetic coupling," Electrical Design of Advanced Packaging and Systems Symposium (EDAPS), 2012 IEEE , vol., no., pp.5,8, 9-11 Dec. 2012

[4] J. Zhang, Le Yu, H. Yang, Y.L. Xie, F.B. Zhou, W Wang, "Self-test method and recovery mechanism for high frequency TSV array," VLSI and System-on-Chip (VLSI-SoC), 2011 IEEE/IFIP 19th International Conference on , vol., no., pp.260-265, 3-5 Oct. 2011

[5] L. Jiang, Q. Xu, B. Eklow, "On effective TSV repair for 3D-stacked ICs," Design, Automation & Test in Europe Conference & Exhibition (DATE), 2012 , vol., no., pp.793-798, 12-16 March 2012

Modeling and Analysis of Open Defect in Through Silicon Via (TSV) Channel

Daniel H. Jung, Heegon Kim, Jonghoon J. Kim and
Joungho Kim
Department of Electrical Engineering
Korea Advanced Institute of Science and Technology
Daejeon, South Korea
Hyunsuk@kaist.ac.kr

Hyun-Cheol Bae and Kwang-Seong Choi
IT Materials and Components Laboratory
Electronics and Telecommunications Research Institute
Daejeon, South Korea

Abstract—**Vertical interconnections of stacked chips through the silicon substrates have enabled higher performance of electronic products with lower power consumption. The advantage of through silicon via (TSV) technique can be maximized by increasing the number of I/Os, which requires fine pitch and smaller diameter. The scale-down of TSVs results in decreased yield level caused by lack of precision in fabrication process. Among various types of possible defects, open defect creates a disconnection in the channel, electrically separating the transmitting terminal from the receiving target. In this paper, the equivalent circuit model for open defect is proposed and inserted as circuit component in a circuit model for defect-free channel. Open defect is analyzed in different locations along the channel to examine the effect in signal transmission characteristics.**

Keywords— Through Silicon Via (TSV), Equivalent circuit model, Open defect

I. INTRODUCTION

Vertical integration of chips has expanded the boundary of power consumption and system density limitations. A technique of stacking multiple chips on top of the other to integrate a single system, called three-dimensional integrated circuit (3D-IC), provides shortest distance between the chips and smallest form factor of the product. The interconnections for communication between vertically placed chips are realized in various types: wire-bonding, ball grid array (BGA), or through silicon via (TSV). Among the mentioned interconnection techniques, TSVs are considered as the most efficient technique to keep up with the trend. Using all available space in the silicon substrate, the I/O count can increase up to the order of tens of thousands for TB/s level data transmission [1].

Despite the absolute advantages of TSV based 3D-IC, the challenges still remain to be solved before application to the commercializing products. Fitting thousands of I/Os in a limited chip space requires fine pitch and reduction of TSV diameter. The reliability issue becomes more critical in satisfying continuously increasing bit rate. Present TSV fabrication technique cannot avoid various types of defects, such as misaligned chips, impurity failure, void and crack formation in TSVs or bumps, and metal shortage between the

channels [2]-[3]. Each of the defects or potential defects contributes in different ways to degrade the signal transfer from the input terminal to the receiving target.

3D-IC comprises multiple layers of chips, which also comprises multiple layers for both vertical and horizontal interconnections. Possible locations for open defect in the channel are in between the layers. Failure to form metal vias between re-distribution layers (RDLs) and TSVs, or non-uniform bump size may lead to open defects as shown in Fig. 1. In addition, thermal and mechanical stress may cause the potential defects, such as voids, to become open defects. This creates a disconnection in the path of the signal, which apparently causes an unexpected transmission performance of the channel. The analysis of the effects of open defect to the signal transmission, according to the cause and location, is vital for further enhancement of TSV fabrication process.

The equivalent circuit models of TSV structure and RDLs are proposed in [4]-[5]. In previous works, the electrical characteristics of TSV channel are analyzed through description of the effect of each circuit component in frequency-domain and the frequency ranges are divided according to dominantly affecting components. The equivalent circuit model for TSV structure provides fast and accurate results for high speed channel characterization. In a similar manner, since it is difficult to acquire test vehicles with known defect types and locations, TSV channel with defects can be analyzed through equivalent circuit model analysis.

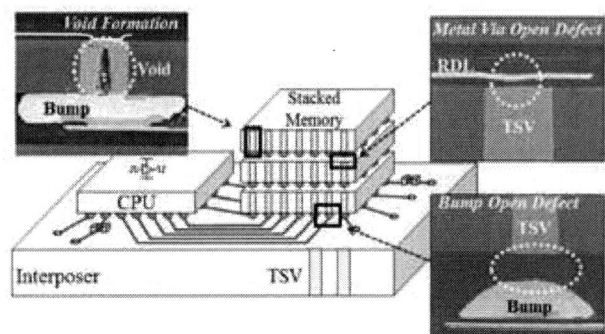

Fig. 1. Possible locations of open defects along the channel

978-1-4799-5004-1/13 $31.00 © 2013 IEEE

In this paper, the effect of open defects in different locations is analyzed. A ground signal ground (GSG) type daisy-chain structure is designed with eight TSVs in each channel and using 2-port analysis, signal transmission and reflection is examined. The results from equivalent circuit model and 3D FEM solver are compared in frequency-domain for model verification. An open defect is represented as circuit components, which are inserted in the defect-free circuit model for analysis and verification. The open defect is analyzed in three cases by reducing the height of one of the bumps in three different locations and calculating the values of the components accordingly.

II. TSV DAISY-CHAIN STRUCTURE DESIGN

The transmitted signals travel through the silicon substrate multiple times when the chips are vertically stacked. For simplification, a daisy-chain structure is designed, in which only two dies are necessary to realize signal paths that pass through the substrate multiple times. The design for 3D FEM solver simulation is illustrated in Fig. 2. Starting from the left contact pad, where port 1 is located, the top RDL leads the signals to first TSV, and approaches the bottom die through the bump. The signals meet the second TSV through RDL in the bottom and the channel continues through six additional TSVs until the other end of the channel, where port 2 is located.

Dimensions of the parameters in Fig. 2 are listed in Table I. The listed values and the constants of known material properties are applied to the equations in [4]-[6], which provide equivalent circuit models for TSVs and RDLs. The models are connected in series for daisy-chain structure modeling. In this section, the equivalent circuit model of GSG-type defect-free daisy-chain structure is presented and verified by comparison with the results from 3D FEM simulation. The ports are located at both ends of the channel for signal transmission analysis. The modeling of defect-free channel has to be accurate in preparation for open defect analysis, which is described in more detail in the following sections.

TABLE I. THE PARAMETERS IN DASY CHAIN STRUCTURE AND THEIR DIMENSIONS.

Parameters	Parameter Description	Dimensions (μm)
h_{tsv}	TSV height	60
d_{tsv}	TSV diameter	10
t_{insul}	TSV Insulator thickness	0.2
p_{tsv}	TSV pitch	100
p_{pad}	Contact pad pitch	250
h_{bump}	Bump height	9
d_{bump}	Bump diameter	20
h_{via}	Metal Via height	1
d_{via}	Metal Via diameter	9
t_{RDL}	RDL thickness	1
t_{IMD}	IMD thickness	4
t_{si}	Silicon Substrate thickness	60

Prior to analyzing open defect in TSV channel, defect-free daisy-chain structure is modeled by alternately connecting the RDL models and TSV models in series. The circuit diagram for GSG-type TSV channel model is shown in Fig. 3. The path of the signal is modeled as series resistance and inductance; the insulation layer and the silicon substrate between the channels are modeled as shunt capacitance and conductance. The schematic in Fig. 3 only includes two TSVs, since the rest of the channel is a repetition of the shown model.

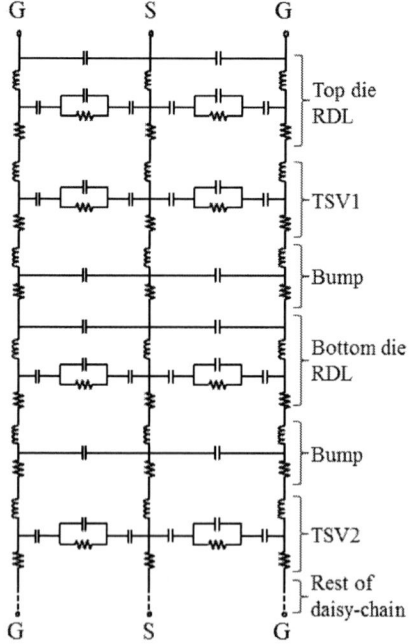

Fig. 3. Equivalent circuit model of GSG-type daisy-chain structure

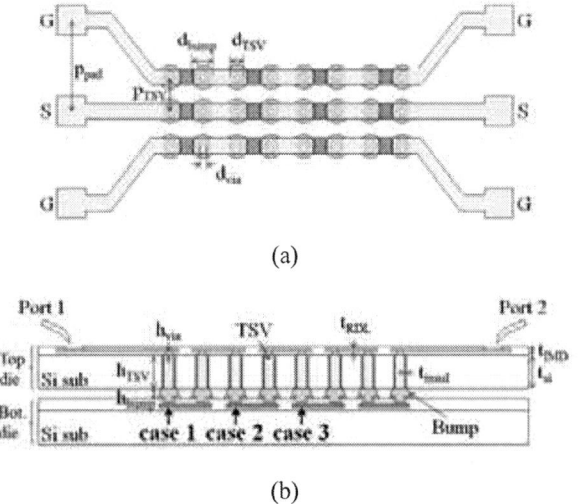

(a)

(b)

Fig. 2. Designed GSG-type TSV daisy-chain structure and three cases of open defect locations. (a) Top view. (b) Cross-sectional view.

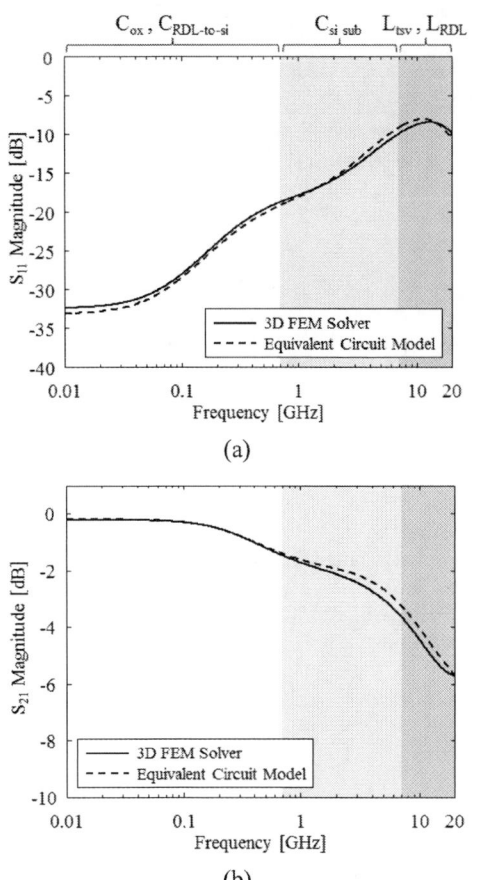

(a)

(b)

Fig. 4. S-parameter results for equivalent circuit model verification. (a) S_{11} magnitude. (b) S_{21} magnitude.

The verification of the defect-free channel model is provided in S-parameter curves. By placing 50-ohm terminations at both ends of the channel, the results are extracted in S_{11} magnitude and S_{21} magnitude. The plots are divided into three regions: low, mid and high frequency ranges. The curvature in each frequency range is dominantly affected by different components as labeled on top of S_{11} plot in Fig. 4(a) [4]. C_{ox} is the capacitance formed in insulation layer between TSVs and silicon substrate and $C_{RDL\text{-}to\text{-}si}$ is the capacitance in between the RDLs and the substrate; relatively higher values of capacitance are the dominant factors in low frequency range. As the frequency increases, the signal is affected by relatively smaller capacitance, such as C_{si}, which is formed between TSVs in silicon substrate. In higher frequency range, the inductance of the channel becomes a dominant factor in signal transmission characteristics. The similarity of the curvatures in all frequency ranges shows that each factor representing different parts in the daisy-chain structure is properly modeled.

III. EQUIVALENT CIRCUIT MODEL OF TSV CHANNEL WITH OPEN DEFECT

Open defect in TSV channel can be modeled as a capacitance as depicted in Fig. 5. Since TSV based 3D-IC is a

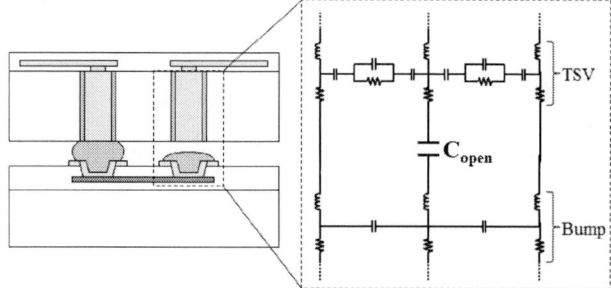

Fig. 5. Representation of bump open defect as a series capacitance.

vertical integration of multiple layers, open defects may be found in between the layers, such as bumps between the stacked dies or metal vias between RDLs and TSVs in inter-metal dielectric (IMD) layer. The effect of open defect is analyzed by intentionally reducing the height of one of the bumps in the channel, creating a 6 μm gap between a TSV and the corresponding bump. The results are obtained in three different locations as labeled in Fig. 2(b).

A. Open Defect Modeling and Verification

A dielectric gap between conductive materials is modeled as a series capacitance, which is calculated by the following equation [7]:

$$C_{open} = \varepsilon_0 \varepsilon_{r,dielectric} \times \frac{\pi \times r_{tsv}^2}{t_{open}} \quad (1)$$

in which ε_0 and $\varepsilon_{r,dielectric}$ are the absolute and relative permittivity values, r_{tsv} is the radius of TSV and t_{open} is the open gap thickness. Applying the calculated capacitance value in series between a TSV and bottom RDL, the effect of open defect can be analyzed. The equivalent circuit model for TSV daisy-chain structure without any defects is verified in the previous section. By inserting the calculated capacitance in the defect-free model, frequency-domain plots for three cases are obtained as shown in Fig. 6.

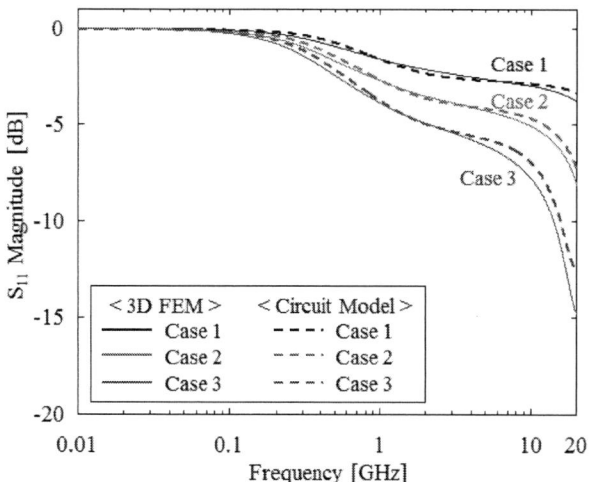

Fig. 6. Comparison of S_{11} magnitude results from 3D FEM solver and the circuit model of TSV channel with three cases of open defect locations

978-1-4799-5004-1/13 $31.00 © 2013 IEEE

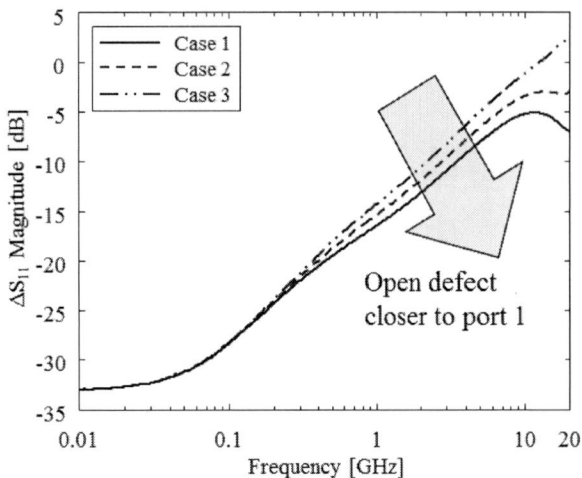

Fig. 7. ΔS_{11} magnitude results from the circuit model of TSV channel with open defect in three different locations: bump under TSV1, TSV3 and TSV5 (from port 1).

As mentioned previously, the open defect for three cases are located under first, third and fifth TSVs from the left side of Fig. 2. For verification of the model with open defect, the height of one of the bumps in the signal channel is reduced to 3 μm instead of 9 μm to form a 6 μm gap, which is calculated as a capacitance of 0.46 fF. Although the ports are located at both ends of the channel, only S_{11} magnitude results are presented for verifying open defect, since reflection of the signal is more of a concern in a disconnected channel. S_{21} magnitude curves are dependent to the thickness of open gap; if the series capacitance value stays the same, S_{21} curve is not affected regardless of the location. The correlation between the results from 3D FEM solver with small-sized bump and equivalent circuit model with series capacitance shows that the components are accurately calculated.

B. Open Defect Detection and Isolation

The results in Fig. 6 are obtained by varying the location of the defect. The pattern in curvature change in S_{11} magnitudes can be further analyzed for open defect detection and isolation. The reason behind designing the structure with 2-port probing is to find the location of open defect. A disconnection of the channel is easily detected by 2-port probing since the transmitted signal cannot be accurately received by the port at the other end in low frequency range. Once the defect is detected, the test vehicle has to be further processed for fault isolation. The frequency-domain results in Fig. 6 can be modified as ΔS_{11}, which is a plot of the curvature difference between S_{11} magnitude and the reference. The S_{11} level drop in higher frequency range is dominantly determined by the amount of shunt capacitance that the signal experiences in approaching the open defect.

The reference may be chosen according to the applications. The plot in Fig. 7 is obtained by curvature difference between S_{11} magnitudes from the model of defect-free channel in Fig. 4(a) and that of three cases in Fig. 6. The curves in low frequency range are indistinguishable, since shunt capacitance

in a channel with open defect is not a dominant factor. As the frequency increases, the effect of signal loss in silicon substrate appears. The signal transmitted from port 1 experiences 1, 3 and 5 TSVs before the reflection. ΔS_{11} for case 1 is lower than case 3, which proves that the amount of shunt capacitance in case 3 is higher, resulting in greater signal loss in the substrate. Taking account of the design and applications, the method can be further developed for connectivity testing of chip level interconnections.

IV. CONCLUSION

In this paper, the equivalent circuit model for TSV daisy-chain structure with open defect is proposed. A designed GSG-type daisy-chain structure is modeled by RLGC calculations. By comparing the results from the model and 3D FEM solver the values for the components is verified. In a prepared defect-free model, bump open defect is represented as a capacitor and the calculated value is inserted in the defect-free model for analysis. The results with series capacitance are also verified by an identical procedure. The effect of open defect according to its location is analyzed in three locations: bump open under first, third and fifth TSVs from port 1. The curvature difference of each case is caused by the amount of shunt capacitance experienced by the signal before approaching the open defect. The level difference can be further developed as a connectivity test method for detection and isolation of open defect in TSV based 3D-IC.

ACKNOWLEDGMENT

This work was supported by the Smart IT Convergence System Research Center funded by the Ministry of Education, Science and Technology as Global Frontier Project (STRC-2011-0031863) and by the R&D program of ISTK [Development of an image-based, real-time inspection and isolation system for hyperfine faults].

REFERENCES

[1] A. Papanikolaou, D. Soudris and R. Radojcic, Three Dimensional System Integration, Springer: New York, 2011.

[2] Ming-Che Hsieh, "Energy release rate investigation for through silicon vias (TSVs) in 3D IC integration," Thermal, Mechanical and Multi-Physics Simulation and Experiments in Microelectronics and Microsystems (EuroSimE), 2011 12th International Conference on , vol., no., pp.1/7,7/7, 18-20 April 2011

[3] Dunne, R., "Development of a stacked WCSP package platform using TSV (Through Silicon Via) technology," Electronic Components and Technology Conference (ECTC), 2012 IEEE 62nd, vol., no., pp.1062,1067, May 29 2012-June 1 2012

[4] Joohee Kim, "High-Frequency Scalable Electrical Model and Analysis of a Through Silicon Via (TSV)," Components, Packaging and Manufacturing Technology, IEEE Transactions on , vol.1, no.2, pp.181,195, Feb. 2011

[5] Kihyun Yoon, "Modeling and analysis of coupling between TSVs, metal, and RDL interconnects in TSV-based 3D IC with silicon interposer," Electronics Packaging Technology Conference, 2009. EPTC '09. 11th , vol., no., pp.702,706, 9-11 Dec. 2009

[6] S. H. Hall, G. W. Hall and J. A. McCall, High-Speed Digital System Design, Wiley Interscience: New York, 2000

[7] Jung, D.H., "Disconnection failure model and analysis of TSV-based 3D ICs," Electrical Design of Advanced Packaging and Systems Symposium (EDAPS), 2012 IEEE , vol., no., pp.164,167, 9-11 Dec. 2012

978-1-4799-5004-1/13 $31.00 © 2013 IEEE

The Direct RF Power Injection Method up to 18 GHz for Investigating IC's Susceptibility

Yin-Cheng Chang[1,2], Shawn S. H. Hsu[2], *Member, IEEE*, Yen-Tang Chang[3], Chiu-Kuo Chen[3], Hsu-Chen Cheng[1], and Da-Chiang Chang[1]

[1]National Chip Implementation Center, National Applied Research Laboratories, Hsinchu, Taiwan
[2]Institute of Electronics Engineering, National Tsing Hua University, Hsinchu, Taiwan
[3]Bureau of Standards, Metrology and Inspection, M.O.E.A, Taipei, Taiwan

Abstract—**The direct RF power injection (DPI) measurement up to 18 GHz is proposed to investigate the IC immunity. The DPI method is reviewed and the consideration of extending frequency range is discussed. Furthermore, the details of the measurement setup are depicted in this work. The critical part, on-board injection network in the power injection path with a 3 dB bandwidth of 18.7 GHz is realized. A low dropout regulator (LDO) is used to demonstrate the test setup. The proposed DPI test with the experimental results shows the significance up to 18 GHz.**

Keywords—integrated circuit; EMC; immunity; DPI

I. INTRODUCTION

The continuous miniature of the feature size in integrated circuit (IC) technology, as known as Moore's law, increases the significance of the electromagnetic compatibility (EMC) of IC. Scaling down the size of the devices as well as the increasing transistors amount allow IC to be operated at high-speed with low power consumption. The consequently desired high performances not only produce noise but also make the IC itself sensitive to interference. This situation leads the demand of characterizing their behaviors of emission and immunity. To investigate these problems, several measurement methods have been developed as the standards.

The technology subcommittee 47A of International Electrotechnical Commission (IEC) published a series of IC level test methods on EMI (61967 series) [1] and EMS (62132 series) [2]. They are widely adopted as the comparative evaluation for choosing the best candidate of product from different designs. Among them, a popular method to characterize the immunity of IC is called DPI [3] as shown in Fig. 1. The straightforward test setup in this method is helpful to observe the conducted emission behavior of a certain IC pin/pins. The DPI was utilized to differentiate the improved susceptibility levels of an IC with several embedded on-chip EMI protection [4]. Similarly, the electrostatic discharge (ESD) protection strategy also used the DPI as the EMI aggression to demonstrate the impact of interference on EMC performance [5]. These important findings reveal the fact that DPI is recommended when the experimental results were desired to be reproducible, repeatable, and confident. In addition, the models of DPI test setup were built for simulations which agree well with the measured results [6-7]. Based on the

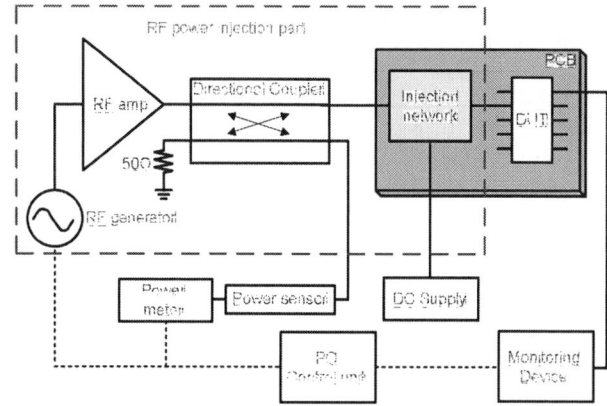

Fig. 1. The test setup of DPI method for ICs.

mature modeling, the differences between DPI and BCI (bulk current injection) [8] tests are analyzed with high conformity [9]. These previous studies provide valuable information regarding DPI. However, its applicable frequency bandwidth remains an issue which needs further investigation.

Most released standards such as IEC series have the frequency range below 1 GHz. Sometimes it is insufficient to evaluate the EMC behaviors while the modern circuits operate higher than 1 GHz. Therefore, some measurement method like GHz transverse electromagnetic (GTEM) cell [10] was proposed which has the frequency range up to 18 GHz. As a trend, the DPI is also expected to having the capability of a wider bandwidth. A new methodology was proposed by using the edge coupled transmission line as a part of the injection network [11]. The result showed the requirement above 1 GHz can be achieve, but the frequency bandwidth was limited by the narrow band nature of coupler. A significant study [12] was proposed to perform the DPI up to 20 GHz which focused on the PCB fixture design for a SOIC8-packaged IC, while the injection network is off-board.

This paper integrates the injection network into the test board with the bandwidth up to 18 GHz and is organized as follows. At first, the DPI method is revisited and some principles are emphasized. In section III, the feature of DPI test up to 18 GHz is proposed. The measurement setup and components used in the injection path with the test board are discussed in details. Finally, a LDO used widely in communication module is tested as the DUT. The result of immunity level from the conducted RF disturbances is measured, and the measurement setup is validated.

II. REQUIREMENTS OF WIDEBAND DPI TEST SETUP

The DPI method is defined to characterize the immunity of IC in the presence of conducted RF disturbances. This delivered conducted forward power ($P_{forward}$) injects into a IC through the cable harness or the traces on a PCB. To characterize the immunity of an IC, the $P_{forward}$ which causes malfunction of IC is recorded. The general test setup according to the IEC 62132-4 standard is shown in Fig. 1. It contains the DC power supply, RF power injection part, test PCB with injection network and DUT, monitoring device, and a control unit.

Several elements in the power injection part become critical while the measurement frequency extends above 1 GHz. In order to deliver enough power into DUT, a 50Ω characteristic impedance (Z0) system has to be implemented for effective power injection with less path loss. Besides, the power level from signal generator is often insufficient at high frequency. Therefore, a power amplifier is needed for driving enough power level into the pin under test which often presents high degree of mismatch. Because of the wide bandwidth, several amplifiers are needed to cover the whole frequency range. Besides, the level of harmonics has to be 20dB lower than the interference according to the standard.

The directional coupler is employed to monitor the $P_{forward}$ injected to the port of test PCB from power amplifier. The $P_{forward}$ can be measured by a power meter with a power sensor. Therefore, the dynamic range and frequency range of power sensor should be taken care. Also the VSWR is desired to be smaller than 1.15. Notice that the $P_{forward}$ measured by power meter has to be corrected by adding the coupling factor of the directional coupler. Accompanying with various applications at high frequencies, most devices in power injection parts can be found with expected performance. But the components in the on-board section which will be discussed in the next section are not common.

An oscilloscope, test receiver or other monitoring device is used to monitor the malfunction of the DUT during the experiment. The injected power has to be recorded when DUT becomes susceptible. A control unit or program can be used to control these instruments which will save time.

III. ON-BOARD INJECTION NETWORK DESIGN

The RF power from amplifier is expected to transmit onto the pin of DUT for a wide bandwidth of 18 GHz. Therefore, all the elements which form this injection network on PCB have to be designed carefully and supposed to have excellent performance. So the traces on PCB have to be design as short as possible with a characteristic impedance of 50Ω. And a DC block capacitor is inserted to prevent DC current destroying the amplifier. 6.8nF capacitors as an example mentioned in the standard gives the lower frequency limit around 150 kHz. The larger capacitance can achieve lower 3dB bandwidth and

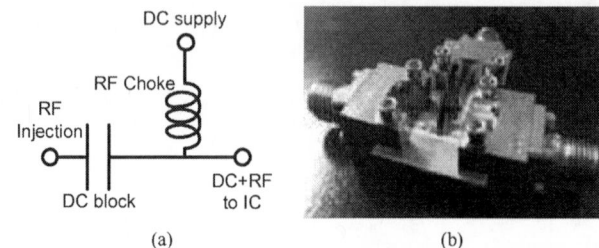

(a) (b)

Fig. 2. (a) The configuration and (b) photograph of on-board injection network.

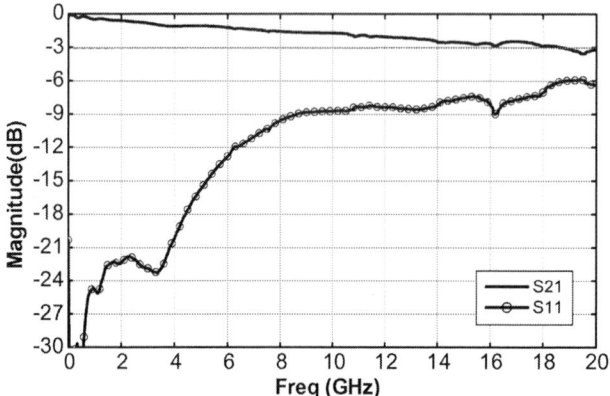

Fig. 3. The S-parameters of on-board injection network.

present a high pass response. The problem is the parasitic effect makes the resonance happen and transforms the capacitor become inductive which limits the upper 3dB roll-off bandwidth. Therefore, a capacitor with wide bandwidth, flat frequency response, and low insertion loss is preferred. In this work, a 100nF capacitor (ATC 545L) is chosen with its S-parameters can be obtained for estimation in advance.

If the pin under test is also supplied by a DC source, a decoupling component is necessary to avoid the injected RF power heading to the DC source where presents a low AC impedance path. Generally, a RF choke like the inductor is a good candidate to have the AC impedance over 400Ω in the test frequency range without causing too much DC voltage drop on the path. Again, the parasitic has to be minimized to guarantee higher operating frequency without resonance. A 2uH inductor (ATC 506WLS) is chosen in this work. The capacitor and inductor are formed as an injection network as shown in Fig. 2.

Except for the selection of lump components with wide bandwidth, the performance of PCB and connectors should also be concerned. The capacitor and inductor are mounted on a 0.254mm double side high frequency PCB (RO4350B) with SMA end launch connectors (Southwest Microwave 292-04A-5) as shown in Fig. 2(b). In this work, the vector network analyzer used to measure the S-parameters is Agilent PNA N5230A with the measurement capability of 300 KHz to 20 GHz. The standard four-port short-open-load-thru (SOLT) calibration was performed before testing.

In the standard, a 3dB insertion loss of the on-board injection network is permitted to perform the DPI test. The measured 3dB bandwidth reveals the DPI measurement

978-1-4799-5004-1/13 $31.00 © 2013 IEEE

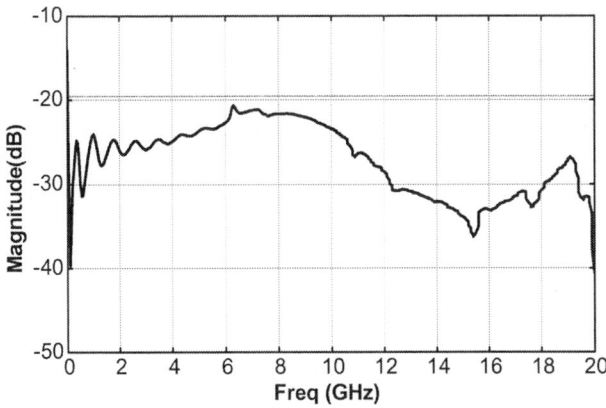

Fig. 4. The attenuation contributed by the decoupling inductor on the DC path.

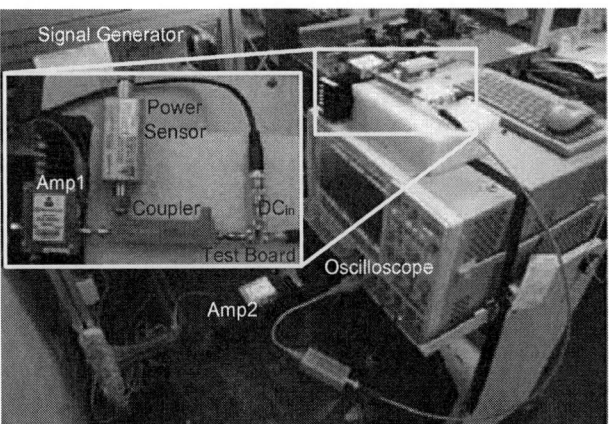

Fig. 5. The photograph of 18 GHz DPI test setup.

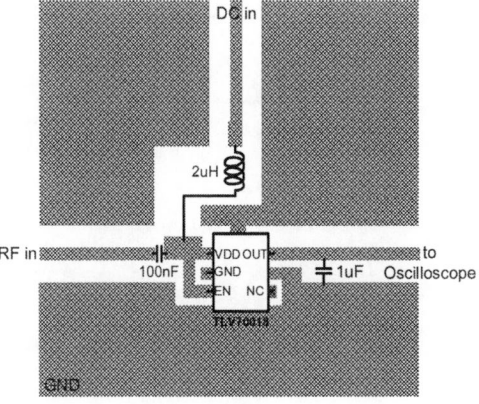

Fig. 6. The test board composed of on-board injection network and DUT.

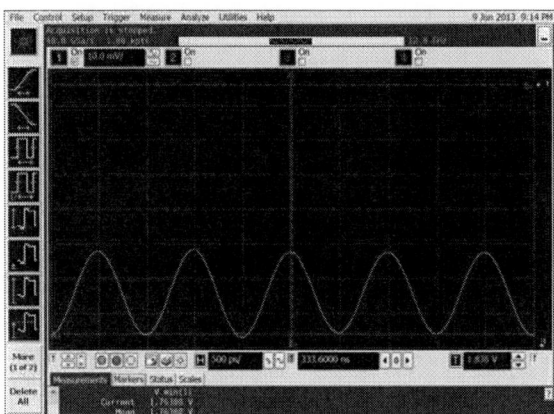

Fig. 7. The measured waveform when DUT reach the immunity criterion.

frequency range can be extended to 18.7 GHz as shown in Fig. 3. Regarding to the AC impedance over 400Ω recommended in the standard, the corresponding insertion loss of DC path is 19.6 dB. Fig. 4 shows the validated results with the insertion loss greater than 19.6 dB over the whole frequency range. The following section will apply this injection network to achieve a DPI testing.

IV. EXPERIMENT OF DPI UP TO 18 GHz

The test setup in this work for DPI measurement similar to Fig. 1 is shown in Fig. 5. It contains DC power suppliers, power meter, RF power injection part (RF generator, RF amplifier, directional coupler, and on-board injection network), test board with DUT, and the oscilloscope as monitoring device, but lack of the control unit.

A LDO (Texas Instruments TLV70018) for portable devices is chosen as the IC under test. The function of LDO is to provide an accurate and stable DC voltage to the whole system. This component is used widely in the modern communication modules which operate at the frequency range from hundred MHz to several GHz. So it is a good candidate to perform DPI test. Fig. 6 shows the test board design with on-board injection network and the typical configuration of IC operation. The injection point is set at the VDD pin to emulate

the interference injection which may cause malfunction of IC and lead the failure of the whole system.

To observe the failure of DUT, the output pin is connected to an oscilloscope (Agilent DSA91204A 12GHz real time oscilloscope) through a 1MΩ probe. The immunity criterion has to be defined to tell if the DUT fail or not when subject to an interference. A failure is determined when the output voltage reach ±2% tolerance by referring to the data sheet. In other words, once the output voltage lower than 1.764V or higher than 1.836V, it fails. Neither the average voltage nor the AC ripple reaches limits; the $P_{forward}$ is recorded and represents the immunity level at that frequency. Fig. 7 shows a failure occurs at 1 GHz while the ripple of waveform hits the lower boundary. Besides, the highest injected power level has to be set based on the performance of facilities. In this work, a maximum $P_{forward}$ of 30dBm is defined because of the restriction of instrumentation. The DPI test was demonstrated from 1 GHz to 18 GHz with a frequency step of 500 MHz as shown in Fig. 8. The immunity remains at 30 dBm limit in the frequency range between 4 GHz to 11.5 GHz which indicates the pin under test is immune to the electromagnetic aggression. And the immunity degrades again at higher frequency band. It demonstrates the interference at the band which is out of operating frequency of IC may cause malfunction. This finding shows the significance that the DPI method with

978-1-4799-5004-1/13 $31.00 © 2013 IEEE

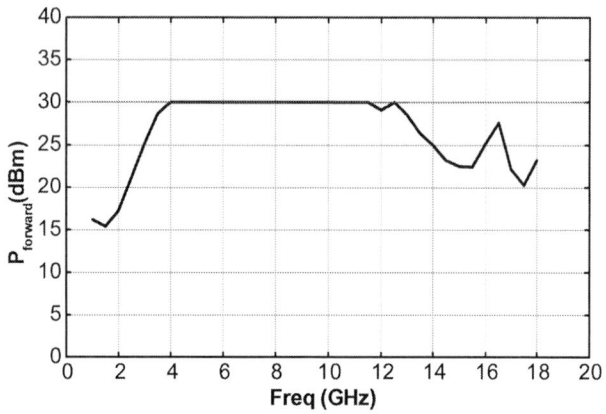

Fig. 8. The immunity of LDO by applying DPI measurement up to 18 GHz

extended bandwidth is needed. All the details of used instruments and components are listed in Table. I.

TABLE I. INSTRUMENTS AND DEVICES USED IN DPI SETUP UP TO 18 GHZ

Instrument/Device	Vendor / part	Feature
RF generator	Agilent/E8247C	250k~20GHz
RF amp1	Mini-Circuits ZVE-3W-83	2G~8GHz
RF amp2	Mini-Circuits ZVE-3W-183	5.9~18GHz
Directional Coupler	Agilent 87300B	1G~20GHz
Power sensor	Agilent E4413A	50M~26.5GHz
Power meter	Agilent E4416A	20M Sa/sec
Oscilloscope	Agilent DSA91204A	12GHz, 40G Sa/sec
DC supply	Agilent E3615A	0~20V, 0-3A
Capacitor	ATC 545L	100nF, 16kHz~40GHz
Inductor	ATC 506WLS	2uH, 400kHz~40GHz
LDO	TLV70018	1.8V output

V. CONCLUSION

In this paper, the setup of establishing DPI measurement up to 18 GHz for ICs is proposed. To achieve such a wide bandwidth measurement, all the components, instruments, and Z0 of the PCB traces are carefully considered. The on-board injection network in the power injection path is designed and verified. By employing a LDO as the DUT, this proposed DPI measurement is demonstrated. The experimental result shows the capability of investigating the immunity of ICs up to 18 GHz. A significant finding that IC could be affected by the interference at such high frequency is observed.

ACKNOWLEDGMENT

The authors would like to thank research group from BSMI (Bureau of Standards, Metrology and Inspection) for their technical support. The original research work presented in this paper was made possible in part by the BSMI under Contract No. 2C101011222-05, grant from BSMI, Taiwan.

REFERENCES

[1] *Integrated Circuits, Measurement of Electromagnetic Emission, 150 KHz to 1 GHz: General Conditions and Definitions—Part 1*, International Electrotechnical Commission Standard IEC61967-1, Mar. 2002.

[2] *Integrated Circuits, Measurement of Electromagnetic Immunity, 150 KHz to 1 GHz: General Conditions and Definitions—Part 1*, International Electrotechnical Commission Standard IEC62132-1, 2007.

[3] *Integrated Circuits, Measurements of Electromagnetic Immunity 150 kHz to 1 GHz—Part 4: Direct RF Power Injection Method*, Standard IEC 62132-4, 2006.

[4] A. Alaeldine, N. Lacrampe, J. L. Levant, R. Perdriau, M. Ramdani, and F. Caignet, "Efficiency of embedded on-chip EMI protections to continuous harmonic and fast transient pulses with respect to substrate injection," in *Proc. IEEE Symp. EMC 2007*, Jul. 9–13, pp. 1–5.

[5] K. Abouda, P. Besse, and E. Rolland "Impact of ESD strategy on EMC performances," in *Proc. International Workshop on Electromagnetic Compatibility of Integrated Circuits EMC Compo*, Nov. 2011, pp. 224-229.

[6] Loeckx J., Georges G. "Assessment of the DPI standard for immunity simulation of integrated circuits," *IEEE Symp. on EMC*: Workshop and Tutorial Notes, Honolulu, HI, USA, 2007, vol. 1–3, pp. 813–817.

[7] A. Alaeldine, R. Perdriau, M. Ramdani, J. Levant, and M. Drissi, "A direct power injection model for immunity prediction in integrated circuits," *IEEE Trans. Electromagn. Compat.*, vol. 50, no. 1, pp. 52–62, Feb. 2008.

[8] *Integrated Circuits, Measurements of Electromagnetic Immunity 150 kHz to 1 GHz—Part 3: Bulk current injection (BCI) method*, Standard IEC 62132-3, 2007.

[9] S. Miropolsky, S. Frei "Comparability of RF Immunity Test Methods for IC Design Purposes," in *Proc. International Workshop on Electromagnetic Compatibility of Integrated Circuits EMC Compo*, Nov. 2011, pp. 59-64.

[10] D. Konigstein and D. Hansen, "A new family of TEM-cells with enlarged bandwidth and optimized working volume," *Proc.7th Int. Zurich Symp. and Tech. Exh. on EMC*, pp. 127 – 132, March 1987.

[11] J. Catrysse, D. Pissoort, and F. Vanhee, "Expanding the frequency range for DPI testing of IC's above 1 GHz: an alternative proposal," in *Proc. Int. Symp. Electromagn. Compat. (EMC Europe)*, Sep. 2011, pp. 400–404.

[12] Sjoerd Op 't Land et al., "Design of a 20 GHz DPI Method for SOIC8," in *Proc. Int. Symp. Electromagn. Compat. (EMC Europe)*, Sep. 2012, pp. 1–6.

Anti-resonance Peak Frequency Control by Variable On-die Capacitance

Wataru Ichimura, Sho kiyoshige, Masahiro Terasaki Ryota Kobayashi,
Genki Kubo, Hiroki Otsuka, and Toshio Sudo
Shibaura Institute of Technology, 3-7-5 Toyosu, Koto-ku, Tokyo, Japan
{ma13009,toshio}@shibaura-it.ac.jp

Abstract— **Power integrity design has been becoming important in the advanced CMOS digital systems, because power supply noise induces logic instability and electromagnetic radiation. Especially, anti-resonance peaks in power distribution network (PDN) due to the chip-package interaction induce the unwanted power supply fluctuation, and result in large electromagnetic radiation. In this paper, power supply noises and total impedances of power distribution network (PDN) for the variable structure of on-die capacitances have been examined. In addition, power supply noise and total PDN impedance have been examined by changing the number of power supply terminals. As a result, it has been proved that anti-resonance peaks could be controlled by on-die capacitance and the number of power supply terminals. Simulated anti-resonance peak frequencies were well correlated with the peak frequency spectra of measured power supply noise.**

Keywords—Anti-resonance peaks, Co-design, Power supply noises, Power integrity

I. INTRODUCTION

As the CMOS LSI systems operate at higher clock frequencies and at lower supply voltage, power integrity is becoming a critical issue to maintain digital electronic systems more stable. Especially, chip-package anti-resonance in the power distribution network (PDN) occurs due to the parallel combination of on-die capacitance and package inductance. In order to estimate the peak frequency and peak level of anti-resonance exactly, on-die PDN impedance must be properly designed along with package, and board PDN design [1]-[3].

Power supply noises on the core or I/O circuits have been almost studied in the time domain. Total PDN impedance which consisted of chip PDN, package PDN, and board PDN has become an important approach in the frequency domain to understand the noise phenomena more clearly and optimize the PDN characteristics properly.

In this paper, we have evaluated the total PDN and the power supply noise by using an evaluation chip designed by rohm180um process technology. Evaluation chip is a variable capacitance chip capable of varying the additional capacitance Cadd. This chip, it is possible to verify the power supply noise and integration of PDN when changing the additional capacity. Further, confirmed the correlation

power supply noise integration and PDN from the results of the FFT power supply noise

II. PARALLEL RESONANCE OF TOTAL PDN

Fig.1 shows a simplified total PDN model consisted of chip, package and board. In this model, anti-resonance peak occurs by the parallel resonance of on-die capacitance (Cdie), on-die resistance (Rdie), and package inductance (Lpkg), because the board impedance (Zpcb) is normally smaller than the package inductance (Lpkg) for the conventional board with several decoupling capacitors. This parallel resonance occurs at around the cross point between package inductance and on-die capacitance. Fig.2 shows an example of the anti-resonance peaks seen from chip.

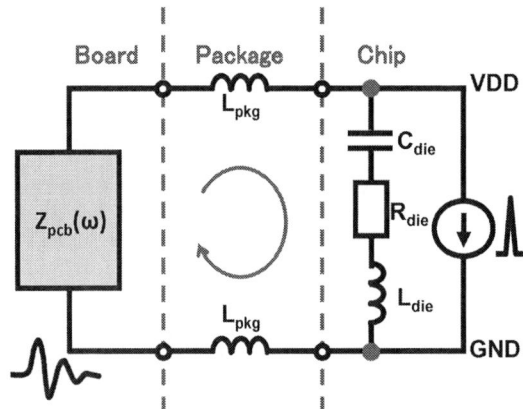

Fig.1 Schematic of PDN system and simplified model

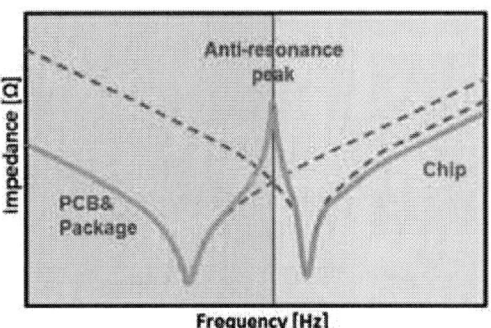

Fig.2 Anti-resonance peak seen from chip

III. CHIP CIRCUIT DESIGN

Test chip with variable on-die PDN properties were designed as shown in Fig.3. The size of the test chips was 2.5 by 2.5 mm. test chip was designed to have both noise generating circuits and on-chip noise monitoring circuits. This test chip has a noise generating circuit using shoot-through current (STNG: shoot-through noise generator) with different current drivabilities by changing the number of CMOS inverter stages. Noise monitoring circuits were designed utilizing CMOS output buffer circuits. They were located at the four corners of each chip. Power supply noise was observed by fixing the signal output at the high level, and ground noise was observed by fixing it at the low level.

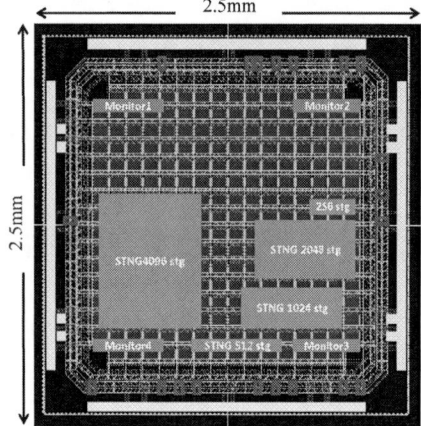

Fig.3 Circuits implemented in test chip

IV. PDN DESIGN WITH VARIABLE CDIE

Fig.4 shows a detailed total PDN model with additional C circuit in a chip which connected to the intrinsic RC circuit in parallel. This chip can be changed additional capacitance of chip C_{add} in step 8. Fig.5 shows eight capacitor groups switched by external switches. In addition, this figure shows the sequence of changing the P/G number of pairs for the supply of the core power. This chip has eight P/G pair in total. The first P/G pair is the lower left. It is possible to reduce Lpkg with increasing P/G number of pairs and increasing parallel number.

Table 1 shows the capacitance values of C_{add} for the eight switching groups. Package inductance was assumed a conventional QFP whose size was 14 by 14 mm with 80 pins.

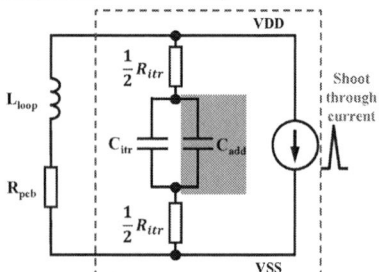

Fig.4 Detailed PDN model with additional RC circuit in parallel in a chip

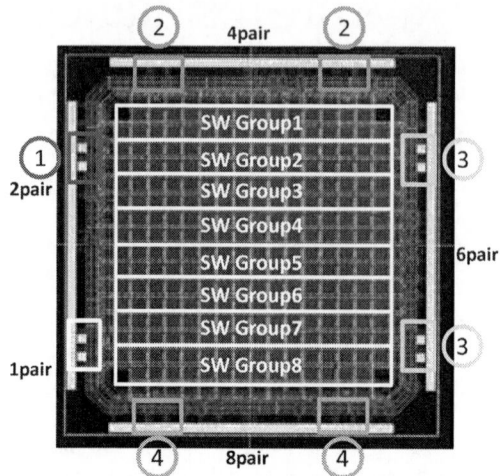

Fig.5 Eight switching groups of decoupling capacitance

Table 1 Switched group and corresponding cell capacitance values

	Group Name	Capacitance Value[pF]
SW1	SW Group1	69
SW2	SW Group1-2	143
SW3	SW Group1-3	216
SW4	SW Group1-4	290
SW5	SW Group1-5	364
SW6	SW Group1-6	437
SW7	SW Group1-7	511
SW8	SW Group1-8	580

V. MEASUREMENT OF POWER SUPPLY NOISES

Fig.6 shows a external appearance of test board. The power and ground lines were excited by shoot-through current noise generators (STNG) with different current drivabilities, and the power and ground fluctuations were observed by noise monitoring circuits placed at four corners of the chip.

Figs.7 to 9 shows the measured power supply noise when changing the Cadd by switching ON / OFF of CMOS SW. Fig.11 shows a power supply noise which P/G pairs were 2 pair. Fig.12 shows a power supply noise which P/G pairs were 4 pair. Fig.13 shows a power supply noise which P/G pairs were 8 pair. At this time, the STNG with the maximum strength of 2048 was excited at a frequency of 10 MHz. These measured on-chip power supply noises were monitored at the monitor 4 which was located at the bottom left corner of the test chip.

As a result, the peak-to-peak amplitude of the power supply noise and the ringing frequency becomes low as the Cdie increases. In the same manner, the peak-to-peak amplitude of the power supply noise and the ringing frequency becomes low as the number of P/G pairs reduces.

978-1-4799-5004-1/13 $31.00 © 2013 IEEE

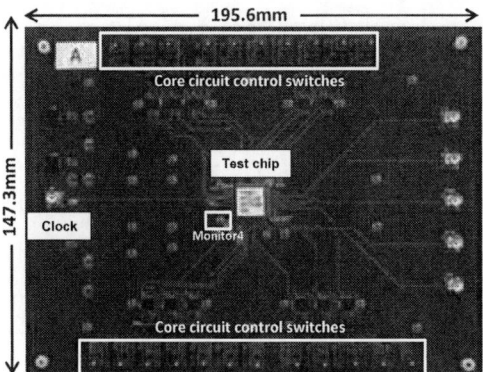

Fig.6 External appearance of test board

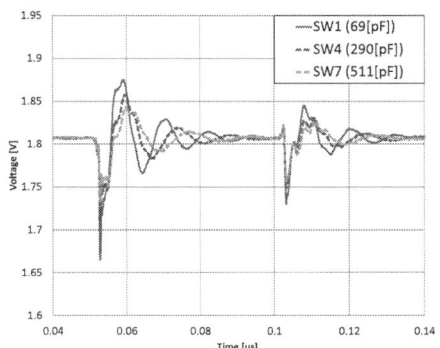

Fig.7 Measured power supply noises (P/Gpair2)

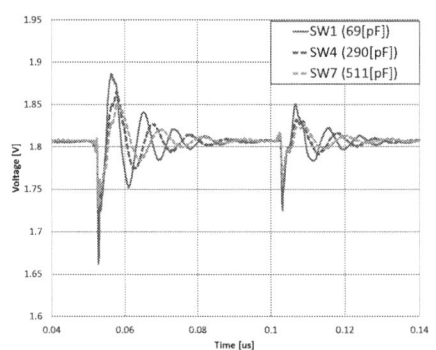

Fig.8 Measured power supply noises (P/Gpair4)

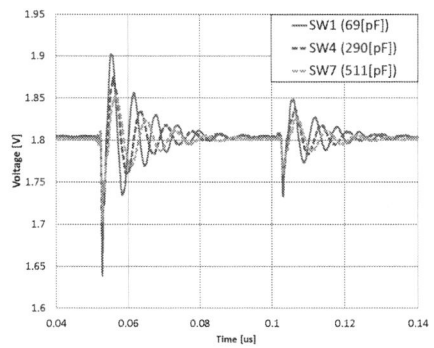

Fig.9 Measured power supply noises (P/Gpair8)

VI. FREQUENCY SPECTRA OF POWER SUPPLY NOISE WAVEFORMS

Figs.10 to 12 shows frequency spectra of power supply noise waveforms for Figs.7 to 9. As a result, peak frequency was shifted to the low frequency side by decreasing Lpkg.

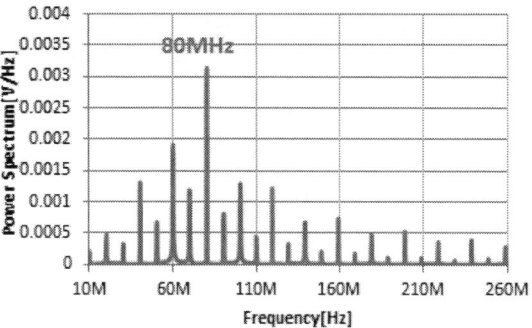

Fig.10 Frequency spectra of power supply noise waveforms (SW4, P/G pair2)

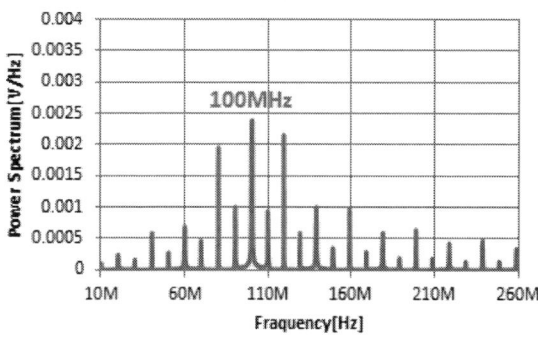

Fig.11 Frequency spectra of power supply noise waveforms (SW4, P/G pair4)

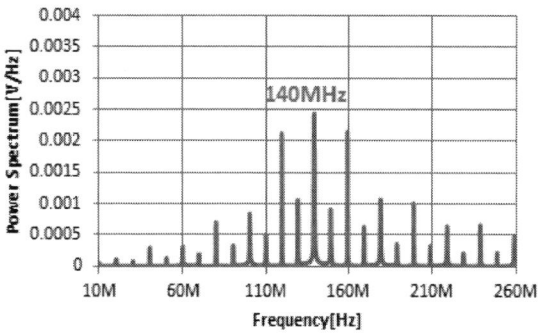

Fig.12 Frequency spectra of power supply noise waveforms (SW4, P/G pair8)

978-1-4799-5004-1/13 $31.00 © 2013 IEEE

VII. MEASURED ANTI-RESONANCE PEAKS

The on-chip PDN impedances were measured by using a vector network analyzer (Agilent E5017C). Fig.13 shows measurement setup to observe on-chip PDN impedances by directly contacting a high-frequency probe. Figs.14 to 16 shows the measured anti-resonance peaks. Anti-resonance peaks were clearly and definitely observed. These peak frequencies were almost the same as the peak frequencies of the spectra of the power supply noise waveforms as shown in Figs.10 to 12.

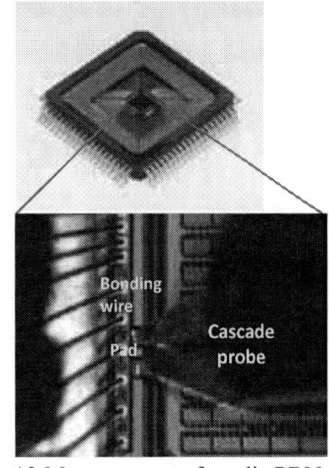

Fig.13 Measurement of on-die PDN

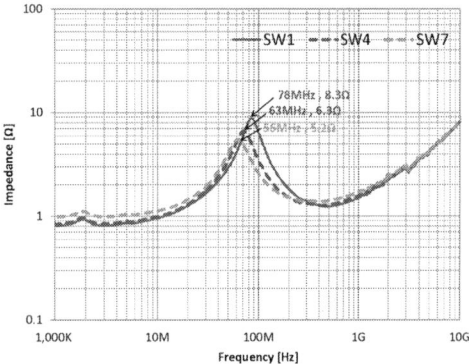

Fig.14 Measured anti-resonance peaks (P/G pair2)

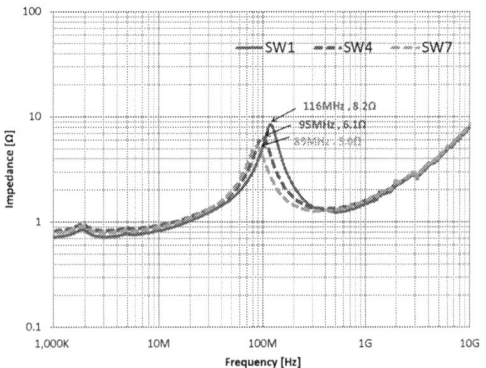

Fig.15 Measured anti-resonance peaks (P/G pair4)

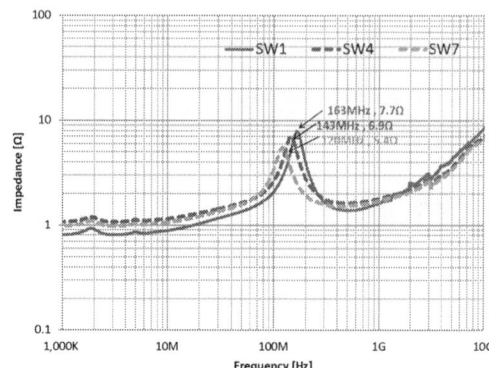

Fig.16 Measured anti-resonance peaks (P/G pair8)

VIII. SUMMARY

In this paper, we have developed a test chip with the variable structure of on-die capacitance. The power supply noises were observed under the various on-die capacitance and the various number of the power supply terminals. It has been found that anti-resonance peaks were shifted to the lower frequency range by increasing the on-die capacitance Cadd, while anti-resonance peaks were shifted to the higher frequency range with decreasing package inductance Lpkg.

Furthermore, it has been found that measured anti-resonance peaks were well coincident with the peak frequencies obtained by the FFT of the power supply noises.

ACKNOWLEDGMENT

This work was supported by STARC. The test chips were fabricated through the chip fabrication program of VDEC.

References

[1] P. Larsson, "Resonance and Damping in CMOS Circuits with On-Chip Decoupling Capacitance," *IEEE Trans. on Circuits and Systems-1* vol.45, no.8, pp.849-858, Aug. 1998.

[2] R. Kobayashi, et al., "Effects of Critically Damped Total PDN Impedance in Chip-Package-Board Co-Design," Proc. of *IEEE EMC Symposium*, Pittsburgh 2012.

[3] W. Kim, "Estimation of Simultaneous Switching Noise From Frequency-Domain Impedance Response of Resonant Power Distribution Networks," *IEEE Trans. on CPMT*, vol.1 no.9, pp. 1359-1367, Sept. 2011.

Estimation of Data-Dependent Power Voltage Variations of FPGA by Equivalent Circuit Modeling from On-Board Measurements

Kengo Iokibe [#1], Yoshitaka Toyota [#2]

Graduate School of Natural Science and Technology, Okayama University
3-1-1 Tsushima-naka, Kita-ku, Okayama 700-8530 Japan
[1] iokibe@okayama-u.ac.jp
[2] toyota@okayama-u.ac.jp

Abstract—An equivalent circuit model was evaluated in simulating data-dependent power voltage variations of a field-programmable gate array (FPGA). The equivalent circuit model was Linear Equivalent Circuit and Current Source (LECCS) model representing dynamic switching current inside the FPGA with an equivalent current source. The current source was supposed to depend on input data for the FPGA on which a cryptographic circuit was implemented. Model identification was based on the procedure of LECCS model identification from on-board measurements and the current source was identified for all values of input data used in this work. The identified current source was investigated in accordance with the operation process of the cryptographic circuit and found an excellent correlation to the operation process. The identified LECCS model was combined with an equivalent circuit of the power distribution network for the FPGA core circuit to simulate power voltage variations for the 1,000 input texts. The simulated variation waveforms were compared to the corresponding measured ones to evaluate the LECCS model. Results indicated that the simulated and measured power variations matched excellently for all input data with high cross-correlation coefficients from 0.7 to 0.9. LECCS model is, therefore, able to predict the data-dependent power voltage variation by combining a PDN equivalent circuit.

I. INTRODUCTION

Modern integrated circuit (IC) technology is still advancing in terms of operating clock rate, power consumption, and functional integration. This trend remains designing power distribution networks (PDNs) for modern ICs a challenging issue. To suppress power voltage variation, or power bounce, is growing of importance as the voltage and temporal noise margin is getting smaller.

There are similar equivalent circuit model developed for simulating the dynamic power current and electromagnetic emission, such as LECCS and ICEM models[1], [2], [3], [4]. The equivalent circuit model was also utilized for simulating the power voltage variation of microcontrollers[5] and field-programmable gate arrays (FPGAs)[6]. These simulations used simple circuit configurations not designed for a practical use, such as toggling flip-flops or binary counters. Besides the simulation results were validated with no consideration of data dependency of the PND noise.

Regarding the PDN design of FPGA, the equivalent circuit models can be difficult to be identified accurately from design information of the circuit, because several types of chip information are inherently unobtainable such as implementing circuit locations and logic gate connections. The equivalent circuit model, therefore, can help PND designers to suppress the power voltage variation if they obtain accurate models identified form measurements.

This work evaluate LECCS model in terms of data dependency of the power voltage variation by use of a FPGA on which an encryption circuit has been implemented. A data dependent LECCS model of the FPGA is extracted for 1,000 input data and applied to simulate the power voltage variation at an on-board location. Simulated power voltage variations are compared with corresponding measured ones in the time-domain to evaluate the simulation. The power voltage variations were also simulated for the same FPGA in our previous works[7], [8], in which simulation accuracies were evaluated with respect to security estimation of the cryptographic circuit implemented in the FPGA. In other words, the simulation accuracies were evaluated in indirect ways in the previous works. In this work, simulated power voltage variations will be validated directly by investigating their cross-correlation to measured profiles. In addition, the current source of LECCS model will be improved by reducing background noise in model identification measurements.

The following sections are composed as: Section II reviews LECCS model briefly and introduces the data dependent LECCS model. Section III represents the configuration of PDN for the FPGA under test, and then extraction of the LECCS model and an equivalent circuit model of the PDN are shown in Section IV. Finally, the power voltage variation is simulated with the extracted model to validate the model.

II. LECCS MODEL

A short review of LECCS model is presented and the data dependent LECCS model is introduced in this section.

A. Short Review of LECCS Model

LECCS model has been developed to predict IC switching current behavior in a PDN for the RF power noise reduction

978-1-4799-5004-1/13 $31.00 © 2013 IEEE

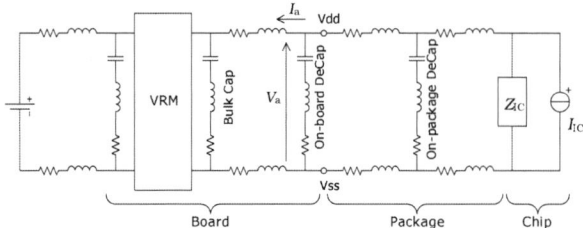

Fig. 1. LECCS model combined with a typical PDN

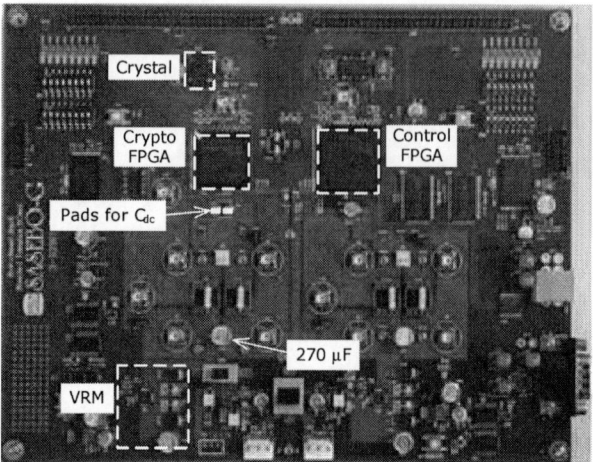

Fig. 2. Top view of SASEBO-G

design. LECCS model is composed of an equivalent current source and package impedances. The equivalent current source expresses the IC switching current generation in an interesting circuit area. The package impedances include impedances of bonding wires, lead frames, and the interconnecting board and are usually inductive. Combined with an equivalent circuit of the board PDN as shown in Fig. 1, LECCS model produces predictions of RF currents and voltage fluctuations at any location of the PDN. Since LECCS model is much simpler than the transistor level model, the combined equivalent circuit is so simple that designers obtain the prediction with a short calculation time. The short calculation time allows designers to make an iteration of try-and-errors until their product meet noise regulations.

A practical methodology to identify the LECCS model parameters was based on on-board measurements[3], [4]. Chip and package impedances was identified from an impedances measurement at a V_{dd}-V_{ss} port close to the target IC. The ideal current source I_{IC} was calculated by the definition,

$$I_a = K I_{IC}, \qquad (1)$$

where I_a is a current spectrum measured at an arbitrary port on board PDN, and K is the current transmittance from I_{IC} to I_a and depends on the PDN impedances. These on-board measurements can also be carried out on commercial tools with design information of an IC and printed board is available to use. Besides, an equivalent circuit like LECCS and ICEM was extracted by a commercial tool specified for simulating the chip power impedance and dynamic current activity of a circuit[9].

B. Data Dependent Current Source

To express the data dependency on the RF power noise, LECCS model is supposed that the ideal current source is depend on data, and the impedances are not. In other words, the switching current generated in the circuit changes with data though the chip impedance does not. It is reportedly that the chip impedance changes slightly with data. On the contrary, the switching current varies with data. The switching current is occurred as logic gates transit their states and becomes large when the number of logic gates transiting their states is large. It becomes small when a small number of logic gates switches. The number of logic gates to switch depends on data input to the circuit. Therefore the switching current depends on the

data.

C. Identification of Current Source

The model parameters, impedance and current source, are identified from on-board measurements in the following section. Although chip and package impedances are determined from a measured impedance as described above, the current source is calculated from a dynamic power voltage V_a instead of the dynamic power current I_a for ease of measurement. The equation (1) is rewritten as

$$V_a = Z_K I_{IC}, \qquad (2)$$

where Z_K represents the transmitting impedance from I_{IC} to V_a and obtained from a circuit simulation with the PDN equivalent circuit.

The current source I_{IC} is identified for all input data to the circuit. After measuring V_a for all the input data, I_{IC} is calculated by Eq. (2) for all the input data.

III. Printed Circuit Board under Test

A commercial printed circuit board developed for evaluating cryptographic devices, SASEBO-G, was used here It has two FPGAs on it as shown in Fig. 2: one for operating encryption processes and the other for controlling the encryption operation. Circuitry composition of the PDN of the encryption FPGA is drawn in Fig. 3. Between the cryptographic FPGA and the voltage regulator module (VRM), only an electrolytic capacitor is mounted as a decoupling capacitor of 270 μF, labeled C_{blk} here, and a single pair of pads for a chip decoupling capacitor is prepared near the FPGA. In this study, a 2012 sized chip capacitor of 10 nF was mounted on the pads, as C_{dc}.

In the following experiments, the cryptographic FPGA processed a standardized encryption algorithm, Advanced Encryption Standard (AES)[11], with a 128-bit key of (2B 7E 15 16 28 AE D2 A6 AB F7 15 88 09 CF 4F 3C)$_{16}$. The AES-128 encryption process was composed of 10 round-operations as

978-1-4799-5004-1/13 $31.00 © 2013 IEEE

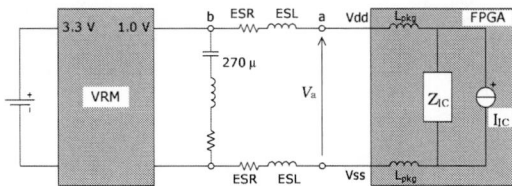

Fig. 3.   PDN composition for modeling

TABLE I
EQUIPMENT USED IN MEASUREMENTS

Impedance measurement	
Vector Network analyzer	E5071A, Agilent Technologies
Freq. range	300 kHz–2 GHz for FPGA
	30 kHz–500 MHz for VRM
No. of Points	1601
No. of Averaging	16
IFBW	70 kHz for FPGA
	10 kHz for VRM
Microprobe	FPC-SG-1250, Cascade Microtech
Voltage variation measurement	
Digital oscilloscope	54845A, Agilent Technologies
Bandwidth	1.5 GHz
Sampling rate	4 GSa/s
Coupling	AC
Passive probe	1161A, Agilent Technologies
Bandwidth	500 MHz

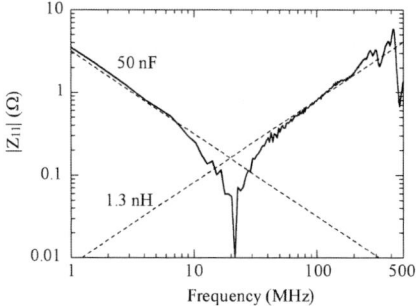

(a) FPGA core circuit

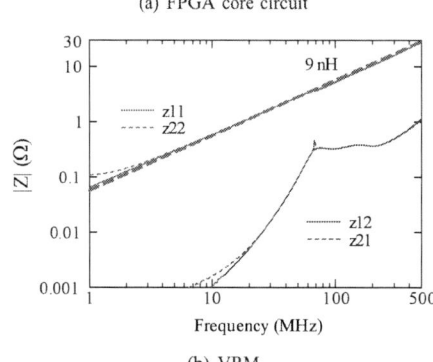

(b) VRM

Fig. 4.   Impedances of components

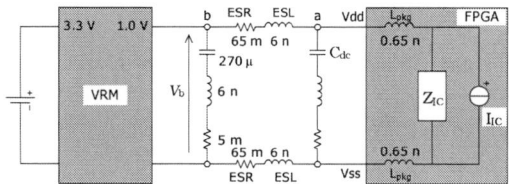

Fig. 5.   Equivalent circuit identified

well as a pre-operation including preparation of subkeys used in the round operations. Each operation began synchronized to the clock signal of 24 MHz that was supplied from the crystal oscillator mounted by the cryptographic FPGA, see Fig. 2.

IV. MODEL IDENTIFICATION

A. Impedance Measurements

The FPGA impedance Z_{IC} was measured at the pads for a decoupling capacitor C_{dc} by using a microprobe (FPC-SG-1250, Cascade Microtech) and a vector network analyzer (E5071, Agilent Technologies), as the FPGA was disconnected with the PDN and not biased. Obtained S parameters were converted into the driving point impedance at the measurement port, plotted in Fig. 4(a), found the impedance composed of a capacitance of 50 nF, a resistance of 10 mΩ, and an inductance of 1.3 nH. Each half of the inductance was given as the parasitic inductance of V_{dd} or V_{ss} interconnects and wirings, L_{pkg}. The chip impedance Z_{IC} was as the series connection of the 50 nF capacitance and 10 mΩ resistance.

The remaining impedance of on-board traces and components were determined from measurements. The capacitors of 270 μF and C_{dc} were measured to obtain their ESLs and ESRs as well as their capacitances with an impedance analyzer. ESLs and ESRs of board traces and vias were determined from V_{a} waveforms that contains a damping oscillation induced by the switching current of the FPGA. The time constant and oscillating frequency of the damping oscillation gave an estimation of the parasitic impedances. Extracted impedance values are written in the schematic in Fig. 5. The impedance parameters of VRM are plotted in Fig. 4.

B. Current Source

According to Eq. (2), the current source of the LECCS model can be calculated from V_{a} and Z_{K}. At first, V_{a} was measured at Port a in the PDN configuration of Fig. 3 with a digital oscilloscope and a passive probe. The PDN configuration is a non-decoupling configuration in which the nodes a and b were shunt with short-bars of which ESLs and ESRs were taken into account. Specifications of experimental equipment are listed in Table I. Waveforms of V_{a} were obtained for all the input data values.

Next the system constant Z_{K} was calculated by simulating the linear equivalent circuit of the PDN on a commercial circuit simulator, AWR Microwave Office, using the impedances determined in Sec. IV-A. Substituting Z_{K} and V_{a}, the current sources I_{IC} was obtained for all the 1,000 input data values.

An example of the calculated current sources are indicated in Fig. 6. A series of periodic sharp peaks are seen in each of the time-domain traces. The period of 41.7 ns was matched

978-1-4799-5004-1/13 $31.00 © 2013 IEEE

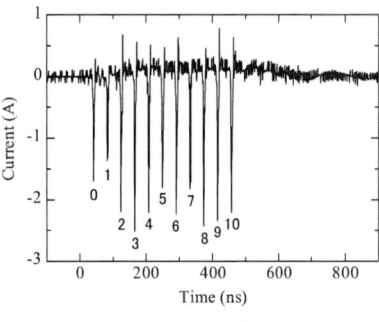

(a) Time-domain

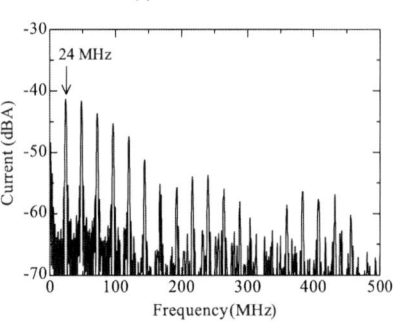

(b) Frequency-domain

Fig. 6.   Current source identified

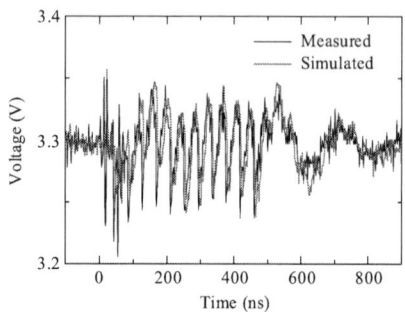

Fig. 7.   Simulated power voltage variation waveform with measured one

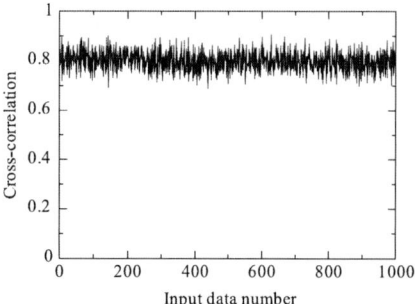

Fig. 8.   Cross-correlation between simulated and measured traces

with that of the 24 MHz clock by which the encryption circuit was operated. The number of the periodic peaks is 11 that agreed with that of sub operations of AES-128: the pre-operation and 10 round operations. The orientation of the sharp peaks was negative, that is, the currents directed from V_{dd} to V_{ss}. This is very reasonable as the direction of the IC switching current. In the frequency-domain, harmonics of the clock rate of 24 MHz were observed clearly up to 500 MHz.

V. SIMULATION OF POWER VOLTAGE VARIATION

Power voltage variation on the PDN for the 1.5 V core circuit of the cryptographic FPGA was simulated by LECCS model at an on-board location. The location was at the Port b indicated in Fig. 5. The PDN configuration was changed from the model identification configuration by inserting a decoupling capacitor C_{dc}, the value of which was 0.1 μF. The power voltage variation simulation was carried out for 1000 different texts of input data.

Simulated waveforms of power voltage variation were matched with measured ones. An example of simulated waveforms was plotted in Fig. 7 and superposed on the corresponding measured one. The input text for the example was $(00\ 00\ 00\ 00\ 00\ 00\ 00\ 00\ 00\ 00\ 00\ 00\ 00\ 00\ 00\ 00)_{16}$. The time origin is the moment at which the AES operation started. There is a periodic oscillation in the simulated waveform composed of 11 periods of 41.7 ns, corresponding to the clock frequency of 24 MHz. The simulated waveform agreed with the measured one in their shapes including the period and phase of the periodic oscillation. The amplitude of the estimated power

voltage variation also agreed with that of measurement in the last 6 rounds, $t > 250$ ns although it was smaller than measured one in the former rounds.

Simulated waveforms also agreed with measured ones for all the other input data tested. To investigate whether simulated and measured results agreed for all the input data, cross-correlations between simulated and measured waveforms were calculated, as shown in Fig. 8. The horizontal axis represents the data number and the vertical the cross-correlation. Obtained cross-correlation ranged from 0.7 to 0.9. These high correlation coefficients confirm that simulated power voltage variation waveforms agreed with measured ones for all the input data.

VI. CONCLUSION

An equivalent circuit model of an FPGA was verified whether it simulate data-dependent power voltage variations. As the equivalent circuit model, Linear Equivalent Circuit and Current Source (LECCS) model was applied to the FPGA on which a cryptographic circuit was implemented. The LECCS model of the FPGA was identified from on-board measurements by an LECCS modeling procedure presented in a previous work. The identified LECCS model was evaluated in two aspects: consistency of current source with the encryption operation and accuracy of simulation results. First, the identified LECCS model itself was found that it obtained a reasonable current source waveform of which involved a periodic oscillation that was consistent with the encryption operation in terms of the period and the number of oscillation. Second, the LECCS model was combined with an

978-1-4799-5004-1/13 $31.00 © 2013 IEEE

equivalent circuit of the PDN for the cryptographic circuit and applied to simulate the data-dependent voltage variations, and then simulated power voltage variations were compared with corresponding measured ones for all input data texts. Results of the comparison showed that the simulated and measured power voltage variations agreed excellently producing high cross-correlation coefficients from 0.7 to 0.9. LECCS model was, therefore, confirmed that it is available to predict the data-dependent power voltage variation.

ACKNOWLEDGMENTS

The authors would like to thank the Strategic Information and Communications R&D Promotion Programme (SCOPE) from the Ministry of Internal Affairs and Communications (MIC) for its financial support. They would also like to acknowledge Ms. Kana Shimizu for her generous contribution to this work.

REFERENCES

[1] S. D. Dhia, M. Ramdani, and E. Sicard, *Electromagnetic Compatibility of Integrated Circuits*. New York: Springer, 2006.

[2] C. Labussiere-Dorgan, S. Bendhia, E. Sicard, Tao Junwu, H.J. Quaresma, C. Lochot, B. Vrignon, "Modeling the Electromagnetic Emission of a Microcontroller Using a Single Model," *IEEE Trans. Electromagn. Compat.*, vol. 50, no. 1, pp. 22-34, 2008.

[3] K. Nakamura, T. Toyota, O. Wada, R. Koga, and N. Kagawa, "EMC Macro-Model (LECCS-Core) for Multiple Power-Supply Pin LSI," in *Proc. IEEE Int. Symp. Electromagn. Compat.*, Aug. 2004, pp. 493-496.

[4] K. Iokibe, R. Higashi, T. Tsuda, K. Ichikawa, K. Nakamura, Y. Toyota, and R. Koga, "Modeling of Microcontroller with Multiple Power Supply Pins for Conducted EMI Simulations," *2008 Electrical Design of Advanced Packaging and Systems Symposium (EDAPS 2008)*, Dec. 2008, pp. 135–138.

[5] S. Li, H. Bishnoi, J. Whiles, P. Ng, H. Weng, D. Pommerenke, and D. Beetner, "Development and Validation of a Microcontroller Model for EMC," in *Proc. 2008 Int. Symp. Electromagn. Compat. (EMC Europe)*, Sep. 2008.

[6] I. Zamek, P. Boyle, Z. Li, S. Sun, X. Chen, S. Chandra, T. Li, D. Beetner, and J.L. Drewniak, "Modeling FPGA Current Waveform and Spectrum and PDN Noise Estimation," *DesignCon*, 2008.

[7] K. Iokibe, T. Amano, K. Okamoto, and Y. Toyota, "Equivalent Circuit Modeling of Cryptographic Integrated Circuit for Information Security Design," *IEEE Trans. Electromagn. Compat.*, vol. 55, no. 3, pp. 581–588, 2013.

[8] K. Iokibe, T. Amano, K. Okamoto, Y. Toyota, and T. Watanabe, "Improvement of Linear Equivalent Circuit Model to Identify Simultaneous Switching Noise Current in Cryptographic Integrated Circuits," in *Proc. IEEE Int. Symp. Electromagn. Compat.*, Aug. 2013, pp. 834–839.

[9] H. H. Park, S.-H. Song, S.-T. Han, T.-S. Jang, J.-H. Jung, and H.-B. Park, "Estimation of Power Switching Current by Chip-Package-PCB Cosimulation," *IEEE Trans. Electromagn. Compat.*, vol. 52, no. 2, 2010.

[10] AIST. Side-channel attack standard evaluation board (sasebo). [Online]. Available: http://staff.aist.go.jp/akashi.satoh/SASEBO/en/index.html

[11] *Advanced encryption standard (AES)*, NIST FIPS publication 197, Nov. 2001.

Microcontroller Emission Simulation based on Power Consumption and Clock System

Thomas Steinecke [1]

[1] *Infineon Technologies AG, Germany*

thomas.steinecke@infineon.com

Abstract — **Various approaches exist to build simulation models for the electromagnetic emission (EME) of digital integrated circuits [1-6]. However, several drawbacks constrain their configuration and usage. A new modelling approach is described in this paper. It is based on the main EME-relevant parameters of digital circuits, i.e. dynamic power and clock rates. This information should be even available before the IC design phase starts. With this approach, the models can be created without any additional information about module size, functional patterns or embedded decoupling capacitors. Thus it can be used to perform design studies wrt. clock rate selection, clock modulation and impact of dynamic power consumption. The modelling and simulation software can even handle very complex ICs like high-end 32-bit microcontrollers. The program is named EMISoC which stands for "EMI simulation of systems-on-chip". It has been validated for existing 65 nm CMOS designs and is currently used for emission estimations of future 40 nm microcontrollers for automotive applications. This paper describes the EMISoC modelling approach, features and limitations.**

I. INTRODUCTION

The electromagnetic emission of digital SoCs (systems on chip) is determined by a few parameters:

- Complexity of its functional modules,
- Clock rates of its functional modules,
- RLC parasitic of the SoC design,
- RLC parasitic of the IC package,
- Decoupling concept on the PCB.

State-of-the-art modelling approaches [1]-[4] and past reported approaches [5] [6] struggle with some important drawbacks:

- High model complexity leads to long simulation time,
- Late availability of IC design data makes simulation usefulness questionable,
- Prediction quality for planned new ICs.

The vast majority of existing EME simulation tools focus on the analysis of existing designs, either on netlist or layout level, or even on physical measurement level. This means that the models are available late in the IC design cycle and very special for this IC, i.e. difficult to use for EME prediction of future ICs. Table 1 gives an overview of these model key parameters.

The modelling approach was straight forward:

- Identification of EME key determining parameters,
- Modeling on architectural level,
- Post-processing for parasitic parameters.

Table 1 Drawbacks of different EME modelling approaches

		Previous tools					EMISoC
Hardware	Transistors+RLC (layout)	X					
	Standard cells (netlist)		X	X			
	Tile matrix (floorplan)				X		
	Modules (specification)						X
	Package RLC	X	X				
	"Smoothing"						X
Software	Functional pattern (netlist)	X	X				
	Statistical vectors (netlist)			X			
	Clocks (specification)						X
	Dyn. current (measurement)				X	X	
	Average current (specific.)						X
Drawback	Enable simul. (design cycle) 0=late ... 2=early	0	0	1	1	0	2
	Simulation accuracy 0=inaccurate ... 2=accurate	2	2	1	1	0	1
	Simulation speed 0=slow ... 2=fast	0	0	1	1	2	2
	Prediction for virtual ICs 0=bad ... 2=good	0	0	0	0	0	2

The motivation to go for an alternate modelling approach was the request to estimate the EME potential for microcontrollers in the next technology node, i.e. 40 nm CMOS technology. The ICs of interest are "virtual", i.e. not yet designed. There exists only an architectural specification which includes however information about the current consumption and the target system clocks. These are (luckily) the EME-determining parameters and determined the chosen modelling approach. The new modelling approach was implemented in an Infineon-proprietary software called EMISoC. It is aiming at a good EME prediction of not yet designed microcontrollers during its specification phase. Furthermore, it should have fast runtimes to be able to compare different microcontroller configurations in a ahort time. It should at least generate a detailed dynamic current profile of the complete microcontroller and visualize the frequency domain representation (i.e. emission spectrum) of the dynamic current.

II. EMISoC EME SIMULATION SOFTWARE

A. Modeling Approach

Table 1 indicates that simulations on transistor level, considering the parasitics of signal and power traces, lead to the most accurate results. On the other hand, the design data is available very late in the design cycle, i.e. close to tapeout. And the simulation times are quite long because of the device-level analog simulation. The more abstract the model gets (transistors → standard cells → tiles → modules), the earlier the model data is available and the faster runs the simulation.

An exception is the case when no design data is available and a "black box IC needs to be measured to extract the model data, e.g. by measuring the dynamic supply current. In this case, although the model is the most abstract one (i.e. "black box"), it is available very late (i.e. after IC production). This case is not interesting for IC vendors, but only for IC users. IC vendors should always care to build IC EME models based on design data (for EME-related design sign-off prior to tapeout). Unfortunately this focus does not allow early estimations of the EME behaviour of new ICs.

The EME of not yet designed ICs must be estimated based on general architecture information of the new planned IC. The EMISoC software considers this requirement by selecting the following EME-relevant IC parameters, see also Table 1:

- Current consumption of functional modules,
- Clocking scheme of functional modules,
- List of functional modules in the IC of interest,
- Result post-processing ("smoothing") to consider parasitic filter or resonance effects.

B. Modeling Preparation

Now let's see what information is required to build a microcontroller EME simulation model with EMISoC.

The first EME-determining parameter is the *switching current*. It is part of the overall power consumption.

Since microcontrollers mostly use standard QFP or BGA packages without additional cooling, their power consumption is limited. The design flow includes power consumption estimations which are based on previously measured data. If a new microcontroller re-uses functional modules in the same technology, its total SoC power budget can be easily estimated by adding the power contributions of all single modules. Therefore, current consumption lists are required which indicate dynamic and static currents for existing functional modules. Static currents are leakage currents which are determined by the used technology. They are PVT (process, voltage, temperature) dependent and rise significantly with the die temperature, but they are not EME-relevant. The dynamic currents are the switching currents. They are PVTC (process, voltage, temperature, clock frequency) dependent.

The second EME-determining parameter is the *time variation* of the switching current. Because of their clock frequency dependency, the dynamic switching currents are the EME-relevant parts of the supply current. They are given by the clock frequency or data rate, respectively, and the rise/fall times of the clock or data signal of a functional module. Both

parameters determine the envelope of the resulting emission spectrum. Figure 1 shows three spectra of a trapezoid clock signal with different frequencies $f=1/T$ and rise/fall times t. The signal amplitude is 2 A in all three cases.

- Figure 1a: f=100 MHz → T=10 ns; t=2.5 ns
- Figure 1b: f=100 MHz → T=10 ns; t=0.5 ns
- Figure 1c: f=500 MHz → T=2 ns; t=0.5 ns

Figure 1a-c indicate that a faster rise/fall time leads to higher harmonics towards high frequency, and that a faster frequency (i.e. shorter clock period) leads to less damping at higher frequency. The peak amplitudes of the nth harmonics stay the same, but their distance in frequency domain increases with shorter period.

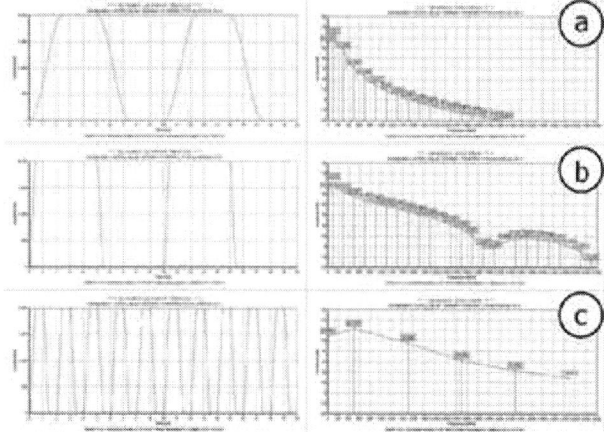

Figure 2 Emission spectrum for a trapezoid waveform

A real SoC consists of many functional modules which do not operate necessarily at the same clock frequency. Furthermore, the modules' current consumptions are different. To calculate realistic dynamic current waveforms and from there realistic frequency spectra, the dynamic currents of all functional modules in the SoC have to be added, and a Fast Fourier Transformation (FFT) has to be applied on the current waveform.

C. Features of the EMISoC Software

A microcontroller model for EMISoC [7] consists of some general clocking information for the SoC and a module list for the SoC. The module list contains individual current consumption and clocking information for every functional module on the SoC.

The general clocking information describes the system-PLL which provides the main system clock. Clock frequency, clock modulation shape and amplitude are the most important system-PLL parameters. Currently supported are triangular, clipped [8] and random modulation.

Every functional module is characterized by its average dynamic current, i.e. the integration of the dynamic clock over time. The calculation of a realistic current shape over time is done considering a rising and falling clock edge which causes a current pulse by switching transistors in the module. This can be logic or I/O transistors. It is assumed that the transistors charge or discharge capacitive nodes, following an exponential function. Mathematically, the starting current peak is reached immediately with the clock edge. In reality, the starting current increases stepwise when the signal pattern released by the falling clock edge ripples through the combinatorial logic. A four-stage logic depth is considered by EMISoC. The propagation delay of one logic depth is given by the technology- and library-dependent parameter *pd*. The preceding rising clock edge stored the new signal pattern in the master flipflops. The rising edge of this current peak takes one *pd*, according to the fact that all master flipflops are active simultaneously. The decay times of both current pulses can be scaled individually. Pad drivers will take a longer time to charge discharge their nF loads than logic gates to drive their fF or pF loads. Furthermore, the ratio between the current content in the "rising" current pulse (i.e. the one storing the new signal pattern) and the total current can be varied. This allows to select e.g. similar current distribution for core logic modules and to select asymmetric current ratios for pad drivers.

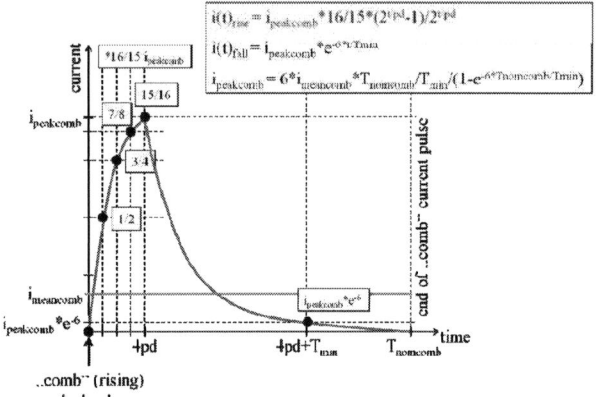

Figure 2 Dynamic current pulse for the rising clock edge

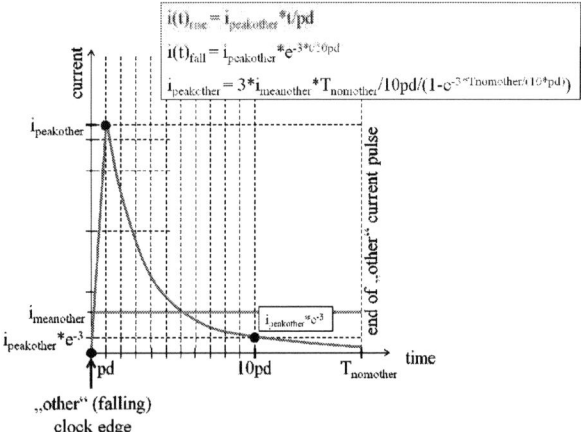

Figure 3 Dynamic current pulse for the falling clock edge

The peak value for the "rising" and "falling" current pulse in every module is calculated in a way that – considering the decay and current share parameters – the average current over

time for this module is met. In this calculation, also the operating clock and the maximum (i.e. design target) clock frequency are considered. Figures 2 and 3 show the dynamic current pulse parameters for the rising and falling clock edges.

The duty cycle, i.e. the position of rising and falling clock edges within one clock period, can be configured in EMISoC. Therefore, for each functional module the input clock (from a PLL), the clock divider, and the duty cycle are specified. Example: The VCO of a system-PLL runs at 600 MHz, is divided by 3 and distributed as a 200 MHz clock to the functional modules. A module takes this 200 MHz clock and divides it by 2, i.e. operates at 100 MHz. But this division is not done symmetrically, but instead by masking every second 200 MHz pulse. This is resulting in a 1:4 duty cycle of the module's 100 MHz clock. Figure 4 shows how this configuration is implemented in EMISoC.

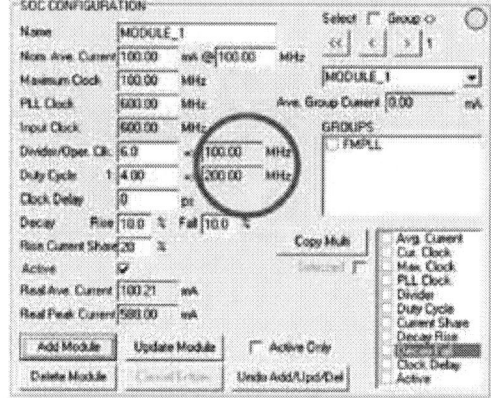

Figure 4 Clock timing control for a functional module

Smoothing effects on the dynamic current caused by on-chip capacitors and trace inductances are considered by a related parameter which integrates the current values over a selectable time interval.

Damping effects caused by resonances in the chip/package/ PCB system are not yet considered, but planned to be implemented into the EMISoC software.

All general and module-specific modelling data can either be typed in manually or prepared in a csv-formatted file and imported by the program. Vice versa, the SoC configuration can be exported in csv-format.

To simplify the configuration of related SoC modules (e.g. all core modules connected to a high-speed bus, all peripheral modules connected to a lower-speed bus, all I/O modules), up to eight "groups" can be defined. A "copy multi" command allows to set the respective parameter values simultaneously by one mouse click.

To simplify the overall execution of simulations for multiple SoC configurations, a batch language has been introduced. The batch commands reflect all manually possible commands. Batch files in csv-format can be prepared and checked by EMISoC prior to execution for syntax or systematic errors like illegitimate clock configurations. Batch jobs perform automatically the following tasks:

978-1-4799-5004-1/13 $31.00 © 2013 IEEE   182

- Configure the SoC,
- Calculate the total dynamic switching currents,
- Calculate the resulting emission spectra,
- Store all result diagrams in gif-format (the smallest graphic file size).

D. EMISoC Graphical User Interface

EMISoC is written in Delphi ("Visual Pascal") [9]. The graphical EMISoC user interface consists of one window displaying the control panels and the result panels simultaneously, see Figure 5.

The left part of the screen contains the control and edit panels for the program flow, the general settings, SoC configuration, FFT envelope handling, and a statistics display. The statistics panel lists all used module clocks which are calculated from the PLL and local clock divider settings.

The right part of the screen shows the dynamic waveform, the emission spectrum of the SoC, and the input field for a diagram title.

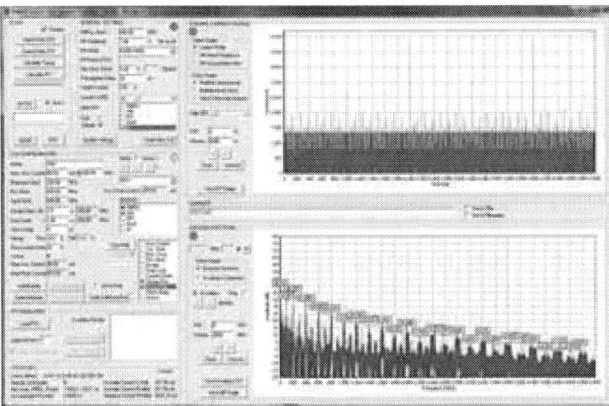

Figure 5 EMISoC graphical user interface

The frequency spectrum contains an optional envelope curve which can be saved in csv-format. Up to eight envelopes can later be loaded and displayed in one diagram for comparison purposes; an example is shown in Figure 6. The envelope shows the peak emission or frequency values; the window to "catch" the emission peaks is configurable.

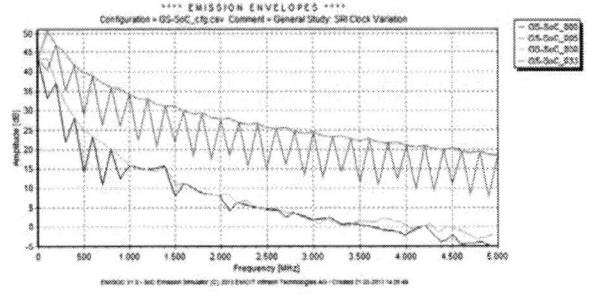

Figure 6 FFT envelope overlay

All filenames follow a special syntax, i.e. they contain a string token which identifies the meaning of the file contents,

e.g. *_cfg.csv is the filename of a configuration file, *_bat.csv denotes a batch file, *_cur.gif contains the dynamic current waveform, *_fft.gif the frequency spectrum, and *_env.csv an FFT envelope data file. * stands for a user-determined string which reflects the SoC configuration.

E. EMISoC Simulation Performance

The time-consuming procedures in EMISoC are the current waveform and the FFT calculation. The duration of the current waveform calculation increases with every additional module. Batch files contain several EXECUTE commands, each of which calculates the dynamic current waveform and the frequency spectrum. Furthermore, it stores the FFT envelope data and an individual configuration file for every executed SoC configuration to simplify detailed analysis of certain configurations by just importing the corresponding config files. Prior to starting the batch file, a quick check can be performed to ensure that the batch file contains no syntax or clock system configuration errors. Progress bars indicate the execution state of the program.

Table 2 lists the duration of execution tasks, measured on a laptop computer with Intel Core i5, CPU M540 running at 2.53 GHz, equipped with 4 GB memory.

Table 2 Run times of different EMISoC jobs

#Modules / #Configs	Dyn. Current	PWL Export / Timestep	FFT Spectrum	Batch Check	Batch Run
1/1	1s	1s/213ps		-	-
16/1	10s	4s/13.3ps	12s	-	-
139/10	1m30s	25s/1.7ps		1m35s	18m

III. EMISoC Test Cases

The purpose of EMISoC is the prediction of electromagnetic emission for future, not yet designed microcontrollers. Because it needs no simulation pattern, the dynamic current waveform of a yet non-existing microcontroller can be created without having any netlist or circuit design. Since EMISoC generates the dynamic current profile based on module-specific clock settings, this approach is more realistic than a random-pattern approach offered by some of the tools referred in chapter I. Furthermore, EMISoC can reveal the impact of changes in the clocking configurations on the resulting dynamic current profile and thus on the EME. Following this motivation, 2 test cases have been executed up to now. They are described in the next sections.

A. Test case 1: Debug port configuration

A debug port transfers internal microcontroller data to an external debugger for further analysis and actions. With increasing performance, the data transfer rate rises. Instead of the previous two bit channels, up to 8 bit channels are considered in future products. EMISoC was used to judge the impact of bit clock modulation (this is allowed because the bit clock is generated by the microcontroller) and bit line skewing on EME. EMISoC offers a built-in pad driver configuration based on driver on-resistance and external capacitive load. The results of this study are summarized in Figures 7 and 8.

The pad drivers were modelled in EMISoC with respect to their toggle rate, driver transistor "on" resistance and external capacitive load.

Figure 7 shows the effect of bit skewing on the debug port's dynamic current – skew implies edge smoothing while reducing the peak current.

- Fig. 7a: Unskewed dynamic current
- Fig. 7b: Skewed dynamic current

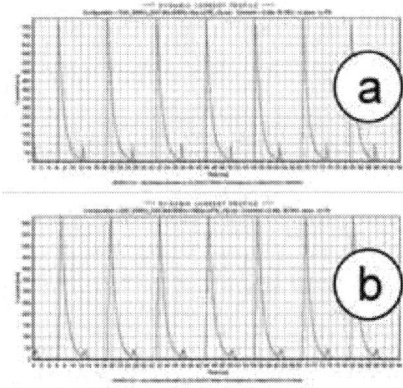

Figure 7 EMISoC dynamic current results for debug port study

Bit skewing leads to minor emission reduction and is closely related to the skew step. This is indicated by the Figure 8 picture "skew, no FM" which shows skew-time dependent (sin x)/x artefacts in the emission envelope. Frequency modulation leads to a smooth emission decrease towards higher frequency. The most effective solution would be – as expected – the combination of FM and skewing, see Figure 8 picture "skew, FM".

- Fig. 8a,b: EME without FM, without/with skew
- Fig. 8c,d: EME with FM, without/with skew

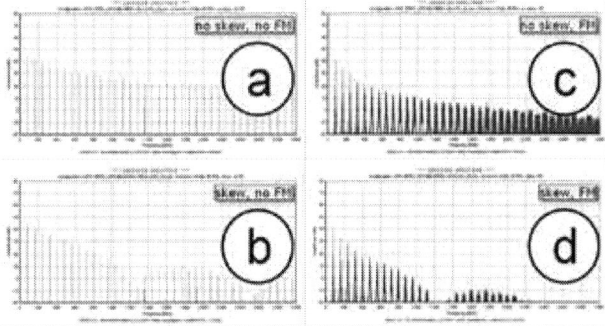

Figure 8 EMISoC emission results for debug port study

B. Test case 2: New 40 nm microcontroller family

EMISoC was used to find optimal solutions for the challenge to identify the most emission-friendly clocking concept. At this time, no design data of the new microcontroller was available, except some ideas and proposals for the clocking scheme, and the power

consumption estimation for the functional modules running at certain clock rates. The EMISoC model consists of 139 modules, thereof 80 I/O drivers. We know that the most relevant parameters for electromagnetic emission are clock rates, clock modulation and clock skew. Clock rates imply the frequency and coincidence (i.e. whether to use same or different clock rates to distribute on SoC level) of clocks used in different functional modules. Since in a high-speed synchronous design the clock skew must be minimized towards zero to ensure the full performance, clock skewing applies mainly to switching I/O groups or bits.

Consequently, a test plan for this study may look like:

- Apply clock modulation to functional modules.
- Variation of modulation amplitude.
- EME impact of disabling clock modulation for one or more modules.
- Apply clock skewing to I/O groups and single I/Os.
- Variation of clock skew.
- EME impact of I/O clock skew variation.

Figure 9 shows some results of this study. The two diagrams on the left side indicate the EME reduction implied by clock modulation of the functional modules. For modulation amplitudes greater than 0.5%, the I/O switching noise harmonics start to dominate in the spectrum.

The two diagrams on the right side indicate the EME reduction implied by I/O clock skewing. Even a clock skew of 1 ns does not lead to an EME peak reduction of I/O switching noise below the envelope given by the clock modulation peaks.

- Fig. 9a: EME w/o FM, w/o skew (reference for Fig. 9b)
- Fig. 9b: EME reduction from module clock modulation
- Fig. 9c: EME w/o FM, w/o skew (reference for Fig. 9d)
- Fig. 9d: EME reduction implied by I/O clock skewing

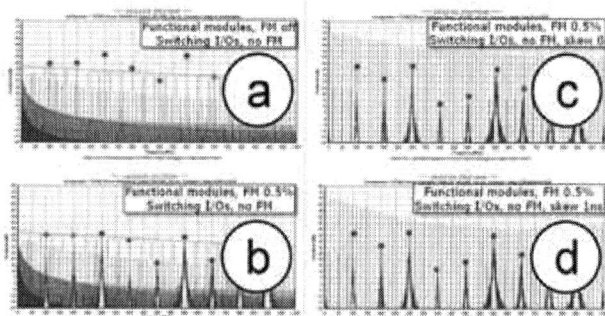

Figure 9 EMISoC emission results for microcontroller study

The specification of clocking systems for complex chips requires a thorough investigation of module and I/O clocking options under special consideration of I/O characteristics, such as driver strength, toggle rate and activity levels. EMISoC helps the designers to compare the expected emission of various clocking configurations. The most promising configurations should be evaluated according to their impact on SoC performance, system restrictions and design tool limitations, in order to select the best configuration for future designs.

C. Test case 3: Layout-based voltage drop simulation

EMISoC can precisely calculate a dynamic current over time which reflects the complete microcontroller activity of core and I/O in an "ideal" way. It is not intended to consider any damping or resonance effects caused by parasitic package or board RLC characteristics. Due to lacking RLC-based noise propagation path data, the EMISoC simulation results differ from real emission measurement results according to the BISS IC EMC test specification [10] and the underlying IEC 61967-4 IC emission test standard [11]. Results for the TC277B microcontroller without and with system clock modulation are shown in Figure 10.

- Fig. 10a/b: Measured core supply without/with modulation
- Fig. 10c/d: Simulated core supply without/with modulation

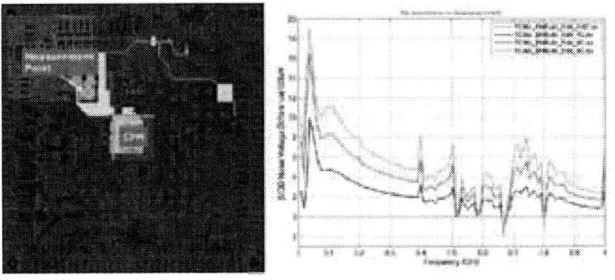

Figure 10 Simulation vs. measurement comparison

But EMISoC can co-operate with other EDA tools. In our case, we used the same SoC data base as described in test case 2. In addition, a preliminary floorplan layout was available. This floorplan was partitioned into four rectangles, each of which contained a subset of the 139 modules. EMISoC was used to create an individual PWL file for each of the four subsets, and each PWL stimulus was connected to the center of a floorplan partition.

Figure 11 Test case 3 model setup on PCB level (left) and EME results (right)

This approach allows considering different switching activities in different parts of the SoC layout, leading to more realistic voltage drop or emission simulation results. Finally, Spice models of the IC package and a printed circuit board

were added. The ideal voltage source was connected to the supply connector on the PCB model. For the simulation, Apache and Sigrity tools were used [1] [3]. RedHawk simulated the dynamic current distribution over the floorplan layout, based on the four PWL files delivered by EMISoC. Furthermore, RedHawk generated four different chip power models (CPM) for different on-chip capacitor configurations. PowerSI used these CPMs as stimuli and simulated the EME appearing at the measurement point on the BISS emission test board. Figure 11 shows the test board layout and the comparison EME spectra for the four on-chip capacitor configurations.

IV. EXPERIENCE, OUTLOOK AND CONCLUSIONS

EMISoC should be used to calculate and compare the emission spectra of "ideal" dynamic currents, allowing to identify preferable clock configurations minimizing the EME.

A simulation versus measurement study will start in the frame of a master thesis in December 2013, aiming at the extraction of "typical" noise coupling paths over the IC package, and adding these scalable transfer functions into EMISoC. Based on this new feature, correlations between EMISoC simulation and BISS measurement results are planned on the existing TriCore 32-bit microcontroller Aurix TC277.

Another planned enhancement are programmable activity sequences on module-level, i.e. by pre-defined macros like "memory read/write access" or "I/O data bitstream".

EMISoC was proven to be a valuable addition to our microcontroller design flow to estimate different microcontroller clock architectures for future not yet designed products. Together with simple on-chip power-grid and package models, the dynamic current profiles provided by EMISoC allow design optimization studies on the layout level as well.

REFERENCES

[1] Ansys/Apache: *EDA software Totem and Redhawk, CPM emission models*, http://www.apache-da.com/products

[2] Ansys: *EDA software HFSS and Q3D*, http://www.ansys.com/ Products/Simulation+Technology/Electromagnetics

[3] Cadence/Sigrity: *EDA software PowerSI*, http://www.sigrity.com/products

[4] E. Sicard, INSA: *IC-EMC simulation software*, http://www.ic-emc.org/

[5] D. Hesidenz, T. Steinecke, *Chip-Package EMI Modeling and Simulation Tool "EXPO"*, EMC Compo 2005, Munich, Germany

[6] A. Gstöttner, T. Steinecke, M. Huemer, *High Level Modeling of Dynamic Switching Currents in VLSI IC Modules*, EMC Compo 2005, Munich, Germany

[7] T. Steinecke, *EMISoC Electromagnetic Emission Calculator for Systems on Chip – User's Manual*, Infineon internal document, 2013

[8] T. Steinecke, *Low-Jitter Frequency-Modulated PLL*, 2012 Asia-Pacific Symposium on Electromagnetic Compatibility, Singapore

[9] Embarcadero: Delphi software development suite, http://www. embarcadero.com/de/products/Delphi

[10] Bosch, Continental, Infineon, ZVEI, *Generic IC EMC Test Specification*, http://www.zvei.org/Verband/Publikationen/Seiten/ Generic-IC-EMC-Test-Specification-english.aspx

[11] IEC 61967-4, *Integrated Circuits – Measurement of electromagnetic emissions, 150 kHz to 1 GHz – Part 4: Measurement of conducted emissions – 1 Ω/150 Ω direct coupling method*, http://www.iec.ch

A Microcontrller Conducted EMI Model Building for Software-level Effect

Shih-Yi Yuan
Dep. Communication Engineering,
Feng Chia University
Taichung, Taiwan, R.O.C.
syyuan@fcu.edu.tw

Abstract—**This paper proposes a model building process for conducted electromagnetic interference (cEMI) model of microcontroller (μC) considering software effect. Due to the fast advances of embedded system design technologies, software now is capable of controlling nearly all the features of electronic modules, which means software can actually affect EMI characteristics of target modules. Thus, a software-level EMI model is essential for electronic modules. Due to intellectual property (IP) considerations, IC designers seldom expose the internal architecture details of their IC products to EMI modelers. Because the internal module behaviors are unknown, it makes EMI modeling very difficult. This paper proposes a block-box cEMI modeling procedure for μC. The concept is based on a set of block-box impulse response (BBIR) functions. BBIR modeling method is based only on measurement information and treats the target as a block-box. After the model building process, the cEMI behavior of a new testing boards (or modules) with the same μC can be estimated. This model is verified by a case study. From the experiment results, it shows that the proposed method can estimate different machine code cEMI behaviors. The estimated result is in good accordance with the measurements both in time-domain and frequency-domain. The results also shows the internal impedances of a μC are quite different among machine codes executed by the μC.**

Keywords—*black-box EMI modeling; software-level EMI modeling; microcontroller EMI modeling*

I. Introduction

Electromagnetic Interference (EMI) researches are undergoing radical changes by electronic technologies. Electronic devices are now essential to many daily life applications. Among all these devices, the central part is a microcontroller (μC).

Due to the fast advances of μC design and VLSI technologies, the Platform-Based Design Methodology (PBDM) is now the major design trend for modules with μCs. PBDM can improve the design reuse percentages, reduce the hardware/software errors, and thus, increase the robustness of such modules. Thus, PBDM actually controls the EMI behavior of the electronic modules. The phenomena are observed both in literature [1] and practical EMI industry [2]. Thus, an EMI model for μC is essential for the research and industry.

The IEC 62433 [3] standards are a family of EMI modeling method developed for integrated circuit (IC). The standard has been successfully applied to model the power

and ground signal fluctuations for many μCs, ASICs, and programmable devices within the range between 1MHz–2GHz.

Although IEC-62433 is successful in many application areas, there are several difficulties coming from the model building process:

1. Due to the intellectual property (IP) considerations, IC design companies seldom expose the internal architecture details which are necessary for an accurate estimation of IEC-62433 model.

2. The clues for estimating gross Internal Impedance (IntZ) [3] are few. Thus, a rough conjecture to build a presumed internal static netlist relies on modeler's domain knowledge. The estimation of the Internal Current Activity (IntCA) also faces the same difficulty.

3. After the initial guessing, IntZ and IntCA in the primitive model are simulated and, then, iteratively fine-tuned for the measurement information. As it can be expected, the trial-and-error modeling processing is both tedious and error prone.

4. Even a detail internal architecture (IntZ) is supported by IC designer, the IntCA is still an estimation and is modeled by several fixed-periodic linear (periodic triangle or periodic trapezoidal) current waveform. As the waveform can only estimate the first-order accuracy of the current activities, it also decreases the model accuracy [4] even the IntZ is accurate.

5. Since the model building process of IEC-62433 is based on fine-tuning the primary model to fit the measurement data. The effects of IntZ and board-level impedance (Z_{PCB}) cannot be separated and discriminated. Thus, the model can only be applied to the same board (module). Any other boards, even with the same μC, are not applicable and a new model should be built from the beginning.

6. Since EMI behaviors can be affected by software, a software-related EMI model should be built based on the program (or machine code) sequence and the effect of the hardware driven by the program [1][2][3]. This means both the IntZ and IntCA can be dynamic and depend on the machine codes executed [5].

In reference [6], we propose a method using a set of black-box impulse response (BBIR) functions to automatically build a conducted EMI model. The method is based on a black-box mathematical deduction for the EMI

978-1-4799-5004-1/13 $31.00 © 2013 IEEE

measurement data. Neither the internal architecture of μC nor the IntZ/IntCA is to be conjectured. Thus, the first and second difficulties are greatly alleviated. The method is automatic procedures without any fine-tune steps. Thus, it solves the third difficulty. The method can use BBIR to deduce both IntCA and IntZ and the accuracy is greatly increased, so the fourth difficulty is solved. BBIR functions can discriminate board impedance and IntZ by our multi-board constrains, the fifth difficulty is improved.

However, the last difficulty is still a problem. In this paper, we try to apply the BBIR concept to build a software-level dynamic cEMI estimation model of a target μC.

The paper is organized as following: section II briefs the BBIR concept and the model building procedures based on it. Section III describes the software-level dynamic cEMI model building method. Experiment results are described in section IV and followed by conclusions.

II. BLACK-BOX IMPULSE RESPONSE (BBIR) FOR CEMI BUILDING

Example of a circuitry netlist and its reduction form are shown in Fig. 1. The netlist is based on IEC 61967 EMI measurement standard [7]. The blue blocks represent the controllable part of the cEMI modeler which is the board-level impedance. The red parts represent the IEC-62433 model for IC. It is not controllable and not known to the cEMI modeler.

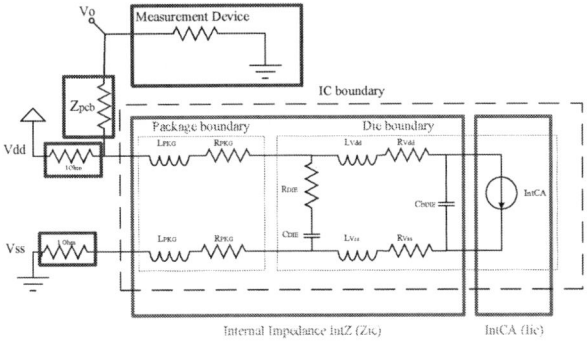

Fig. 1. (a) Simple netlist example of IEC 61967-4

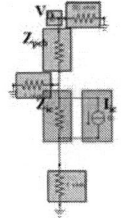

Fig. 1. (b) The reduction form of Fig. 1. (a)

Fig. 1. Simple IEC 61967-4 model and its reduction form

From Fig. 1, $V_O(f)$ can be deduced as:

$$\frac{-50 \times I_{ic} \times Z_{ic}}{\left(101 + 51 \times Z_{ic} + 2 \times Z_{pcb} + Z_{pcb} \times Z_{ic}\right)} = V_O \quad (1)$$

where Z_{PCB} is the board-level impedance.

Since there are 2 variables, at least 2 constrains should be given to solve the equation. We use 2 testing boards (PCB1 and PCB2) with different board level impedance for the BBIR model building.

Assume the different board-level impedance $Z_{PCB1}(f)$ and $Z_{PCB2}(f)$ are designed and 2 boards are embedded with the same μC, we can reasonably assume the unknown $Z_{IC}(f)$ and $I_{IC}(f)$ are generally the same with different Z_{PCB1} and Z_{PCB2}. The board-level IR function $Z_{PCB1}(f)$ and $Z_{PCB2}(f)$ can be easily measured and the unknown variables be solved by [6] as (2) and (3). These equations (IR functions) can be more than 1000 equations for different frequencies (in our case, 2048).

$$Z_{ic} = \frac{\left(-101 + (V_{o1} - V_{o2}) - 2 \times Z_{pcb1} \times V_{o1} + 2 \times Z_{pcb2} \times V_{o2}\right)}{\left(51 \times (V_{o1} - V_{o2}) + Z_{pcb1} \times V_{o1} - Z_{pcb2} \times V_{o2}\right)} \quad (2)$$

$$I_{ic} = \frac{V_{o1} \times \left(101 + 51 \times Z_{ic} + 2 \times Z_{pcb1} + Z_{pcb1} \times Z_{ic}\right)}{-50 \times Z_{ic}} \quad (3)$$

A third board (verification board or PCBv) is designed for the verification purpose. When the Z_{PCBV} is given, the $V_O(f)$ of PCBv ($V_{OV}(f)$) is estimated by the proposed model. The estimated $V_O(t)$ can be derived from IDFT [8]. The real PCBv responses can be measured and compared by high speed oscilloscope.

III. SOFTWARE-LEVEL DYNAMIC CEMI MODEL BUILDING

From the deduction above, any groups of "Vo-pairs" (the output of the two testing boards) can be used to estimate the IntZ and IntCA. We consider a quasi-static condition in a short period of time that the IntZ and IntCA are assumed to be static.

Generally, all PBDM-style μC is a digital circuit and clock-driven. The quasi-static period can be safely estimated by the external clock-cycle or internal CPU machine-cycle (the integer multiple or fraction of the clock cycle). In the case study, the external clock-cycle is 4MHz and the machine-cycle is 1 MHz (a quarter of the external clock-cycle). The period is set to 1μs in the case study.

The detail software-level quasi-static IntZ and IntCA extraction processes are briefly described in Session II. The extraction and identification process of the dynamic IntZ/IntCA is similar to the Instruction current extraction process [4] and is briefed in Fig. 2 for 1 machine code per IntZ/IntCA.

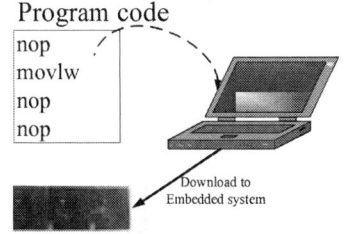

(a) Program code preparation and target DUT download

978-1-4799-5004-1/13 $31.00 © 2013 IEEE   187

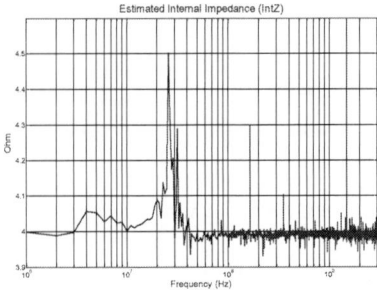

(b) IntZ(f) estimation by BBIR

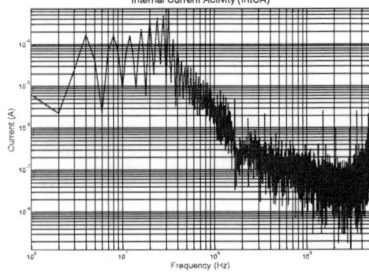

(c) IntCA(f) estimation by BBIR

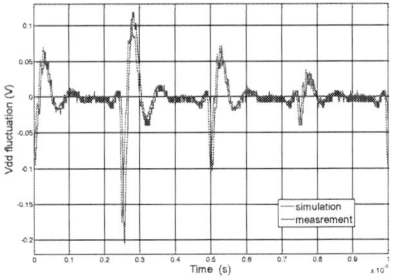

(d) PCBv Vo(t) BBIR estimation and measurement

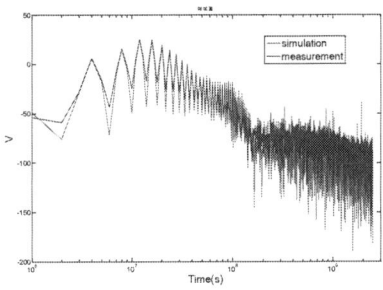

(e) PCBv Vo(f) BBIR estimation and measurement
Fig. 2. The proposed method for one machine code IntZ/IntCA calculation and cEMI estimation

The IntZ and IntCA are not observable from outside world. However, the cEMI behavior Vo(f) of the target board PCBv is observable and comparable. Thus, the comparison result can be used to show the accuracy of the IntZ and IntCA estimation.

IV. EXPERIMENTAL RESULTS

PCB1, PCB2, and PCBv are specially designed that their board-level impedance (Z_{PCB}, or the blue part in Fig. 1) are

NOT the same. The Z_{PCB} of PCB1 and PCB2 are NOT following IEC 61967 standard. The selection of Z_{PCB} can be found in [6]. The Z_{PCBv} is designed according to the standard.

TABLE I shows the Z_{PCB} of the 3 boards. The DUT to be estimated is a commercial μC PIC12F629 [9]. The oscilloscope is DPO72004 (Fig. 4).

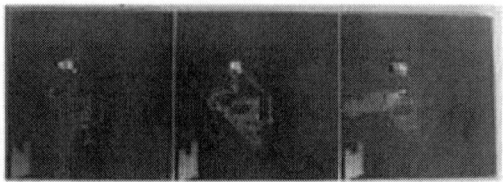

Fig. 3. Three testing boards for physical verification (PCB1, PCB2, and PCBv)

TABLE I. Testing board Z_{PCB} (Definition of Z_{PCB} can be found in Fig. 1)

	Z_{PCB}
PCB1	$1//(s \times 470e\text{-}6)$
PCB2	$1000 + 1/(s \times 6.8e\text{-}9)$
PCBv	$51 + 1/(s \times 6.8e\text{-}9)$

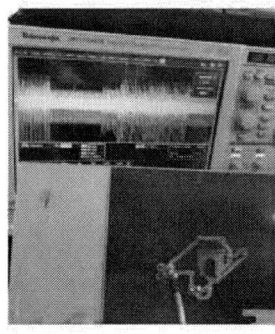

Fig. 4. Testing setups

Fig. 2 (d) and (e) shows one machine code comparison result between the measurement and the estimated result. Fig. 2 (d) is the time-domain comparison; and Fig. 2 (e) is the frequency-domain comparison. The estimated and measured waveform is very accurate within the range DC to 100 MHz. From the comparison result, the estimation of the new board (PCBv) is very accurate both in time-domain and frequency-domain. This implies that the estimation of the internal current impedance and internal current activity (IntZ and IntCA) are accurate.

By using the same procedure, different machine codes' EMI can be estimated. In the same time, the IntZ/IntCA can be also calculated by BBIR method. According to Section III the quasi-static period is set to 1μs, the IntZ and IntCA of different machine code is show in Fig. 5.

From Fig. 5, it is clear that the estimated IntZ(s) are intrinsically different and are dynamic depending on the machine code executed by μC. The IntCA(s), although with some variations, are generally static at low frequencies (DC-80MHz). The different machine codes' fluctuations of IntZ below 10MHz are almost negligible. The fluctuations of

978-1-4799-5004-1/13 $31.00 © 2013 IEEE

different machine codes between 11MHz-300MHz are very different.

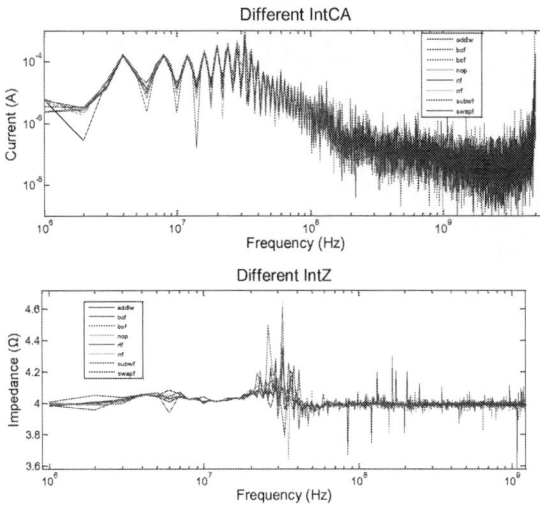

Fig. 5. IntCA/IntZ of different machine codes

V. CONCLUSIONS

This paper proposes a black-box methodology for software-level microcontroller µC conducted EMI (cEMI) modeling procedures. Owing to the black-box impulse response (BBIR) method, the cEMI characteristics of different machine codes executed in the target µC can be estimated. The automatic procedure and non-conjecture property of BBIR solves or alleviates many difficulties when IEC 62433 procedure is applied to a µC.

Because of no easy way to observe the IntCA and IntZ from outside world, the paper proposes the BBIR method to estimate these 2 characteristics by the effects that is observable from outside world – the cEMI behaviors of different machine codes. The estimation of the IntZ suggests that IntZ are very different among different machine code executed in µC. This result can be further extended to estimate the EMI behaviors of a code sequence or even an entire program which will pave the way to software-level EMI optimization.

ACKNOWLEDGMENT

This work was supported by grants from the Bureau of Standards, Metrology and Inspection (BSMI 0121136A) and the National Science Council (NSC) (NSC 101-2221-E-035-050-), Taiwan, Republic of China.

REFERENCES

[1] Martin O'Hara, "The EMC Impact of Embedded Software", Conformity, Sep. 2007, pp. 36 – 45.
[2] Osami Wada, Yoshiyuki Saito, Katsuya Nomura, Yukishige Sugimoto, and Tohlu Matsushima, "Power Supply Current Analysis of Micro-controller with Considering the Program Dependency," 8th Workshop on Electromagnetic Compatibility of Integrated Circuits (EMC Compo 2011), pp. 93-98, Nov. 6-9, Dubrovnik, Croatia.
[3] IEC 62433 "Models of Integrated Circuits for EMI behavioral simulation," [Online]. Available: http://www.iec.ch
[4] S. Y. Yuan, H. E. Chung, and S. S. Liao, "A Microcontroller Instruction Set Simulator for EMI Prediction," IEEE Trans on EMC, vol. 51, pp. 692-699, 2009.
[5] Shih-Yi Yuan, Huai-En Chung, Chiu-Kuo Chen, and Shry-Sann Liao, "Time varying instruction current EMC simulation improvement," IEEE International Symposium on EMC, Aug., pp. 18-22, Detroit, USA, 2008.
[6] Shih-Yi Yuan and Shry-Sann Liao, "Automatic Conducted-EMI Microcontroller Model Building," submitted to EMC Compo 2013.
[7] IEC 61967 "Integrated circuits - Measurement of electromagnetic emissions, 150 kHz to 1 GHz," [Online]. Available: http://www.iec.ch
[8] Alan V. Oppenheim, Alan S. Willsky, and S. Hamid Nawab, "Signals and Systems," 2nd ed., Prentice Hall, 1996.
[9] 8-Bit CMOS Microcontroller (PIC12F629) Microchip, 2003.

Characterization and Modeling of Electrical Stresses on Digital Integrated Circuits Power Integrity and Conducted Emission

A. Boyer, S. Ben Dhia

LAAS-CNRS

Université de Toulouse ; UPS, INSA, INP, ISAE ; UT1, UTM, LAAS

Toulouse, France

alexandre.boyer@laas.fr

Abstract - **Recent studies have shown that integrated circuit aging modifies electromagnetic emission significantly. The proposed paper aims at evaluating the impact of aging on the power integrity and the conducted emission of digital integrated circuits, clarifying the origin of electromagnetic emission evolution and proposing a methodology to predict this evolution. On-chip measurements of power supply voltage bounces in a CMOS 90 nm technology test chip and conducted emission measurements are combined with electric stress to characterize the influence of aging. Simulations based on ICEM modeling modified by an empirical coefficient to model the evolution of the emission induced by device aging is proposed and tested.**

Keywords: Integrated circuits, power integrity, conducted emission, accelerated aging, ICEM modelling

I. INTRODUCTION

Recently, many publications have forecast a decrease of new CMOS technology device lifetime down to few years [1], with anticipated appearance of hard or soft failures due to wear-out mechanisms (e.g. hot carrier injection (HCI), negative bias instability (NBTI), electromigration...) [2] [3]. The drift of the electrical characteristics of semiconductor devices can have direct consequences on integrated circuit electromagnetic emission (EME) and power integrity (PI). Recently, some publications have shown that accelerated aging tests such as high or low temperature operating life, thermal cycling, or electrical overstress induce a significant variation of EME produced by power supply units [4], high side switch devices [5] or I/O buffers [6]. However, these studies do not clarify the origins of EME changes and do not address modeling and prediction issues.

This paper intends to clarify the impact of IC aging on PI and conducted emission (CE) experimentally and by simulation. A test chip designed in Freescale CMOS 90 nm technology, which includes a digital core and on-chip sensors dedicated to the measurement of power supply voltage fluctuations, has been developed to characterize the evolution of PI vs. time. In order to explain the effect of IC aging on PI and CE, an equivalent model of the circuit based on the Integrated Circuit Emission Model (ICEM) approach [7] is developed. A simple empirical coefficient is introduced in the model to simulate the effect of aging. This modeling methodology could constitute a method to predict the evolution of EME of ICs with time.

The paper is organized as follows: after a description of the circuit under test, on-chip sensors and experimental set-up, the measurements of the evolution of the power supply voltage bounces and CE are presented. Then, the proposed modeling approach of the circuit PI and CE is described and, finally, simulation and measurement results are compared in order to validate both our hypothesis about PI and CE evolution with time and the proposed modeling method.

II. EXPERIMENTAL SET-UP DESCRIPTION

A. Test chip description

A dedicated test chip has been designed in Freescale CMOS 90 nm technology for IC emission and susceptibility modeling, and characterization of aging impact on EMC of ICs. In particular, the test chip includes a digital core with a dedicated power supply voltage equal to 1.2 V. This structure is a basic 100-stage shift register, synchronized by a 40 MHz clock.

B. On-chip voltage sensor

In order to monitor the voltage drops on the power supply of the digital core, an on-chip voltage sensor is placed along the power supply rail of the core. This sensor is able to measure the waveform of voltage bounce across non accessible nodes with a precise time resolution (up to 15 ps). Its analog bandwidth is equal to 10 GHz. Its principle and its implementation in the test chip are explained in [8]. In order to prevent noise coupling with the digital core, the designed sensor has a dedicated power supply and is isolated from the IC substrate by a buried N layer. Due to its small input capacitance (4 fF), the sensor is not intrusive and does not alter the performance of the tested block. Although the electrical stress applied to the circuit under test can partially alter the performances of the sensor, it can be recalibrated before any measurements. The calibration process is necessary to compensate effects of imperfections due to non ideal behavior and mismatch on voltage and frequency responses.

C. Description of the experimental set-up

During their lifetime, CMOS transistors are affected by intrinsic failure mechanisms such as NBTI or HCI, mainly activated by harsh environmental conditions such as high or low temperature and electrical overstress. In this study, DC electrical stresses are applied on the power supply pin of the digital core to accelerate the damage rate for the relevant wear-out failure mechanisms and thus the circuit aging. The experiment is based on a measure-stress-measure flow, which consists in applying electrical stress on the circuit under test and interrupting the stress during short periods regularly for characterization purpose. With this procedure, the characteristics of the circuit are monitored at various degrees of aging. The choice of stress conditions (stress voltage and duration) is based on a preliminary failure analysis of MOS devices developed in this technology [9]. For DC stress voltages ranging from 3 to 4 V and applied between drain and source of NMOS and PMOS devices, significant degradations of threshold voltage and carrier mobility are induced due to HCI and/or NBTI after several hundreds of seconds. The amount of degradation is related to stress voltage, duration, transistor geometry and gate oxide thickness. A 3.6 V electrical stress is applied during 120 minutes in order to validate the influence of the stress voltage on power integrity change.

Sensor measurements of power supply voltage fluctuations are performed after each stress interval. The sensor is not supplied during stress phases in order to reduce its aging. Recalibration is done after each stress period, although experimental characterizations have not shown any significant degradation of the sensor. In addition, conducted emission measurements according to the "1 Ω" method [10] are performed after each stress interval, in order to monitor the variations of the transient current that returns to the ground. The conducted emission measurements are done on a dedicated board which is not stressed to prevent from any changes of the 1 Ω probe characteristics. All the experiments have been repeated on several samples to ensure that similar results are obtained. Nevertheless, the uncertainties linked to process dispersion are not taken into account in this study.

III. EXPERIMENTAL RESULTS: EVOLUTION OF POWER INTEGRITY AFTER ELECTRICAL STRESS

A. Initial measurement of power integrity

Fig 1 presents the power supply voltage fluctuations produced by the digital core activity measured by the on-chip sensor. Positions of clock edges are indicated. Rapid drops appear at each clock switching. These events are linked to the rapid current demand from every gates and latches of the core. A damped oscillation with a pseudo-period equal to 4.5 ns follows the first rapid current impulsion. It is linked to the anti-resonance produced by on-chip capacitor and package inductor (see part V for more details). This type of noise is also measured on Vss node. Similar voltage fluctuations are also measured on the ground node of the circuit.

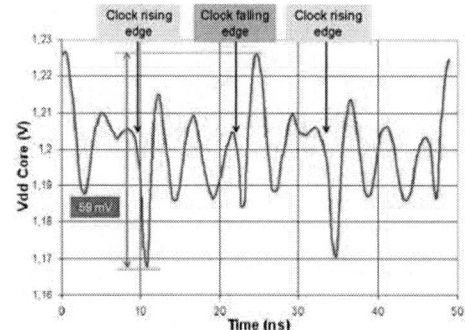

Figure 1. Measurement of the power supply voltage bounce of the digital core

B. Impact of electrical stress on digital core power integrity

After stress, the core remains operational and its quiescent current has not changed. However, the core timing characteristics have evolved. A specific part of the core has been designed to measure the propagation delay through one D-latch and 100 inverters. The propagation delay through the core after stress increases up to 31 % depending on the stress duration. The electrical stress applied on core power supply accelerates wear-out mechanisms such as HCI and/or NBTI. They increase the propagation delay of each gate of the core.

Fig. 2 presents the power supply voltage bounces measured by the on-chip sensor before and after 3.6 V stresses. A -30 % reduction of the peak-to-peak amplitude is observed. The waveform of the signal is also modified. The first peak is less steep and the period of the resonance oscillation has not changed. Its amplitude is reduced, but it is still damped at the same rate. This observation indicates that the package inductance, the equivalent capacitance of the digital and resistance linked to the power distribution network (PDN) have not been affected by the electrical stress significantly.

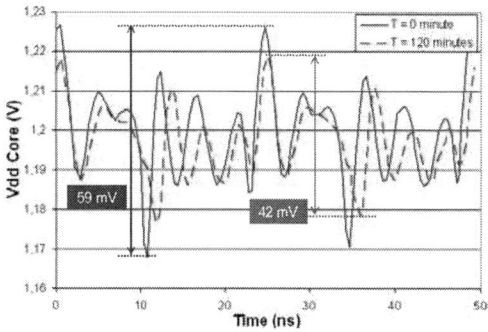

Figure 2. Evolution of the core power supply voltage bounce after 120 minutes of 3.6 V electrical stress

This hypothesis is confirmed by impedance measurement of the PDN with a vector network analyzer, as shown in Fig. 3. Whatever the electrical stress amplitude and duration, the impedance of the circuit power distribution network remains constant over a large frequency range.

This study has been done with the financial support of French National Research Agency (project EMRIC JC09_433714), the regional council of Midi-Pyrénées

978-1-4799-5004-1/13 $31.00 © 2013 IEEE

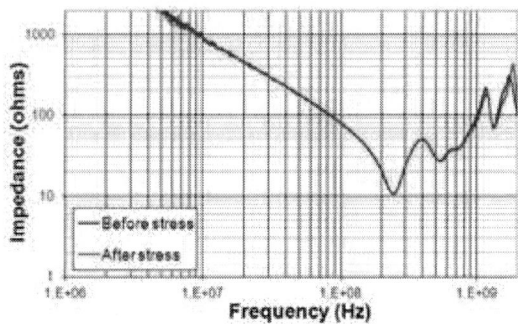

Figure 3. Measurements of the impedance of the power distribution network of the digital core before and after electrical stress (the core is not biased)

From these experimental results, the following hypothesis can be proposed to explain the evolution of power integrity of the circuit: the reduction of power supply voltage bounce is only linked to the change of transient current produced by the core activity. Wear-out mechanisms accelerated by electrical stress have reduced mobility of carrier in MOS devices so the switching transient current has been spread.

Fig. 4 presents the evolution of the spectrum of conducted emission of the digital core during the exposition to the 3 V electrical stress. Only the spectrum envelop is shown to facilitate the comparison between the different curves. The result shows a time-dependent gradual reduction of the conducted emission spectrum, which is more important at high frequency. The emission level at the circuit LC resonance (240 MHz) is decreased after stress. Above 400 MHz, the emission level decrease exceeds 5 dB after a 3 V stress applied during 240 minutes. Above 1 GHz, the reduction of the CE level reaches 15 dB.

This experimental result confirms the previous hypothesis about the effect of the electrical stress on the PI: the degradation mechanisms accelerated by the electrical stress not only reduces the amplitude of the transient current produced by the circuit activity, but also spread it because of a slowing down of the circuit. The next part intends to validate the proposed explanation about PI and CE evolution by simulation and thus to propose a method to predict the amount of variation of emission due to circuit aging.

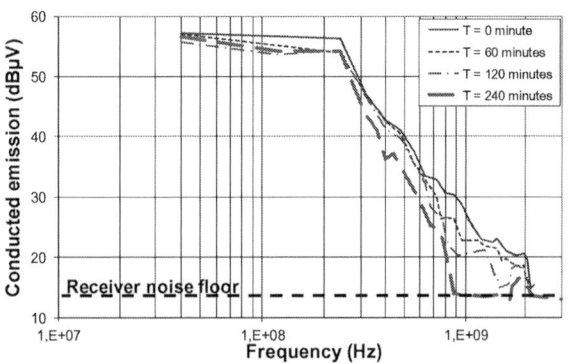

Figure 4. Evolution of the core conducted emission after 240 minutes of 3 V electrical stress

IV. MODELING OF DIGITAL CORE AGING EFFECT ON POWER INTEGRITY AND CONDUCTED EMISSION

A. Emission model construction and validation

The EME model is based on ICEM approach [7]. The model includes two main parts: the internal activity (IA) block which models the transient current produced by circuit operation, and the power distribution network (PDN) which models the filtering effect of the transient current due to IC and package. Fig. 5 presents a simplified structure of the ICEM model of the digital core.

A linear model based on an equivalent RLC circuit is proposed for the PDN. The parameters of the model are extracted from measurements. The passive element values are fitted from an impedance measurement between Vdd and Vss pins of the digital core made with vector network analyzer, as shown in Fig. 3. A very simple approach is used for IA modeling: two triangular waveform current sources describe the current produced at each clock edge. Three parameters describe the current pulse waveform: the amplitude Δi, the rise and fall times Tr and Tf. Even though such a waveform is quite simplistic, it provides a good estimation of the actual current waveform which can be tuned without a precise analysis of the circuit power consumption. Initially, the IA parameters are extracted from a SPICE transient simulation on the core model. Then, they are tuned in order to fit the simulation results on PI measurements done on a "fresh" device.

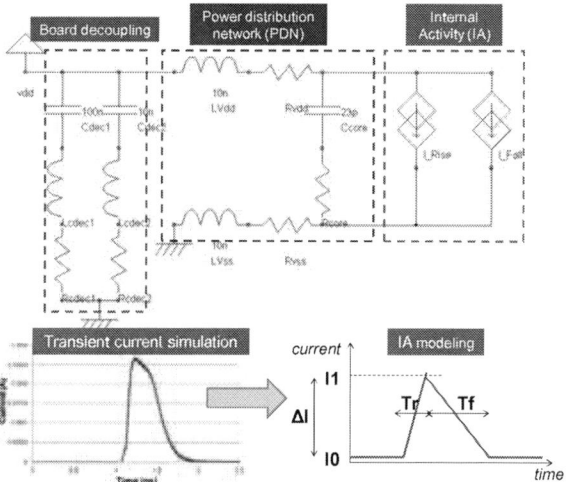

Figure 5. ICEM model of the digital core

SPICE transient simulations are performed to compute the waveform of the power supply and the spectrum of the conducted emission of the digital core from the ICEM model. Fig. 6 presents a comparison between the measurement and the simulation of the power supply voltage bounce before electrical stress. The simulated waveform is similar in term of peak-to-peak amplitude, pseudo-oscillation period and damping. Although the correlation between measured and simulated curves is not totally perfect, the model offers a sufficient accuracy for the aim of the study. A better correlation would rely on a more complex transient current waveform, extracted

978-1-4799-5004-1/13 $31.00 © 2013 IEEE

from a precise analysis of the power consumption of the digital core.

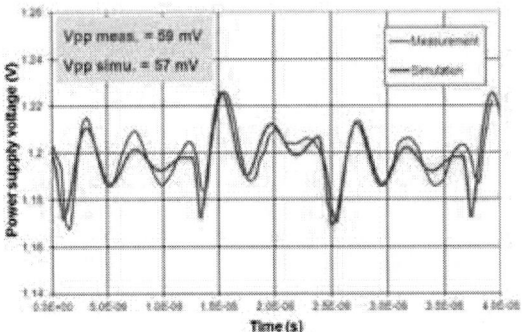

Figure 6. Comparison between measurement and simulation of the digital core power supply voltage bounce before electrical stress

Fig. 7 presents a comparison between the measured and simulated spectra of the core CE before electrical stress. The correlation between measurement and simulation is also acceptable up to 1 GHz. Extending the validity range of the model requires a more complex model of the PCB and IC substrate coupling.

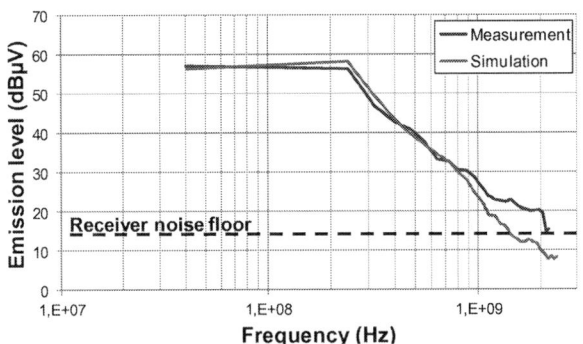

Figure 7. Comparison between the measurement and simulation of the digital core conducted emission before electrical stress

B. Modeling of degradation mechanisms induced by electrical stress

Continuous electrical stresses have been carried on 90 nm transistors and have shown that HCI and NBTI were activated on NMOS and PMOS transistors respectively, and were dependent on the stress voltage [9]. Both degradation mechanisms are activated in the digital core by the large voltage amplitude applied on the power supply. However, the contribution of the PMOS degradation on the PI and CE evolution is predominant. This fact can be verified by a SPICE simulation of the digital core with a modified model of PMOS transistors which takes into account NBTI effect. In the rest of the study, only the NBTI on PMOS transistor will be considered. NBTI is activated when a negative gate-source voltage is applied and leads to an increase of the threshold voltage V_{TH} of PMOS transistor and thus a reduction of the saturation current. Fig. 8 presents the experimental characterization of the PMOS transistor V_{TH} evolution with time when a constant gate-source voltage is applied. Two stress voltages ware applied: 1.2 V and 3.2 V. The time evolution of

V_{TH} can be estimated by a power law model given by equation (1), where A and γ are fitting coefficients [11].

$$\Delta V_{TH}(\%) = A \times t^{\gamma} \quad (1)$$

Figure 8. Measurement and modeling of V_{TH} evolution of 90 nm low voltage PMOS device exposed to negative gate-source voltage

The effect of NBTI can be taken into account in a transistor model such as BSIM4 by changing parameters which deal with the threshold voltage. The modeling of threshold voltage is complex and depends on numerous parameters, such as the substrate bias, channel geometry or doping profile [12]. Changing the parameter V_{TH0}, which defines the threshold voltage for a long channel without substrate bias, provides a simple method to model NBTI effect. Thus SPICE simulations can be done to predict the evolution of IC transient current vs. the V_{TH} drift. If V_{TH} time evolution models such as shown in Fig. 8 are known, the evolution of IC transient current vs. stress time can be estimated. For example, with this CMOS technology, a 3 V stress applied during 4 hours induces an increase of V_{TH} of 19 %. SPICE simulations done on the digital core model with a modification of PMOS V_{TH} model leads to a division by 2 of the transient current amplitude and a spreading of the current peak.

Information about transistor degradation is useful to modify the content of an ICEM model in order to predict the evolution of EME. In the next part, experimental data about PMOS degradation will be disregarded in order to propose and test a simple method to simulate aging effect on EME with ICEM

C. Integration of aging in ICEM model

In order to take into account the aging of the circuit in the ICEM model, the following simple methodology is proposed. The methodology relies on the assumption that the charge transfer associated to each gate switching does not change after stress. Although degradation mechanisms can increase leakage current and thus modify the dynamic current consumption, the measurement of average current consumption of the circuit does not evolve after stress, which confirms the validity of the hypothesis.

Only three parameters of the equivalent current sources in IA block are changed to spread the current pulse created by the digital core: the rise and fall times Tr and Tf of the current pulse are increased while its amplitude Δi is reduced. The evolution of these parameters must ensure that transient current integration over time remains constant. Their evolution is governed by an empirical coefficient δ called "degradation ratio", as shown by equations 2 and 3. It makes the link between the intrinsic degradation mechanisms which affect the circuit and the impact on the dynamic current consumption. δ equal to 0 means that the digital core is not degraded and the current pulse is not spread. If δ increases, a larger amount of degradation is considered and the current pulse is spread. In our ICEM model, the degradation ratio is applied on both current sources of the IA block symmetrically.

$$\Delta i_{stress} = \Delta i_{initial}(1-\delta), \quad 0 \le \delta \le 1 \quad (2)$$

$$t_{r\,stress} = \frac{t_{r\,initial}}{1-\delta}, \quad 0 \le \delta \le 1 \quad (3)$$

D. Simulation of electrical stress impact on power integrity

By simulation, the effect of δ on the PI can be studied. Fig. 9 presents the evolution of the peak-to-peak amplitude of the power supply bounce when δ is increased in both current sources. The result confirms that the power supply bounce is reduced when the current pulse is spread. However, we need to confirm that our approach is able to model the evolution of the power supply voltage waveform by a comparison with the on-chip measurements done in stressed cores.

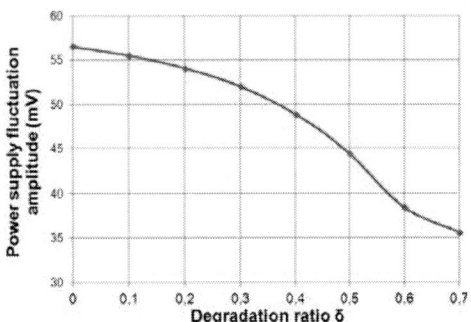

Figure 9.   Simulated evolution of the peak-to-peak amplitude of the power supply voltage bounce vs. degradation ratio

Except if information about transistor degradation is available, the coefficient δ value for a given amount of stress is unknown initially. However, the curve shown in Fig. 9 can help us to choose a reasonable value to fit with EME measurement. From on-chip measurement results, a value of δ is selected to obtain the same voltage fluctuation amplitude. Fig. 10 presents the comparison between the measurements and simulations of the power supply voltage bounces before and after a 120 minutes 3.6 V electrical stress. The coefficient δ is set to 0.58 to model the effect of the stress in the ICEM model. The correlation between measurement and simulation curves is acceptable. The ICEM model modified by the empirical degradation ratio is able to reproduce the evolution of the power supply voltage bounce with a reasonable accuracy.

E. Simulation of electrical stress impact on conducted emission

In order to evaluate the relevance of our modeling methodology, the conducted emission of the core after a 120

minutes 3.6 V electrical stress is simulated with the modified ICEM model. The degradation ratio value is also set to 0.58. Fig. 11 presents the comparison between measurement and simulation results which are in a good agreement up to 1 GHz. The simulation reproduces the observed reduction of the CE spectrum at high frequency.

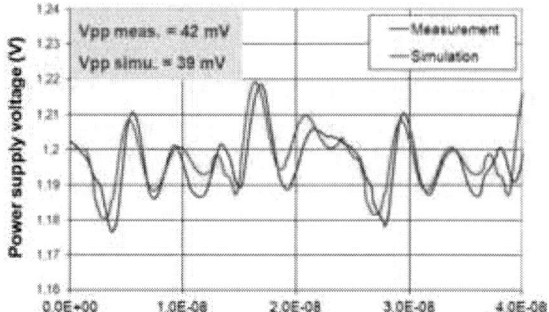

Figure 10.   Comparison between measured and simulated power supply voltage bounce after 120 minutes of 3.6 V electrical stress

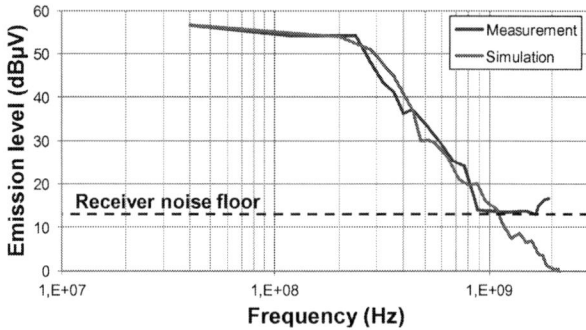

Figure 11.   Comparison between the measurement and simulation of the digital core conducted emission after 120 minutes of 3.6 V electrical stress

The ICEM model is reused to simulate the effect of a different value of stress voltage. The following figures compare the measurement and the simulation of the evolution of the CE spectrum with time when the core is exposed to a 3 V stress. According to the measured power supply voltage bounces and the graph presented in Fig. 9, the degradation ratio values are 0.4 and 0.5 for 180 and 240 minutes respectively. The model is able to reproduce with a quite good accuracy the time-dependent reduction of the CE spectrum up to 1 GHz.

F. Evaluation of the degradation ratio

The prediction of the evolution of EME of an IC with ICEM relies on an accurate evaluation of the degradation ratio. It can be extracted from data about transistor degradation mechanisms and an electrical model of the circuit to simulate how the transient current is spread. In this case study, the experimental data about NBTI helps us to evaluate the threshold voltage increase vs. stress time, for a given electrical stress condition. SPICE simulations on the digital core netlist are performed to simulate the evolution of transient current consumption and to estimate the degradation ratio evolution vs. the stress time, as shown in Fig. 13. The evolution of δ is correlated to the change of threshold voltage and follows a power law as given by equation (1).

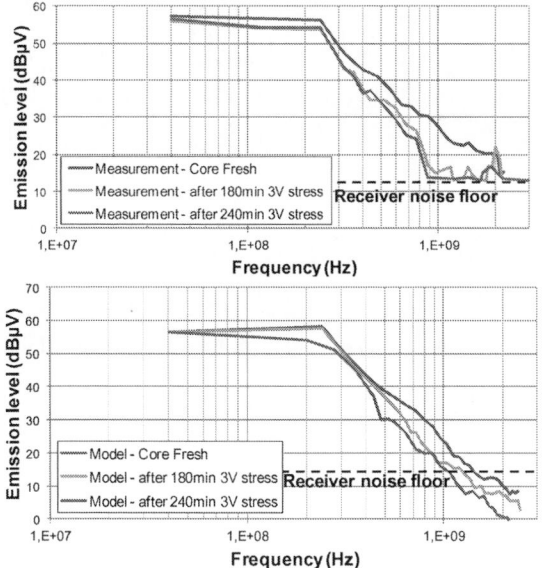

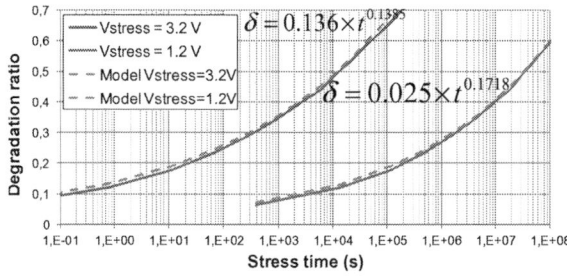

Figure 12. Measurement (top) and simulation (bottom) of the evolution of the CE spectrum of the digital core exposed to a 3 V electrical stress

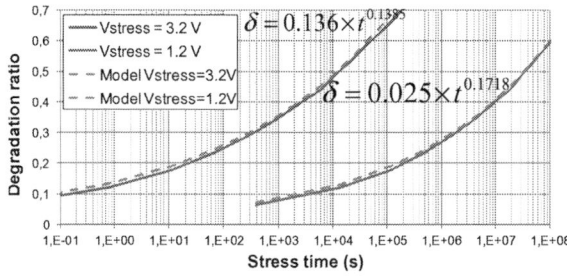

Figure 13. Simulated evolution of the degradation ratio vs. stress time

For example, according to Fig. 8, a 3 V stress applied for 4 hours leads to an increase of V_{TH} of 19 %. SPICE simulation of the digital core shows that the transient current amplitude is divided by 2 while its duration is nearly multiplied by 2. This current spread can be characterized by a degradation ratio equal to 0.5, which correlates with the choice made in the previous part to simulate the evolution of CE. According to V_{TH} evolution equation, under a nominal power supply voltage equal to 1.2 V, the same variation of V_{TH} and, thus the same PI and CE change, would be obtained after nearly 2 years.

V. DISCUSSION AND PERSPECTIVES

The simulation results confirm our hypothesis about the origin of the evolution of the PI and CE after circuit aging. The degradation mechanisms induced at MOS device level, especially the NBTI mechanism, tend to spread the current pulses produced by the switching of the gates of the digital circuit. It leads both to a decrease of the power supply voltage bounce and a reduction of the conducted emission spectrum, especially the contribution of high frequency harmonics. Moreover, these results demonstrate that the evolution of the power integrity can be predicted by a correct modeling of the change of the circuit current consumption. A complex

methodology based on an accurate prediction of the power consumption from the circuit netlist combined with the modeling of MOS device degradation mechanisms could provide an accurate estimation of the evolution of the power supply voltage bounce. However, in this paper, it has been demonstrated that a more simple approach based on an ICEM model can also a reasonable estimation of the evolution of the parasitic emission and power supply voltage bounce. A single empirical parameter called degradation ratio is introduced in the model to take into account the current spread induced by the IC aging. This parameter can be estimated from data about degradation mechanisms in the considered technology and the electrical model of the tested circuit.

Modeling IC emission and integrating the effect of aging is fundamental in order to predict the emission at a larger scale, e.g. a board which integrates several ICs and numerous passive devices, whose characteristics may change with time. The perspectives of this work are threefold: firstly, validating the proposed method on more complex circuit and verify if a unique degradation ratio is enough to predict EME evolution with ICEM model; secondly, the development of a simple method based on simulation aiming at extracting the degradation ratio according to the stress conditions and duration without a complete simulation of the tested circuit; thirdly, the integration of technological process dispersion which adds a non negligible uncertainty to the prediction of the evolution of the EME.

REFERENCES

[1] J. Srinivasan, S. V. Adve, P Bose, The impact of technology scaling on lifetime reliability", The International Conference on Dependable Systems and Networks, (2004).

[2] J. W. McPherson, "Reliability trends with advanced CMOS scaling and the implications for design", 2007 IEEE CICC Conference.

[3] M. White, Y. Chen, "Scaled CMOS technology reliability user's guide", NASA WBS 939904.01.11.10, 2008, http://nepp.nasa.gov.

[4] I. Montanari, A. Tacchini, M. Maini, "Impact of thermal stress on the characteristics of conducted emissions", IEEE Int. Symp. on EMC, 2008

[5] A. Boyer, A. C. Ndoye, S. Ben Dhia, L. Guillot, B. Vrignon, "Characterization of the evolution of IC emissions after accelerated aging", IEEE Trans. on EMC, vol. 51, no 4, Nov. 2009, pp 892 – 900.

[6] S. Ben Dhia, A. Boyer, B. Li, A. C. Noye, "Characterization of the electromagnetic modelling drifts of a nanoscale IC after accelerated life tests", Electronic Letters, 18th February 2010, Vol. 46, no. 4.

[7] IEC62433-2, "Models of Integrated Circuits for EMI behavioral simulation – ICEM-CE, ICEM Conducted Emission Model", International Electrotechnical Commission, Geneva, Switzerland, 2006

[8] A. Boyer, S. Ben Dhia, C. Lemoine, B. Vrignon, "Characterizing circuit susceptibility with on-chip sensors", AP-EMC, May 2012, Singapore.

[9] B. Li, "Study of aging effects on electromagnetic compatibility of integrated circuits", Thesis, University of Toulouse, December 2011.

[10] IEC 61967-4 Ed. 1.1, "Integrated circuits - Measurement of conducted emissions - 1 Ω/150 Ω direct coupling method", IEC, Geneva, Switzerland, 2006.

[11] S. Chakravarthi, A. T. Krishnan, V. Reddy, C. F. Machala, S. Krishnan, "A comprehensive framework for predictive modeling of negative bias temperature instability", IEEE Proc. IRPS, pp. 273-282, 2004.

[12] M. V. Dunga, et al., « BSIM46.1 MOSFET Model – Suser's Manual », 2007 UC Berkeley.

System-ESD Validation of a Microcontroller with External RC-Filter

Thomas Steinecke [1], Markus Unger [1], Stanislav Scheier [2], Stephan Frei [2], Josip Bačmaga [3], Adrijan Barić [3]

[1] *Infineon Technologies AG, Germany;* [2] *Technical University of Dortmund, Germany;* [3] *University of Zagreb, Croatia*

thomas.steinecke@infineon.com

Abstract — **Although microcontrollers are generally well separated from ESD events happening on a fully equipped and mounted electronic control unit, special configurations expose some microcontrollers to these system-ESD events. In the BISS IC EMC Test Specification [1], several system-level disturbance tests are referenced. One of them is the unpowered system-ESD test according to the international standard ISO 10506 [2]. Automotive companies request that microcontrollers and other ICs shall withstand e.g. 6 kV system-ESD stress applied to IC-pins either directly or via discrete protection components. This paper describes the experience made with a 65 nm CMOS 32-bit microcontroller including an external ESD protection filter when exposed to normative system-ESD pulses. Although not expected, discrete SMD protection capacitors degraded or even showed short-circuits after being exposed to several ESD events.**

I. INTRODUCTION

While many integrated circuits (IC) are designed to withstand high ESD voltages, this is not the case for automotive microcontrollers. A major reason for this situation is the expensive additional chip area required for large protection structures. According to the Q100 standard, microcontrollers fulfil the common HBM and CDM ESD requirements which are linked to packaging and handling.

System-ESD describes an electrostatic discharge event which happens when the microcontroller is already soldered on an application board inside an electronic control unit (ECU), which is mounted inside a vehicle. The ESD pulse is generated either by a human or a tool touching an exposed wire or metallic piece which is electrically connected to a microcontroller pin. A typical application is a sensor somewhere in the car, whose signal line is connected via a cable to an analog input of the microcontroller. Such analog/digital converter (ADC) input pins are often connected in this manner. Also digital input pins may be used as battery monitor or other functions which require that cables from the harness are connected to microcontroller or ASIC pins which are not designed to withstand such high discharge voltages.

The solution is to connect one or more discrete external protection elements to the IC pin exposed to ESD. A cheap filter solution is the low-pass combination of a capacitor and a series resistor, see Figure 1. The capacitor is placed very close to the respective PCB connector pin and is expected to conduct the majority of the ESD energy to ground. Nevertheless, a significantly high voltage will still remain for a short time over this capacitor. A current-limiting resistor will care for further energy reduction. The energy which appears finally at the microcontroller pin shall then be low enough to avoid any over-voltage or thermal damage of the IC.

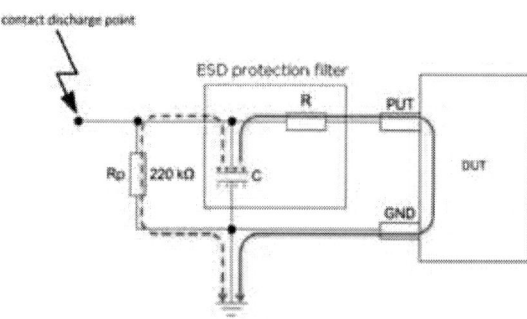

Figure 1 Electrostatic discharge path over RC-filter to IC-pin

Originally, components like capacitors and resistors are not designed to protect against ESD. The majority of capacitors used on PCBs have voltages rated between 50 V and 200 V. In case of ESD the capacitor has to withstand voltages up to several kV and capacitors operate far above the absolute maximum ratings specified in their datasheets.

It is generally known that MLCCs reduce their capacitance at higher voltages. It is not always possible to counteract this phenomenon by a MLCC of higher capacitance, because often the IC pins to be protected have to deal with high bitrate signals. A compromise is a lower capacitance which provides higher speed but less ESD protection. In [10] an improvement of the trade-off using an antiferroelectric capacitor as ESD-protection is proposed. The dielectric constant of the hand-made capacitor increases with rising voltage.

Repeated exposure of MLCC to ESD can cause irreversible parameter shift on capacitors. The effects of derating on RF properties were investigated in [4]. A worst case out of four X7R 4.7 nF capacitors has shown a capacitance drop to 75 % of the nominal value, measured at 13.5 MHz. However, no modeling approaches for observed degradation effects were proposed. Physical damage to MLCCs exposed to ESD stress was examined by authors in [11] [12]. It was shown that ESD may cause a permanent damage to dielectric material. Significant differences in ESD performance of MLCCs from several manufacturers have been measured. The metallization of conductive plates within a capacitor results in a non-recoverable shift of insulation resistance of several MΩ to a lower value in the kΩ range. This will have an effect on the energy consumption of a system, and low current communication systems could be disturbed.

978-1-4799-5004-1/13 $31.00 © 2013 IEEE

II. SYSTEM-LEVEL ESD TEST SETUP ACCORDING TO BISS

A. ESD test place

To handle the huge variety of integrated circuits, the "Generic IC EMC Test Specification" – also known as "BISS paper" [1] – provides all information necessary to perform IC-level electromagnetic compatibility tests. The new, yet unpublished version 2.0 addresses two system-level tests: automotive pulses according to ISO 7637 and system-level ESD tests according to ISO 10605.

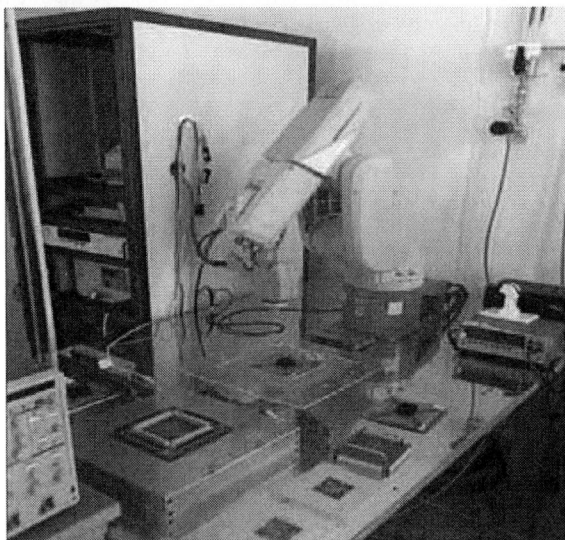

Figure 2 Automatic robot-controlled ESD test place

Microcontroller system-ESD measurements were performed in the frame of a master thesis [2]. The setup for conducted system-level ESD tests consists of an ESD gun, a high-ohmic discharge resistor, and the possibility to either monitor the IC function during the test (in case of powered measurement) or to perform a pre-stress and a post-stress read-out test to check if the device under test's (DUT) AC/DC parameters and functions are still in spec. Figure 2 shows an automatic robot ESD test place, holding an ESD gun and a platform for the DUT test board. Pogo-pins can connect the DUT pins at the bottom test board side with channels for U/I-characteristics measurements. This way, a pin degradation can immediately be detected.

B. Microcontroller test boards

The test boards contain the microcontroller (DUT) itself, plus a lot of discrete ESD protection filters. The trace width is selected to match real-life designs. The board can be assembled with different package sizes of SMD capacitors and resistors. Figure 3 shows the system-ESD test board designed for the 65 nm CMOS 32-bit microcontroller Aurix TC275L in an L-QFP-176 package.

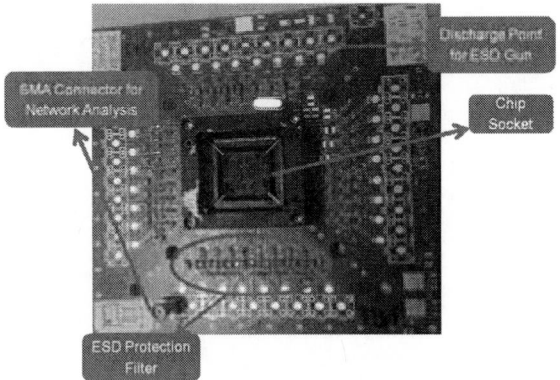

Figure 3 Top view of microcontroller ESD test board

We decided to go for a socketed test because soldered DUTs cause big trouble when they need to be removed and replaced by a new device. Since the ESD pulses first propagate through the RC low-pass, the higher frequency content is already suppressed before the remaining pulse reaches the socket pin. Therefore, the microcontroller socket parasitics will not affect the pulse shape on its way from the socket pin to the IC pin.

III. TEST RESULTS

Time domain simulations of the ESD test setup showed a sufficient grounding of the energy injected by the ESD gun over the SMD capacitor, such that no spark-over or filter component degradation was expected. The simulation model consisted of sub-models for the ESD gun, the filter capacitor and resistor, and transmission lines as well as via holes along the discharge path. The gun model according to the international system-ESD test standard ISO 10605 (ESD-gun with 150 pF charging capacitor and 330 Ω discharge resistor) [3] and the resulting discharge pulse are shown in Figure 4. The SMD components are modelled including their parasitic behaviour, such as ESR and ESL.

Figure 5 shows the simulation results for the maximum voltage which occurs during an ESD event over the capacitor. We use the simple simulation model consisting only of the ESD-gun model [4, 5] and the capacitor model (including parasitics) instead of a full ESD path model as shown in Figure 1. This simplification is valid because the additional discharge path through the series resistor and the microcontroller pin is high-ohmic and limits its partial discharge current significantly.

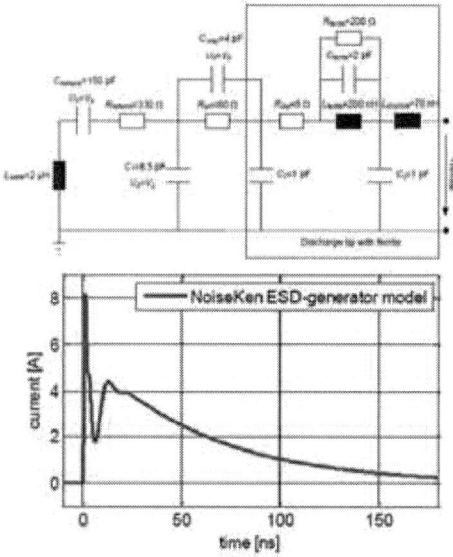

Figure 4 ESD gun model and resulting discharge pulse

Values from 1 nF up to 10 nF have been considered; typical values used for ESD protection capacitors are 4.7 nF or 10 nF. For these values, the simulated voltages stay well below 300 V and thus should neither lead to a sparkover nor to degradation.

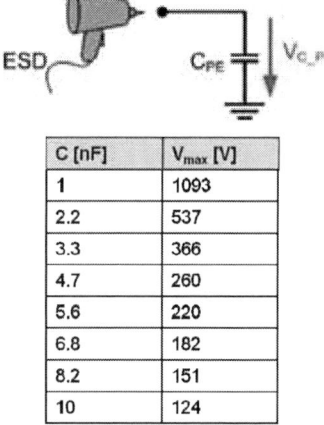

C [nF]	V_{max} [V]
1	1093
2.2	537
3.3	366
4.7	260
5.6	220
6.8	182
8.2	151
10	124

Figure 5 Simulated peak voltages across ESD protection capacitors

Virgin SMD capacitors have a series DC resistance far above 10 MΩ. After being ESD-stressed several times in the robot setup, the filter capacitors showed a significant degradation. This was verified by measurements which indicated a decreased DC resistance of ca. 3 kΩ after a few zaps, or even a conducting channel of <1 Ω after several zaps, equivalent to an EOS damage. Additionally, sparkovers were observed.

Since significant differences in the ESD performance of SMD capacitors were reported and no systematic characterization and modelling approaches were proposed, we started a detailed research on the behavior of the selected samples when exposed to high voltages, i.e. electrical overstress. Previous publications reported irreversible damages on capacitors [6] [7].

First, we measured the required voltage to cause a sparkover, as a function of contact distance. Therefore, a test PCB was built which contained different shapes and distances of sparkover contacts to cover the range of SMD component sizes. Depending on the geometry and the voltage sequence, sparkover started at voltages >2 kV, see Figure 6.

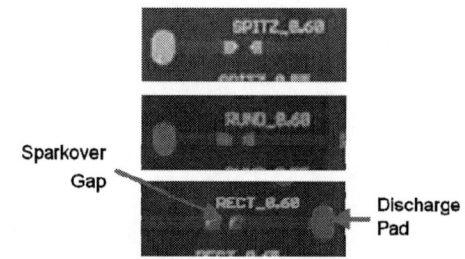

Pad geometrie	0603 (0.6mm)			1206 (1.15mm)		
	V_{BD_up} [kV]	V_{BD_down} [kV]	V_{BD_rand} [kV]	V_{BD_up} [kV]	V_{BD_down} [kV]	V_{BD_rand} [kV]
Rect	2.5	2.3	2.6	2.5	2.5	2.6
Round	2.7	2.4	2.7	2.9	2.9	3.4
Peak	2.5	2.7	3.0	2.8	3.0	3.4

Figure 6 Measured sparkover voltages through air

In a second step, stand-alone SMD capacitors were exposed to conducted discharge pulses, see Figure 7. The sparkover gaps on the PCB contained no solder resist material to avoid any discharge bypass.

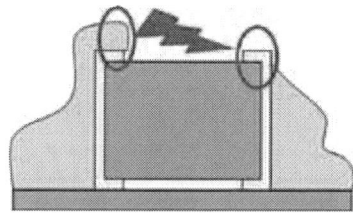

Figure 7 Measured sparkover voltages over SMD capacitors

With this test setup, the previously after the robot ESD tests observed SMD capacitor degradation and damage were confirmed. Sparkovers were observed on a 0603-sized 10 nF / 200 V ceramic SMD capacitor for discharge voltages down to 650 V. In addition, ESD pulse sequences caused insulation defects, resulting in approximately 3.5 kΩ insulation resistance, i.e. 1.5 mA leakage current at 5 V. Furthermore, multiple ESD zaps caused EOS defects, i.e. fully conducting channels (resistance less than 1 Ω) within the capacitors' insulation regions.

978-1-4799-5004-1/13 $31.00 © 2013 IEEE   198

IV. CERAMIC CAPACITOR BEHAVIOUR DESCRIPTION

A. Ceramic capacitor characteristics

These results led to the conclusion that the common SMD capacitor model, consisting of the capacitance plus equivalent series resistance and inductance is no more valid in voltage regions above the capacitor's maximum operating voltage. Unfortunately, we do not know any ceramic SMD manufacturer who specifies electrical parameters above the components' absolute maximum ratings. Nevertheless, it is very common to use standard SMD capacitors in all kinds of electronic systems to protect logic ICs from ESD damage, i.e. operate them for short time intervals above their regular operating conditions.

The strange consequence is that multiple ESD zaps do not destroy the IC, but the protection capacitor itself. When becoming low-ohmic, the capacitor conducts a higher amount of the ESD energy towards ground and thus protects the IC even better. On the other hand, a low-ohmic connection of a signal line to GND will likely prohibit proper function because of signal distortion. Thus the risk of a non-reversible system failure is high.

Since electronic units are commonly ESD-stressed only a few times, the protection capacitors have to withstand typically only one to three zaps. There is a chance that this low number of ESD events may not cause any severe capacitor degradation or damage.

To be on the save side, alternate component selections should be considered. For example, larger package sizes reduce the risk of capacitor ESD damage. We found that 0805 SMD sizes can be used up to 6 kV ESD events without causing any degradation.

Nevertheless, we were interested to find the explanation for this capacitor degradation when operated above the specified maximum voltages.

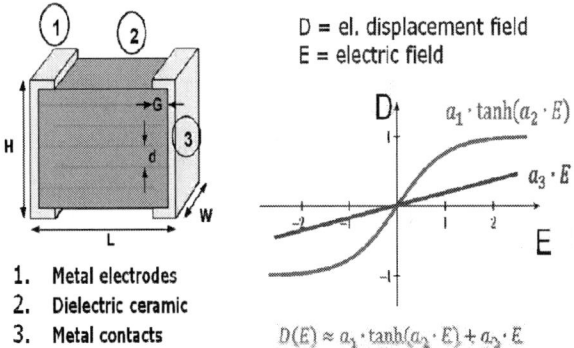

1. Metal electrodes
2. Dielectric ceramic
3. Metal contacts

Figure 8 Multilayer ceramic SMD capacitor and its field characteristics

A literature study [8, 9] revealed the following facts:

Typical ceramic SMD capacitors use a multi-layer ceramic insulator technology (MLCC). The insulator material is Barium titanate ($BaTiO_3$) plus vendor-specific additives. Pure Barium titanate is ferroelectric; the additives remove those ferroelectric properties. The spontaneous polarization is zero, and it shows no hysteresis.

If the electric field across the capacitor increases, the resulting electric displacement field rises, but soon reaches a saturation state. An MLCC and its displacement field are shown in Figure 8.

B. Simulation model improvement

These findings led to the following conclusion:

- The MLCC capacitance is voltage dependent; i.e. the voltage over the capacitor depends on the collected charge, and the same charge on a smaller capacitance leads to a higher voltage.

From this capacitor behaviour, some assumptions were deducted:

- The real capacitance is smaller than the ideal capacitance.
- The voltage across the capacitor gets higher than expected.
- The observed sparkover may be confirmed by an improved capacitor model.

This advanced capacitor model is shown in Figure 9. We changed the fixed capacitance into a voltage-dependent variable capacitance. We added an insulation resistor in parallel to the variable capacitor. Furthermore, a spark gap path was introduced.

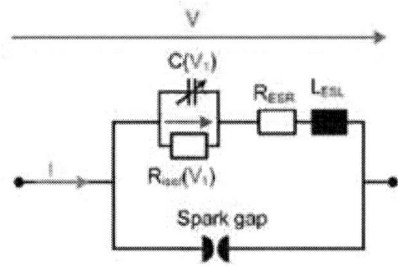

Figure 9 Advanced SMD capacitor model

In a next step, the parameter values for the advanced capacitor model had to be calculated. This was done by a measurement setup according to Figure 10, using a transmission line pulser, a current clamp to measure the time-variant ESD current, and a 6 GHz oscilloscope to measure the time-variant ESD voltage. A voltage divider is used to protect measurement devices.

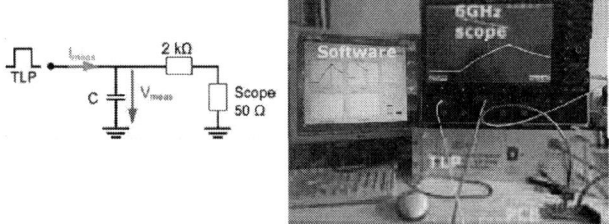

Figure 10 Test place for the voltage-dependent capacitance

978-1-4799-5004-1/13 $31.00 © 2013 IEEE

The calculation of the voltage-dependent capacitance took place in two steps:

- Step 1: Calculate the voltage-dependent charge.
- Step 2: Fit the capacitor model parameter values for the variable voltage according to the measurement results.

Basis for the parameter calculations were the measured time-variant voltages and currents. Voltage over time was directly measured (Figure 11a). The charge on the capacitor was calculated as $Q=\int I\,dt$ (Figure 11b). Taking voltage and charge from the same time stamps, the charge over voltage diagram (Figure 11c) was calculated.

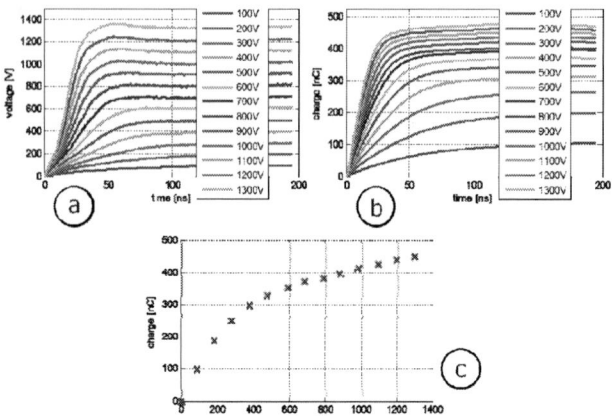

Figure 11 Charge calculation from measured voltages and currents over time

The link between the paraelectric polarization of the ceramic material (Figure 8) and the DC-bias curve given in a capacitor's data sheet can be explained with a parallel plate capacitor model. Assuming a constant layer distance, the electric field is defined as:

$$E = \frac{V}{d} = \frac{voltage}{layer\ distance} \propto V \tag{1}$$

and the electric displacement field as:

$$D = \frac{Q}{A} = \frac{charge}{area} \propto Q \tag{2}$$

where Q is the accumulated charge. Therefore the polarization curve is proportional to the QV-curve of a capacitor. Using the capacitance definition:

$$C = \frac{dQ}{dV} \tag{3}$$

it can be seen that the DC-bias curve or the CV-curve is proportional to the derivative of the QV-curve, or to the polarization curve of the ceramic. That means all three description options are qualitatively equivalent, and may be

used for high-voltage characterization of MLCCs using equation 4b.

$$Q(V)\big|_{nC} = b_1 \cdot tanh\left(\frac{V}{b_2}\right) + b_3 \cdot V \tag{4a}$$

$$C(V)\big|_{nF} = \frac{dQ}{dV} = \left(b_3 - \frac{b_1 \cdot \left(tanh^2\left(\frac{V}{b_2}\right) - 1\right)}{b_2}\right) \tag{4b}$$

Figure 12 shows the voltage dependencies of 10 nF SMD capacitors in a 0603 package (Fig. 12a) and in a 0805 package (Fig. 12b).

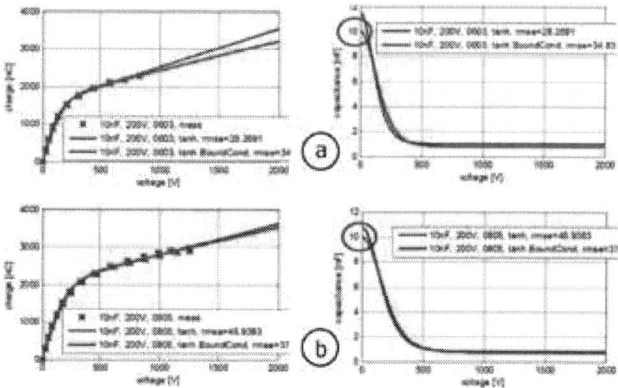

Figure 12 Capacitance over voltage for different SMD package sizes

Using these voltage-dependent capacitance characteristics, for an ESD event with its measured voltage/current characteristics, the maximum occurring voltage can be calculated. The table in Figure 13 lists these maximum voltages for a set of SMD capacitors which are widely used as ESD protection capacitors.

Capacitor	ISO10605 ESD Network: 150pF, 330Ω	
	V_{cap} for 8kV [V]	V_{cap} for 8kV [V] 25% derating
4.7nF 0603 50V	1450	2080
4.7nF 0603 200V	670	1170
4.7nF 0805 50V	325	530
4.7nF 0805 200V	320	500
10nF 0603 50V	700V	1100
10nF 0603 200V	150	260
10nF 0805 50V	177	360
10nF 0805 200V	140	205

Figure 13 Maximum voltages over SMD capacitors during ESD events

Detailed information about the presented characterization and modelling approach was published in [13].

978-1-4799-5004-1/13 $31.00 © 2013 IEEE

C. Microcontroller ESD test results

After knowing about the reliability weaknesses of ceramic SMD capacitors under over-voltage conditions, we repeated our microcontroller system-ESD tests with robust 0805-sized protection capacitors. The device under test was the 65 nm CMOS 32-bit microcontroller Aurix TC275L in an L-QFP-176 package. Up to 6 kV ESD stress according to ISO 10605, the microcontroller did not fail, and the ESD protection capacitors did neither degrade nor were they damaged.

V. CONCLUSIONS

According to Figure 13, SMD package sizes smaller than 0805 lead to high voltages over the protection capacitor which can cause sparkover, degradation or EOS damage at least after several zaps. Therefore, we recommend to use only 0805 or larger SMD package sizes for ESD protection capacitors. Alternatively, two 0603 capacitors can be connected in series, or special high-voltage ESD capacitors can be chosen. However, reliable SMD capacitor behaviour can only be achieved by spending PCB space or cost.

If that space or cost cannot be spent, it must be clear that multiple ESD events will likely degrade or even short SMD capacitors, which will increase the system's current consumption and distort the signals which are transmitted over the nets to be protected.

VI. ACKNOWLEDGMENTS

The work described in this paper was conducted in the frame of the European Catrene funding project EM4EM by the project partners Infineon Technologies AG and the Technical University of Dortmund. The project is funded on national level by the German BMBF (Infineon: Fkz 16M3092F; TU Dortmund: Fkz 01M3092I).

REFERENCES

[1] Bosch, Continental, Infineon, *Generic IC EMC Test Specification. Ed. 2.0*, not yet published.

[2] J. Bačmaga, *Influence of Board Geometries and External Protection Network Configurations on the Automotive Microcontroller Behaviour when Exposed to System-related Electrostatic Discharge*, Master Thesis, Zagreb, September 2012.

[3] *ISO 10605 2nd Ed. 2008-07-15: Road vehicles - Test methods for electrical disturbances from electrostatic discharge.*

[4] F. Streibl, *Electrostatic Discharge Performance of Passive Surface-Mount Components*, Institut für Energieübertragung und Hochspannungstechnik, Stuttgart, 2011.

[5] L. Müller and K. Feser, *Untersuchung von Gleitentladungen und deren Modellierung durch Funkengesetze im Vergleich zu Gasentladungen*, EMV Düsseldorf, 2000.

[6] F. zur Nieden, B. Arndt, J. Edenhofer, and S. Frei, *Vergleich von ESD-System-Level Testmethoden für Packaging und Handling*, ESD-Forum 2009, 2009.

[7] F. zur Nieden, Y. Cao, B. Arndt, and S. Frei, *Vergleichbarkeit von ESD-Prüfungen auf IC- und Systemebene oder welchen Einfluss hat eine Reduzierung der ESD-Festigkeit auf die Systemfestigkeit?*, EMV-Düsseldorf, 2010.

[8] Gordon R Love, *Energy Storage in Ceramic Dielectrics*, Journal of the American Ceramic Society, February 1990.

[9] Stanislav Scheier, Stephan Frei, *Analysis of Passive ESD-Filters*, Technical Report, Technical University of Dortmund, February 2013.

[10] Hongyu Li; *Nonlinear capacitors for ESD protection*, Electromagnetic Compatibility Magazine, IEEE, 2012

[11] Rostamzadeh, C.; Dadgostar, H.; Canavero, F. *Electrostatic Discharge analysis of Multi Layer Ceramic capacitors*, Electromagnetic Compatibility, 2009. EMC 2009. IEEE International Symposium

[12] Demcko, R.; Ward, B. *MLCC ESD characterization*, CARTS 2007 Symposium Proceedings, Albuquerque

[13] Stanislav Scheier, Stephan Frei, *Characterization and Modeling of ESD-Behavior of Multi Layer Ceramic Capacitors*, EMC Europe, Brugge, 2013

Automatic verification of EMC immunity by simulation

B. Vrignon, P. Caunegre, J. Shepherd
Freescale Semiconductor
Toulouse, France
bertrand.vrignon@freescale.com

Jianfei Wu
School of Electronic Science and Engineering
National University of Defense Technology
410073 Changsha, Hunan, China

Abstract—**Immunity of analog circuit blocks is becoming a major design risk. This paper presents an automated methodology to simulate the susceptibility of a circuit during the design phase. More specifically, we propose a CAD tool which determines the fail/pass criteria of a signal under direct power injection (DPI). This contribution describes the function of the tool which is validated by a LDO regulator.**

Keywords: Circuit simulation, immunity, susceptibility, direct power injection

I. INTRODUCTION

The knowledge about electromagnetic compatibility (EMC) issues is a key element to be successful with design challenges of integration and advanced technology. Nowadays, EMC is one of the main reasons for integrated circuit "re-design" because it is not sufficiently taken into account in the design flow. Moreover, the designer knowledge of noise generation and coupling mechanisms are not always sufficient.

Little information exists about disturbance behavior carried through an integrated circuit [1] and designers are not able to predict easily and rapidly circuit immunity thresholds at the design stage. Generally, characterization of immunity is realized during product qualification according to customer specifications and/or international standards like IEC-62132-4 [2]. When weaknesses are detected by measurements on the manufactured device, the cost of redesign and manufacture may be prohibitive and external solutions may be preferred, even if the customer is unwilling to add additional components to his application.

To cope with this issue, it is necessary to estimate the immunity level of ICs during the design phase. Simulation of EM immunity of ICs allows potential weaknesses to be detected before the product is first manufactured. Short simulation times are therefore of utmost importance. For this challenge, an automated simulation process which determines the critical frequencies and powers is essential. The goal of this design flow is to run and optimize automatically all the simulations, evaluate the pass/fail criteria, and generate a graph with the immunity curve. Moreover, for use in industrial IC design, standard models and tools have to be employed.

In the past, some approaches for simulation-guided EMI predictions are published in [3], [4], [5] or [6]. Reference [3]

uses standard models within design environment. It gives the most sensitive nodes of the circuit by using periodic steady-state methods or harmonic balance. But, this method does not give directly the immunity level of the circuit like in measurements. Reference [4] applies to the device a disturbing signal whose amplitude is modulated by a ramp. The amplitude of the disturbing signal is therefore continuously changing and there is no well defined dwell-time. In [5], the in-house tool needs EMI-specifically upgraded models and in [6], the approach uses AC-DC simulation that is coupled with transient analysis. The validation of these analyses is reported only for small excitations where strong non-linear effects don't appear.

This paper introduces a method of simulation of an immunity test with the automatic default detection. The immunity levels are simulated in the similar way to the immunity measurement methods. First, the challenge to run immunity simulation and the flow developed in this new tool are described. Then, we detail the different parameters and options of the tool and also the possible outputs for result analysis. Finally, we study the immunity of a LDO regulator and compare the simulation and the measurements.

II. DEFINITION OF AN EM IMMUNITY SIMULATION TOOL

A. Challenge

Today, standards exist for the measurement of EM immunity. For example, in the case of integrated circuits, IEC-62132-4 is applied to measurements using continuous wave (CW) and amplitude modulated (AM) sine wave disturbing signals. Disturbing signal is applied to the device under test (DUT) and its amplitude is increased stepwise until a default is detected. A default may be a change in an output signal (e.g. amplitude, DC level, phase shift, jitter, frequency, etc), a change of state (e.g. a digital signal passing from a low to a high state, or vice-versa) or any other indication of a malfunction of the device. As soon as a default is detected, the level of the disturbing signal is noted and the measurement is carried out again by changing a parameter such as frequency of a sine wave. The signal level is reset to a suitable minimum level. In the case of measurements on a real device a maximum disturbing signal level is specified and the measurement is stopped when this level is reached. It is easily understood that the level of the disturbing signal cannot be further increased without a risk of damage to the device. The figure 1 presents

978-1-4799-5004-1/13 $31.00 © 2013 IEEE

the flowchart of immunity measurements as described in IEC62132-4.

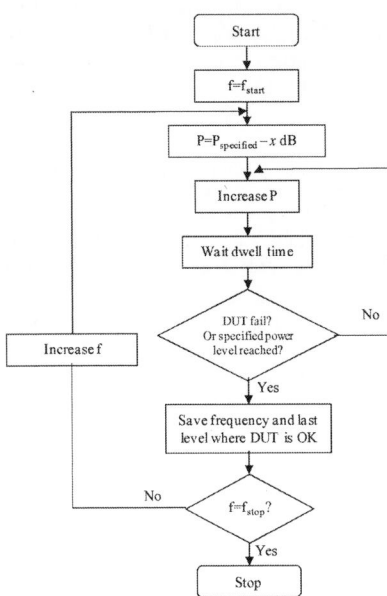

Figure 1. Flowchart of a direct power injection (DPI) test procedure [2]

The default may be detected by the simple measurement of a signal level (amplitude or DC), a frequency, etc. In cases where the default is a difference between the nominal output signal (i.e. the signal when no disturbing signal is applied) and the disturbed signal, limits are placed around the nominal signal. Fig. 2 shows the limits around a typical signal.

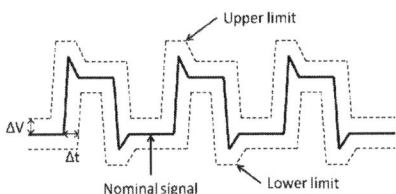

Figure 2. Example of limits around a typical signal

In this example the amplitude of the observed signal may not vary by more than ΔV and its timing by more than Δt. Fig. 3 presents the detection of a failure when the disturbed signal goes outside the limits.

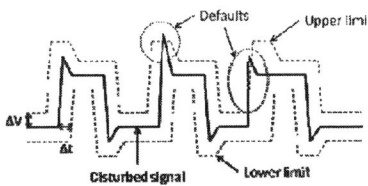

Figure 3. Example of failure detection

EM immunity measurements are generally automated by controlling the measurement equipment remotely with a controller, such as a personal computer. In order to be certain that a default condition will be detected, a dwell-time is specified during which the controller permanently checks for a default at a given disturbing signal level. The dwell-time may last from a fraction of a second to several seconds. This leads to excessively long measurement times. In order to reduce the measurement time, it is usual practice to stop the measurement as soon as a default is detected, rather than continuing to the end of the dwell-time. In the case of a measurement according to IEC 62132, the level of the disturbing signal for the next frequency is not reduced to the minimum level, but to a level only slightly below the level at which the default was previously detected (x dB in Fig. 1). This avoids having to measure at levels where there is little chance of finding a default, thereby reducing the measurement time.

In the case of the simulation of EM immunity, it is usual practice to run a complete simulation before looking for a default condition in the resulting data. The simulation is therefore very time consuming. The resulting amount of data is enormous and the post processing is also very long. This is worsened when the frequency of the disturbing signal is many times that of the disturbed signal, or vice versa. In a transient (time domain) simulation the number of points to be simulated depends on the resolution required (i.e. number of points for one cycle of the highest frequency to be simulated). For example, with a disturbing signal at 1GHz and an observed signal at 1MHz, 1000 cycles of disturbing signal must be simulated for one period of observed signal. Moreover, if a resolution of 10 points per cycle of disturbing signal is required and ten cycles of observed signal are needed to ensure good detection of a default (dwell-time), the number of points reaches 100 000. This simulation must be run for each level and frequency of disturbing signal. In many modern simulators, the time step is automatically reduced as the slope of a signal increases. This avoids calculation when the signal is changing slowly, but in the case of a sine wave, this is not particularly advantageous.

The aim of the tool [7] presented in this paper is to apply to EM immunity simulations similar methods to those used to reduce the time of EM immunity measurements. In order to achieve this, the simulator must include functions allowing default conditions to be detected "on-the-fly" and allowing the simulation to be stopped or paused and restarted with new conditions.

B. Methodology

The tool [8] provides a method of simulation of an immunity test with the automatic default detection. The immunity levels are simulated in the similar way to the immunity measurement methods. To go faster, all steps are automated and when a default is detected, the immunity simulation is stopped immediately and then goes to the next frequency. The method can be applied for all the immunity tests: radiated, conducted and impulse. But, it requires an equivalent model of each injection method. For radiated and impulse immunity (near-field scan method), [9] proposes several electrical model of the injection setup that can be reuse in the simulation test bench. Refinements to optimize and reduce automatically the simulation time as a function of the previous simulations are included.

978-1-4799-5004-1/13 $31.00 © 2013 IEEE

Figure 4 shows the automated flowchart. Before applying the disturbance to the device, a first simulation is run in order to save the signal waveform in nominal conditions. During this simulation, the conditions are saved at a predefined time when the simulation is considered to have reached a steady state (settle time). The simulation is then continued to the predefined dwell time. After that, an algorithm calculates automatically the test limits according to the specified tolerances (e.g. ΔV and Δt). Then, the conditions at the settle time are loaded and the simulations are re-run with the disturbing signal up to the dwell time. Finally, an algorithm compares the disturbed signal with the test limits. If a default is detected or if the level reaches the target level, the simulator goes to the next frequency. Otherwise, the level is increased. The increase in level can be adapted to the nearness of the disturbed signal to the limits.

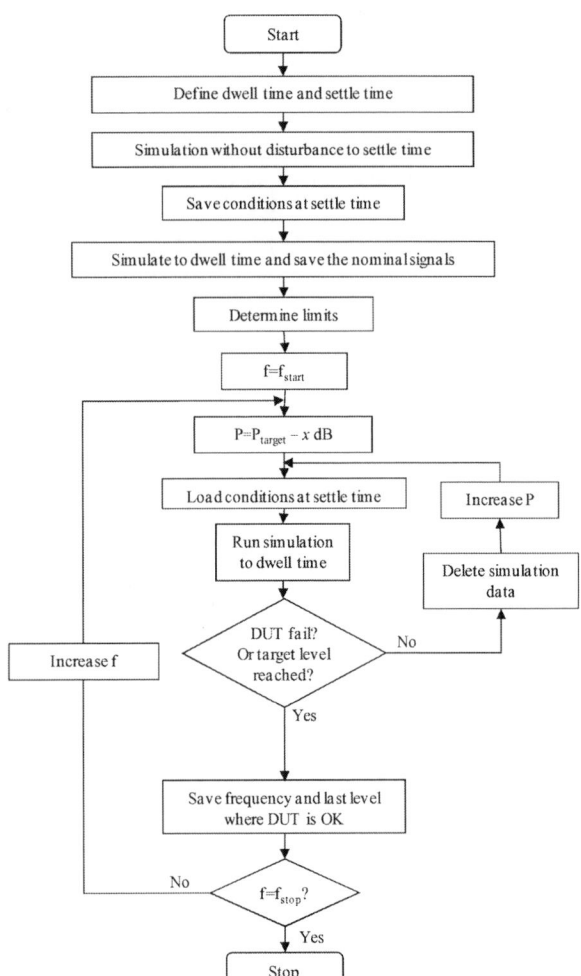

Figure 4.   Automated immunity simulation flowchart

This process can be completely automated, avoiding a tedious manual post-processing. For each level at the end of a simulation run, the disturbed signal is compared to the test limits. If there is not a default the waveform is deleted avoiding storing all the simulation results. So, the process keeps only the last simulation results where the signal goes outside the limits and deletes all the other simulations for a given frequency. This automated simulation flowchart can be easily improved in order to optimize automatically the simulation runtime and the iteration number.

Many improvements can be added to this process to improve the simulation run-time. Firstly, the first applied level can be optimized according to the knowledge of the previous tendencies. For example, if the level at which the default occurs is very similar at each frequency, x can be smaller. On the contrary, when the levels are varying considerably between successive frequency points, x can be increased automatically. Another improvement concerns the simulation dwell-time. It is possible to run the simulation only for n points and not all the dwell time. The disturbed signal may be compared with the limits after that n points have been simulated instead of waiting until the dwell time has been reached.

A last improvement can be implemented by adjusting a delay to the disturbing signal so as to advance the time at which the default occurs. Generally, a default is created by a certain combination of the disturbing signal, disturbed signal (i.e corresponding to the critical states) and in some cases internal conditions of the device. By applying a delay to the disturbing signal it is possible to generate earlier in the simulation the default on the device. An algorithm can change automatically the value of the delay according to the frequencies of the disturbing signal and disturbed signal.

C. Implementation

EMC simulation flow is performed through an EMC toolbox inside an internal in-house tool [8]. This tool is meant to perform automatic verification that an analog and mixed signal circuit meets its specifications over the combined variation of process, temperature, power supply and any other user defined parameters. The description of the simulation flow is performed in a graphical way through a very user-friendly interface that will allow the designer to define each characterization and verification task by adding functional boxes.

The EMC box behaves as a "Loop" box: the simulation (type of analysis) and measurement flow (failure criteria) must be nested inside this box as shown in Fig. 5. The EMC simulation flow consists in iterating over frequency and power of a sinusoidal wave representing the EM input stimuli, and finds the limit where the circuit is no longer immune.

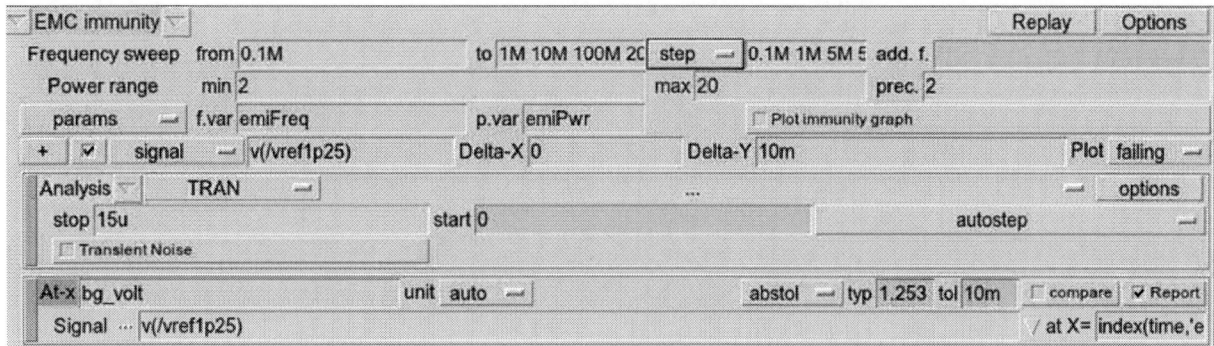

Figure 5. EMC immunity box

III. EMC FLOW SETUP

A. Stimuli definition

The sinusoidal wave generator can either be specified by the designer inside the test-bench or be inserted on the fly during the execution. Power must be expressed in dBm, and this will have to be taken into account when specifying the input stimuli.

First, a conversion between a power expressed in dBm and a voltage sine source magnitude is needed (Fig. 6). It is necessary to take into account that the port model assumes the circuit input impedance is matched with the one of the source, thus introducing a voltage divider bridge.

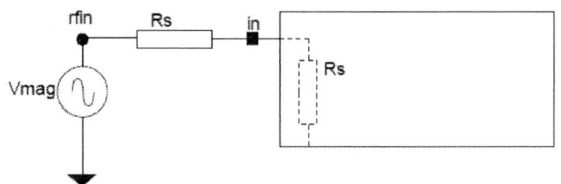

Figure 6. Voltage source model

Voltage delivered by the source must thus be twice the one that will be at the input. In addition, one needs to take into account the fact that converting the power to voltage leads to an *rms* value, whereas the voltage source must be defined with a voltage magnitude (peak value).

This leads to (1):

$$V_{mag} = 2 \times V_{in} = 2 \times \sqrt{2} \times \sqrt{Rs \times P_{in}}$$
$$V_{mag} = \sqrt{\frac{8 \times Rs \times 10^{(PdBm/10)}}{1000}} \qquad (1)$$

B. Sweeping frequency and power

Frequency sweep (Fig. 7) is defined through a from-to range and a step. Additional frequencies can be specified on the "Additional frequencies" field. There are several frequency sweep modes:

- dec: this is a logarithmic sweep where one specifies the number of points per decade

- step: this is a linear sweep where one specifies the frequency step value

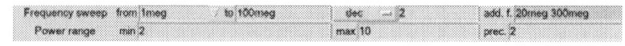

Figure 7. Frequency sweep definition

It is also possible to specify different frequency steps between different ranges.

Power range is defined as a min max range, and is swept using a dichotomous approach in the present version of the tool. The improvements, presented in part II-B, will be implemented later if the run-time is too long. Note that it is possible to define the power range dependent upon frequencies, as a list of frequency power pairs.

C. Immunity criteria

Immunity is computed by analyzing signal deviation or any types of measurements (average voltage, rising/falling time...) compared to the nominal case.

For measure results, their specification limits are being used to check if the results are valid or not as illustrated in Fig. 8.

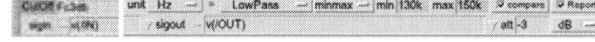

Figure 8. Specification of measurement limit

For signals, a comparison with the nominal case is performed using a 'halo' function analysis in signal match within a given tolerance on X and Y axis. In addition, it is possible to control when the signal must be plotted: 'never', 'always' or 'only when failing'.

D. Execution and result analysis

For the execution, all frequency iterations are run in parallel. For one given frequency, all power dichotomous runs are sequentially executed. The replay mode allows enriching the report with additional measurement data performed at additional frequency or power values.

Finally, a graph (Fig. 9) showing the maximum power on Y axis that can be injected without causing circuit to fail the specifications, for each one the frequency values on X axis.

978-1-4799-5004-1/13 $31.00 © 2013 IEEE   205

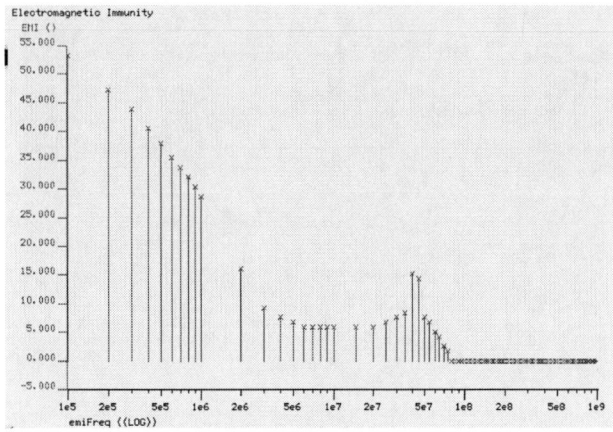

Figure 9. Immunity graph example

A report containing the various reported measures can be also generated.

IV. APPLICATION EXAMPLE: LDO REGULATOR SUSCEPTIBILITY

A. Testchip description

The LDO regulator under test was implemented with CMOS 90nm process and aims at providing a regulated power supply voltage to a small digital core. Fig.10 illustrates the internal structure of the LDO voltage regulator which contains a Kuijk bandgap reference circuit and an output amplifier designed to work with a +3.3 V supply, applied on terminal V_{IN}. The nominal voltage of the bandgap reference voltage is +1.2 V. The gain of the output amplifier is 1. Bandgap reference circuit and regulator outputs are monitored through the terminals V_{REF} and V_{OUT}.

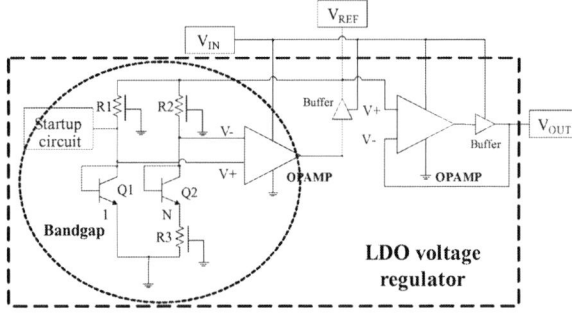

Figure 10. LDO structure and general setup

B. Conducted susceptibility test setup

Measurements are carried out in accordance with IEC standard 62132-4 [2]. The test bench for the DPI measurements is described in Fig. 11. The DC offsets induced on both bandgap reference circuit and regulator outputs were monitored. A deviation of +/- 0.1 V from the RMS output voltage V_{REF} and V_{OUT} is tolerated.

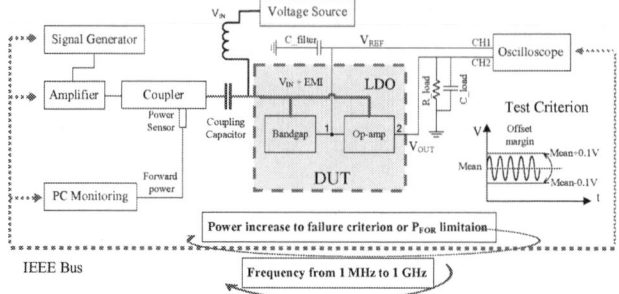

Figure 11. Direct Power Injection test bench

The DPI measurement result is illustrated in Fig. 12. When the EMI is below the op-amp unity gain frequency, the regulator works in the normal mode and V_{OUT} is controlled by a negative feedback circuit. Above this frequency, the op-amp no longer works in the negative feedback regime, which explains the decrease in susceptibility level. With the parasitic effects of the test chip, more EMI noise is transferred to the output and induces more distortion and offset. But at frequencies above 700 MHz, the parasitic effects of bonding, packaging and the PCB will work as filters improving the immunity level.

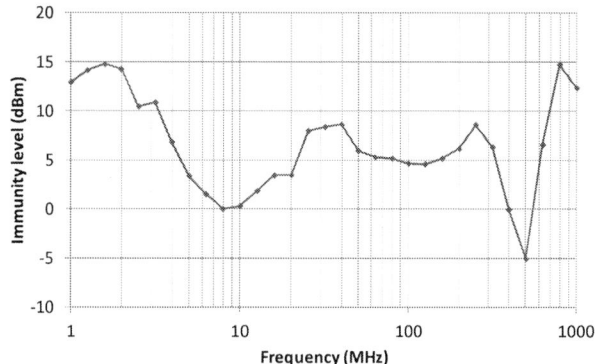

Figure 12. DPI measurements of LDO regulator on Vout

C. Simulations

First of all, we need to model the package and board to compare the measurement with the simulation. For the IC, we use directly the netlist of the regulator without any simplification. Between Vddreg pin and the SMA connector, there is an injection path (about 5 cm) for EMI propagation. As the characteristic of the injection path is very important to DPI simulation, the characteristic of the path is extracted from the PCB board design information. Electrical parameters as shown in Fig. 13 are extracted from analytical formulations [10] or by using electromagnetic simulators. The parameters are used for PCB track model building in ADS.

Fig. 14 describes the model of the PDN which includes the Vddreg/Vssreg package, bonding and coupling capacitor, decoupling capacitor and resistor, substrate resistor and capacitor. This model predicts the amount of noise which is coupled into the LDO circuit because PDN acts as filtering element ADS is used to simulate the model and compare the results with S-parameter measurements.

978-1-4799-5004-1/13 $31.00 © 2013 IEEE

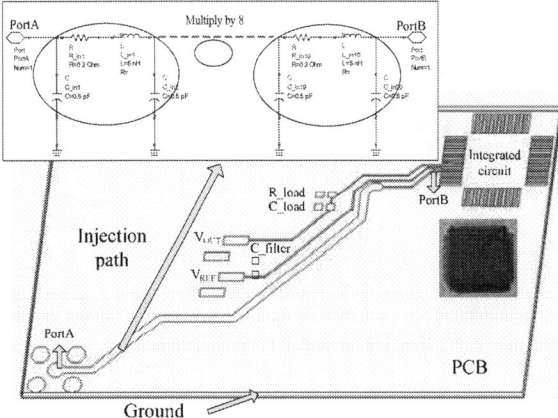

Figure 13. The PCB track model

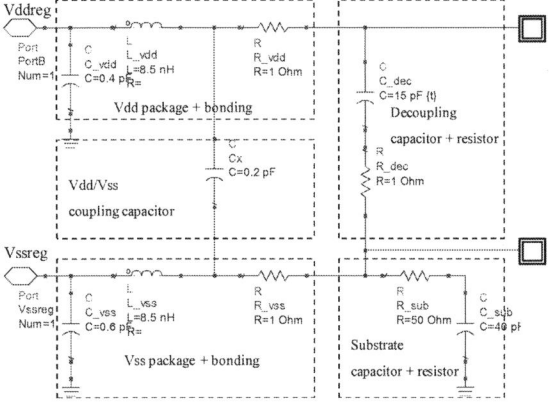

Figure 14. The PDN model of Vddreg and Vssreg for LDO regulator

Decoupling capacitors were removed from the power supply V_{IN} to increase coupling of EMI disturbance to the LDO input. In the test chip design, the filtering capacitor is crucial for op-amp output stability. As integrating a large capacitor on-chip takes up a very large area, an off-chip 47 nF capacitor (C_filter) is included to improve the output stability of the bandgap reference V_{REF}. The regulated output V_{OUT} is loaded by a 330 Ω resistor (R_load) and a parallel 100 pF capacitor (C_load). However, due to the physical presence of the leads, the load capacitors have an equivalent series inductance (ESL) and an equivalent series resistance (ESR) as shown in Fig. 15. These parasitic elements are important for high frequency immunity simulation up to 3 GHz.

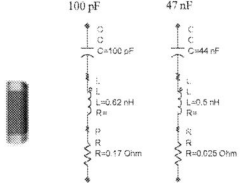

Figure 15. The equivalent load capacitor models with ESL and ESR

Then, all the PDN as presented previously and LDO schematic are assembled and simulated thanks to the EMC tool box. Fig. 16 presents the comparison between measurement and simulation. The difference is below 5dBm for all the frequency range between 1MHz and 1GHz. The simulation time takes only one hour for all frequencies and powers, which is acceptable for a block of this size. The simulation up to 100MHz takes only few minutes. But after as the dwell time is 5μs, we need to simulate several hundred periods of the disturbed signal.

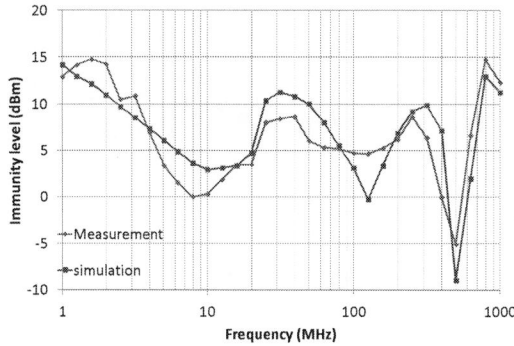

Figure 16. Comparison between DPI measurement and simulation of LDO regulator

V. CONCLUSION

This paper proposes an EMC tool box that simulates automatically the immunity level of a circuit. The methodology and the parameters of the tool are presented. The simulation time of the tool is not yet optimized, but several improvements are also detailed.

Finally, an application case is presented and the simulation is compared with the measurement.

REFERENCES

[1] A. Boyer, S. Ben Dhia, E. Sicard, "Modeling of a direct power injection agression on a 16-bit microcontroller input buffer", *EMC Compo, Torino, Italy, 2007*

[2] IEC 62132-4: Integrated Circuits – Measurements of Electromagnetic Immunity, Direct RF Power Injection Method; International Elektrotechnical Comission, 2006.

[3] U. Stürmer, B. Zhang, O. Jovic and W. Wilkening, "Direct Power Injection Sensitivity Analysis Tool", *EMC compo conf., Toulouse, France, 2009*

[4] E. Sicard, A. Boyer, "IC-EMC, User's Manual, part 7: Immunity simulation", version 2.0, INSA Toulouse, ISBN 978-2-97649-056-7, 2009, pp. 162-183

[5] S. B. Worm, "Simulation of the RF immunity property of analog circuits", *EMC Zürich, 1995.*

[6] J. Loeckx, G. Gielen, "Efficient identification of major contributions to EMI-induced rectification effects in analog automotive circuits", *EMC Zürich 2006.*

[7] B. Vrignon, J. Shepherd, M. Deobarro, "Method, Computer Program Product, and Apparatus for Simulating Electromagnetic Immunity of an Electronic Device", Patent n° WO 2012/143749 A1

[8] DesCoVer reference guide, version 2.5.2, Freescale Semiconductor, Inc., 2012

[9] A. Alaeldine, J. Cordi, R. Perdriau, M. Ramdani, J. Levant, "Predicting the immunity of integrated circuits through measurement methods and simulation models", *EMC Zürich conf., Zurich, 2007*

[10] N. Delorme, M. Belleville, J. Chilo, "Inductance and capacitance analytic formulas for VLSI interconnects", Electronic letters, vol 32, n° 11, pp. 996-997, May 1996.

978-1-4799-5004-1/13 $31.00 © 2013 IEEE

Electro-Magnetic Robustness of Integrated Circuits: from statement to prediction

S. Ben Dhia, A. Boyer

LAAS-CNRS
Université de Toulouse ; UPS, INSA, INP, ISAE ; UT1, UTM, LAAS
Toulouse, France
sbendhia@laas.fr

Abstract - **EMRIC project, a new research activity mixing integrated circuits electromagnetic compatibility (EMC) and integrated circuits (ICs) reliability, provides methods and guidelines to circuits and equipment designers to ensure EMC during lifetime of their applications. In order to improve the ICs electromagnetic robustness (EMR) this project studies the effect of ICs ageing on electromagnetic emission and immunity to radio frequency interferences, clarifies the link between IC degradations and related EMC drifts and develops prediction models and propose "time insensitive" EMC protection structures.**

Keywords: Integrated circuits, electromagnetic compatibility, reliability, robustness, accelerated aging, EMC modelling

I. INTRODUCTION

Introducing new high performance electronic modules in automotive, aeronautic and aerospace applications forces system manufacturers to optimize system reliability and reduce time to market delivery and manufacturing costs. This trend has triggered off an increasing demand for conclusive statements about future lifetime and function of the product already at the design stage, ranging from electromagnetic effects (EMC/RF) to thermal management issues and thermo-mechanical reliability forecasts.

Technological evolution of CMOS technologies brings a shrink of dimensions, a better integration rate and an enhancement of performances such as switching speeds. However it also leads to an increase of dissipated power and leakage current that have a direct impact on ICs parasitic emissions. Moreover reduced supply voltages and an increased number of interfaces tend to decrease the immunity to radio-frequency interference. EMC has become a major concern and a key differentiator in overall IC performance [1] [2].

During their lifetime, integrated circuits may be affected by failure mechanisms mainly activated by harsh environmental conditions (high or low temperature, humidity, vibration, electrical overstress...). The acceleration of intrinsic degradation mechanisms in nanoscale devices threats the reliability of circuits [3]. Even if failure mechanisms do not compromise the circuit operation, IC intrinsic degradations can have a significant impact on EMC performances. The need to predict EMC of ICs after several years of operating life is driven by the trend towards extended warranty. Ensuring the electromagnetic robustness of nanoscale integrated circuits (extension of the electromagnetic compatibility for the full lifetime of the product) has become a key challenge.

EMRIC project aimed at developing a new research activity that mixes EMC and IC reliability to improve the electromagnetic robustness of integrated circuits, with a special emphasis on deep submicron technology. This research topic is still under-explored as research communities on "IC reliability" and "IC electromagnetic compatibility" have often no overlap. The effect of natural aging of ICs on their electromagnetic behavior is still misevaluated and has to be clarified. Highlighting the EMC drift issues induced by IC natural aging, understanding the links between the physical degradation mechanisms induced by ICs aging and EMC drifts, predicting the extent of aging induced EMC drifts and providing guidelines to design robust EMC protection structures are the main objectives of this project.

The first part of the paper presents the methodology set up to study the evolution of ICs electromagnetic behavior over the full lifetime of the component. The second part, dedicated to experimental results, exhibits the drifts of a test chip electromagnetic emission and immunity levels and the link between degradation mechanisms and transistors physical parameters drifts that affect EMC. In the last section, a modeling and simulation approach is proposed.

II. EMR METHODOLOGY

The characterization of the EMR of a circuit consists in measuring its emission and/or susceptibility level before and after ageing procedures that accelerate intrinsic degradation mechanisms. The main objective of this study is to evaluate electromagnetic drifts as function of stress types, stress conditions and stress duration. Figure 1 presents the methodology used to extract EMC level drifts induced by ageing as accurately as possible [4].

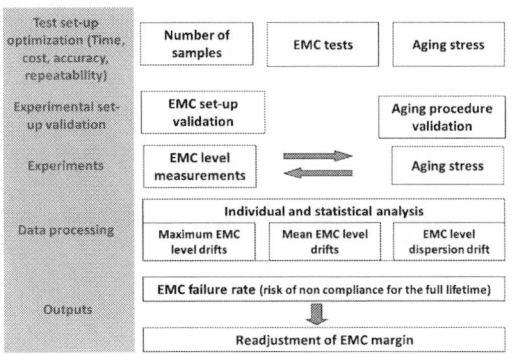

Figure 1.   ICs Electromagnetic robustness experimental metodology

Test set-up optimization is a critical step of the proposed methodology. A compromise should be found between a sufficient number of samples to obtain a precise evaluation of aging impact on EMC and reasonable tests duration and experimental costs.

Accelerated life tests, required to qualify integrated circuits and guarantee their quality and robustness, consist in applying an overstress conditions (extreme temperatures, strong humidity, vibration, shocks, over voltage...) during a short time in order to accelerate the damage rate for relevant degradation mechanisms.

ICs electromagnetic behaviour is strongly linked with the electrical characteristics of semi-conductor active devices such as MOS transistor electrical parameters. Recently, some publications have shown that accelerated aging tests such as high or low temperature operating life (HTOL, LTOL), thermal cycling or electrical overstress induce a significant variation of electromagnetic behaviour [5]. In the following experimental studies, HTOL, LTOL and electrical stress have been selected to accelerate electrical degradation mechanisms. For each experiment, a batch of ten components are tested and placed during 408 h in a climatic chamber which regulates the temperature at 150°C (HTOL) or -40°C (LTOL). An external source provides a power supply 10% higher than the normal supply voltage.

As far as concerned EMC tests, IEC standards methods adapted to IC conducted emission and immunity characterization are applied on the batches. Uncertainties of these methods are acceptable to ensure consistency of EMR results.

Characterization of conducted emission of components under test consists, according to IEC 61967-4 [6], in sensing the parasitic current produced by the IC and flowing through the ground pin.

Characterization of components immunity consists, according to IEC 62132-3 [7], in superimposing RF disturbances to a low frequency signal through a decoupling network, to couple a conducted disturbance to a pin of a circuit (figure 2). The direct power injection (DPI) is an efficient method to couple disturbances to the circuit. Susceptibility is characterized over the band 10 MHz to 1 GHz.

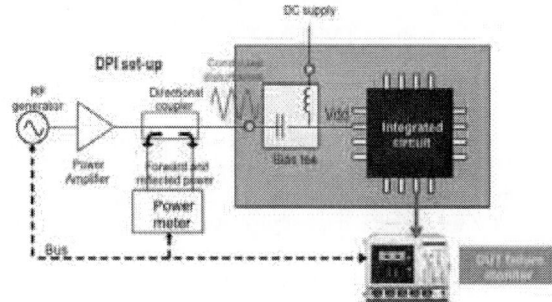

Figure 2.   Direct Power Injection test set-up

Once EMC and ageing procedure set-ups have been completely defined, they have to be validated experimentally. A validation step is necessary to ensure that ageing procedure is the only responsible of DUT EMC drifts. The EMC validation step must ensure that the EMC measurements are enough repeatable to extract precise information concerning the EMC drifts related to ageing effects. Measurement repeatability and uncertainties set a limit for the consistency of EMC level drifts. For all experiments, the components should be still functional in nominal condition after aging. At the end of the measurement campaign, the EMC measurement results obtained before and after ageing are processed to extract statistical data concerning the EMC level drifts, such as the worst case and the mean drifts. These data are required to predict the risk that a component becomes incompliant after ageing and, if necessary, readjust the EMC margins at the design level.

Variations of emission or susceptibility levels are given in terms of mean drift ΔM over all the components:

$$\Delta_M = \frac{1}{N} \sum_{i=1}^{N} \left(X_{Ai} - X_{Bi} \right)$$

X_{Ai} : emission or immunity level of sample 'i' after aging
X_{Bi} : emission or immunity level of sample 'i' before aging
N: number of samples

Emission and susceptibility levels cannot be accurately known as they are subject to statistical distributions due to measurement errors and variability between components. Aging can affect the variability of component characteristics so the dispersion σ of emission or susceptibility levels must be computed before and after aging:

$$\sigma = \sqrt{\frac{\sum_{i=1}^{N} \left(X_{(A\,or\,B)i} - \overline{X_{(A\,or\,B)}} \right)^2}{N-1}}$$

X: mean emission or immunity level of all samples.

The measured drifts between samples are also affected by repeatability errors owing to the measurement equipment or test bench variations.

III. EXPERIMENTAL RESULTS

A. Initial measurements

In order to exhibit aging effect on IC EMC, the EMR methodology has been applied to a 65 nm low power CMOS technology test chip developed by ST-Microelectronics and dedicated to the characterization of several I/O structures [8]. Standard conducted emission is characterized by measuring the I/O current consumption on ground pin (figure 3). As far as concerned I/O immunity, RFI are injected on I/O power supply pin (figure 5). These two tests are applied on a batch of 10 components, before and after aging. Two separated batches of 5 components have been placed during 408H in a climatic chamber which regulates the temperature respectively at 150°C (HTOL) and -40°C (LTOL) (AECQ100 protocol) combined with high supply voltage (10% higher than the nominal).

The mean difference (over 5 samples) between the emission levels before and after aging (figure 4) and immunity levels before and after aging (figure 6), for both HTOL and LTOL accelerated life tests have been computed. Experimental results are reported in Table 1.

Emission levels decrease clearly after both aging tests over almost all the frequency range. LTOL induces a greater emission level reduction (with a mean value about -1.48 dB) than high temperature stress (-0.3 dB on average).

Conversely, aging seems to have a negative impact on IO immunity when RF injections are performed on power supply pins. Both aging tests induce a noticeable reduction of the immunity level over a large frequency range. HTOL induces a greater immunity level reduction. Even if the average global fading of immunity level remains acceptable, the large reductions of immunity level observed on some samples at several frequencies (up to -8.75 dB) can seriously affect their EMC compliance.

Weak measurement uncertainties are taken into account. Process dispersion computed characterized before and after stress shows that aging stress doesn't affect each component on the same way. EMC level dispersion is higher after aging stress than before.

These first experiments that exhibit some potential issues due to EMC levels evolution after aging, clearly indicate that there is a direct link between internal degradation mechanisms and EMC behavior.

This study has been done with the financial support of French National Research Agency (project EMRIC JC09_433714), the regional council of Midi-Pyrénées

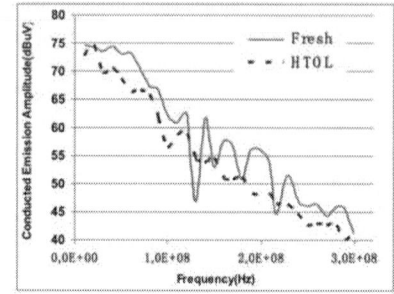

Figure 3. HTOL aging impact on emission level (measured on one sample)

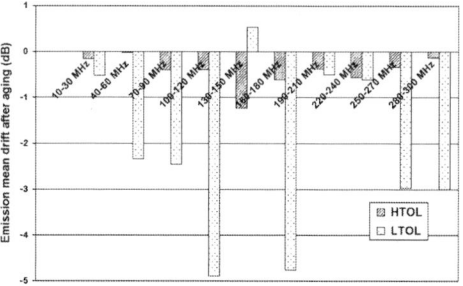

Figure 4. LTOL and HTOL aging impact on emission level (mean drift for all samples)

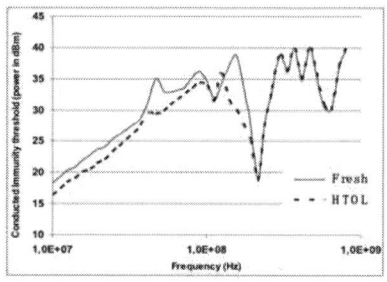

Figure 5. HTOL aging impact on IO immunity threshold (RFI injected on the IO power supply of one sample)

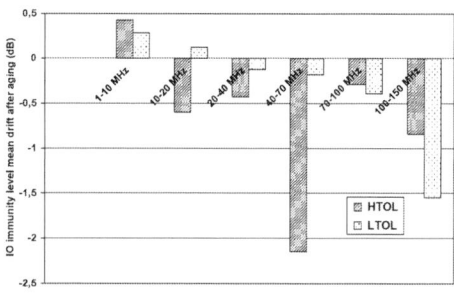

Figure 6. LTOL and HTOL aging impact on immunity level (mean drift for all samples)

EMC test	Aging Test	Max. aging drift above all frequencies (for the most affected sample)	Mean aging drift (entire batch)	Standard deviation of the drift (entire batch)
Emission	HTOL	2.2 dB / -9.5 dB	-0.3 dB	1 dB
	LTOL	15.3 dB / -12 dB	-1.48 dB	1 dB
Immunity	HTOL	2 dB / -8.75 dB	-0.8 dB	0.7 dB
	LTOL	1.26 dB / -8.25 dB	-0.6 dB	1 dB

Table 1. Summary of LTOL and HTOL aging impact on emission and immunity levels

B. Impact of semiconductor device parameters drifts

During their lifetime, CMOS transistors are affected by intrinsic failure mechanisms such as time dependent dielectric breakdown (TDDB), negative bias temperature instability (NBTI) or hot carrier injection (HCI), mainly activated by harsh environmental conditions such as high or low temperature and electrical overstress. The failure origins and modelling are studied for years and active researches are still on-going. Among them, HCI and NBTI are the major contributors to the device performance degradation in advanced CMOS technologies. The degree of degradation of a device and, hence, its lifetime depend on the stress level and duration (Table 2). Applying a high level stress (e.g. high temperature or high voltage) for a short period accelerates the damage rate for relevant wear-out failure mechanisms. This principle is commonly used in accelerated-life tests to extrapolate semiconductor device lifetime or in screening tests.

Failure mechanism	Cause of failure	Failure modes	Stress conditions
Hot carrier injection (HCI)	Trapped carrier in gate oxide due to the impact ionization of channel carriers near the drain region.	Threshold voltage shift and mobility degradation. Serious issue in short channel NMOS transistor.	LTOL High positive voltage V_{DS} Medium positive voltage V_{GS}
Negative Bias Temperature Instability (NBTI)	Formation of fixed-oxide charge in PMOS gate oxide.	Threshold drift with partial recovery phenomenon. Serious issue in thin oxide PMOS transistor.	HTOL High negative voltage V_{GS} Low negative voltage V_{DS}

Table2. CMOS transistor intrinsic failure mechanisms and failure modes

A dedicated test chip has been developed in Freescale CMOS 90 nm technology to characterize the failure mechanisms and the associated MOS transistors electrical parameter drifts. The test chip contains MOS devices with various geometrical and gate options (thin and thick oxide options, with nominal operating voltages equal to 1.2 and 3.3 V respectively). The transistor terminals (gate, drain, source, bulk) are accessed with DC probes. Ids characteristics of transistors have been measured with a Keithley 2601A source meter and an acquisition board after several electrical stress periods. The tests aim at characterizing NBTI in PMOS and HCI in NMOS devices specifically. More details about the measurement set-up and results can be found in [9]. Temperature is not considered in our experiments to simplify the set-up. All the tests are performed at ambient temperature.

The experimental characterizations show that HCI and NBTI effects arise in NMOS and PMOS devices respectively when electrical stresses are applied. Threshold voltages V_{TH} and saturation currents I_{dsat} are strongly affected depending on the stress voltage and duration for both transistors. Experimental results on NMOS transistors are given in Table 3. Changes of transistors physical characteristics such as threshold voltage and mobility could affect the emission and susceptibility levels of integrated circuits by modifying electrical behavior of internal functions such as noise margins, Jitters, current consumptions, delays…

Stress conditions	ΔIdsat	ΔVth
After 2000 s @ 3 V	- 40 %	+ 25 %
After 500 s @ 4 V	- 75 %	+ 40 %

Table 3: Characterization of HCI impact on thin oxide NMOS transistors

To illustrate the effect of degradation mechanisms on IC EMC behavior figure 7 presents the power supply voltage fluctuations produced by the 90nm digital core activity measured by an on-chip sensor before and after accelerated aging stress (over voltage stress). A -30 % reduction of the peak-to-peak amplitude is observed. The waveform of the signal is also modified. This modification of power supply voltage bounce shape has a direct impact on the component emission spectrum [10].

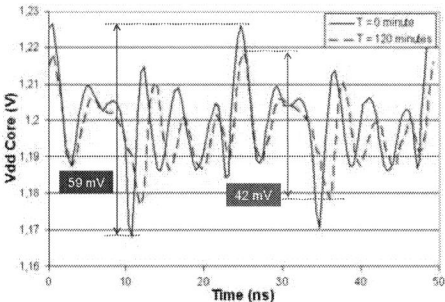

Figure 7. Evolution of the core power supply voltage bounce after 120 minutes of 3.6 V electrical stress

IV.   METHODOLGY APPLIED TO THE PREDICTION OF A PLL ELECTROMAGNETIC ROBUSSTNESS

In this section, we focus on the evolution of the susceptibility of a CMOS phase-Locked-loop (PLL) to EMI coupled on its voltage controlled oscillator (VCO) power supply after accelerated ageing. The developed methodology is applied to model and explain variations of the VCO electrical behavior and consequently the variation of the PLL susceptibility level. The device under test is a test chip developed in the 0.25μm SMARTMOS technology from Freescale semiconductor dedicated to automotive applications.

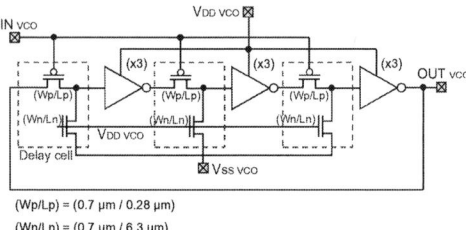

(Wp/Lp) = (0.7 μm / 0.28 μm)
(Wn/Ln) = (0.7 μm / 6.3 μm)

Figure 8.   VCO schematic view

The PLL is made of three sub-blocs that have their own separated power supply pins: a phase detector, a VCO and a frequency divider. The VCO is a delay-controlled ring oscillator as shown in figure 8, designed to operate nominally at 112 MHz. The delay introduced by the three delay cells depends on the voltage applied on the gate of their PMOS transistors. The voltage on NMOS gates is constant.

As ICs embedded in automotive applications often suffer from high temperature, High Temperature Operating Life (HTOL) stress is applied. Ten components are tested and placed during 408 h in a climatic chamber which regulates the temperature at 150 °C. An external source provides a power supply 10% higher than the normal supply voltage. DPI tests are conducted before and after aging stress. Susceptibility is characterized over the band 10 MHz to 1 GHz and the maximum forward power is set to 45 dBm. This value, larger than usual maximum power level used in DPI tests (30dBm), has been chosen to induce failure on a large frequency range. Three susceptibility criteria are considered: a static margin of the output amplitude set to 20% of the power supply V_{DD}, a dynamic margin set to 20% of the output signal period (8ns) and a power current limited to 40 times the nominal current.

A significant evolution of the susceptibility for conducted injection on VCO power supply has been observed on the 10 samples of the batch after aging. Immunity level reduction can reach up to 10 dB over a large frequency range on some samples. A statistical analysis shows that among 10 samples after HTOL the maximum reduction is 15.88dB; while the mean reduction for all 10 samples over all frequency is 2.6 dB (figure 9) with a standard deviation 3.64dB. These results indicate that ageing and induced intrinsic degradation mechanisms affect considerably the PLL immunity to power supply ripple.

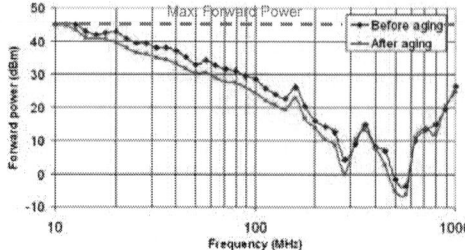

Figure 9.   Evolution of the average susceptibility level measured on a lot of 10 samples before and after accelerated aging test

HTOL mainly accelerates NBTI that can degrade transistor parameters, which could constitute the source of the immunity level variation after ageing. Degradation of transistor transconductance (gm) and mobility (l) could directly affect the power supply rejection ratio and delay time of gates composing the VCO. The prediction of immunity level change induced by aging relies on a model able to simulate the susceptibility level and a physical model that simulates the impact of aging induced degradation mechanisms. To take into account both specifications, a transistor based model is used for the VCO modeling in order to integrate the effect of device aging. The degraded transistors can be identified and their characteristics can be changed according to an aging model.

The structure of the PLL susceptibility model relies on ICIM standard proposal [11]. The model presents two parts:

- The Internal Behavior (IB) block which describes the nominal operation of the PLL and detects failures induced by coupling of EMI

- The Power Distribution Network (PDN) that describes the coupling path of the EMI to the sensitive nodes of the PLL. The PDN includes the DPI system, board tracks, package and on-chip interconnects.

Once PLL Immunity model has been validated, aging model has to be included. A quantitative prediction of the effect of aging on susceptibility relies on the identification of critical transistors and the extraction of variation laws of transistor electrical parameters vs. stress conditions and time. Using CAD simulations, different scenarios of transistor degradation can be tested. In the following analysis, we assume that threshold voltage and mobility are the only model parameters affected by aging. Experimental results have shown that the susceptibility increase is correlated with VCO slowing down so the degradation can be located on the transistors which have a major contribution on VCO oscillation frequency: the delay cell. Degradation mechanisms such as NBTI can increase the channel resistance of the delay cell PMOS transistor and the delay. NBTI applied on the PMOS transistor also decreases the carrier mobility and increase the threshold voltage. The variation of the VCO oscillation frequency is simulated for an arbitrary change on the carrier mobility and threshold voltage values. A change of 10% on a transistor parameter is usually considered as a failure. The worst-case degradation scenario is now considered: the mobility is decreased while the threshold voltage is increased by 10% from their nominal value. The PLL susceptibility level is simulated between 10MHz and

1GHz. Figure 10 presents the comparison between the simulated susceptibility level of fresh and aged PLL.

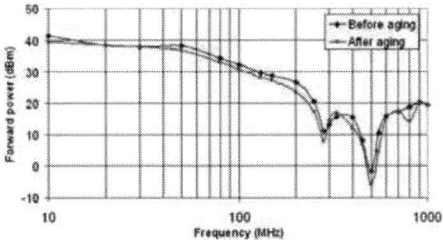

Figure 10. Comparison between simulated susceptibility level of fresh and aged PLL

The simulation predicts a global reduction of several dB of the immunity level over a large frequency range (Fig. 10). This trend is similar to the immunity level decrease observed in measurement (Fig. 9). The degradation of the mobility and threshold voltage of the delay cell PMOS transistor is responsible for a large part of the reduction of the PLL robustness to conducted interferences. The proposed model reproduces qualitatively the measured evolution of the susceptibility level of the PLL to conducted disturbances. A better transistor model would certainly improve immunity level drift prediction.

V. CONCLUSION

This paper gives an overview of EMRIC project that aimed at developing a new research activity that mixes electromagnetic compatibility (EMC) and integrated circuits (ICs) reliability. The long-term EMC has to be ensured at circuit design level to guarantee a high level of safety of the final electronic applications. This paper has araised the fact that natural aging in harsh environment condition may affect EMC of electronic systems. The link between ICs internal degradation mechanisms, electrical parameters drifts and EMC evolution has been exhibited. A simulation methodology to predict the impact of circuit aging on conducted susceptibility has been presented. This method has been applied to model and explain the variation of susceptibility level of a phase-locked loop measured after an accelerated-life test. An electrical model of the PLL has been developed to predict the susceptibility level of the PLL to conducted disturbances coupled on the VCO power supply pin. A hypothesis has been proposed about the origin of the susceptibility level drift. The degradation of carrier mobility and threshold voltage of the

PMOS transistor of the VCO delay cell slow down the VCO, modify the capture range of the PLL and thus increase the susceptibility. Including the change of transistor parameters in the susceptibility model has induced a similar reduction of immunity level as the measured one. Although the presented simulation results remain qualitative, the characterization and the accurate modeling of transistor degradation mechanisms will allow a quantitative prediction of susceptibility level drifts.

REFERENCES

[1] Ben Dhia, S., Ramdani, M., Sicard, E., 2005, EMC of Ics: Techniques for low emission and susceptibility, Springer, ISBN 0-387-26600-3

[2] Ramdani, M. Sicard, E. Boyer, A. Ben Dhia, S. Whalen, J.J. Hubing, T.H. Coenen, M. Wada, O., "The Electromagnetic Compatibility of Integrated Circuits—Past, Present, and Future", IEEE Transactions on Electromagnetic Compatibility, Volume 51, Issue 1, pp 78 - 100, February 2009

[3] M. White, J. B. Bernstein, "Microelectronics Reliability: Physics-of-Failure Based Modeling and Lifetime Evaluation", NASA WBS 939904.01.11.10, 2008, nepp.nasa.gov.

[4] B. Li, A. Boyer, S. Ben Dhia, C. Lemoine, "Ageing effect on electromagnetic susceptibility of a phase locked loop", Microelectronic Reliability, Vol. 50, Issues 9-11, September – November 2010, pp. 1304-1308.

[5] I. Montanari, A. Tacchini, M. Maini, "Impact of thermal stress on the characteristics of conducted emissions", 2008 IEEE Int. Symposium on Electromagnetic Compatibility, EMC 2008, Detroit.

[6] IEC 61967-4, "Integrated Circuits, Measurement of Electromagnetic Emissions, 150 KHz - 1 GHz: Measurement of Conducted Emissions, 1 Ω/150 Ω Method", IEC standard, 2006.

[7] IEC 62132-3, "Direct RF Power Injection to measure the immunity against conducted RF-disturbances of integrated circuits up to 1 GHz" , IEC standard, 2007.

[8] Ben Dhia, S., Boyer, A., Li, B., Ndoye, Characterization of the Electromagnetic behavior drifts of a nanoscale IC after Accelerated Life Tests", Electronic Letters, 18th February 2010, Vol. 46, no. 4

[9] N. Berbel, R. Fernandez-Garcia, I. Gil, B. Li, A. Boyer, S. Ben Dhia, "Experimental verification of the usefulness of the nth power law MOSFET model under hot carrier wear out", Microelectronics Reliability, vol. 51, no 9 -11, pp. 1564-1567, September 2011

[10] A. Boyer, S. Ben Dhia , " Characterization and Modeling of Electrical Stresses on Digital Integrated Circuits Power Integrity and Conducted Emission ", 9th International Workshop on electromagnetic Compatibility of Integrated Circuits, EMCCompo 2013, Dec. 15 – 18, 2013, Nara, Japan.

[11] A. Boyer, S. Ben Dhia, C. Lemoine, B. Vrignon, "Construction and Evaluation of the Susceptibility Model of an Integrated Phase-Locked Loop", 8th International Workshop on electromagnetic Compatibility of Integrated Circuits, November 6 – 9, 2011, Dubrovnik, Croatia.

EMC Immunity of Integrated Smart Power Transistors in a non-50Ω Environment

Hermann Nzalli,
Wolfgang Wilkening
Robert Bosch GmbH (AE/EID)
Reutlingen Germany
Hermann.Nzalli@de.bosch.com

Rolf H. Jansen
Chair of EM Theory (RWTH Aachen University)
52072 Aachen, Germany
jansen@ithe.rwth-aachen.de

Abstract— **The Direct Power Injection (DPI) standard, widely used for the susceptibility analysis of integrated circuits (IC), specifies an ideal 50Ω-environment for the investigations. This constant load assumption does not fully cover latter stages or the IC final operating environment, where ICs are subjected to various load impedances, especially at pins which are connected to wiring harnesses. We present variable-load DPI measurements and large-signal simulations for new circuit blocks, namely a simplified high-side driver and an ESD structure. The results extend the applicability of small-signal simplification methods beyond a low-side driver formerly reported by the same authors.**

Keywords—EMC (Electromagnetic compatibility); Direct Power Injection (DPI); Load-pull ; S-parameter; Variable-load DPI

I. INTRODUCTION

In the last years, the extensive use of electronics in modern cars, much of which controls safety-related functions, raise the EMC (Electromagnetic Compatibility) concerns in the automotive industry. Especially in what concerns EMS (Electromagnetic Susceptibility), a lot of attention is paid to ICs (Integrated Circuits). In fact, ICs have not only increased in number, but they are now biased by lower supply voltages. On top of this, they have to operate in the very harsh vehicle environment polluted by interfering sources from within the automobile or by outside sources. These disturbance sources generate EMI (Electromagnetic Interference) whose amplitude and variety are steadily growing. Therefore, automotive IC designers have invested a lot of efforts to tackle EMC immunity issues.

In this context, the DPI standard [1] (Direct Power Injection) has been established as the most appropriate method to evaluate the conducted susceptibility of ICs. The literature reports several studies where the immunity of custom automotive analog chips has been analyzed by use of DPI measurements and simulations [2] [3] [4] [5].

As stipulated by the standard, DPI investigations have to be performed in an ideal 50Ω environment. However, this constant load assumption is rarely fulfilled as soon as the chip is mounted / installed in its final destination, for example in an IC package or in the ECU (Electronic Control Unit). In fact, because of the IC immediate and intermediate environment,

which might consist of package, PCB traces or cable harnesses, the IC terminals are subjected to a wide range of load impedances which are difficult to predict. Hence, from an impedance point of view, the final ICs locations can be rigorously considered as non-50Ω environments. Prior to this work, the authors introduced variable-load DPI measurements for the example of a low-side driver in [6], in order to study the circuit immunity while the DPI signal was injected through various RF loads. The investigations enable identifying worst-case load impedances leading to 5dB less immunity in comparison to the standard 50Ω DPI, thereby underlining the significant impact of the load alteration. Yet, the equipment required for such investigations is expensive, and the calibration and measuring procedure laborious. On top of this, the load-pull tuner bandwidths only allow measurements above several hundreds of MHz, making the DPI frequency band only partly coverable.

This paper evaluates the load dependency of the EMC immunity by way of variable-load DPI measurement and large-signal simulations. Furthermore, two simplified S-parameter based methods, recently validated for a low-side driver in [7], are applied to a high-side driver and an ESD-protection.

Section II presents variable-load DPI measurement results for the investigated ICs, and computes corresponding large-signal simulations. Subsequently, the results are interpreted and the need for simplification is addressed, in view of the complexity and limitations of both measurements and large-signal simulations. Accordingly, section III introduces two small-signal approaches which draw upon S-parameter to ease variable-load DPI analysis. Subsequently, Section IV discusses the impact of the load-dependency on the EMC immunity within the frequency range from 1MHz up to 4GHz. Finally, section V draws concluding statements.

II. VARIABLE-LOAD DPI MEASUREMENTS AND LARGE-SIGNAL SIMULATION RESULTS

A. DUTs, Variable-load DPI setup description and results

In this work, three DUTs are investigated: a low-side driver, a high-side driver and an ESD-protection diode. The circuits as well as the variable-load DPI measurements setup are respectively illustrated in Fig. 1, Fig. 2 and Fig. 3.

978-1-4799-5004-1/13 $31.00 © 2013 IEEE

The low-side driver, already studied in [6] and [7], consists of a 30V LDMOS transistor diffused in a 0.35μm smart-power technology. A 3kΩ poly-Si resistance is joined to its drain, and both a high-impedance resistive voltage divider and protection diode are connected at the drain side. The driver is operated in OFF-mode ($V_{GATE}=0V$) and supplied by a typical battery voltage ($V_{DRAIN}=14V$), as an interfering signal is applied at its output where the DC voltage V_{OUT} is also monitored. The failure criterion is set as a 10% drop relatively to the initial V_{OUT} value ($V_{OUT}<12.6V$).

The main component of the high-side driver is a 65V HV-PMOS diffused in the same technology as the low-side driver. The 270kΩ poly-Si resistance connected to the transistor drain enables monitoring the output voltage V_{OUT}. The transistor is also driven in cut-off region ($V_{GATE}=14V$, $V_{SOURCE}=14V$), and the DPI signal is superimposed with the DC supply in this case. A 10% increase of the output voltage V_{OUT} is set as failure criterion ($V_{OUT}>1.4V$).

The ESD-protection is a diode-connected BJT transistor, composed of a 100 cells array and diffused in a 0.18μm smart-power technology. Its susceptibility in reverse region is investigated, as the EMI-signal is overlaid with the cathode DC voltage ($V_{CATHODE} = 14V$). A current flow above 1mA is considered as malfunction.

As principal extension to the standard DPI setup, a load-pull tuner is integrated into the test-bench. It is placed between the RF-source and the disturbed DUT pin (see Fig. 1, Fig. 2 and Fig. 3). From the 50Ω of the RF-source, the tuner can nearly synthesize any desired load Z_L, which is as well the impedance of the EMI-injection. During variable-load DPI measurements at 1GHz, for each interfering power level P_{AVS}, the DUT susceptibility is monitored while all impedances Z_L of the Smith-chart are scanned. Starting from small values, P_{AVS} is increased until impedances which lead to the malfunction are found.

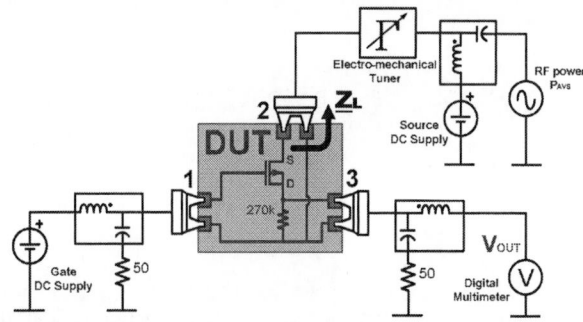

Fig. 2 Standard-DPI measurement setup for the high-side driver

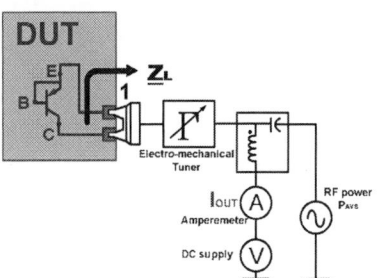

Fig. 3 Standard-DPI measurement setup for the ESD-protection

Furthermore, in order to provide a foundation for verification, large-signal transient simulations have been computed with the Spectre simulator [8], on the basis of extended BSIM3v3 and Ebers-Moll models.

Fig. 4, Fig. 5 and Fig. 6 depict the results obtained for each DUT at 1GHz, starting from the smallest P_{AVS} where load impedances Z_L causing a DUT failure were found. The identified critical impedances have been exemplary shown for measurements of the low-side driver with $P_{AVS}=20dBm$ (see red circles in Fig. 4.c). For the sake of clarity and comparability with other simulation-guided methods, only the contours of the variable-load DPI regions will be displayed in the remainder of this paper. Regardless of DUT, it could be noticed that these regions expand and move towards the center of the Smith-Chart as P_{AVS} increases. From similar crescent RF-power sweep measurements performed at 1GHz with $Z_L = 50Ω$, the defined failure was triggered at 23dBm for the low-side driver, 10dBm for the high-side driver and 25dBm for the ESD-protection. Therefore, the DUTs respectively exhibit 5dB, 8dB and 7dB less immunity in the presence of worst-case impedances.

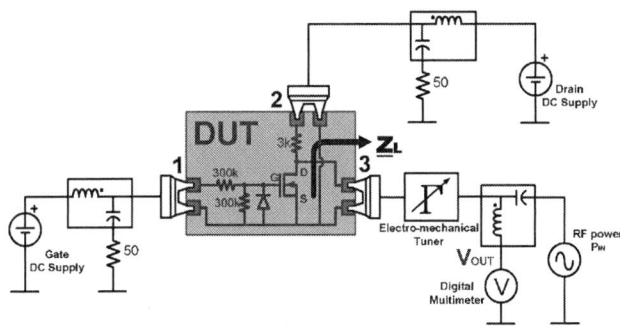

Fig. 1 Standard-DPI measurement setup for the low-side driver

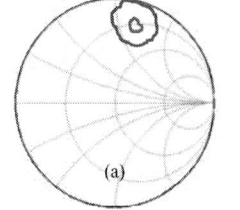 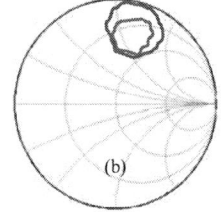

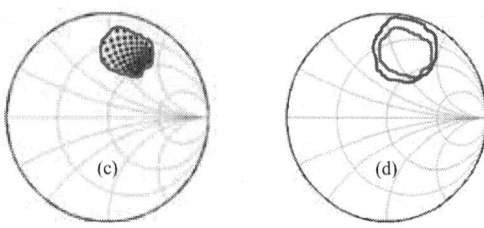

Fig. 4 Variable-load DPI contours obtained for the low-side driver from measurements and large-signal simulations with P_{AVS} equal to 18 dBm (a), 19 dBm (b) and 20 dBm (c, d) at 1 GHz. Red circles: Measured critical impedances leading to $V_{OUT} < 12.6$ V; Red line: measurements; Blue line: large-signal simulations

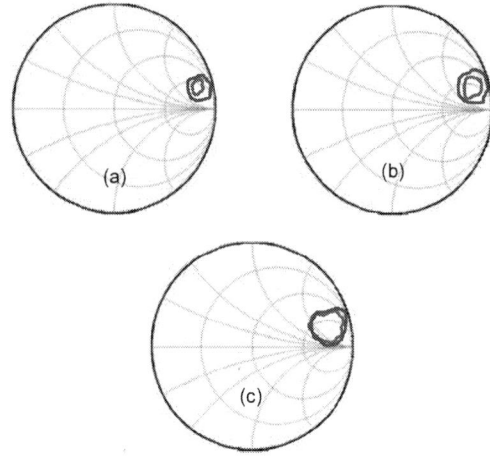

Fig. 5 Variable-load DPI contours obtained for the high-side driver from measurements and large-signal simulations with P_{AVS} equal to 2dBm (a), 3dBm (b) and 4dBm (c) at 1 GHz. Red line: measurements; Blue line: large-signal simulations

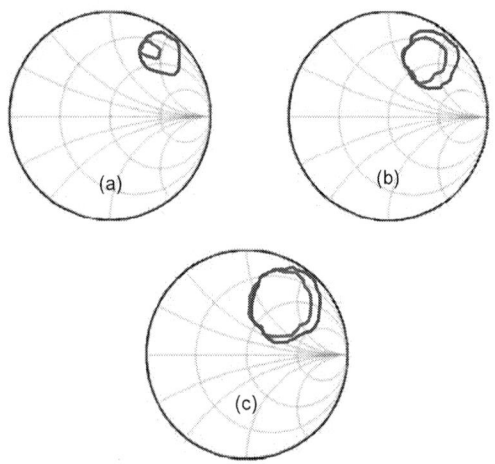

Fig. 6 Variable-load DPI contours obtained for the ESD-protection from measurements and large-signal simulations with P_{AVS} equal to 18 dBm (a), 19

dBm (b) and 20 dBm (c) at 1 GHz. Red line: measurements; Blue line: large-signal simulations

B. Results interpretations and limitations of measurements and large-signal simulations

1) Interpretation of variable-load DPI results

ICs interact with the outer world trough their externally connected pins which are joined to cables. These act as antennas and collect RF-disturbances, which consequently appear at IC pins from impedances which are hardly predictable. These can be interpreted as RF impedances which load IC pins. In order to investigate how these loads impact the susceptibility behavior, it is beneficial to consider the 50Ω RF-power source and the load-pull tuner as a single variable-load DPI source (see Fig. 7). Since the tuner is an ideally lossless impedance network, it conserves power (note: voltage or current may be altered). Therefore, we decide to continue the discussion in terms of RF power. Its transfer is analyzed and eventually, i.e. at the failure relevant node, power will be converted in RF voltage or current, depending on which quantity is relevant for the failure criterion. The tuner entirely transfers the interfering power P_{AVS} (which is actually generated by the 50Ω DPI-source with matched load $Z_0=50Ω$) while synthesizing an impedance Z_L at the DUT-side. Hence, the variable-load DPI source is an RF-power source with internal impedance Z_L and maximum available power P_{AVS}.

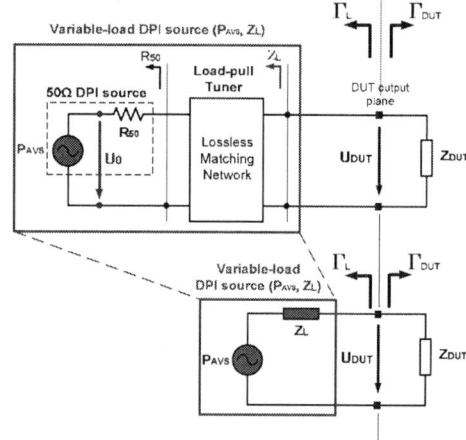

Fig.7 Variable-load DPI source

With V_0 being the open-circuit voltage of the variable-load DPI source, P_{AVS} is expressed as [9]:

$$P_{AVS} = \frac{|V_0|^2}{8Z_0} \times \frac{|1-\Gamma_L|^2}{1-|\Gamma_L|^2} \qquad (1)$$

In this consideration, Γ_L is the reflection coefficient which corresponds to the impedance Z_L. Besides, the driver being initially driven in OFF-state, it acts as a passive device representable by its impedance Z_{DUT} or the equivalent reflection coefficient Γ_{DUT}. Thus, the active power P_{DUT}

978-1-4799-5004-1/13 $31.00 © 2013 IEEE   216

absorbed by the DUT at the drain-side can be expressed as follows:

$$P_{DUT} = \frac{(1 - |\Gamma_{DUT}|^2) \times (1 - |\Gamma_L|^2)}{|1 - \Gamma_{DUT} \times \Gamma_L|^2} \times P_{AVS} \qquad (2)$$

The worst-case scenario in terms of EMC immunity occurs in the presence of an impedance $Z_{L,WORST\_CASE}$ when the maximum possible RF-power couples into the DUT, or in other words, when P_{DUT} is equal to the available source power P_{AVS}. This occurs when power matching is given, that is when the load impedance is conjugately matched at the frequency of operation to the DUT impedance Z_{DUT}, meaning $Z_{L,WORST\_CASE}$ = Z_{DUT}*, or $\Gamma_{L,WORST\_CASE} = \Gamma_{DUT}$*. A closer look at equation (2) also confirms the previous statement. Therefore, the DUT impedance Z_{DUT} behaving mostly capacitive in the investigated frequency range, the load impedances to which the DUTs are particularly susceptible are located in the upper half of the Smith-chart.

From RF linear circuit theory, this matching in linear mode is even known to be reasonably close to matching for maximum power transfer under large-signal conditions. For all studied ICs, it could be verified trough S-parameter measurement that the first identified critical impedance were not far away from the DUT impedance at the disturbed DUT pin. Accordingly, in the standard DPI test, a large portion of the incoming power might be reflected, whereas in the presence of the load impedance $Z_{L,WORST\_CASE}$ during variable-load DPI, the DUT absorbs the entire available power source. As a consequence of this power, RF-voltage levels appear at the "EMI-hotspot" of the DUT, that is, the most critical circuit node which directly triggers the DUT malfunction in the presence of EMI-disturbances. Therefore, the higher the power transfer to the DUT, or formulated differently, the better the matching, the higher RF-induced disturbance of the EMI-hotspot, and consequently, the earlier the malfunction occurs. Hence, from this point of view, the standard DPI measurement at 50Ω-level represents a power mismatching condition, and the discrepancy expressed in dB quantifies the deviation from the matched loading situation.

2) Limitations of the variable-load DPI measurements and simulations

Unfortunately, the variable-load DPI calibration and measuring procedure is a laborious task. On top of this, the setup exhibits inescapable losses in the test-bench which raise the power insertion loss. This restricts the measurement capabilities when moving close to the boundary of the Smith chart. Moreover, the bandwidth of common load-pull tuners is limited to frequencies from several hundreds of megahertz upwards (in our case 900MHz-6.9GHz). On the other hand, large-signal simulations are computed in an ideally lossless environment, and thus allow a complete scan up to the outer rim of the Smith chart. This is the reason why simulated variable-load DPI contours are larger than their measured counterparts. However, the lengthy simulation time is a serious drawback.

To bridge these deficits and assess with less complexity the impact of the load alteration on the EMC immunity, two small-signal simplification approaches are presented in the next chapters. Of course, they are only applicable if the linearity assumption is valid. After being recently validated for a low-side driver in [7], they are now extended to the high-side driver and the ESD-protection.

III. SMALL-SIGNAL APPROXIMATION TECHNIQUES

A. Linear standard-DPI-related approach

This technique necessitates a prerequisite knowledge of the standard DPI immunity levels. The idea behind this method is to quantify the load dependence of the susceptibility by computing, for each load Z_L, the offset ΔP regarding the absorbed RF-power P_{DUT} in comparison to the 50Ω reference case. For this purpose, a linear model containing measured S-parameter data in a simulator compatible format (Touchstone, Citi, etc...) can be generated for the DUT. On this basis, by means of transient simulations, the active power P_{DUT} absorbed by the circuit output from the variable-load DPI source can be calculated as following:

$$P_{DUT} = \frac{1}{2}\left(V_{DUT} \times I_{DUT}* + V_{DUT}* \times I_{DUT}\right) \qquad (3)$$

Alternatively, ΔP can also be calculated by employing equation (2) to compare P_{DUT} between the case of a particular load Z_L and the 50Ω case. This yields:

$$\Delta P[dB] = 10 \times \log_{10}\left(\frac{1 - |\Gamma_L|^2}{|1 - \Gamma_L \times \Gamma_{DUT}|^2}\right) \qquad (4)$$

Hence, because the susceptibility threshold for a 50Ω load is known in advance, the immunity level can be calculated for each load Z_L (see Fig. 9, Fig.10 and Fig. 11).

B. Linear analytical approach

This method is instead self-contained and independent of prior DPI testing. Its principle resides in identifying the circuit EMI-hotspot and linearizing the circuit in its initial operating point in order to analytically express the EMI-caused RF-voltage level $V_{HOTSPOT}(s)$ at this node. Beforehand, the failure criterion should preliminary be translated in terms of a threshold DC value $V_{THRESHOLD}$ related to the EMI-hotspot. If $v_{HOTSPOT}(s)$ reaches $V_{THRESHOLD}$, the DUT is expected to behave susceptible at the tested frequency. Then, the transfer function $H(s) = v_{HOTSPOT}(s)/v_{DUT}(s)$ between the rms-value of the voltage $v_{DUT}(s)$ at the disturbed IC pin and the voltage $v_{HOTSPOT}(s)$ at the EMI-hotspot is to be calculated. This is done by making use of small-signal equivalent circuits illustrated in Fig.8. Finally, taking into account that the failure criterion is fulfilled in the case v_{HOT_SPOT} reaches $V_{THRESHOLD}$, the susceptibility level $P_{SUSCEPTIBILITY}(s)$ can be obtained from:

$$P_{SUSCEPTIBILITY}(s) = \frac{V_{THRESHOLD}^2}{4 \times |H(s)|^2 \times \left|\frac{Z_{DUT}}{Z_{DUT} + Z_L}\right|^2 \times Z_0 \times \frac{1 - |\Gamma_L|^2}{|1 - \Gamma_L|^2}} \qquad (5)$$

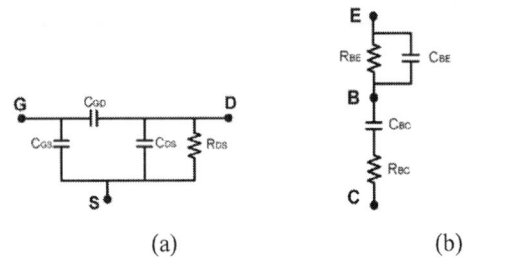

(a) (b)

Fig. 8 Small-signal cut-off models for MOS (a) and diode-connected bipolar transistor (b)

1) Application for the high-side driver

Because of the non-linearity properties of the HV-PMOS, the RF-induced symmetric variation of V_{GS} and V_{DS} around the operating point results in an asymmetric change of the drain current. The resulting DC current offset causes an increase of the output voltage V_{OUT} across the poly-Si resistor.

The EMI-hotspot of the circuit is the transistor gate, and the transfer function H(s) is found to:

$$H(s) = \frac{v_{gs}}{v_{dut}} = \frac{\left(\dfrac{Z_0}{R_{DS}}-1\right)+s\,\xi}{\left(1-\dfrac{Z_0}{R_{DS}}\right)+s\left[\alpha\left(1-\dfrac{Z_0}{R_{DS}}\right)-\xi-Z_0\dfrac{\beta}{R_{DS}}\right]+s^2\left[\beta^2-(\alpha+\beta)(\beta+\xi)\right]}$$

(6)

, with the abbreviations:

$$\alpha = Z_0 \times C_{GS} \tag{7}$$

$$\beta = Z_0 \times C_{GD} \tag{8}$$

$$\xi = Z_0 \times C_{DS} \tag{9}$$

$V_{THRESHOLD}$ is estimated from the output characteristics, and is the gate-source voltage leading to a drain current of magnitude $I_D=1.4V/270k\Omega=5.185\mu A$. The results, shown in Fig. 9, are in good agreements with the measurements and large-signal simulation results.

2) Application for the ESD-diode

Regarding the ESD-protection, the EMI-disturbances applied at the anode or, in this case, at the pnp-emitter, generate an RF-voltage across the parasitic basis-emitter resistance R_{BE}. If this voltage rises above the built-in potential of the basis-emitter diode, the transistor switches into forward-active mode, leading to a current flow. Thus, the EMI-hotspot is the emitter of the pnp-transistor, which is as well the disturbed DUT pin. However, the relevant parameter is the voltage v_{be} across R_{BE}. Hence, by voltage division, H(s) takes the form:

$$H(s) = \frac{v_{be}(s)}{v_{dut}(s)} = \frac{s\,\zeta}{1+s(\alpha+\beta+\xi)+s^2\alpha\beta} \tag{13}$$

, with:

$$\alpha = R_{BE} \times C_{BE} \tag{10}$$

$$\beta = R_{BC} \times C_{BC} \tag{11}$$

$$\xi = R_{BE} \times C_{BC} \tag{12}$$

With $V_{THRESHOLD}$ being the built-in potential of common silicon diodes (typically 0.7V), the variable-load DPI contours can be reconstructed. Here also, the results corroborated with measurements and large-signal simulation results, as the variable-load DPI contours are well reconstructed (see Fig. 10).

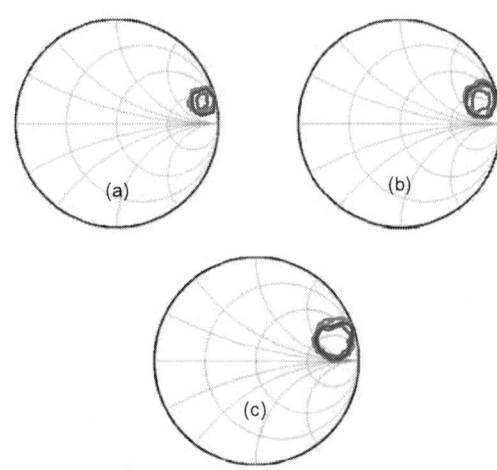

Fig. 9 Variable-load DPI contours obtained for the high-side driver from measurements and large-signal simulations with P_{AVS} equal to 2dBm (a), 3dBm (b) and 4dBm (c) at 1 GHz. Red line: measurements; Blue line: large-signal simulations; Green line: Linear Standard-DPI-related approach; Brown line: Linear Analytical approach

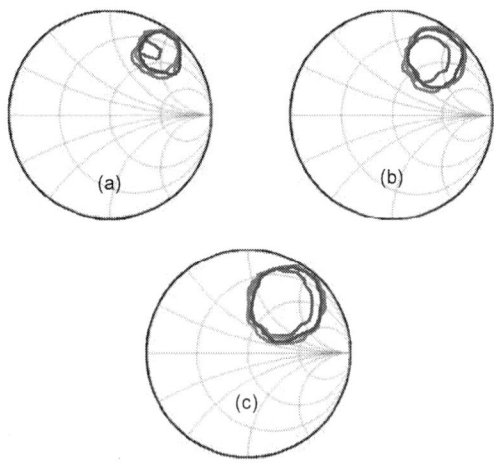

Fig. 10 Variable-load DPI contours obtained for the ESD-protection from measurements and large-signal simulations with P_{AVS} equal to 18 dBm (a), 19 dBm (b) and 20 dBm (c) at 1 GHz. Red line: measurements; Blue line: large-signal simulations; Green line: Linear Standard-DPI-related approach; Brown line: Linear Analytical approach

978-1-4799-5004-1/13 $31.00 © 2013 IEEE

IV. VARIABLE-LOAD DPI INVESTIGATIONS IN THE FREQUENCY SPECTRUM 1MHZ-4GHZ

The results discussed in the previous chapters are dedicated solely to 1GHz. This section now studies discrete worst-case loading scenarios for frequencies from 1MHz to 4GHz, for power levels up to 30dBm. However, instead of the standard 50Ω reference, the worst-case impedances of selected matching frequencies $f_{MATCHING}$ were set before each sweep, so that $Z_L(f_{MATCHING}) = Z_{DUT}*(f_{MATCHING})$. Measurements on chip level are not done because, on the one hand, the tuner bandwidth is limited to frequencies above 900MHz, and on the second hand, the LC values necessary to reproduce the worst-case impedances at lower frequencies have values which are too high to allow integration on chips. Instead, we compare small-signal analysis to large-signal simulations, in order to find out whether there are systematic differences or large signal is validated. The results are illustrated in Fig. 11, and Fig. 12. As expected, the susceptibility profiles respectively reach their minimum at the matching frequencies, thereby indicating a deviation up to 10dB from the 50Ω immunity level. By making use of equation (4), the worst-case EMC risk at the matching frequency is evaluated to:

$$\Delta P(f_{matching})[dB] = 10 \times \log_{10}\left(\frac{1}{1 - \left|\Gamma_{DUT}{}^*(f_{matching})\right|^2}\right)$$

(13)

This clearly highlights the significant impact of the load alteration on the EMC immunity.

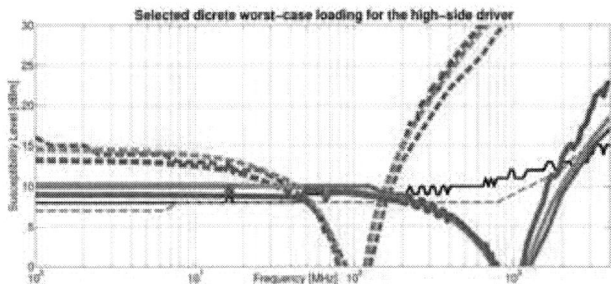

Fig. 11 Variable-load DPI susceptibility profile of the high-side driver with selected discrete worst-case loading at $f_{matching}$=100MHz (dashed line) and $f_{matching}$=1GHz (full line), obtained from large-signal simulations (blue), linear standard-DPI-related approach (green) and linear analytical approach (brown). Comparison is to be made with standard-DPI profile from measurements (continuous black line) and large-signal simulations (dashed gray line).

V. CONCLUSION

This paper employs variable-load DPI measurements and large-signal simulations to investigate the susceptibility of ICs in the presence of load impedances other than the 50Ω-reference. This more realistic consideration is particularly relevant, as it is not covered by the established DPI-standard, but yet critical for externally connected IC pins. The investigations show that the investigated DUTs exhibit

significantly less immunity in the presence of worst-case impedances. These deviate by up to 10dB from standard DPI susceptibility. the

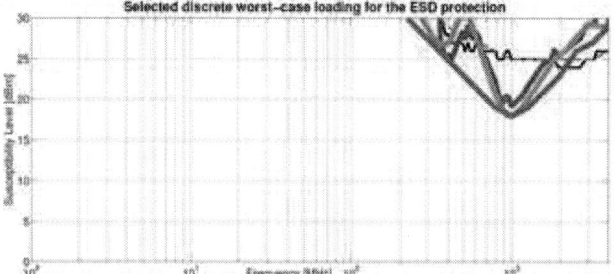

Fig. 12 Variable-load DPI susceptibility profile of the high-side driver with selected discrete worst-case loading at $f_{matching}$=1GHz, obtained from large-signal simulations (continued blue line), linear standard-DPI-related approach (continued green line) and linear analytical approach (continued brown line). Comparison is to be made with standard-DPI profile from measurements (continued black line) and large-signal simulations (dashed gray line).

Two small-signal approaches, which accelerate the load-dependency assessment of the immunity, have been successfully validated.

The proposed techniques might prospectively help to investigate load-dependent design margins and support the definition of EMC requirements at the interface between IC and ECU.

REFERENCES

[1] Integrated Circuits-Measurements of Electromagnetic Immunity 150 kHz to 1 GHz—Part 4: Direct RF Power Injection Method, Standard IEC 62132, Dec. 2004

[2] E. Sicard, S. Ben Dhia, M. Ramdani, T. Hubing, " EMC of Integrated Circuits : A Historical Review", IEEE International Symposium on Electromagnetic Compatibility, EMC 2007, vol., no., pp.1,4, 9-13 July 2007

[3] O. Jovic, W. Wilkening, U. Stuermer, A. Baric, "Susceptibility of a Brokaw bandgap to high electromagnetic interference" in Proc. 20th Int. Zurich Symposium on EMC, Zurich 2009, pp. 401-404.

[4] F. Fiori, and P. S. Corvetti, "Nonlinear Effects of Radiofrequency Interference in Operational Amplifiers," IEEE Trans. On Circuits and Systems I: Fund. Theory and Applications, Vol. 49, Issue 3, March 2002, pp. 367 – 372.

[5] J-M. Redoute, C. Walravens, S. Van Winckel, M.S.J. Steyaert, "An externally trimmed integrated DC current regulator insensitive to conducted EMI," IEEE Transactions on Electromagnetic Compatibility., Vol. 50, Feb. 2008, pp. 63 – 70.

[6] H. Nzalli, C. Lautensack, W. Wilkening, R.H. Jansen, "EMC Immunity of an Integrated Low Side Driver Circuit under Varying RF Loads", 8th Workshop on Electromagnetic Compatibility of Integrated Circuits, EMC Compo 2011, pp.1-6, Nov. 2011

[7] H. Nzalli, W. Wilkening, R.H. Jansen, "Load Dependency Assessment of the EMC Immunity for Integrated Low Side Drivers", accepted for publication in IEEE Transactions on Electromagnetic Compatibility

[8] Cadence Virtuoso Spectre Circuit Simulator, Cadence Design Systems, Inc., [Online]. Available: www.cadence.com

[9] D.M. Pozar, "Microwave Engineering", 3rd edition, Wiley India Pvt. Limited, 2009

Discrete low-frequency transistors subjected to high-frequency CW and pulse-modulated sine signals.

S. Jarrix, J. Raoult, A. Doridant
Institut d'Electronique du Sud
Montpellier, France
name@ies.univ-montp2.fr

C. Pouant, P. Hoffmann
CEA
DEA/SERE/LMFP
Gramat, France
Patrick.Hoffmann@cea.fr

Abstract—**Discrete low-frequency bipolar transistors are subjected to two types of interferences: CW (continuous wave) and pulsed modulated sine signal. In the goal to study the electromagnetic immunity of integrated circuits, devices are biased at low current level. Specific interference frequency bands induce changes in the transistor output voltage, even with frequency values out of band of operation of the devices. Analysis of results obtained under CW signal injection highlights the presence of physical phenomena of rectification and ac current crowding. Pulse-modulated sines show that the amplitude of the interference mean power influences the value of the output voltage offset. Parameters of the pulse interference can be changed to modify the transient response of the transistor.**

Keywords—*high frequency interference, bipolar transistor, rectification, current crowding, pulse-modulated signal, interference.*

I. INTRODUCTION

Study of wave propagation and electromagnetic compatibility (EMC) in a more general way is today one of the major concerns in electronic systems and circuits [1][2]. Indeed EMC may be an important cause of re-design in integrated circuits. Indeed high frequency interference and electromagnetic compatibility have become more important as the speed of processors and clocks have increased. Moreover the drastic rise in high-frequency communications leads to a multiplication of sources of interference in the environment. The goal today is to understand the physical mechanisms occurring during high frequency parasitic signal propagation, so as to propose optimised circuit design, reduced costs and enhanced reliability. In this paper, experiments focus on the behaviour of single bipolar BCW72 transistors subject to high-frequency interferences. These devices can be found in the early stages of low-frequency amplifiying circuits. Transistors are biased at low-level of currents, to be representative of what is found in integrated circuits. Interference is first of a CW-type, to analyse the steady-state operation of the device under disturbance. Then pulse-modulated sine signals are considered, more typical in shape of what can be found in radars. This enables the measurement of the transient behavior of the devices. Results are analysed with regard to intrinsic transistor physics but also with considerations about the

external elements such as bias tees or transistor package. The knowledge gained on discrete transistors should in the future be helpful in understanding basic phenomena occurring in integrated ciruicts under interference.

II. DEVICE SUNDER TEST AND TEST-BENCH

To gain in-depth physical knowledge of the impact of high-frequency interferences on low - frequency circuits, single silicon BCW72 bipolar transistors are under test. They are mounted in a SOT-23 package. This type of package is meant for radio-frequency operation. Transistors have a lateral-type structure. The transistor is placed upon a FR4 substrate, with no particular focus on track widths, leading to no specific impedance matching which could favor the propagation of a particular frequency value. The maximum transition frequency value of these transistors is around $f_T = 200$ MHz, hence interference frequencies, comprised between 100 MHz and 1.5 GHz, will be way out of band of operation.

Devices are biased through bias tees, at very low current level, with a maximum collector current fixed at $Ic = 10$ µA. These bias levels are characteristic of those met in integrated circuits. The mean forward collector gain $\beta = 250$. A bias collector resistance of $Rc = 1$ MΩ is placed on the collector. The value of the base bias resistance is set to $Rb = 20$ kΩ.

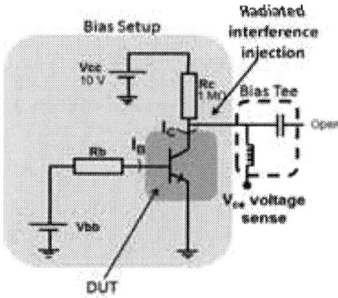

Fig. 1: Scheme of the transistor under test in a common emitter configuration, with bias set-up under contactless interference injection.

The transistors under test are in a common emitter configuration. High frequency bias tees meant for the 100 MHz – 1.5 GHz frequency band (cf. Fig. 1) are used when interference is injected. The high frequency part of the bias tee

978-1-4799-5004-1/13 $31.00 © 2013 IEEE

is left open circuited. Indeed within the frequency bands used in this study, we may consider that whatever circuit placed after the transistor, its impedance would be close to an open circuit. For a contactless mode of interference injection, the disturbance is injected through a near-field probe above the tracks leading to the collector. The mean collector voltage Vce is under observation during interference injection. This voltage is inversely proportional to the collector current, which would be a key parameter to observe, but less easier and accurate to measure. A CW interference is first injected in a radiated contactless mode through a magnetic probe, as in a classical EM scanner [3]. The choice of a magnetic probe came from the fact that it induced more dysfunctions on transistor behavior than an electric probe. This kind of interference injection is representative of an environmental interference wave leaking into a system, coupling itself on the tracks and reaching the transistor. The CW - type of disturbance allows an analysis of the transistor operation in an steady-state regime under interference. For injection of a pulse modulated sine signal, a conducted mode will be used. Indeed this time the transient behavior of the device under test is observed. The probe, if used, would induce a slight distortion in the pulse-modulated signal. In all cases, the interference power Pi mentioned in text will be the power value set on the high-frequency generator. For the contactless injected mode, it has been established that about 20 dB loss is observed at the output of the probe. All changes observed in Vce under CW and pulse-modulated interference shown here, are reproducible on the whole batch of BCW72 transistors under test.

III. TRANSISTOR BEHAVIOR UNDER A CW INTERFERENCE

A. Response of the transistor to CW interference frequency

The nominal quiescent point is given by Vce = 4.5 V, collector current Ic = 5.5 μA. Power of interference is set on the generator at Pi = 23 dBm. Interference frequency Fi varies from 150 MHz to 1.5 GHz. The mean output voltage measured is given on Fig. 2.

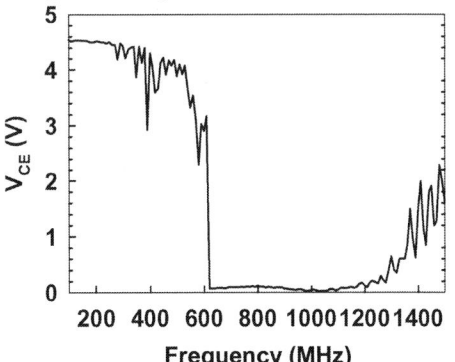

Fig. 2: Output voltage Vce under CW interference, Pi = 23 dBm, as a function of interference frequency.

The most striking fact observed is the drastic change in the value of Vce for an interference frequency band ranging from 600 MHz to 1200 MHz. For these values the output voltage

reaches Vce = 0 V. The transistor has switched from normal operation to a saturated mode. For the lower frequency band, i.e. Fi < 600 MHz, Vce values tend to decrease slowly, till the abrupt fall at Fi = 600 MHz. For Fi values above 1200 MHz, Vce values tend to increase. At Fi = 1400 MHz it still has not reached its initial value of Vce = 4.5 V. When the injection of the interference is stopped, Vce recovers its initial value without interference. These experiments show that a sinusoidal interference with frequency close to maximum frequency f_T, but still out of band of operation of the bipolar transistor, will induce mean voltage and current offsets. Now, in order to understand the existence of interference frequency bands leading to the largest disruptions in device operation, S-parameter measurements were performed.

B. S parameter measurements

The S_{22} parameter, i. e. the collector reflection coefficient, is measured with the transistor biased at the nominal operating point chosen. A Vector Network Analyzer (VNA) is linked to the output of the transistor. Values of $|S_{22}|$ are given on Fig. 3.

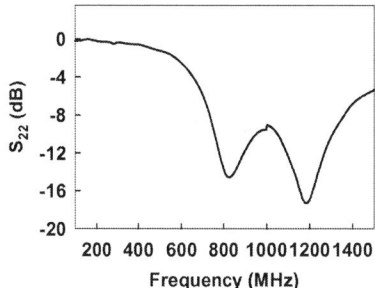

Fig. 3: Amplitude of the S_{22} parameter on the output of the system comprising transistor+bias set-up + package.

The S parameters are measured on the whole system comprising the transistor in package and the bias tees. On Fig. 3 one can notice the irregular shape of the curve as a function of frequency. Considering a frequency band ranging between 600 MHz and 1200 MHz, $|S_{22}|$ (dB) values are small. Indeed $|S_{22}|$ values are comprised between -5 dB and -18 dB. Thus, all the power of a high frequency signal is not reflected by the system, but part of it passes through the system. For frequencies over 1200 MHz, $|S_{22}|$ values increase. For the same frequency bands (600 MHz - 1200 MHz) for which a certain amount of power is not fully rejected, Vce decreases (Fig. 2), putting the transistor in a saturation mode. All impedances involved in the system under test, from the capacitances of the bias tees to the parasitics of the package and of the transistor itself, favor particular frequency values for which a high frequency signal will be less reflected. However no distinction can be made between the roles played by each element of the system with regard to the amount of reflected power. S parameters will only give an indication if a small power, high frequency signal will reach, or not, a system. It does seem interesting however to notice the link between interference frequency bands leading to operation disruptions and those for which a certain amount of high frequency power reaches the system under test.

978-1-4799-5004-1/13 $31.00 © 2013 IEEE

C. Physical phenomena involved in the response of the transistor under CW interference

It has already been established that the injection of a low-level high frequency signal on a low-frequency device leads to the phenomenon of rectification [4]. This phenomenon induces a modification in the nominal quiescent operating point during interference injection on a non-linear junction. The high frequency signal induces a small change in the mean static current of the junction, in turn leading to a small change in the mean voltage. This effect will be the more drastic if the junction is forward-biased. Bipolar transistors comprise two non-linear junctions, the base-emitter and base-collector junctions, hence the rectification phenomenon will occur. The value of the offset following rectification depends on the amount of power reaching the junction.

In addition to rectification, Richardson [4] mentions a possible ac current crowding effect which can occur under high frequency injection. To explain this physical phenomenon we first consider a small DC current. In a bipolar transistor the current lines will flow through the active base region under the emitter (Fig. 4a). A model of the emitter-base junction can be established [5]. This model represents the active area as several transistors in parallel, each in a unit zone, with common emitter and collector, and bases linked through resistances (Fig. 4b).

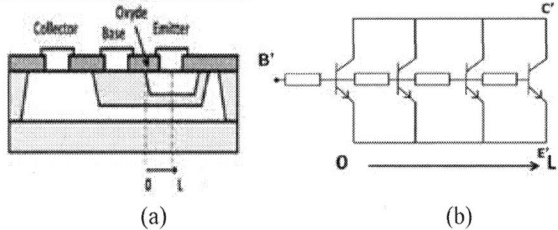

(a)   (b)

Fig. 4: (a) Section of the transistor with active base-emitter zone (b) Model of the base-emitter junction.

This batch of resistances represents a distribution of the base resistance across the active area, hence leading to different voltage potentials in each unit zone.

The DC current flowing through the junction tends to crowd around the edges of the emitter, where the voltage potential difference is the highest. Current crowding will therefore lead to less current in the center of the active zone. This phenomenon would occur mainly for high current injection. In our case the effect will be less as current levels are low. Indeed we chose to bias the devices for base currents Ib around 10 nA. However in the case of a high frequency injection, one can consider again the distributed model of the junction, based on distributed series of diodes. For frequencies above the maximum operation frequency of the diodes, they exhibit mostly a capacitive behavior [6]. This leads to the model presented in Fig. 5.

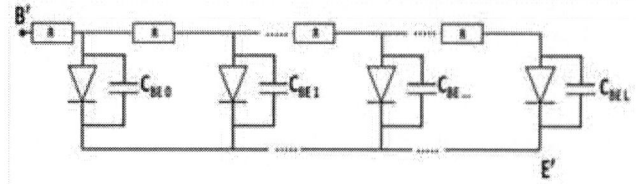

Fig. 5: ac model of the emitter –base junction.

The small amplitude ac voltage is therefore distributed though the base-emitter junction. It is shown that the maximum high frequency voltage potential occurs near the edge of the emitter and decreases in the center of the base-emitter zone.

Hence there is an ac current crowding effect, proportional to the power of the high frequency signal reaching the junction. This phenomenon adds itself to the always present rectification phenomenon. The ac current crowding is perhaps difficult to observe through our experiments, nonetheless it has been particularly put in evidence in dose-irradiated devices [7].

Now that the different physical phenomena occurring in transistors under interference have been presented, we will in the following sections analyze the transient behavior of the transistor with another type of interference.

IV. TRANSISTOR BEHAVIOR UNDER A PULSE MODULATED SIGNAL

In this section the interference is injected in a conducted mode. We first consider a single pulse, so as to gain knowledge on the transient behavior of the bipolar transistors under interference. A high frequency sine signal with frequency Fi is modulated by a pulse signal, with pulse width Wi. This type of signal, schemed on Fig. 6, has a typical shape of the signal emitted by a radar system. Usually UWB radars satisfy the UWB standard with pulse widths of 1 ns or less [8]. However in our study we will enlarge the range of possible values for pulse widths Wi to get an exhaustive analysis of the device behavior.

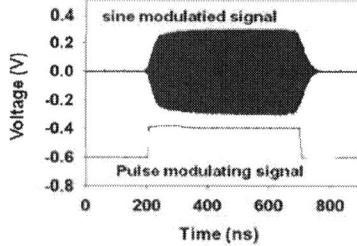

Fig. 6: Example of a pulsed-modulated sine signal, with Fi = 1.2 GHz, Wi = 500 ns, Pi = 19 dBm.

The mean output voltage of the transistor Vce is measured on an oscilloscope. The setup used is given Fig. 7.

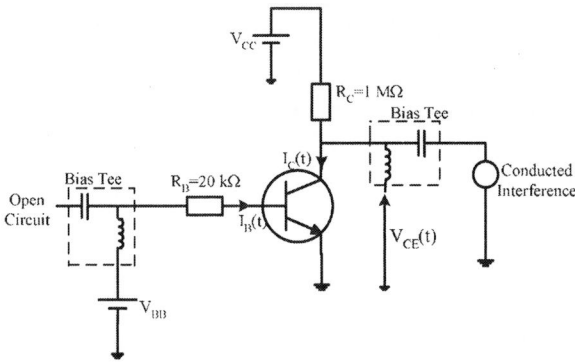

Fig. 7: Scheme of the transistor under test with bias tees, measurement oscilloscope and conducted interference injection.

The nominal quiescent point chosen is such that the collector current keeps at Ic = 5μA. Bias and measurement set-up then lead to a collector voltage Vce = 7 V. Sine frequency Fi = 860 MHz, value comprised in the frequency bands leading to the most disruptions in transistor behavior.

A. Response of a transistor to a single pulse modulated sine interference

We first choose a large pulse width Wi = 10 ms. Power of the sine signal set on the high frequency generator is Pi = 19 dBm. The interference is injected on the output of the transistor. Results are shown on Fig. 8(a) and (b).

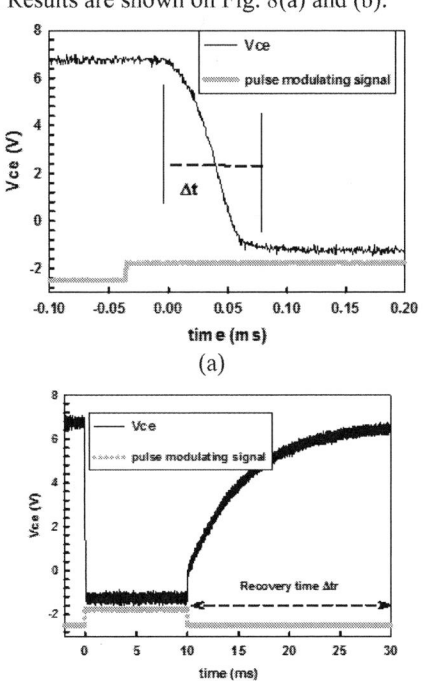

Fig. 8: Mean output voltage Vce as a function of time under a single pulse-modulated sine interference. (a) zoom on the establishment time, (b) establishement and recovery time.

First of all, one can see that the intererence induces an offset in the collector voltage with value ΔVce = 8 V. The

offset does not take place immediately, a certain amount of time, noted here "establishment time Δt", is necessary. Here Δt = 0.07 ms. Once the offset is established, Vce keeps its new value for all the remaining time the interference continues to be injected. This kind of device behavior, with ΔVce remaining at 7 V, can be compared to that of the transisors subject to a CW interference. Then, once the interference injection is stopped, the collector voltage comes back to its initial value, after a certain amount of time noted "recovery time" Δtr. Here Δtr = 20 ms.

Fig. 9 presents the collector voltage for three different values of sine power Pi : Pi = 2.5 dBm, Pi = 0 dBm, Pi = -4dBm.

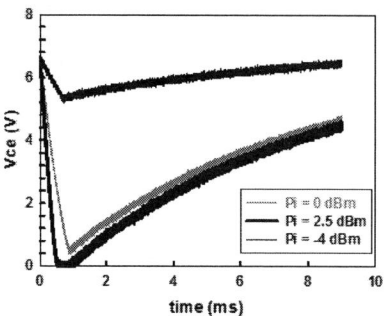

Fig. 9: Mean output voltage Vce under single pulse-modulated sine, for three different sine powers Pi, and Wi = 1 ms.

First the amount of offset ΔVce depends on power Pi. For example for Pi = - 4 dBm transistor does not reach the saturation mode as ΔVce = 1 V. Second, the large recovery time observed seems to be about the same whatever interference sine power.

It was mentioned in section III-C that under high frequency interference the capacitances of the transistor play an important role in the behavior of the device. In the case of pulse-modulated sine signal interferences, with observation of the transient behavior of the system under test, establishemnt and recovery times may be linked to capacitances. Both the base-emitter and the base-collector junctions comprise junction and diffusion capacitances. These internal capacitances will be changed under interference. Indeed it was mentioned that the high frequency interference induces a change in the quiescecnt operation point, known as the rectification phenomenon. A modification in bias will induce a change in the space charge widths, hence the junction capacitance will vary. The amount of stocked minority carriers in the base will also change. All these modifications are not immediate. Capacitance values associated with resistances lead to time constants necessary for the device to put itslef into a steady state under interference. On Fig. 8 and Fig. 9 the recovery time is very large and nearly independent on sine power. Hence, one can think that bias resistances and capacitances of bias tees and parasitics of the package also greatly intervene in the time constants. In the same way, though perhaps with less importance, the establishment time Δt can be linked to time constants themselves due to the system (bias tees, package) and to the intrinsic capacitances of the transistor.

B. Electrical response of a transistor to a train of pulse modulated sine interferences

In this section we focus on the effect of a train of pulses. This time we will take into account the pulse repetition frequency, noted P_{RF}. One can define a mean power by:

$$P_{mean} = Pi \ Wi \ P_{RF}$$

Output voltage is kept to Vce = 7 V. Sine frequency is again fixed to Fi = 860 MHz.

First, the pulse width is set to Wi = 5 µs and pulse repetition frequency P_{RF} = 1 kHz. Peak power of the sine is Pi = 19 dBm, leading to a mean power P_{mean} = -4 dBm. On Fig. 10 one can see that the voltage offset ΔVce increases with the number of pulses. After each end of pulse the voltage tends to go back to initial value of Vce. Time between two pulses is probably much shorter than the necessary recovery time Δtr measured with a single pulse at the same interference power. ΔVce reaches its maximal value, here ΔVce = 7 V, after a certain amount of time measured to be Δt = 28 ms.

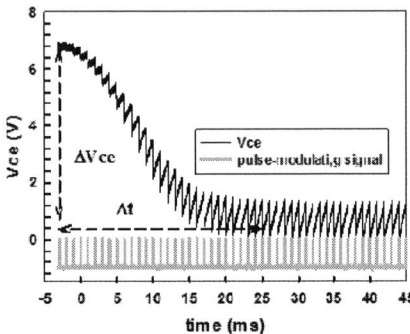

Fig. 10: Mean output voltage Vce under a train of pulse-modulated sine interference with P_{RF} = 1 kHz, Pi = 19 dBm, Wi = 5 µs.

We can observe in Fig. 10 that the transistor reaches the saturation mode. Let's now consider what happens for different interference parameters.

We first change the value of the pulse width, it is decreased to Wi = 1 µs. Mean output voltage measured is given on Fig. 11.

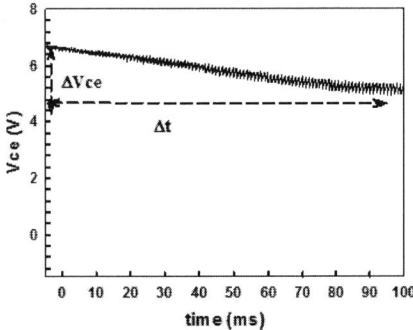

Fig. 11: Mean output voltage Vce under a train of pulse-modulated sine interference P_{RF} = 1 kHz, Pi = 19 dBm, Wi = 1 µs.

The offset voltage value is much less than for interference with Wi = 5 µs, indeed this time ΔVce ≈ 1.5 V. In this case P_{mean} = -11 dBm. One can think that there is not enough power

for a big change in the quiescent point to occur. Compared to what is obtained for higher values of pulse widths, the establishment time Δt is much longer.

Then the power Pi of the sine carrier is decreased from Pi = 19 dBm to Pi = 0 dBm. Pulse width is kept to Wi = 5 µs (cf. Fig. 9). Mean power is then even lower, P_{mean} = -23 dBm. With respect to results obtained with a single pulse at different powers, it can be assumed that the transistor will not reach the saturation mode.

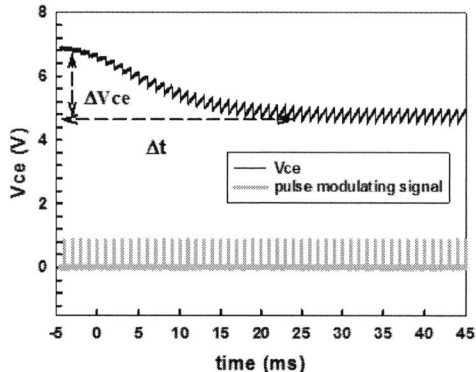

Fig. 12: Mean output voltage Vce under a train of pulse-modulated sine interference with P_{RF} = 1 kHz, Pi = 0 dBm, Wi = 5 µs,.

Compared to Fig. 10 where sine power is large (Pi = 19 dBm), the offset induced is much smaller, ΔVce = 1.9 V. The transistor does not reach the saturation mode, as expected. Not enough power reaches the intrinsic transistor to induce a complete switch in behavior. The time needed to reach the maximum value for ΔVce is however equivalent to the Δt measured with a higher Pi value.

Now let's consider what happens if the pulse repetition frequency P_{RF} is changed. We choose a larger value, P_{RF} = 10 kHz. Again we keep Wi = 5 µs and Pi = 19 dBm. The mean power P_{mean} = 5.9 dBm.

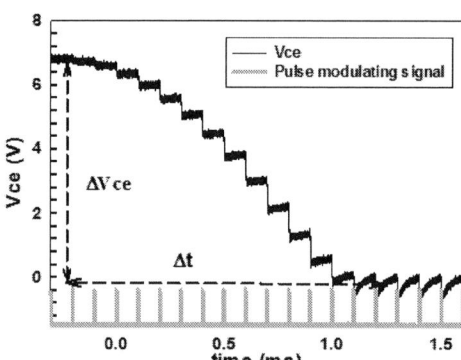

Fig. 13: Mean output voltage Vce under a train of pulse-modulated sine interference with P_{RF} = 10 kHz, Pi = 19 dBm, Wi = 5 µs,.

Mean power is sufficient for the transistor to reach the saturation mode. In the same time Δt = 1 ms, which is much shorter than what was measured for the same interference with P_{RF} = 1 kHz. With P_{RF} = 10 kHz the repetition period between two pulses is 0.1 ms. This value must be much smaller than the recovery time necessary for the whole system to come back to its initial value. The value of the voltage offset therefore increases with each pulse.

CONCLUSION

Low-frequency bipolar transistors, of reference BCW72, were submitted to high-frequency interferences. Interference was first a CW sine and then a pulse-modulated sine. For all interferences, an offset ΔVce in the mean output voltage was observed. The process of rectification of a high frequency signal by a semiconductor junction can explain part of the offset observed. An ac current crowding effect can occur in addition to rectification. A model of a distributed base resistance is given to highlight this last physical phenomenon. The importance of internal transistor capacitances in the change of Vce value is discussed. However, it is shown, particularly under pulse-sine interference, that the global response measured comes from the whole system, transistor in package with bias set-up. All elements comprise parasitic capacitances and resistances playing a role in the different time constants measured during transient behavior of the device. Parameters such as the mean power or pulse repetition frequency will influence the amount of voltage offset taking place during interference injection.

REFERENCES

[1] M. Ramdani, E. Sicard, A. Boyer, S. B. Dhia, J. J. Whalen, T. H. Hubing, M. Coenen, and O. Wada, "The electromagnetic compatibility of integrated cuits—Past, present, and future," *IEEE Trans. Electromagn. pat.*, vol. 51, no. 1, pp. 78–100, Feb. 2009.

[2] E. Sicard, S. B. Dhia, M. Ramdani, J. Catrysse, and M. Coenen, "Towards an EMC roadmap for integrated circuits," Electromagn. pat. (EMC) Compo., 2007, Tourino, Italy.

[3] A. Boyer, S. Bendhia, E. Sicard, "Characterisation of electromagnetic susceptibility of integrated circuits using near-field scan", Electron. Letters, vol. 43, n°.1, pp. 15-16, January 2007.

[4] R. E. Richardson "Quiescent Operating Point in Bipolar Transistors with AC Excitation", IEEE Journal of Solid-State Circuits, vol. sc-14, n°6, p. 1087-1094, Dec. 1979.

[5] J.C.J Paasschens, "Compact modeling of the noise of a bipolar transistor under DC and AC current crowding conditions", IEEE Trans. on Electron. Devices, vol. 51, n° 9, pp 1483-1495, 2004.

[6] H. Ghosh, "A distributed model of the junction transistor and its application in the prediction of the emitter base diode characteristic, base impedance, and pulse response of the device", IEEE Trans. on Electron. Device, vol. ED-12, pp 513-531, 1965.

[7] A. Doridant, S. Jarrix, J. Raoult, A. Blain, N. Chatry, P. Calvel, P. Hoffmann, L. Dusseau, "Impact of Total Ionizing Dose on the Electromagnetic Susceptibility of a single bipolar transistor, "IEEE Transactions on Nuclear Science, vol. 59, n°4, pp. 860-865, août 2012.

[8] IEEE STD 1762 Standard for Ultrawideband radars definition.

Noise Analysis using On-Chip waveform Monitor in Bandgap Voltage References

Akitaka MURATA, Shuji AGATSUMA, Daisaku
IKOMA, Kouji ICHIKAWA, Takahiro TSUDA
DENSO CORPORATION
Kariya, Aichi 448-8661, Japan
akitaka_murata@denso.co.jp

Makoto NAGATA, Kumpei YOSHIKAWA, Yuuki
ARAGA, Yuji HARADA
Graduate School of System Informatics
Kobe University
Kobe, Hyogo 657-8501, Japan

Abstract— **In this paper, the susceptibility of a CMOS bandgap voltage reference (BGR) to external noise was investigated using an on-chip waveform monitor circuit in conjunction with circuit simulations. A Direct RF Power Injection method was employed for the immunity test of the BGR. Also, we evaluated the performance of the on-chip waveform monitor and analyze the BGR immunity using the on-chip monitor. As the results, we have clarified the mechanism of the BGR malfunction. The output voltage drop of the BGR was caused by the offset of operational amplifier in BGR.**

Keywords-vehicle,immunity, on-chip waveform monitor, bandgap voltage reference

I. INTRODUCTION

In the worldwide automotive market, a number of new products are emerging markedly in the fields of safety, environmental friendliness, and telecommunications. They are adopted new technologies such as ISS (Idling Stop System), HV (Hybrid Vehicle) / EV (Electric Vehicle) technologies and wireless technologies employing smart phones. Under these circumstances, it is foreseen that automotive electronic components are rapidly growing in their systems. Mechanical and electrical systems integration is quickly developing.

An electronic device is a key factor in terms of connection between various systems. Moreover, electronic devices are indispensable to achieve high performance, low cost, and further reduction in size of automotive systems. These devices require high performance and high reliability more than ever. Electronic devices used in vehicles must have high noise immunity. However, as advances in process technology and reduction of power supply voltages continue, to achieve sufficient noise immunity is a quite challenging task for circuit designers.

As for immunity of LSIs, many studies have been carried out through different approaches to ensure noise immunity [1]. For example, it is reported the studies as for on-chip waveform monitors (OCM) [2]-[4] and the immunity of bandgap voltage references (BGR) [5]-[10]. There are a few studies of noise propagation paths on BGRs in mixed signal LSIs.

In automotive electronics, module products such as Electric Control Units （ECU） and sensors are used in many systems. They are equipped with power supply circuitries. Many of these power supplies have BGR core circuits. If LSI chips are sensitive to malfunction of BGRs due to external noise, automotive systems embedded these circuits may experience to serious failures.

Therefore, it is quite important for automotive electronic devices to ensure noise immunity, particularly BGRs. We have to correctly understand the mechanism of malfunction to noise and develop noise models to simulate noise immunity before manufacturing. Ultimately, we need to establish design methodologies for high noise immunity LSIs.

In this paper, the impact of external noise applied to BGR is measured using on-chip waveform monitor. After that we analyze BGR malfunction using circuit simulations.

In section II, we explain the circuit under test, i.e. a CMOS BGR circuit. Next, we show the on-chip monitor circuitry in section III. Section IV describes the experimental set-up. A Direct RF Power Injection (DPI) method is explained. In section V, experimental results are presented. On-chip monitor performance is examined. Section VI depicts the analysis of BGR malfunction and corresponding circuit simulation results. Finally, we summarize this paper and refer to the remaining issues in section VII.

II. CIRCUIT UNDER TEST

Figure 1 (a) shows the outline of our test chip. The test chip has a CMOS BGR circuit and an on-chip waveform monitor circuit. To ensure the power supply of the on-chip monitor to be clean, we separate power supplies for the on-chip monitor and the BGR. And we laid out them with an appreciable distance not to interfere with each other via the substrate. The on-chip waveform monitor circuit measures voltage fluctuations at internal nodes of the BGR. The configuration of the BGR is presented in Figure 1 (b).

978-1-4799-5004-1/13 $31.00 © 2013 IEEE

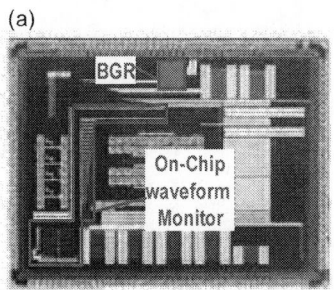

(a)

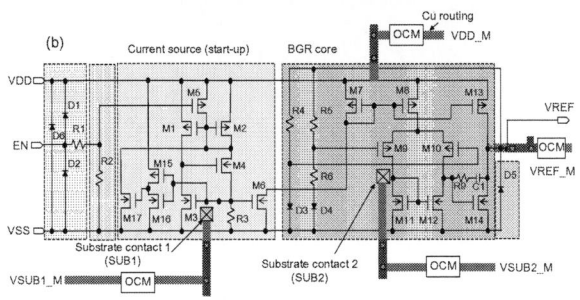

(b)

Figure 1. (a) Test chip photograph, (b) Bandgap reference (BGR) circuit configuration.

The BGR circuit is composed of a constant current circuit for start-up and a BGR core circuit. The BGR contains a standard operational amplifier with a PMOS input differential pair. The power supply voltage is 2.5 V and the BGR output voltage（VREF）is 1.2 V.

Voltage fluctuations at BGR circuit nodes due to the induced electromagnetic interferences (EMI) from the outside environment are observed using the on-chip monitor circuit connected to BGR internal nodes through Cu interconnects.

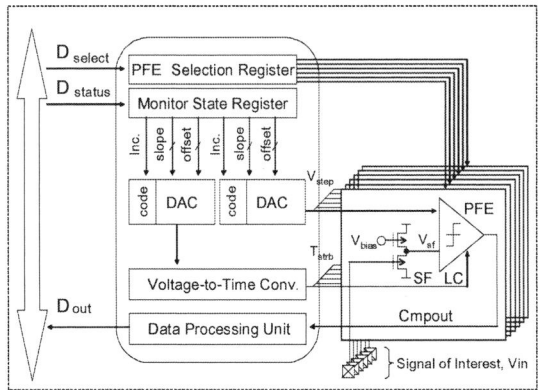

Figure 2. On-Chip waveform Monitor (OCM) architecture.

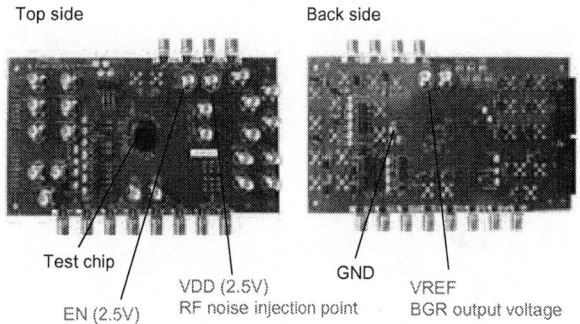

Figure 3. Test board used for DPI experiments.

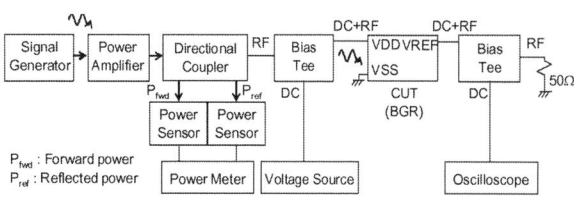

Figure 4. DPI experimental set-up.

The number of IC terminals is four: VDD, VSS, VREF (the output voltage of the BGR), and EN (enable). All four terminals are connected to ESD protection circuits.

III. ON-CHIP WAVEFORM MONITOR

We present the on-chip waveform monitor (OCM) architecture in Figure 2. An internal node signal of interest is captured by the Probing Front-End circuit (PFE), followed by digitalization using the latched comparator (LC). Under the search algorithm proposed in [11], [12], controlling reference voltages applied to LC and sampling timing of LC by DACs, the digitalized time domain voltage waveforms are measured with an on-chip manner.

The monitored waveforms are immediately on-chip digitized within the OCM and therefore immune to the noises and parasitic effects. We can further process the digital on-chip waveform data with digital circuits such as microprocessors. The on-chip waveform monitor is able to measure a variety of noises from the ground bounce to the supply bounce, and cover a wide range of noise amplitudes from a few mV to several hundreds mV. In this study, to measure the voltage fluctuations of the BGR due to the induced EMI, we set up the condition in which the voltage resolution of the on-chip waveform monitor is 0.8 mV, and the voltage range to be covered is 820 mV.

978-1-4799-5004-1/13 $31.00 © 2013 IEEE   227

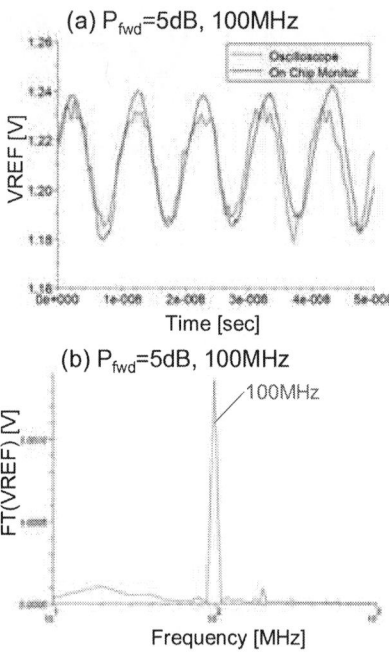

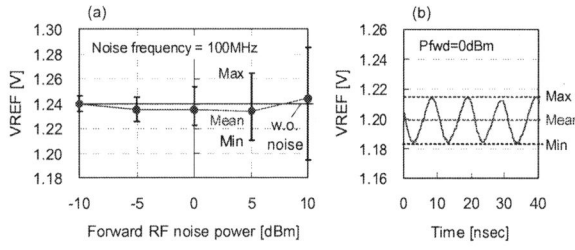

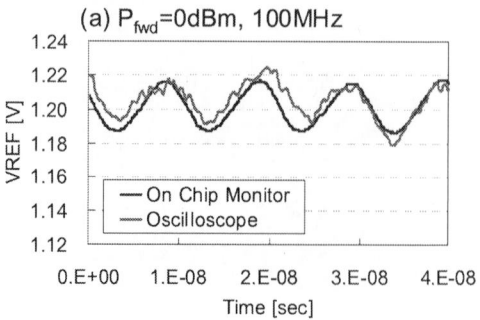

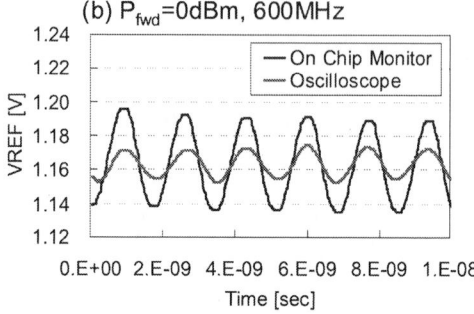

Figure 5. (a) A measured On-Chip Monior waveform at the BGR output voltage (VREF) compared with a oscilloscope waveform, and (b) its FFT spectrum. DPI forward power is 5dBm at 100MHz superimposed on VDD.

Figure 7. Comparison between On-Chip-Monitor waveforms and off-chip oscilloscope results. (a) P_{fwd}=0dBm, 100MHz (b) P_{fwd}=0dBm, 600MHz.

Figure 6. (a) Forward RF noise power dependence of the BGR output voltage (VREF) fluctuation, and (b) the definition of indices using in (a).

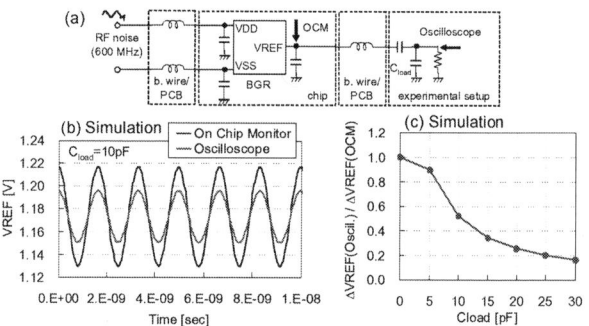

Figure 8. Simulation results of time domain VREF waveforms. Noise frequency is 600 MHz. (a) A simulation circuit diagram. (b) VREF waveforms (C_{load} = 10 pF) at on-chip node and off-chip one. (c) C_{load} dependence of relative amplitudes of VREF fluctuations, off-chip to on-chip.

IV. EXPERIMENTAL SET-UP

Figure 3 illustrates the 6-layer test printed board in which the test chip is placed at the center.

Susceptibility tests of LSIs to externally induced EMI are standardized by IEC 62132. A Direct Power Injection (DPI) test is one of the conducted immunity tests [13]. We show the experimental set-up for the DPI test in Figure 4. In the DPI test, an RF signal generated by the signal generator is amplified by the power amplifier. And the amplified RF signal is injected to the pin of the LSI through the bias-tee. The RF disturbance is monitored through the directional coupler by measuring the forward power (P_{fwd}) and the reflected power (P_{ref}) using the power meters.

DC bias is applied to the DC terminal of the bias-tee and superimposed on the RF signal. In the next section, we explain the DPI test results.

978-1-4799-5004-1/13 $31.00 © 2013 IEEE   228

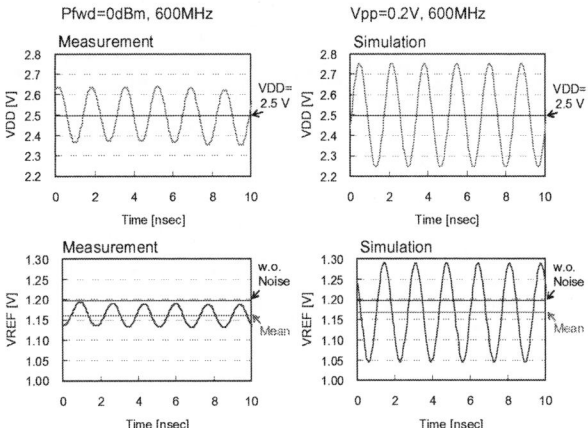

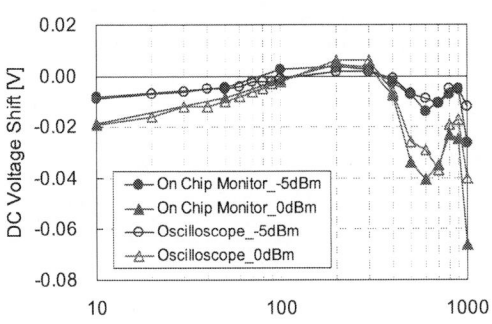

Figure 9. DC voltage shifts of the BGR output voltages under RF noise using DPI. DPI forward powers are -5/0dBm and RF noise frequency range is from 10MHz ~ 1GHz.

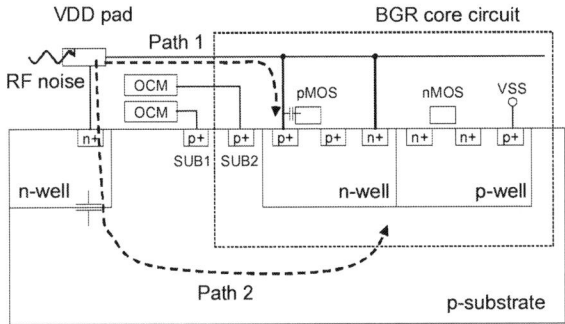

Figure 10. Possible RF noise propagation paths.

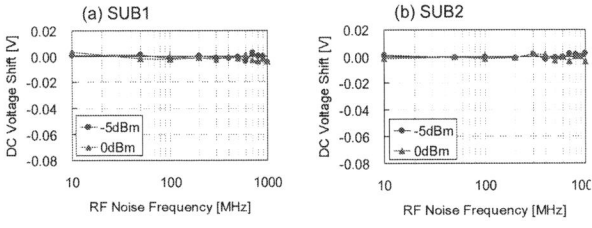

Figure 11. (a) DC voltage shifts of substrate (SUB1), referred to Figure 1, under DPI tests. (b) DC voltage shifts of substrate (SUB2). DPI forward powers are -5/0dBm and RF noise frequency range is from 10MHz ~ 1GHz in both experiments.

Figure 12. Comparison between measurement waveforms captured by the On-Chip Monitor and SPICE simulation waveforms. The upper graphs are correspond to VDD noise waveforms, and the lower ones are VREF waveforms.

V. EXPERIMENTAL RESULTS

Performance parameters to characterize the on-chip monitor circuit are linearity, a bandwidth, input/output voltage swings, and time/voltage resolutions, and so forth.

Figure 5 shows experimental results measured by the on-chip waveform monitor and an off-chip oscilloscope. RF noises are injected to the power supply pin of the BGR using DPI (P_{fwd}=5dBm, RF Noise frequency=100MHz), and the output voltage waveforms are measured. In Figure 5(a), both on-chip and off-chip measurements provide almost the same waveforms. The FFT (Fast Fourier Transform) spectrum of the on-chip monitor waveform has a sharp peak exactly at 100MHz (Figure 5(b)). These results confirm that the on-chip waveform monitor is properly functioning.

On-chip monitor waveforms at output voltages of the BGR are dependent on forward power of DPI experiments (Figure 6). These are the results of mean, max, and min of the waveforms in terms of forward RF noise power. We confirm that the on-chip monitor acquire a broad range of noise amplitudes from 10 mV to a few hundred mV.

We compare the time domain waveforms during DPI experiments at 100 MHz and 600 MHz in Figure 7. At 100 MHz where VREF does not drop, both on-chip monitor and oscilloscope measurements provide almost the same waveforms as mentioned above.

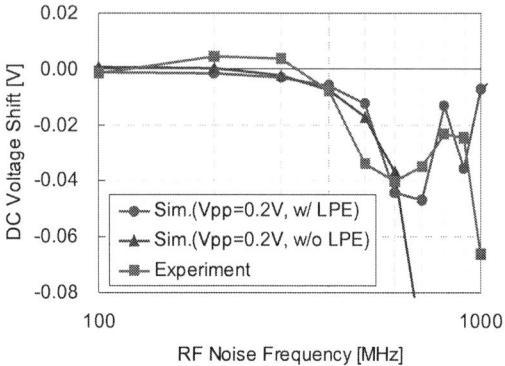

Figure 13. DPI simulation results of DC voltage shift at BGR output voltages.

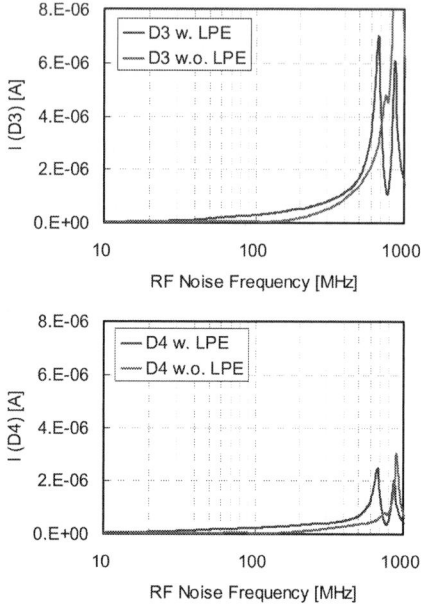

Figure 15. Simulation results of AC currents throu BGR diodes (D3, D4).

On the other hand, at 600 MHz where VREF shows a dip, the noise amplitudes in both cases are different, although the DC shifts are at the same level. It indicates that the waveforms of oscilloscope measurements attenuate in amplitude at high frequencies. In Figure 8, we present simulation results of VREF waveforms measured at an on-chip node and an off-chip node. The off-chip waveform amplitude is smaller than the on-chip one (Figure 8 (b)). This situation is similar to the experimental results of Figure 7 (b). The relative amplitudes of VREF fluctuation is decreased as C_{load} is increased (Figure 8 (c)). From these results, the on-chip monitor has an advantage to the off-chip oscilloscope measurement with respect to capturing high frequency time domain waveforms without attenuation in amplitude.

Figure 9 demonstrates the DC shifts of the BGR output voltage in the DPI experiments. These results are observed with the on-chip waveform monitor circuit where the RF noise frequencies are up to 1 GHz. The output voltage worsens to cause malfunction at 2 points and reduces by 0.04 V at around 600 MHz and by 0.07 V at around 1 GHz. Both the on-chip monitor and the oscilloscope measurement results show the same trend of DC shifts.

VI. DISCUSSION

In this section, we examine the mechanism of malfunction of the BGR to external noise. Firstly we identify the noise path through silicon, secondly we analyze the malfunction mechanism. Figure 10 shows a cross section of the chip.

At the first step, we identify the relevant noise path. We assume that the noise could mainly propagate through Cu interconnects or a silicon substrate. In DPI tests, where the noise is applied to the VDD terminal, the substrate potential waveforms captured with on-chip monitor circuit are shown in Figure 11. The SUB1 is located around the differential inputs of the operational amplifier, the SUB2 is around the constant current circuit. Both DC voltage shifts at the SUB1 node and the SUB2 node are confirmed within ±1 mV. These results indicate that the RF noise from the VDD does not shift the substrate potentials and affects the BGR coupled through Cu interconnects. For the above analyses, the on-chip monitor is quite useful to identify the noise propagation path, since the on-chip monitor is not influenced by the off-chip parasitics.

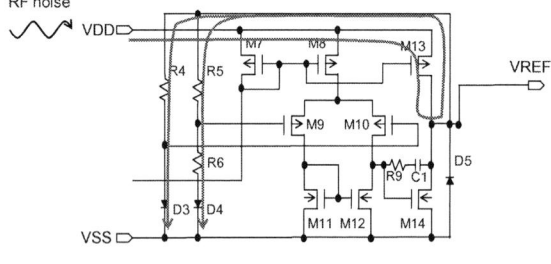

Figure 14. RF noise propagation path model.

At the second step, we examine the mechanism of malfunction to external noise using the on-chip waveform monitor and circuit simulations. Figure 12 depicts the experimental and simulation waveforms at VDD and VREF terminals. In simulations, the VDD fluctuation amplitude is Vpp = 0.2 V leading to the same level of VREF DC shift at 600 MHz compared with the experimental one. Figure 13 is the simulation and the experimental results of the DC voltage shifts of the BGR output voltage. The solid circles present the simulation results using the Layout Parameter Extraction (LPE) netlist at Vpp = 0.2 V. Although the noise amplitude is more than double (Vpp = 0.2 V) compared with the on-chip monitor result (Vpp ~ 0.1 V at 600 MHz), this simulation shows a qualitative agreement with the measurement results. On the other hand, the simulation results without layout parasitics (solid triangles) deviates from experimental results. This means that the LPE modeling accuracy is very important for noise simulation.

The RF noise propagation paths to cause malfunction are shown in Figure 14, and each simulation current of diodes (D3 and D4) is shown in Figure 15. The offset voltage which is the potential difference across the resistance (R6) is increased by resonance current via power source VDD. The voltage of the negative input (M9 gate) of the operational amplifier is higher than that of the positive input (M10 gate). Then, the output voltage or VREF is dropped. In short, the voltage drop is caused by the offset voltage of the operational amplifier.

Also, Figure 15 shows that the simulation currents of D3 and D4 demonstrate the different behavior with or without LPE. We assume that the alteration of the currents depends on the simulation model, the accuracy of LPE, and any other factor.

Considering the discussion mentioned above, we have clarified the mechanism of malfunction to external noise using both the on-chip waveform monitor and circuit simulations where the output voltage drop depends on the offset of the operational amplifier. Remaining problem is to improve the accuracy of circuit simulation, i.e. SPICE simulation models, layout parasitic extraction models, substrate models, experimental-setup parasitics models, and so forth.

VII. CONCLUSION

We have clarified the mechanism of malfunction on a CMOS bandgap voltage reference to external noise using both the On-chip waveform monitor and circuit simulations, and the output voltage drop is caused by the offset of an operational amplifier. The origin of the offset is found out to be the resonant current.

Moreover, we make sure that it is able to measure accurate on-chip waveforms using the on-chip waveform monitor from dc to 1 GHz. The input voltage range of this on-chip monitor is from a few mV to 600 mV.

Future works are to investigate the cause of the resonance current to external noise and to improve the accuracy of circuit simulations for noise analyses.

ACKNOWLEDGMENT

The authors would like to thank Ms. Kanamori of DENSO CORP. for her support.

REFERENCES

[1] P. Schröter, S. Jahn, and F. Klotz, "Improving the Immunity of Automotive ICs by Controlling RF Substrate Coupling," *in proc. Intl. Workshop on the Electromagnetic Compatibility of Integrated Circuits (EMC compo 2011)*, pp. 182-187, Nov. 2011.

[2] K. Ichikawa, Y. Takahashi, Y. Sakurai, T. Tsuda, I. Iwase, and M. Nagata, "Measurement-Based Analysis of Electromagnetic Immunity in LSI Circuit Operation," *IEICE Trans. on Electronics*, vol. E91-C, no. 6, pp. 936-944, 2008.

[3] K. Yoshikawa, Y. Sasaki, K. Ichikawa, Y. Saito, and M. Nagata, "Measurements and co-simulation of on-chip and on-board AC power noise in digital integrated circuits," *in proc. Intl. Workshop on the Electromagnetic Compatibility of Integrated Circuits (EMC compo 2011)*, pp. 76-81, Nov. 2011.

[4] A. Boyer, S. Ben. Dhia, C. Lemoine, and B. Vrignon, "An on-chip sensor for time domain characterization of electromagnetic interferences," *in proc. Intl. Workshop on the Electromagnetic Compatibility of Integrated Circuits (EMC Compo)*, pp. 251-256, No. 2011.

[5] A. Pretelli, A. Richelli, L. Colalongo, and Z. M. Kovacs-Vajna, "Reduction of EMI susceptibility in CMOS bandgap reference circuits," *IEEE Trans. Electromagn. Compat.*, vol. 48, no. 4, pp. 760–765, Nov. 2006.

[6] F. Fiori, and P. S. Crovetti, "Investigation on EMI effects in bandgap voltage references," *in proc. Intl. Workshop on the Electromagnetic Compatibility of Integrated Circuits (EMC compo 2002)*, pp. 35-39, Nov. 2002.

[7] J.-M. Redoute, and M. Steyaert, "Kuijk bandgap voltage reference with high immunity to EMI," *IEEE Trans. Circuits Syst. II*, vol. 57, no. 2, pp. 75-79, Feb. 2010.

[8] E. Orietti, N. Montemezzo, S. Buso, G. Meneghesso, A. Neviani, and G. Spiazzi, "Reducing the EMI susceptibility of a Kuijk bandgap," *IEEE Trans. Electromagn. Compat.*, vol. 50, no. 4, pp. 876-886, Nov. 2008.

[9] Y. Gao, K. Abouda, and A. Huot-Marchand, "Bandgap circuitry with high immunity to harsh EMC disturbances," *in proc. Asia-Pacific Symposium on Electromagnetic Compatibility*, pp. 389-392, May 2012.

[10] O. Jović, U. Stürmer, W. Wilkening, and A. Barić, "Susceptibility of a Brokaw Bandgap to High Electromagnetic Interference," *in proc. Intl. Workshop on the Electromagnetic Compatibility of Integrated Circuits (EMC compo 2009)*, Nov. 2009.

[11] M. Nagata, Y. Kashima, D. Tamura, T. Morie, and A. Iwata, "Measurements and analyses of substrate noise waveform in mixed-signal IC environment," *in proc.* CICC, 2002.

[12] T. Hashida, and M. Nagata, "On-chip waveform capture and application to diagnosis of power delivery in SoC integration," *IEEE J. Solid-State Circuits*, vol. 46, no. 4, Apr. 2011.

[13] IEC 62132-4 Integrated Circuits - Measurement of Electromagnetic Immunity - 150 kHz to 1 GHz - Part 4: Direct RF Power Injection Metho

978-1-4799-5004-1/13 $31.00 © 2013 IEEE

Immunity Evaluation of Inverter Chains against RF Power on Power Delivery Network

Kumpei Yoshikawa[1], Yuji Harada[1], Noriyuki Miura[1], Noriaki Takeda[2], Yoshiyuki Saito[2] and Makoto Nagata[1]

[1]Graduate School of System Informatics, Kobe University 1-1 Rokkodai-cho, Nada-ku, Kobe 657-8501, Japan

[2]Panasonic Corporation, 1006 Kadoma, Kadoma-shi, Osaka, 571-8501, Japan

Email : {kumpei, harada, miura, nagata}@cs26.scitec.kobe-u.ac.jp, {takeda.noriaki, saito.yoshiyuki}@jp.panasonic.com

Abstract—Direct RF power injection on a power delivery network causes timing variations of inverter chains. The amount of period jitter in an inverter chain is strongly dominated by the frequency and amplitude of sinusoidal voltage variations on its internal power supply nodes. The conduction and conversion characteristics of the RF power from an external point of injection to the sinusoidal voltage variation on the node within a chip are modeled with a chip-package-board integrated network. The period jitter is calculated in response to the sinusoidal waveform with the voltage-dependent delay characteristics of an inverter stage. The external RF power is therefore analytically related with the period jitter of an inverter chain. Comparisons are made between the calculation and measurements for a 65 nm CMOS prototype chip featuring on-chip voltage waveform monitoring functions.

Key words : CMOS, jitter, direct power injection, electromagnetic interference

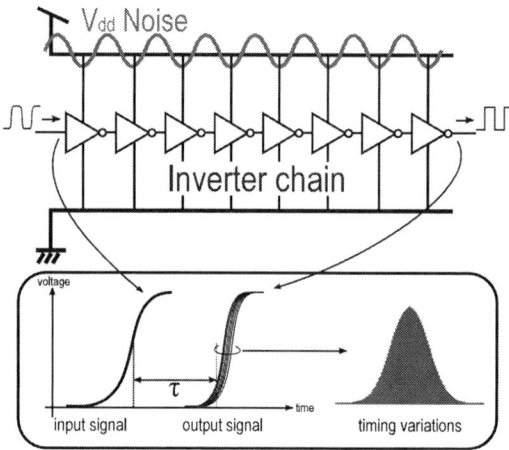

Fig. 1. Timing variations due to voltage variation in inverter chain.

I. INTRODUCTION

Data links with very wide bandwidths and low voltage, low power consumption are demanded in any sort of computation facilities covering from high-end servers to mobile terminals. High-speed data linkage requires precise timing alignments between memory blocks and processing elements in a very large scale integration (VLSI) chip. Clock buffers take a key role in clock and timing delivery of a data link system, by shaping a clock signal waveform and introducing an expected delay for controlling skews. However, it is well known that the delay is very susceptible to the variation of power supply voltage given to buffers. The frequency and amplitude of interference induced on a power delivery network (PDN) influence the amount of jitter. The large amount of jitter can corrupt the data linkage and may halt an entire computation flow.

The frequency generation for clocking often relies on autonomous oscillation of circuits such as ring oscillators or inductor-capacitor (LC) tanks. The frequency components of oscillations inevitably incorporate spurious harmonics due to non-linear mixing or coupling of external periodical signals and introduce phase uncertainties and result in jitter. The analysis of such oscillators deals with their internal feed-back nature [1-5] and different from clock buffers normally in an open loop topology [6]. This paper discusses the period jitter of an inverter chain as a clock buffer.

The susceptibility of jitter in inverter chains against RF power induced on their PDN is discussed in this paper, by using the direct power injection (DPI) method [7-9]. The conductive RF power is injected into the test vehicle by DPI. The transfer characteristics will be thoroughly derived for the sinusoidal voltage variations of PDN caused by DPI into the amount of period jitter in the inverter chains. Theoretical calculations will be compared with measurements on a test vehicle in a 65 nm CMOS technology.

The remaining part of this paper is organized as follows. Section II describes the transfer model of sinusoidal power supply voltage variations into jitter. Section III addresses the chip-package-board integrated PDN model capturing DPI characteristics. Section IV defines a silicon test vehicle and compares measurement results with the analysis. Section V finally gives a brief summary of this paper.

II. JITTER IN INVERTER CHAIN

Inverters in cascade form an inverter chain, as depicted in Fig. 1. It buffers a rectangular shape of an incoming logic signal and also adds an intentional delay time, τ, to an outgoing signal. The length of a chain is a design parameter for adjusting timing, often for ensuring relative time positions between data and clock signal paths. Each inverter stage has a small delay, and the chain levels out the static variation

978-1-4799-5004-1/13 $31.00 © 2013 IEEE

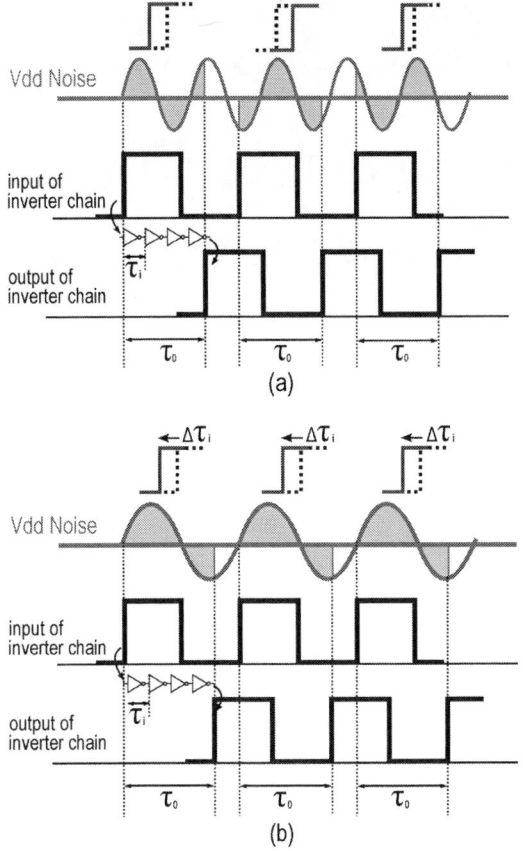

Fig. 2. Dependency of timing variations on frequency of sinusoid. (a) $f_{\text{clk}} \neq f_m$, (b) $f_{\text{clk}} = f_m$

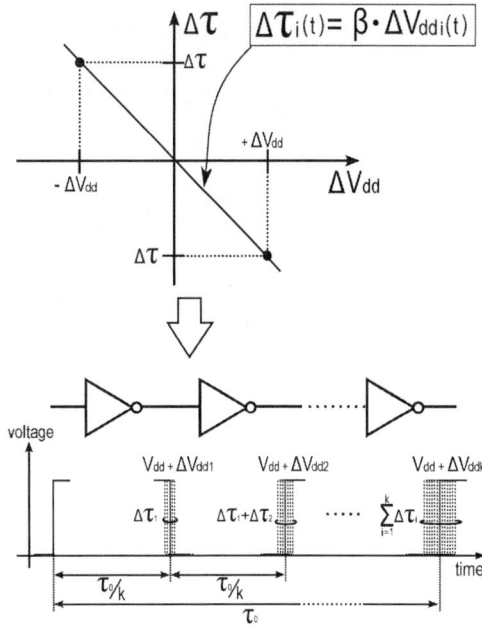

Fig. 3. Delay vs. DC voltage and integral of DC voltage.

$$
\begin{aligned}
\tau &= \sum_{i=1}^{k} \tau_i \\
&= \sum_{i=1}^{k} \left(\tau_0/k + \Delta\tau_i \right) \\
&= \tau_0 + \int_{n/f_{\text{clk}}}^{\tau_0 + n/f_{\text{clk}}} \beta \cdot V_{\text{chip}} \cdot sin\left(2\pi f_{\text{m}} t + \alpha\right) dt \quad (1)
\end{aligned}
$$

The value of V_{chip} represents an amplitude of sinusoid. The deviation of $\tau(n)$ in the nth clock period from the long-time average of τ, that approaches to τ_0 in an ideal case, defines the size of an instantaneous timing variation. The cycle-to-cycle jitter, $J_{\text{p}}(n)$ is defined as the difference of such timing variations between two consecutive clock periods, as (2).

$$
J_{\text{p}}(n) = \tau(n) - \tau(n-1) \quad (2)
$$

The period jitter is calculated as the standard deviation of J_{p} as in (3), where $\bar{J}_{\text{p}}$ is the average of period jitter over the number of continuous clock cycles, normally more than 10,000 in the measurements or simulations.

$$
\sigma = \sqrt{\frac{\sum_{n=0}^{s} \left(J_{\text{p}}(n) - \bar{J}_{\text{p}}\right)^2}{s-1}} \quad (3)
$$

It should be noticed that when the incoming clock signal has the frequency identical with a singular interference of interest, the integration of (1) results in the same delay time for any clock period and therefore the period jitter of (3) becomes zero, as sketched in Fig. 2(b).

of delays among inverter stages and regulates the standard deviation of their timing. However, the amount of τ exhibits a susceptibility to the variation of a power supply voltage, and its variation in a time domain appears as jitter.

We introduce the simplest expression for the calculation of τ in an inverter chain, as given in Fig. 2, assuming that the power supply voltage has a sinusoidal shape at the frequency of the most significant interferences, f_{m}. The voltage then exhibits the nominal DC value of V_{dd} with the time-varying sinusoidal deviation of $\Delta V_{\text{dd}}(t)$. The chain buffers a pure clock signal at the frequency of f_{clk} from an external source and continuously generates a delayed version of the signal with τ. While each inverter stage responds to $\Delta V_{\text{dd}}(t)$ within the small time of τ_i of its operation at t, an inverter chain as a whole integrates the timing variations for the length of inverter stages and provides τ according to (1). Here, it is intuitively considered that each stage leads a nominal delay of τ_0/k with the deviation of $\Delta\tau_i$ that is almost linearly responding to $\Delta V_{\text{dd}}(t)$ with the 1st order coefficient of β, as in Fig. 3. The number of stages in an inverter chain is k.

978-1-4799-5004-1/13 $31.00 © 2013 IEEE   233

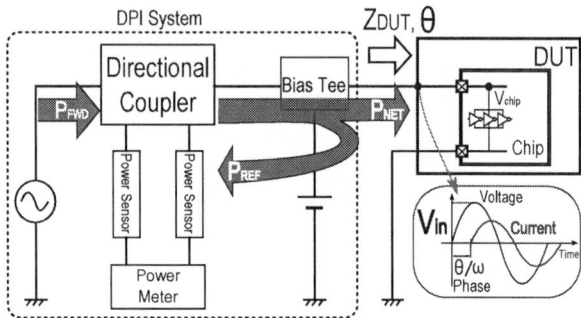

Fig. 4. DPI system diagram.

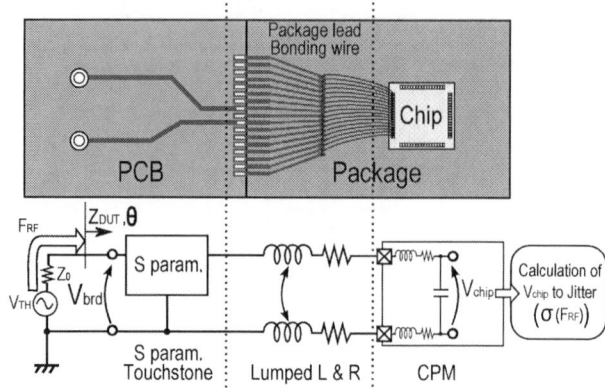

Fig. 5. Chip-package-board integrated PDN model.

TABLE I
INVERTER CHAINS WITH DESIGN PARAMETERS

Name	# of driving inv.	# of loading inv.
inv1	4	40
inv2	94	0

III. POWER SUPPLY VOLTAGE VARIATIONS BY DPI

Direct injection of RF power at the frequency of F_{RF} creates conductive RF currents and induces sinusoidal voltage variations on the on-chip part of the power supply delivery network in a target IC chip, as given in Fig. 4. The IC chip is assembled on an evaluation board and regarded as a whole to be a device under test (DUT). The net RF power (P_{NET}) transmitted into the chip is calculated by (4) from the forwarded power (P_{FWD}) measured at the source side and the reflected power (P_{REF}) at the destination side. A bias-Tee network combines RF power and DC supply voltage of V_{dd} at the power terminal on an evaluation board. The P_{FWD} and P_{REF} become measurable by inserting a directional coupler in series to the power supply path, in combination with power sensors and a power meter.

$$P_{NET} = P_{FWD} - P_{REF} \qquad (4)$$

The input impedance at the power terminal (Z_{DUT}) determines the amplitude of sinusoidal voltage variations (V_{brd}) induced by P_{NET} at the location of the terminal, according to (5) [10]. Here, Z_0 stands for a characteristic impedance of the power path which is designed to be 50 Ω for F_{RF}. The phase angle (θ) leads to the timing difference between the conductive RF current and voltage (V_{brd}) also at the terminal.

$$V_{brd} = \sqrt{\frac{Z_0 + Z_{DUT}}{4Z_0 cos\theta} P_{NET}} \qquad (5)$$

The equivalent circuit network of Fig. 5 integrates the off- and on-chip parts of the PDN. S-parameters representing the power paths on the evaluation board are extracted by a full-wave field solver. The solver uses finite element method (FEM) for calculating S-parameters. The bonding wires and package leads associated with the power paths are lumped to respective inductors and resistors in series. The component values including coefficients of mutual induction are calculated from their physical dimensions. In addition, the on-chip part of PDN incorporated in the circuits of interest (COI) within an IC chip is abstracted in a chip power model (CPM). The CPM contains the entire die power delively network model. The extraction of CPM is closely related with the detailed design data base so that the capacitances parasitic to devices,

wires, and wells that are widely distributed within the COI are to be thoroughly modeled.

Once the entire equivalent circuit is established, the transfer characteristics of (5) are evaluated with a circuit simulator. In addition, the transfer gain of the V_{brd} at the power terminal to the voltage on the power node (V_{chip}) within the COI are defined as in (6) and calculated with the equivalent circuit.

$$Gain(F_{RF}) = 20 \cdot log(V_{chip}(F_{RF})/V_{brd}(F_{RF})) \text{ [dB]} \qquad (6)$$

IV. ANALYSIS AND EXPERIMENTS

A. Test vehicle

A test vehicle of Fig. 6 includes a test chip with inverter chains and an on-chip waveform capture (OCM) [11]. The use of on-chip waveform capturing architecture for immunity test has presented in several works [12][13]. The amplitudes of sinusoidal voltage variations are measured by an oscilloscope for V_{brd} at the power terminal on an evaluation board. Also the voltage variations within the chip are measured by the OCM for V_{chip} at the power supply nodes of inverter chains. Therefore, the transfer gain of (6) associated with V_{chip} and V_{brd}, and the transfer characteristics of (5) with V_{brd} and P_{NET} can be directly evaluated for the RF power at F_{RF}. This is a unique feature of the test vehicle.

An external clock source (DTG) provides a clock signal to the input of inverter chains, in a rectangular wave shape with sufficiently small jitter. The output of the chain is captured by an oscilloscope and stored in a waveform memory. The period jitter is extracted for more than 10,000 adjacent clock cycles from the waveforms.

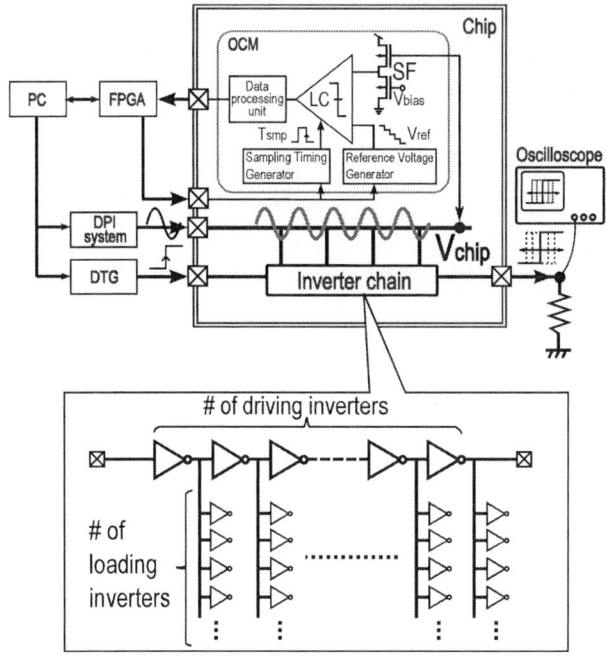

Fig. 6. Test vehicle and measurement system.

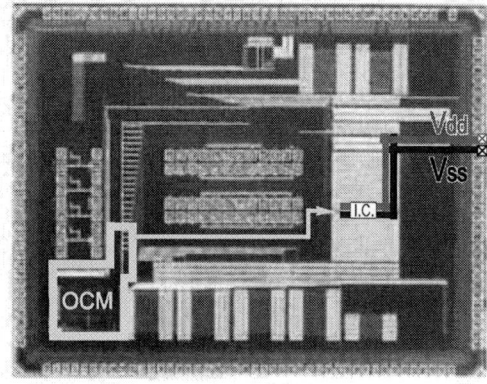

Fig. 7. Chip photograph

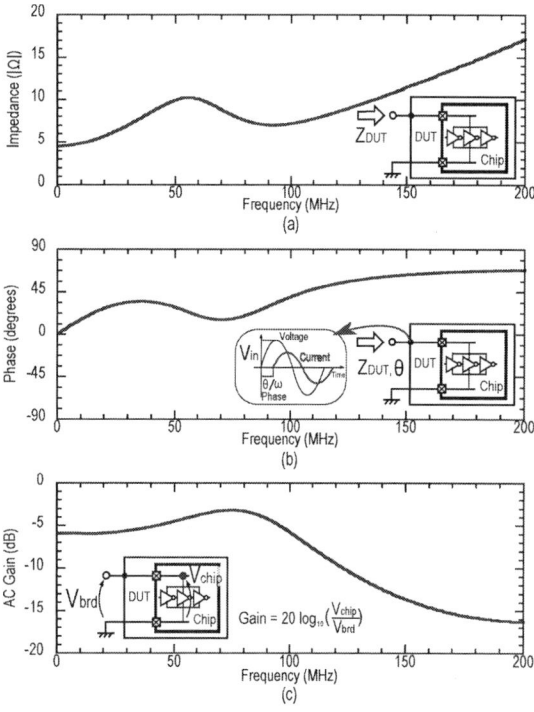

Fig. 8. Derivation of Z_{DUT}. (a) Input impedance of Z_{DUT}, (b) phase angle of Z_{DUT}, and (c) transfer gain between V_{brd} and V_{chip}

Inverter chains with different sets of design parameters are prepared in the test vehicle. The parameters include the number of inverter stages in cascade (driving inverters) and the number of inverters connected in parallel to the output of each driving inverter (loading inverters), as sketched in the inset of Fig. 6. The transistors in the inverters are adjusted in size so as to produce the same amount of τ_0 among the chains. The two examples are shown in Tab. I and tested in this paper.

The test chip is designed in a 65 nm CMOS technology. The inverter chains use the core transistors at 1.2 V while the waveform capture is designed with I/O devices at 2.5 V. A photo of the prototype chip is given in Fig. 7. A unique PDN common to inverter chains is externally supplied by a power source in connection with the DPI system, and internally probed by the OCM for monitoring the time variation of power supply voltages.

B. Sinusoids on power supply voltage

The PDN of the test vehicle as a whole is captured in the equivalent circuit of Fig 5. The input impedance of Z_{DUT} and the phase angle θ are simulated as given in Fig. 8, and the transfer gain is also derived, for the frequency of F_{RF} in the frequency range up to 200 MHz. In Fig. 8 (a) and (b), inductive impedance characteristics are seen even in several MHz frequency range, because of the mounting of on-board decoupling capacitros.

The waveform captured by the OCM on the power supply node of an inverter chain is given in Fig. 9 (a) under the DPI. The amplitude of on-chip sinusoidal voltage variations

reaches approximately 100 mV for external P_{NET} of 0 dBm at $F_{\text{RF}} = 80$ MHz. We confirmed that the sinusoids exhibit voltage amplitudes with two orders of magnitude higher than the power supply noise inherently generated by the inverter chain with the largest design parameters, as given in Fig. 9(b). These waveforms also indicate the capability of OCM for fine resolution experiments [14].

The comparison was made on measurements and simulation of the transfer characteristics of the DPI for the frequency range of 200 MHz, as shown in Fig. 10. The source power (P_{FWD}) is adjusted for P_{NET} to remain constant at 0 dBm among the measurement points for RF injection. It is clear

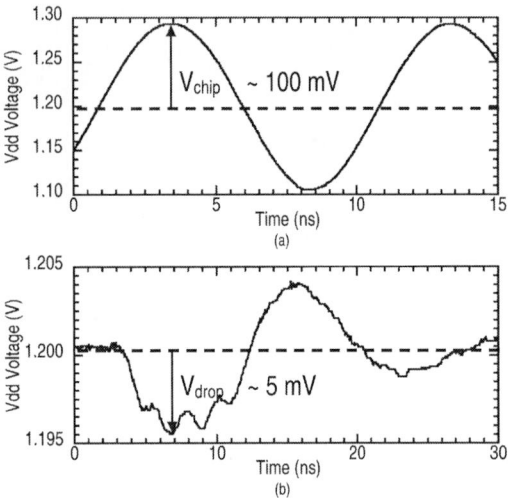

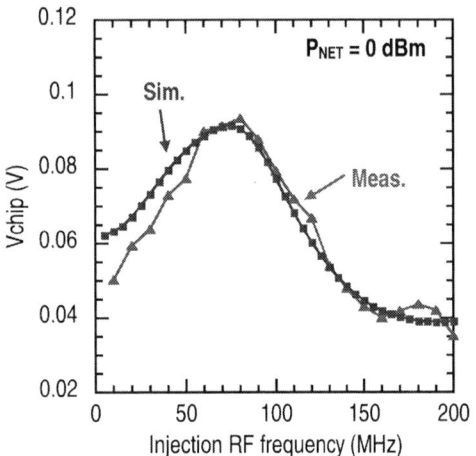

Fig. 9. Voltage waveform measured by OCM. (a) Under DPI and (b) without DPI.

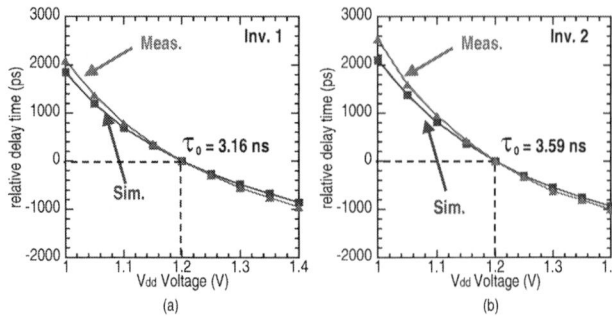

Fig. 11. Measured and simulated delay vs. DC voltage. (a) inv1 and (b) inv2.

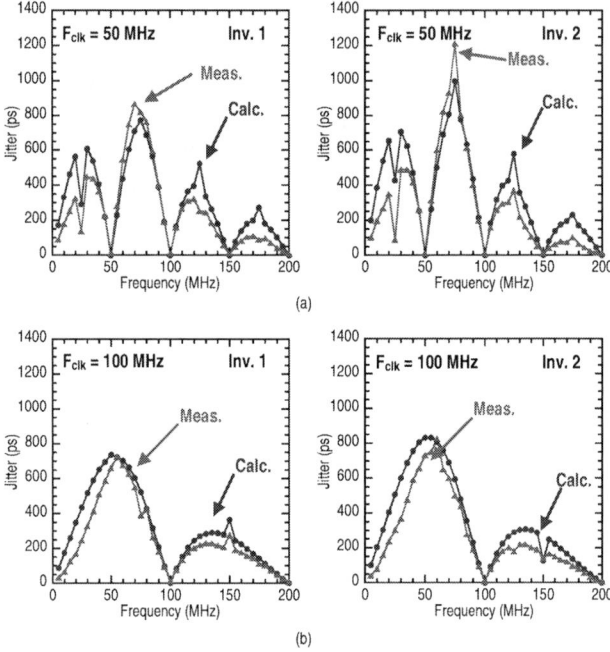

Fig. 12. Period jitter versus frequency of sinusoids among inverter chains (inv1 and inv2). (a) $F_{clk} = 50$ MHz and (b) $F_{clk} = 100$ MHz.

that the proposed equivalent model well predicts the V_{chip} on the inverter chain.

C. Jitter calculation and measurements

The susceptibility of the delay time of an entire inverter chain on the static voltage variation is evaluated as in Fig. 11. The relative delay time indicates the difference of the delay time of an inverter chain compared to the standard 1.2 V voltage function. The observed monotonic response is consistent among circuit simulations with a transistor-level netlist featuring transistor device models and measurements. It is noted that the response is strongly dominated by the properties of transistors forming an inverter, and deviated from the assumed linear response of Fig. 3. However, the analysis of (1) is considered well valid because of the monotonicity of the response.

The period jitter is calculated from (3) for inverter chains

in the presence of RF power injected on its PDN at different F_{RF}, as given in Fig. 12, in comparison with measurements. The two inverter chains in Tab. 1 are tested for the two frequencies of clock signal, f_{clk}, at 50 and 100 MHz. The calculation of period jitter from the derived V_{chip}, that is simulated with the equivalent circuit model for P_{NET} at F_{RF}, matches measurements. The jitter becomes almost zero for every multiple of the frequency of F_{RF} equal to f_{clk}, coinciding with the interpretation given in Fig. 2(b).

V. CONCLUSION

A general expression of DPI with the equivalent circuit model is established and the transfer characteristics from RF power on the power terminal on a board to the sinusoidal voltage variations on the circuit of interest within a chip are derived. The impacts of such sinusoids of power nodes on

Fig. 10. DPI characteristics.

period jitter in an inverter chain is calculated. The equivalent circuit model is established with fundamental design data, therefore the estimation of jitter by interferences on PDN is closed only with design database. The technique proposed in this paper can be applied to the jitter analysis of digital circuits in various noise environments and will help to quantify the electromagnetic immunity of digital systems.

ACKNOWLEDGMENT

This work was in part supported by CREST, JST. The authors would like to thank Yuta Sasaki and Shinichiro Ueyama for technical contributions. The authors are also thankful to Apache Design Inc. for their support on modeling of power delivery networks.

REFERENCES

[1] J.A. McNeil, "Jitter in Ring Oscillators," *in IEEE J. Solid-State Circuits,* vol. 32, no. 6, pp. 870-879, June 1997.

[2] F. Herzel and B. Razavi, "A Study of Oscillator Jitter Due to Supply and Substrate Noise," *in IEEE Trans. on Circuit and Systems,* vol. 46, no. 1, pp. 56-62, Jan 1999.

[3] A. Hajimiri, S. Limotyrakis and T.H. Lee, "Jitter and Phase Noise in Ring Oscillators," *IEEE J. Solid-State Circuits,* Vol. 34, no. 3, pp. 513-519, Aug 2003.

[4] T. Pialis and K. Phang, "Analysis of Timing Jitter in Ring Oscillators Due to Power Supply Noise," *in Proc. 2003 IEEE International Symposium on Circuits and Systems (ISCAS 2003),* pp. 685-688, May 2003.

[5] Asad.A. Abidi, "Phase Noise and Jitter in Ring Oscillators," *IEEE J. Solid-State Circuits,* Vol. 41, no. 8, pp. 1803-1816, Aug 2006.

[6] M.P. Robinson, K. Fischer, I.D. Flintoft and A.C. Marvin, "A Simple Model of EMI-Induced Timing Jitter in Digital Circuits, its Statistical Distribution and its Effect on Circuit Performance," *in IEEE Trans. Electromagnetic Compatibility,* Vol. 45, no. 3, pp. 513-519, Aug 2003.

[7] "Integrated Circuits, Measurement of Electromagnetic Immunity, 150 KHz - 1 GHz: General and Definitions," IEC62132-1 Part 1, 2007, International Electrotechnical Commission: Geneva, Switzerland.

[8] "Direct RF power injection to measure the immunity against conducted RF-disturbances of integrated circuits up to 1 GHz," IEC 62132-4, 2003, International Electrotechnical Commission: Geneva, Switzerland.

[9] T. Sawada, T. Toshikawa, K. Yoshikawa, H. Takata, K. Nii and M. Nagata, "Immunity Evaluation of SRAM Core Using DPI with On-Chip Diagnosis Structures," *in Proc. IEEE 8th International Workshop on Electromagnetic Compatibility of Integrated Circuits (EMC Compo 2011),* pp. 65-70, Nov. 2011.

[10] A. Alaeldine, R. Perdriau, M. Ramdani, J-L. Levant and M. Drissi, "A Direct Power Injection Model for Immunity Prediction in Integrated Circuits," *in IEEE Trans. Electromagnetic Compatibility,* Vol. 50, no. 1, pp. 52-62, Feb 2008.

[11] T. Hashida, and M. Nagata, "On-chip waveform capture and application to diagnosis of power delivery in SoC integration," *IEEE J. Solid-State Circuits,* Vol. 46, no. 4, pp. 789-796, April 2011.

[12] A. Boyer, S. Ben Dhia, C. Lemoine, B. Vrignon, "Construction and Evaluation of the Susceptibility Model of an Integrated Phase-Locked Loop," *in Proc. IEEE 8th International Workshop on Electromagnetic Compatibility of Integrated Circuits (EMC Compo 2011),* pp. 7-12, Nov. 2011.

[13] S. Ben Dhia, A. Boyer, B. Vrignon, and M. Deobarro, "On-Chip Noise Sensor for Integrated Circuit Susceptibility Investigations," *IEEE Trans. Instrumentation and Measurement,* Vol. 61, no. 3, pp. 696-707, March 2012.

[14] Y. Harada, K. Yoshikawa, N. Miura, A. Murata, S. Agatsuma, K. Ichikawa, and M. Nagata, "Power-Noise Measurements of Small-Scale Inverter Chains," *in Proc. 2013 International Meeting for Future of Electron Devices, Kansai (IMFEDK 2013),* pp. 102-103, June 2013.

978-1-4799-5004-1/13 $31.00 © 2013 IEEE

Magnetic Field Coupling on Analog-to-Digital Converter from Wireless Power Transfer System in Automotive Environment

Bumhee Bae, Sunkyu Kong, Joonghoon J. Kim, Sukjin Kim, and Joungho Kim

Terahertz Interconnection and Package Laboratory, Department of Electrical Engineering, KAIST,
373-1 Guseong-dong, Yuseong-gu, Daejeon 305-701, South Korea,
bhbae@kaist.ac.kr, teralab@kaist.ac.kr

Abstract—**There are multiple electrical devices on automotive system, which are control devices, communication devices, and digital devices. Each electrical device can generate magnetic field, one of the critical radiated electro-magnetic interference (EMI) elements. The operating frequency of each device is different and it means that the bandwidth of magnetic field is wide. Therefore, the strong magnetic fields can degrade the performance of diverse semiconductor systems in automotive applications, but it is not well discussed yet, even though the malfunction of electrical devices on automotive system is related to safe issues. So, we focus on strong magnetic field effects of semiconductor system. Among strong magnetic field source, we targeted wireless power transfer (WPT) system, which is spotlighted and promising technology for automotive and mobile charging system and significant magnetic field source. Furthermore, we choose analog-to-digital converter (ADC), sensitive to external noise and critical system involved in control devices related to the safety issues of automobile, as a targeted semiconductor system. In this paper, we discuss the magnetic coupling path and describe how to estimate the magnetic field effects on ADC with WPT. To estimate and analyze the targeted effects on ADC, we designed the ADC using a 0.13um CMOS process and WPT system using printed circuit board (PCB). Consequently, the magnetic field couples to ADC system, and there are three methods to estimate performance degradation of ADC by magnetic field effects, one is modeling, another is simulation, and the other is measurement.**

Keywords— *Strong magnetic field, electromagnetic interference, electromagnetic susceptibility, analog-to-digital converter, wireless power transfer*

I. INTRODUCTION

Recently, with the introduction of electrical automobile for eco-friendly trend and increased needs of convenient functions in an automobile, an automobile system is becoming more of an electrical product, which includes power circuits, communication devices, and control devices, and not just a mechanical product. In harmony with this trend of technological development, the wireless power transfer (WPT) system [1] can also be applied to power suppliers or electrical devices in an automobile. Basically, WPT system generates magnetic field for power transfer, which inevitably means that strong magnetic field can exist nearby critical electrical devices on automotive applications. Fig. 1 shows the importance of electrical control logic, which affects the automotive system

behaviors. It is of utmost importance to analyze the strong magnetic field effects on electrical system, especially if the electrical control logic is sensitive to noise and is tightly related to safety issues on automotive system [2].

Therefore, we analyze strong magnetic field coupling from WPT system, one of the strongest magnetic field source, to analog-to-digital converter (ADC), which is highly sensitive to noise and also plays a critical role in automotive applications, as described in Fig. 1. Prior to actual research, we first need to know which path is most dominantly affected by the magnetic coupling and why. Fig.2 shows the conceptual diagram of strong magnetic field coupling path on ADC in system level. There are several components in system level on the critical coupling path of strong magnetic field, which are PCB metal plane, bonding wire, and on-chip metal line.

To validate the analysis, an ADC is designed and fabricated using a 0.13 um CMOS process. The analysis is verified based on simulation and measurement results from 1 MHz to 1 GHz. The reason we are interested specifically in the frequency range from 1 MHz to 1 GHz is that this region includes both the operating frequency of ADC system and that of WPT system. Consequently, the magnetic field couples to ADC system, and there are three methods used in this paper to estimate the performance degradation of ADC by magnetic field effects; one is modeling, another is simulation, and the other is measurement.

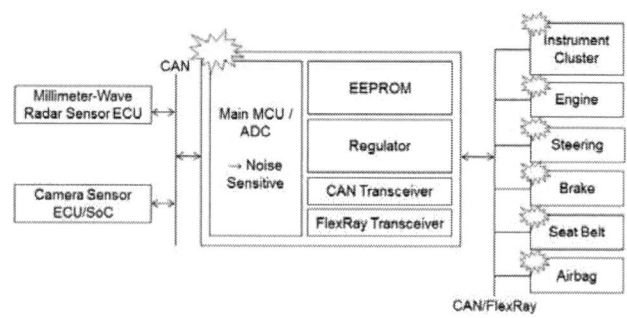

Fig. 1. Block diagram of electrical control logic on automotive system. The malfunction of control logic makes the critical functions of automotive be wrong

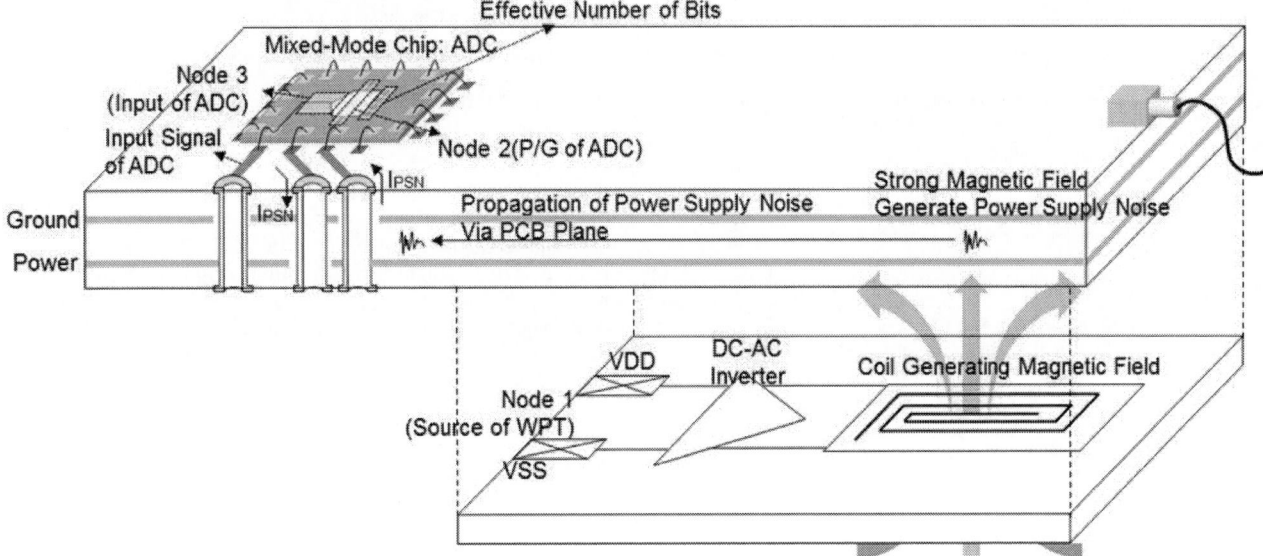

Fig. 2. Conceptual diagram of strong magnetic field coupling path on ADC in system level. The magnetic field is generated by WPT system, and it couples to the ADC with hierarchical power distribution network (PDN)

II. FEASIBILITY STUDY OF MAGNETIC FIELD EFFECTS ON ADC

A. Design of ADC with PCB

For researching magnetic field effects on ADC, it is needed to design the device under test (DUT) of ADC. We fabricated ADC using CMOS 0.13 um process and PCB for considering normal behavior of ADC, as illustrated in Fig.3. There are decoupling-capacitor pads and vias on power/ground plane considering practical design. The designed DUT of ADC works well, of which the performance specification is shown in Table I. The targeted ADC used in this paper is the flash ADC, one of the representative high-speed ADCs. Most high-speed ADCs have nearly the same applied structure as the flash ADC, which means that the analysis of strong magnetic field effects on a flash ADC can be used to understand the strong magnetic field effects on various high-speed ADCs. The performance of ADC is defined using signal-to-noise ratio (SNR) and effective number of bit (ENOB), which are the two main performance factors of the ADC. SNR is the ratio of the input signal amplitude to the output noise level, and ENOB is a parameter most frequently used to express SNR as the bit of an ideal ADC that has only quantization error.

B. Feasibility Test of Magnetic Field Effects on ADC with PCB

We used wide-band near field antenna probe for feasibility test of magnetic field effects on ADC circuit fabricated on PCB. When the 6 MHz sine wave of 25 dBm is applied to the near field antenna probe, the ADC performance is accordingly degraded from 3.7 ENOB to 3.25 ENOB. Furthermore, the effects of performance degradation are easily observed by analyzing the output bits of ADC for following two cases: 1) DC input is applied without external noise and 2) DC input is applied with external noise. When 590 mV DC input without

external noise is applied to the ADC, the output bits are 0111 as shown in Fig. 4 (a). However, when the 590 mV DC input is influenced by the magnetic field radiated from near field antenna probe, the output bits are fluctuated, as can be seen from Fig. 4 (b).

TABLE I.   SPECIFICATION OF DESIGNED ADC

	Specification
Process	Dongbu CMOS 0.13um
Maximum Operating frequency	2GHz
Input Swing Range	0.6Vpp
ENOB	3.7bit
Power Consumption	54.6mW

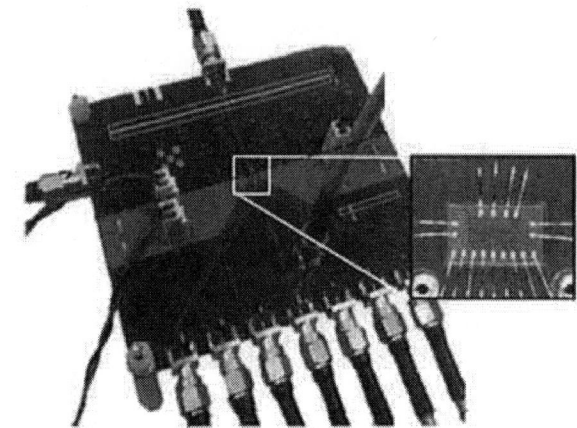

Fig. 3. Designed Device Under Test (DUT) of ADC. It includes chip of ADC, bonding wires, and PCB

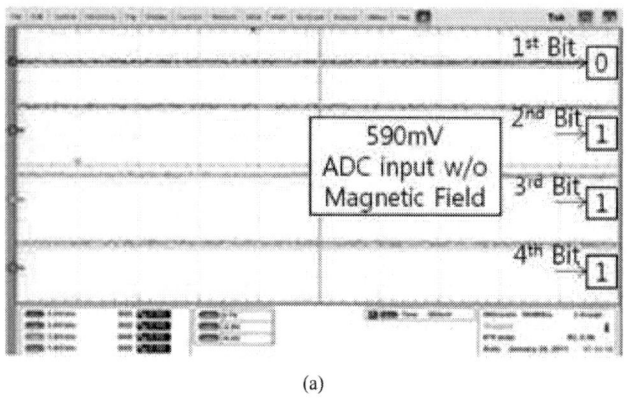

(a)

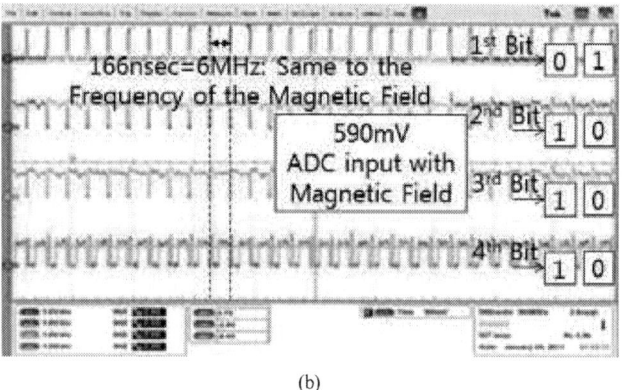

(b)

Fig. 4. Measurement for feasibility study of strong magnetic field effects on ADC. (a) The stable output bits of ADC without external noise source. (b) The fluctuated output bits of ADC with radiated field source

TABLE II.   SPECIFICATION OF DESIGNED WPT COIL

	Value
Total Self-Inductance (Ls)	1 uH
Total Effective Series Resistance (Rs)	10 Ω
Tuning Capacitance (Cs)	10 pF
Targeted Resonant Frequency (fr)	47.3 MHz
Diameter	300 um

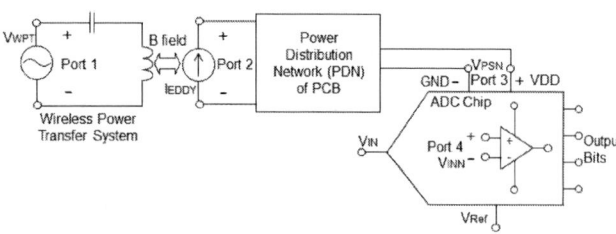

Fig. 5. Block diagram which includes WPT system, power distribution network, and ADC. It defines the position of ports which we will mention

III. POWER SUPPLY NOISE OF ADC FROM MAGNETIC FIELD OF WPT

A. Magnetic Field Generation from Designed WPT Coil

For this research we designed and fabriated WPT coil on PCB with a resonant frequency of 47.3 MHz, which is the frequency practically used in WPT-WG [7]. The specification of the designed WPT system is summarized in Table II. At resonant frequency, power input of WPT coil, port 1, produces high current, which leads to high magnetic field, because magnetic fields are produced by electric currents. The direction of a magnetic field can be determined by using the right hand grip rule [3]. The strength of the magnetic field is inversely proportional to the distance from the wire.

B. Eddy Current Effects, Eddy Current Loss and Power Supply Noise, arise from Magnetic Field Coupling

The amplitude of the current on PCB induced by the magnetic field from the designed coil can be calculated using (1) and (2). There are two main effects caused by the eddy current on PCB plane, one is temperature rise on PCB, and the other is generated power supply noise. The eddy current basically circulates on metal plane, and it leads to eddy current loss. The energy of eddy current loss causes the temperature rise on PCB. In practical PCB design, there are many vias and pads for integrating decoupling-capacitors. When the eddy current is induced to the closed-loop metal line - in this case decoupling-capacitor pad, via, and power/ground PCB plane - then it leads to the production of current from power to ground. Further, the current causes the power supply noise, which degrades the performance of ADC [4]. When the power supply noise gets coupled to ADC, it can be coupled to the input of ADC, and it is the main reason the performance of ADC is degraded by the power supply noise. Therefore, we first have to know how much power supply noise is generated by magnetic field coupling.

We define the power/ground of closed loop metal line and that of ADC chip as port 2 and port 3, respectively, which are depicted in Fig 5. The current of port 2 can be estimated by (1) and (2). Then, we have to consider the power distribution network (PDN) which includes PCB power/ground plane, power/ground bonding wire, de-coupling capacitor, and on-chip power/ground line. Finally, we can estimate the transfer impedance (Z21) from port 2 to port 3. Transfer impedance is the impedance that we encounter when an electrical current passing through one element of the system makes a voltage difference in other element of the same system. Therefore, when we know the current at port 2 and transfer impedance from port 2 to port 3, we can estimate the voltage of port 3, which is the power supply noise of ADC.

$$\nabla \times E = -\partial B_{WPT}/\partial t \tag{1}$$

$$I_{Eddy} = kE + \varepsilon \partial E/\partial t \approx kE \tag{2}$$

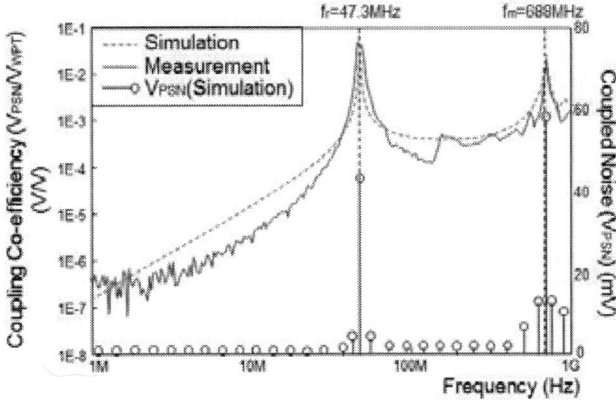

Fig. 6. The noise coupling coefficient from the power input of wireless power transfer to that of ADC. The plot of coupled noise is the result when we assume V_{WPT}=1V

C. Power Supply Noise Generated by WPT

We simulated and measured power supply noise generated from WPT to power/ground of ADC. The results are summarized in Fig. 6. There are two peak points which indicate high coupling ratio; one is the resonance frequency of wireless power transfer system, the other is the cavity resonance frequency of PCB. Therefore, not only do we have to focus on the noise coupling ratio at the resonance frequency of WPT system, we also have to analyze it at the resonance frequency of hierarchical PDN of ADC.

IV. NOISE COUPLING FROM POWER SUPPLY OF ADC (PORT 3) TO INPUT OF ADC (PORT 4)

Fig. 7 shows a simplified schematic of a 4-bit flash-type ADC, the targeted device in this paper. Up to this point, we have discussed the power supply noise which is generated by magnetic field of WPT, as explained in the previous chapters. Now, we have to consider power supply noise effects on ADC. There are two parts to be considered in analyzing the power supply noise effects on ADC; one is the noise coupling from power supply noise of ADC to the input of ADC, and the other is the ADC performance degradation caused by the input noise of ADC.

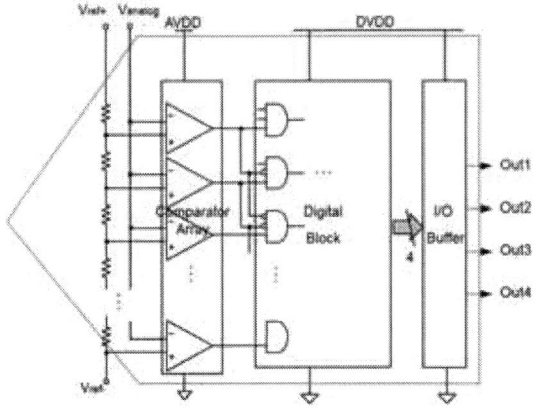

Fig. 7. Simplified schematic of the designed flash type ADC

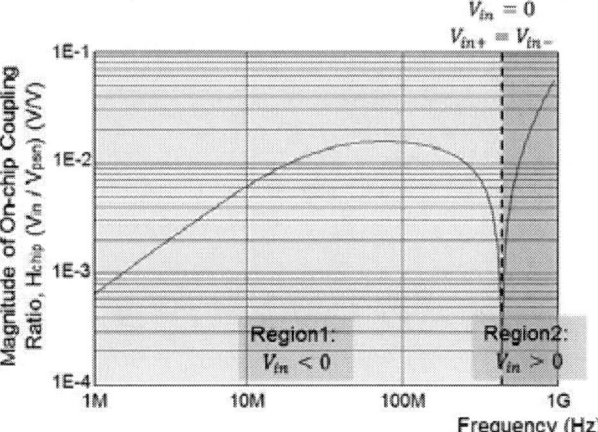

Fig. 8. Noise Coupling from Power Supply of ADC (Port 3) to Input of ADC (Port 4)

In this chapter, we focus on the on-chip noise-coupling from the power/ground of the ADC to the input of the ADC. The comparators are the main circuits that are highly sensitive to power supply noise on ADC. Therefore, we have to consider the on-chip noise-coupling ratio (V_{INN}/V_{PSN}) of comparator in order to estimate power supply noise effects on ADC. The differential inputs of each comparator are the difference between two inputs: one input is connected to a voltage divider, which is designed to be a reference ladder, and the other is connected to an analog input line [5]. Therefore, we can calculate the coupled input noise from power supply noise using circuit theory [6]. It is displayed in Fig. 8.

V. CO-SIMULATION OF MAGNETCIF FIELD EFFECTS ON ADC AND IT'S MEASUREMENT VERIFICATION

A. Co-simulation methodology

We use 3D field solver (HFSS) to estimate noise coupling ratio, and from this ratio, we can obtain the resulting input noise of ADC. We apply the resulting input noise to the spice simulator (Hspice) for estimating the circuit behavior. The HFSS model includes bare PCB board with decoupling capacitor pad/via, wire bond, and on-chip power/ground lines. The Hspice model consists of the net-lists with MOS models, on-chip decoupling capacitor models, and so forth. From these simulation results, we can get the output bits of ADC, which are then converted to obtain the performance factor of ADC (ENOB) using MATLAB code.

B. The result

Fig. 9 shows the decreased ENOB result of ADC, degraded by the magnetic field from WPT system. When we apply on-chip decoupling capacitor whose value is 2.13nF, the effect of noise above the frequency value of 100 MHz, is effectively removed. However, the low frequency noise (below 100MHz) which is generated by the resonance frequency of WPT still significantly affects performance of ADC.

978-1-4799-5004-1/13 $31.00 © 2013 IEEE   241

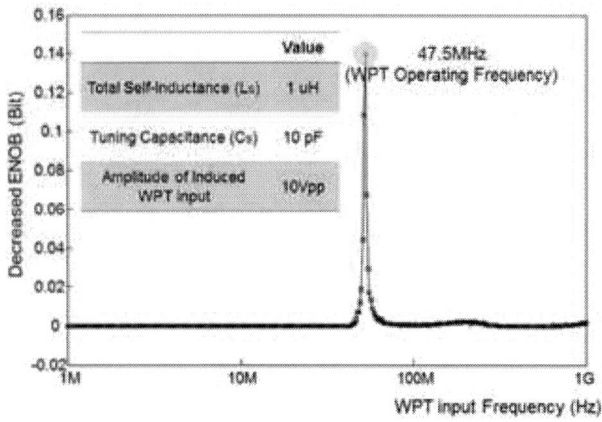

	Value
Total Self-Inductance (Ls)	1 uH
Tuning Capacitance (Cs)	10 pF
Amplitude of induced WPT input	10Vpp

Fig. 9. The decreased effective number of bits (ENOB) of ADC by the magnetic field of WPT system

VI. CONCLUSION

Even though the electro-magnetic interference (EMI) issues are becoming more and more important in recent semiconductor system designs, those issues are not well discussed in depth yet. In this paper, by using wireless power transfer system as an EMI source, and ADC as a victim semiconductor system, we discussed the EMI effects on semiconductor system. Throughout this paper, we demonstrated and experimentally verified that the performance of ADC can be significantly degraded by the strong magnetic field, one of the major EMI types, from WPT system. Furthermore, we proposed how it can be co-simulated using field solver, spice simulator, and MATLAB.

ACKNOWLEDGMENT

This work was supported by the National Research Foundation of Korea (NRF) grant funded by the Korea government (MEST) (No. 2010-0029179).

REFERENCES

[1] Sunkyu Kong; Myunghoi Kim; Kyoungchoul Koo; Seungyoung Ahn; Bumhee Bae; Joungho Kim; , "Analytical expressions for maximum transferred power in wireless power transfer systems," *Electromagnetic Compatibility (EMC), 2011 IEEE International Symposium on* , vol., no., pp.379-383, 14-19 Aug. 2011

[2] Casier, H.; Moens, P.; Appeltans, K.; , "Technology considerations for automotive," *Solid-State Device Research conference, 2004. ESSDERC 2004. Proceeding of the 34th European* , vol., no., pp. 37- 41, 21-23 Sept. 2004

[3] Fleming, John Ambrose. *Magnets and Electric Currents*, 2nd Edition. London: E.& F. N. Spon. pp. 173–174.

[4] Bumhee Bae; Yujeong Shim; Woojin Lee; Kyoungchoul Koo; Woojin Ahn; Joungho Kim; , "Hybrid modeling and analysis of power supply noise effects on analog-to-digital converter considering hierarchical PDNs," *Electrical Design of Advanced Packaging & Systems Symposium (EDAPS), 2010*

[5] Yukawa, Akira; Fujita, Tsuneo; Hareyama, Kyuichi; , "A CMOS 8-bit High Speed A/D converter IC," *Solid-State Circuits Conference, 1984. ESSCIRC '84. Tenth European* , vol., no., pp.193-196, 0-0 Sept. 1984

[6] Agarwal, Anant, and Jeffrey H. Lang. *Foundations of Analog and Digital Electronic Circuits*. San Mateo, CA: Morgan Kaufmann Publishers, Elsevier, July 2005.

[7] Shoki, H., "Issues and initiatives for practical use of wireless power transmission technologies in Japan," *Microwave Workshop Series on Innovative Wireless Power Transmission: Technologies, Systems, and Applications (IMWS), 2011 IEEE MTT-S International* , vol., no., pp.87,90, 12-13 May 2011

Immunity Simulation Method for Automotive Power Module using Electromagnetic Analysis

Yosuke Kondo, Kei Tsunada, Norimasa Oka, Masato Izumichi

Electronics Device Business Unit

DENSO CORPORATION

Kariya, Japan

{yousuke_kondou, kei_tsunada, norimasa_oka, masato_izumichi}@denso.co.jp

Abstract— **This paper provides an immunity simulation method for product-level automotive power modules that contain printed circuit boards (PCBs). A stress–strength model is adopted as an estimation method for the susceptibility. The stress is the value of the RF power injected into a victim circuit, and the strength is the acceptance stress criteria of the victim circuit. By using electromagnetic simulation, the stress of the victim circuit is calculated. The strength of the victim circuit is determined by the DPI test result. By comparing the stress and strength of the victim circuit, the susceptibility thresholds can be predicted. The results show good agreement with the experimental results in the DPI test for a power module containing a complex product-level PCB and exhibit a good correlation with the BCI test for the power module.**

Keywords— *RF immunity; Electromagnetic analysis; DPI; BCI; Stress-strength model;*

I. INTRODUCTION

In recent years, electromagnetic compatibility, especially electromagnetic susceptibility (immunity), has been a critical factor for general IC reliability in automotive electronics. Automotive electronics equipment must pass severe immunity tests, such as Direct RF Power Injection (DPI) [1] and Bulk Current Injection (BCI) [2]. If a device fails to pass the immunity tests, time-consuming and expensive efforts will have to be made to improve the susceptibility. The reason for the failure to pass these tests is difficulty of predicting the results of the immunity test. Hence, a prediction method such as numerical simulation is required.

Several simulation approaches for immunity tests were investigated by multiple research groups [3–8]. SPICE simulations were applied to both DPI and BCI test systems. However, SPICE can predict the abnormal response of ICs only for DPI tests that have simple printed circuit boards (PCBs) and IC packages, and cannot predict BCI test results because of the large-scale test setups that are involved. On the other hand, S-parameter approaches were applied to BCI tests [7, 8], but the Equipments under Test (EUTs) were relatively simple and probably did not contain a complex PCB.

Automotive electronics equipment generally contains complex PCBs, and typical improvements for immunity tests are changing patterns, positions of noise filters, or changing components of noise filters on the PCB. Therefore, a simulation method that can take into account complex product-level PCBs is required. One solution is electromagnetic full-wave analysis, which can take PCBs into account easily. In this paper, a simulation technique using electromagnetic analysis for immunity problems is proposed.

This paper is organized as follow: Section II describes the basic concept of the simulation method, which is the stress–strength model. Section III describes the automotive power module that is used as the EUT. Section IV presents the definition of the stress and the strength, and the methodology of determining the strength by the victim-circuit level DPI (device-level DPI in this power module) test results. Section V describes the detail of the module-level DPI test that contains the PCB, and presents the results compared with the experiments. Section VI shows the BCI test simulation and experiment results. Conclusions are drawn in section VII.

II. BASIC CONCEPT OF THE PROPOSED IMMUNITY SIMULATION METHOD

The basic concept of the proposed immunity simulation method is the stress–strength model. Stress is defined as the value of the RF power injected into a victim circuit. Strength is defined as the acceptance stress criteria of a victim circuit. The victim circuit means the circuit that causes an abnormal response when RF power is injected into the circuit. If the stress exceeds the strength, an abnormal response is considered to have occurred in the simulation. Fig.1 shows the concept of the proposed method.

The stress is calculated by electromagnetic analysis in the frequency domain that takes into account the PCB layout, passive components, and impedances of the IC and victim circuits. The strength is evaluated by applying the DPI test only to the victim circuit. By comparing the stress and strength of the victim circuit, the susceptibility thresholds can be predicted. In the proposed method, electromagnetic analysis is expected to help the designer of the PCB know how to reduce the RF stress in the victim circuit. In this study, CST Microwave Studio is used as the electromagnetic analysis simulator.

III. POWER MODULE UNDER TEST

In this study, a power module that contains a multi-layer PCB (fig. 2) is tested and simulated. This power module is a DC motor driver with an H-bridge circuit, and it is a test piece for this study. It contains a 4-layer PCB made of ceramic. A pre-driver IC, a digital microprocessor, an oscillator, and several chip ceramic capacitors are mounted on the PCB, which is itself mounted on the lead frame. Two N-channel and P-channel power MOSFETs are implemented on the power supply lead frames. These MOSFETs make up the H-bridge circuit. A temperature-sensing diode is fabricated on the P-channel power MOSFET, and the anode and cathode terminals of the diode are connected to the pre-driver IC with bonding wires and patterns of the PCB. These lead frames and components are molded by thermosetting resin.

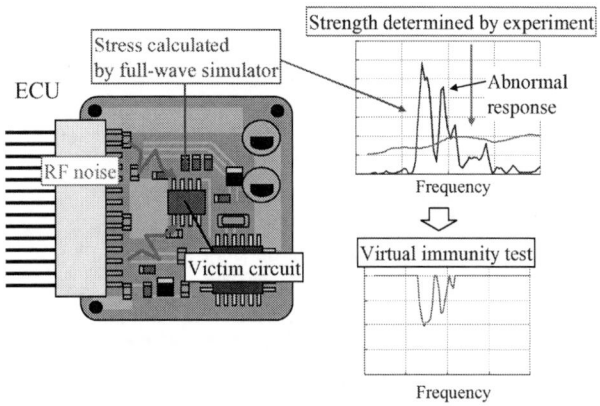

Fig. 1.   Basic concept of the proposed simulation method.

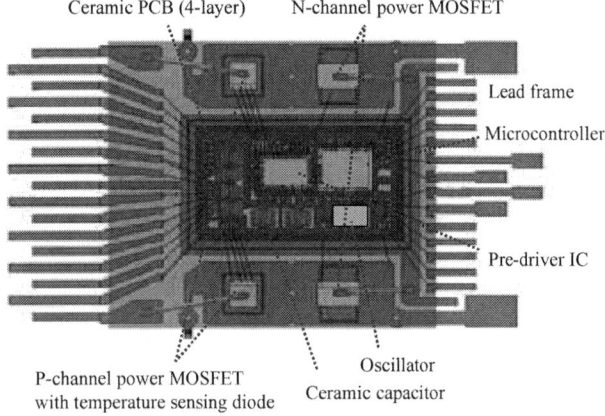

Fig. 2.   Overview of the automotive motor controller.

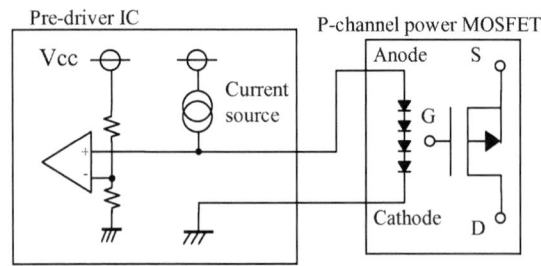

Fig. 3.   The temperature sensing diode and the external drive circuit.

The pre-driver IC powers the temperature sensing diodes on the P-channel power MOSFETs with a constant current, and monitors the voltage of the diode (fig. 3). When the voltage drops below some criteria for the over-temperature protection, the pre-driver IC stops the gate drive signal to avoid thermal runaway. If a RF large signal is injected into the temperature sensing diode, the voltage between the anode and the cathode is dropped, and finally the pre-driver IC stops the output of the power MOSFETs. Therefore, the temperature sensing diode is the victim circuit of the immunity test.

IV. MODELING OF TEMPERATURE SENSING DIODE ON POWER MOSFET AND ABNORMAL RESPONSE

First, the stress and strength of the victim circuit (the temperature sensing diode) need to be defined. This section describes the definition of the stress and the strength. The device-level DPI tests for only the power MOSFET are performed to determine the strength of the temperature sensing diode because the diode is fabricated on the power MOSFET.

A. Device-Level DPI Tests for the Power MOSFET

The measurement of the power MOSFET response is performed in the DPI setup in order to determine the criteria for abnormal response. The power MOSFET is implemented in a 48-pin DIP ceramic package (Kyocera, KD-S78527-G) implemented on the DPI test PCB (fig. 4). The power MOSFET has five terminals including the anode and cathode of the temperature sensing diode. All the terminals are connected to the bias tees. RF power is injected into the RF port of the bias tee of the anode terminal, and the other RF ports of the bias tees are all terminated by 50 ohm resistances (fig. 5). In this test, it is judged that an abnormal response occurs when the voltage at the DC port of the bias tee of the anode terminal drops below the criteria for over-temperature protection. The experimental result is shown in fig. 6.

B. Determination of Abnormal Response Threshold

Electromagnetic analysis is also performed on the device-level DPI test system. The impedances of the drain source (DS), the drain gate (DG), and the gate source (GS) are set as resistances or capacitances by referring to the datasheet. The temperature sensing diode, that is the victim circuit, is coupled with the drain terminal of the power MOSFET by the pF order parasitic capacitances of silicon oxide. The impedance of the victim circuit is very important for the immunity analysis, so the S-parameter of the diode is measured with a vector network analyzer (VNA). In this measurement, the diode is powered at a constant current by using bias tees. The drain is the reference of that S-parameter. The parasitic inductances and capacitances of the measurement system are removed by de-embedding.

The stress, which is used as the abnormal response criterion, is defined as the total net power applied to the temperature sensing diode. The diode is modeled as a 2 port S-parameter. The reference is the drain. The RF noise paths to the diode are anode, cathode and drain, and it is necessary to consider the total power from both ports. Therefore, the net power is adopted as the definition of the stress. The voltages and

978-1-4799-5004-1/13 $31.00 © 2013 IEEE   244

currents at each port can be calculated by electromagnetic analysis, and the net power is calculated from eq. (1):

$$P_{NET} = \text{Re}\left(I_{anode}\overline{V}_{anode} + I_{cathode}\overline{V}_{cahode}\right) \qquad (1)$$

in which the bar means the complex conjugate value. By using this calculated net power and the experimental DPI test results, the strength, i.e., the acceptance diode net power can be introduced. The strength $P_{Strength}$ is calculated from eq. (2):

$$P_{Strength} = T \cdot P_{DPI} \qquad (2)$$

in which T means the transfer function and P_{DPI} means the experimental DPI results, respectively. The transfer function means the transmitted power ratio from the DPI injection port to the temperature sensing diode. Fig. 6 shows the transfer function and the strength.

C. Verification of the Abnormal Response Model

For the verification, electromagnetic analyses and DPI tests in which RF noise is injected into the source and cathode terminals are performed in the same device-level DPI test system. We have confirmed that the results did not depend on

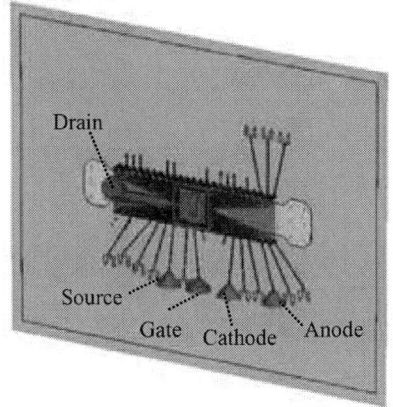

Fig. 4.   DPI test PCB with the power MOSFET in a ceramic package.

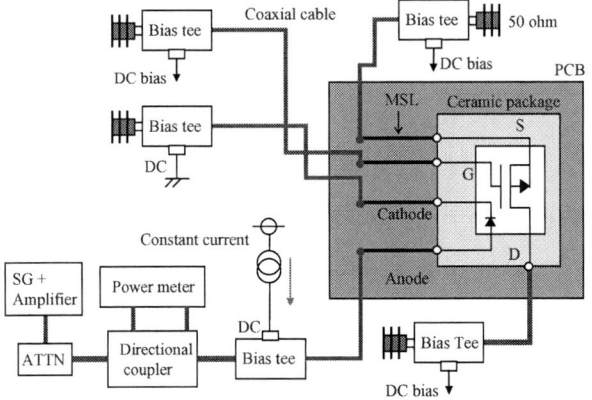

Fig. 5.   DPI test setup for the temperature sensing diode.

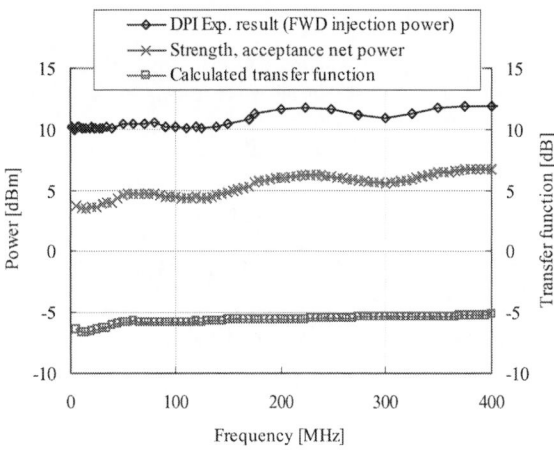

Fig. 6.   DPI test result, calculated stress and acceptance net power for the temperature sensing diode.

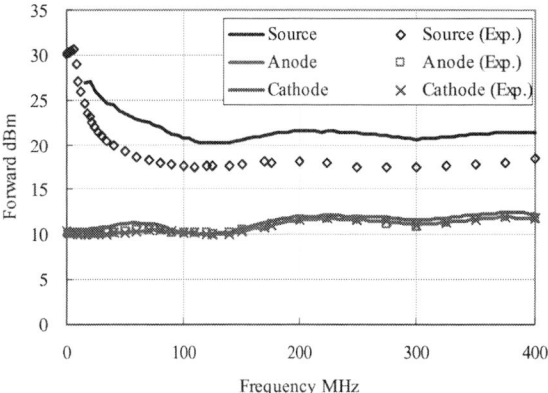

Fig. 7.   DPI tests results: Comparison between simulations and measurements in term of forward power.

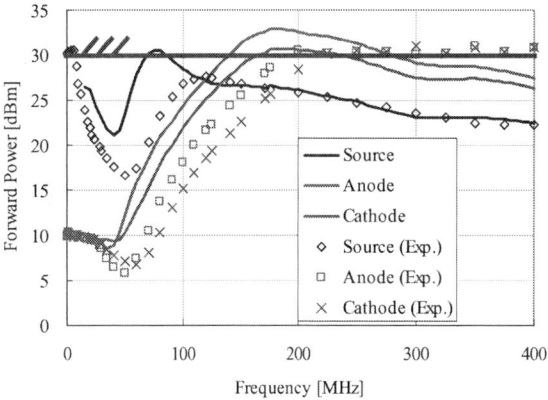

Fig. 8.   Improved DPI tests results: Comparison between simulations and measurements in term of FWD power (with inserted chip ferrite).

the DC gate drive conditions. Fig. 7 compares the simulation and measurement of the susceptibility threshold in terms of forward power. The simulation and experimental results of the anode injection must match because the threshold is determined by the simulation and measurement of the anode injection of. The simulation results of the cathode agree very well with the measurement, and the abnormal response caused by the source injection is reproduced well in the simulation. As stated above, the diode is coupled with the drain of the power MOSFET and the DS impedance is relatively low, therefore the RF noise injected to the source reaches the diode via the drain. That is the reason of the abnormal response when the RF noise is injected to the source. However, the results of the source injection deviate by 4–5 dB. The cause of the deviation is not clear, but a possible reason is the linearization of the diode. S-parameter and frequency domain electromagnetic analysis are restricted to linear systems, so nonlinear behavior of the diode, such as rectification, is ignored in this simulation.

To improve the diode susceptibility, chip ferrites are inserted in both anode and cathode lines on the PCB in both the simulation and the experiment. Fig. 8 shows the comparison between the simulation and measurement. The improvement is verified by both the simulation and the experiment. These results show that the proposed simulation method can verify the effectiveness of an improvement plan.

V. MODULE-LEVEL DPI TEST SIMULATION

Now, a module-level DPI test is discussed. Module-level DPI simulation is difficult for SPICE simulators because the parasitic components of the PCB in the power module are very complex. The proposed electromagnetic simulation method is applied to this module-level DPI test in order to show its capability of dealing with a complex product-level PCB.

A. Modeling of Module-level DPI Test System

Fig. 9 shows the module-level DPI test setup for the power module. The PS1, PS2, PS3, OUT1, OUT2, and IN1 terminals are connected to the external circuit. PS1, PS2, and PS3 are power supply terminals, OUT1 and OUT2 are the signal output terminals, and IN1 is the signal input terminal. These terminals are connected to 50 ohm microstrip lines (MSLs). These MSLs are connected to bias tees, and the RF terminals of the bias tees are terminated by 50 ohm loads. One of the RF terminals of the bias tees is connected to the RF input of the DPI power output.

In the electromagnetic analysis, the lead frame, PCB, adhesive between the PCB and the lead frame, bonding wires, and resin mold are modeled as 3D solids. Passive components such as printed resistors on the PCB and ceramic capacitors are also modeled. The anode and cathode terminals of the temperature sensing diodes are connected only to the pre-driver IC. To include its impedance, the pre-driver IC is modeled as an S-parameter. The S-parameter of the pre-driver IC is calculated with a SPICE simulator. The netlist of the pre-driver IC includes the parasitic components of the IC layout. To take the pre-driver IC S-parameter into consideration, the corresponding ports are set to the electromagnetic simulation model. The references of the ports are the same as for the SPICE simulation, i.e., the substrate of the IC. The digital

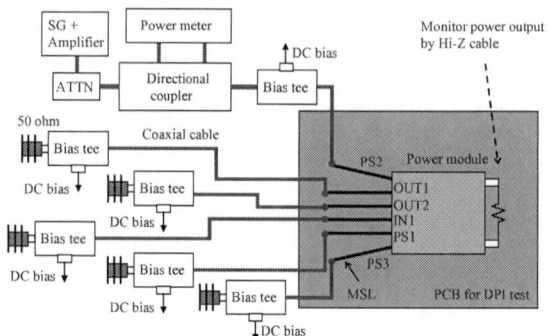

Fig. 9.   DPI test setup for the power module.

(a) Overview of the simulation model   (b) Focus at the pre-driver IC

Fig. 10.   Electromagnetic analysis model of the power module DPI test.

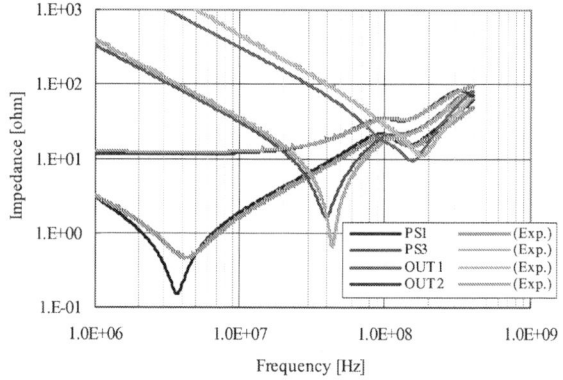

Fig. 11.   Comparison between simulation and experimental results for the input impedance.

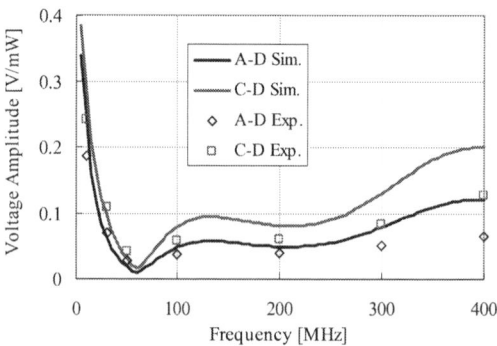

Fig. 12.   RF signal transfer to the temperature sensing diode (RF-injected port: PS2).

microcontroller and the oscillator are ignored because the coupling of the microcontroller and digital nets of the PCB with the victim circuit is weak. The power MOSFETs are modeled as described in section IV. The final simulation model is shown in fig. 10. The S-parameters of the diodes and the pre-driver IC are connected in the high-frequency circuit simulator.

B. Small Signal Response

To confirm the validity of the power module simulation model, the input impedances are measured at the DPI test board by using a VNA, and the measured results are compared with the simulation. The results are shown in fig. 11. The good fit of the curves confirms the accuracy of the simulation model. The OUT1 and OUT2 terminals are only connected to the pre-driver IC; therefore, the results in fig. 11 imply the validity of the pre-driver S-parameter model. However, the input impedance of the OUT1 terminal exhibits a deviation of approximately 40% at lower frequencies. The deviation could be caused by inaccuracy of the S-parameter of the pre-driver IC.

In addition, the small signal transfer functions from the RF-injected port to the temperature sensing diode are observed. The voltages between the drain of the P-channel power MOSFET and the anode or cathode are measured by using an active differential probe and a digital oscilloscope. The injection power at the RF-injected port is about 20 dBm. The result is shown in fig. 12, where the RF-injected port is PS2 and all results are normalized. The simulation results are about 1.5 times larger than the measurements. This means that the calculated stress on the temperature sensing diode may be higher than the experimental stress. However, the frequency dependencies of the transfer functions are well reproduced by the simulation.

C. Module-Level DPI Simulation Results

Fig. 13 shows the results of the module-level DPI tests and the simulation. Each label denotes the RF-injected terminal. The simulation results show a good correlation with the measurements. The differences between the simulation and measurement results are within 3–4 dB. These results show that the proposed method can predict the susceptibility thresholds of complex product-level circuits. One of the reasons of the deviation may be the nonlinearity of the pre-driver IC and the thermal sensing diode and the large injected noise. They might cause the inaccuracy of S-parameters which is only available to linear phenomena.

VI. BCI TEST SIMULATION

The proposed method is applied to the BCI test, which is generally performed for automotive equipment; hence, the BCI test simulation is important. Fig. 14 shows the BCI test setup. The EUT is the same power module as that described in the previous section. The substitution method is applied to this BCI test, and the maximum input current is set at 200 mA in the frequency range of 1–400 MHz. The injection probe is placed 150 mm away from the EUT.

The BCI test system is modeled in the same way as in Reference [8]. The injection probe, the harnesses, the line impedance stabilization networks (LISNs), and the external

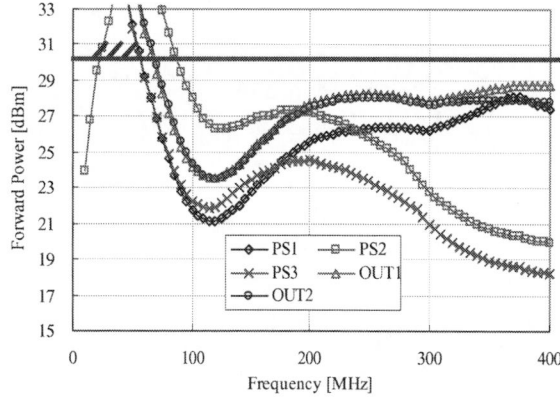

(a) Simulation results

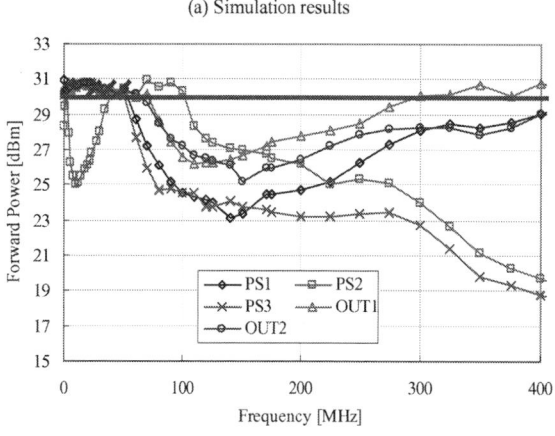

(b) Measurement results

Fig. 13. Comparison of simulation and experimental DPI test results for the susceptibility threshold of each terminal.

electric load are modeled as an S-parameter, which is measured with a VNA. The reference of the S-parameter is the ground plane. The power module is connected to the external PCB, which the connector and two aluminum electrolyte capacitors are mounted on. The simulation model of the EUT is shown in fig. 15. The power module model is the same as that described in the previous section. The PCB, the conductor of the connector terminals, and aluminum electrolyte capacitors are modeled. The measured S-parameter of the BCI test system is imported in the Touchstone file format in a high-frequency circuit simulator, and connected to the S-parameter of the EUT, which is calculated by the electromagnetic simulator. The connection ports between the BCI test system and the EUT are the connector ports whose references are the ground plane. RF noise is injected into the port of the BCI injection probe in the high-frequency circuit simulator. The input RF power is equal to that in the actual BCI test.

The comparison between the simulation and the measurement in the BCI test is shown in fig. 16 in terms of BCI test currents. The simulation results show a good correlation with the measurements in the frequency range ≤200 MHz. However, a shift in the negative peak frequency is observed in the frequency range 150–200 MHz, and the

978-1-4799-5004-1/13 $31.00 © 2013 IEEE

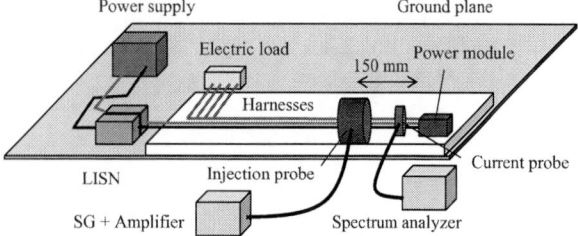

Fig. 14.   BCI test setup.

 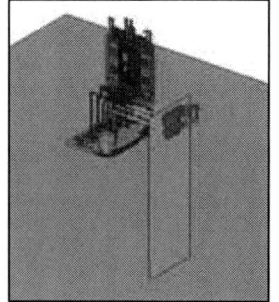

(a) Simulation model of EUT   (b) Simulation model including ground

Fig. 15.   BCI electromagnetic analysis model of the EUT.

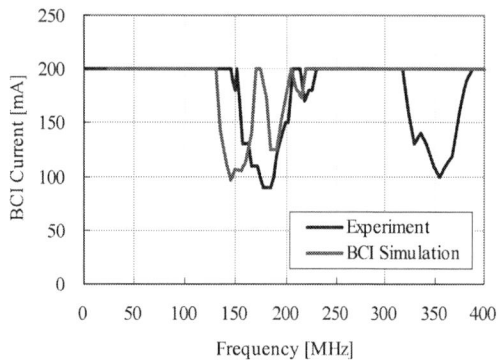

Fig. 16.   Comparison of simulation and experimental BCI test results for the susceptibility threshold.

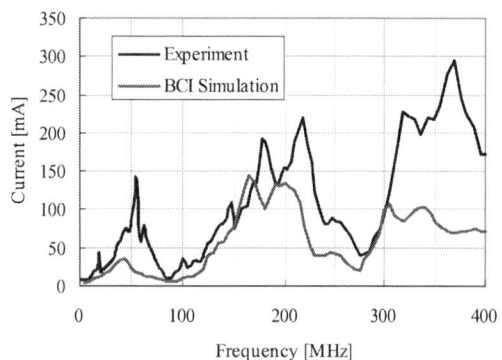

Fig. 17.   Comparison of simulation and experimental BCI test results for the common mode current.

abnormal responses in the frequency range 320–380 MHz are not observed in the simulation. To verify the accuracy of the BCI test simulation, the common mode current at the connector of the EUT is measured with a spectrum analyzer. In this measurement, all harnesses are clamped by a current probe. The result is compared with the simulation results (fig. 17). The calculated current result also shows a good correlation with the measurement, but differs from the measurement in the frequency range of near 50 MHz and over 200 MHz.

The cause of the deviation is unclear, but a possible reason is the inaccuracy of the BCI test system S-parameter. The module-level DPI simulation results in section V show better accuracy than the BCI simulation in the frequency range of over 200 MHz, and the models of the power module are the same. This implies that the accuracy of the BCI test system is not satisfactory in the frequency range of over 200 MHz. Further research is needed to resolve this problem.

VII.   CONCLUSION

This study presents a new simulation approach for estimating the results of immunity tests. The proposed method uses the stress–strength model and electromagnetic analysis. The proposed method is applied to the temperature sensing diode abnormal response in the power module with the multi-layer PCB, and the results show a good correlation with measurements in the DPI and BCI tests. The capability of dealing with the complex product-level PCBs in immunity simulations is also demonstrated. Moreover, the proposed simulation method can verify the effectiveness of an improvement plan for the susceptibility before the experiment is performed. For practical application in the design stage in the future, an improvement in the accuracy of the BCI test system model needs to be investigated.

REFERENCES

[1]   IEC 62132-4 (2006) : Integrated Circuits – Measurement of electromagnetic immunity, part 4, Direct Power Injection.

[2]   ISO 11452-4 (2005) : Road Vehicles – Component test methods for electrical disturbances from narrowband electromagnetic energy – Part 4 : Bulk current Injection.

[3]   F. Grassi, F. Marliani, and S. A. Pignari, "Circuit Modeling of Injection Probes for Bulk Current Injection," IEEE Tran., EMC, Vol.49, no. 3, pp563-576, Aug. 2007

[4]   A. Boyer, S. Bendhia, and E. Sicard, "Modelling of a Direct Power Injection Aggression on a 16 Bit Microcontroller Input Buffer", EMC Compo 2007, Turin, Italy 2007.

[5]   S. Miropolsky and S. Frei, "Comparability of RF Immunity Test Methods for IC Design Purpose", EMC Compo 2011, Dubrovnik, Croatia, 2011.

[6]   O. Jovic, Uwe Sturmer, W. Wilkening, and A. Baric, "Susceptibility of a Brokaw Bandgap to High Electromagnetic Interference", EMC Compo 2009, Toulouse, France, 2009

[7]   A. Durier, H. Pues, D.B. Ginste, M. Chernobryvko, C. Gazda, and H. Rogier, "Novel Modeling Strategy for a BCI set-up applied in an Automotive Application", EMC Compo 2011, Dubrovnik, Croatia, 2011.

[8]   Y. Oguri and K. Ichikawa, "Simulation Method for Automotive Electronic Equipment Immunity Testing", EMC Europe 2012, Rome, Italy, 2012.

Translation of automotive module RF immunity test limits into equivalent IC test limits using S-parameter IC models

Hugo Pues, Ben Briké,
Celina Gazda, Peter Teichmann,
Kristof Stijnen, Christian Peeters
Melexis Technologies N.V.
Tessenderlo, Belgium
hpu@melexis.com

André Durier
Continental Automotive France SAS
Toulouse, France
andre.durier@continental-corporation.com

Dries Vande Ginste
Ghent University, INTEC
Gent, Belgium
vdginste@intec.UGent.be

Abstract—**A method to translate immunity specifications of automotive modules into equivalent requirements at integrated circuit (IC) level, using linear scattering parameter models of the ICs, is presented. A technique is described to determine S-parameters of ICs by simulations based on back-annotated analog schematics. The simulation results are compared with measurement data obtained using a specially designed test board. As an example, simulation and measurement results are given for the input stage of an automotive sensor interface. A good agreement is obtained from the lowest test frequency up to 1 GHz. Above this value, the measured results seem to be dominated by package effects.**

Index Terms—**S-parameters, direct power injection (DPI), RF immunity, automotive EMC.**

I. Introduction

Automotive modules (electronic subassemblies) need to pass many severe electromagnetic compatibility (EMC) tests before they can be integrated into a car. These so-called module-level tests are specified by the vehicle manufacturers (see e.g. [1], which can be freely downloaded from www.fordemc.com). In order to avoid failures, which could require expensive redesigns and increase total time-to-market, it is of utmost importance to predict the results of these tests in advance by means of accurate simulations.

As explained in [2], a correct prediction of the results of module-level radio frequency (RF) immunity tests, such as a bulk current injection (BCI) test [3] or an absorber-lined shielded enclosure (ALSE) test [4], is very challenging. Therefore, IC designers prefer to convert these unpredictable standards to more manageable IC-level ones, such as, e.g., a direct power injection (DPI) test [5] and to compare the results [6]. In [7], a first approach to perform such module-to-IC-level test conversion was presented, successfully translating a BCI test of a pressure sensor assembly to a DPI test. In order to model the IC-under-test, its scattering parameters (S-parameters) were measured. Obviously, such a measurement is only possible if engineering samples of the IC are available.

In this paper we propose to perform the conversion between the tests using simulated S-parameters of the investigated IC.

In this way, module/IC-level requirements can be imposed on the circuit while it is still in its early design stage. We prove that proper simulation of the S-parameters of the IC-under-test in its normal operating point can accurately predict the measurement results, so that it is no longer needed to measure a prototype.

The paper is organized as follows. First, in Section II the principle of the module to IC-level test conversion is explained in detail. Second, in Section III a simulation method to obtain accurate S-parameters is described, which is based on the back-annotated analog schematics of the chip. The next section presents a measurement set-up, which makes use of a specially designed test board with external bias tees and direct current (DC) blocks that was used to validate the simulated data. Section V compares the results for the input stage of the automotive sensor interface described in [7]. Finally, in Section VI conclusions are drawn.

II. Conversion of Module-Level to IC-Level RF Immunity Requirements

In order to avoid unexpected failures of module-level EMC tests, it is appropriate to simulate these tests in advance [8], [9]. For RF immunity tests this becomes very challenging, because a normal model of the full set-up will either be too inaccurate (e.g. a lumped circuit model) or it will require an excessive amount of simulation time (e.g. a full-wave model). However, if the set-up is partitioned into subparts that are each modeled in their own appropriate way, an efficient model of the full set-up can be derived by properly combining all these sub-models in a high-level system model.

Once such a suitably partitioned system model has been set up, one is still faced with the challenge of deriving an adequate behavioral model of the chip (or chips) that is (are) part of the module. Ideally, this should be a nonlinear behavioral model that not only simulates the wanted functional and unwanted EMC behavior of the IC, but that is also efficient enough to be used in a higher-level simulation. However, although techniques have been developed already to derive

978-1-4799-5004-1/13 $31.00 © 2013 IEEE

such models for circuit blocks such as voltage regulators (see [10], [11]), their extension to full chips would still be very laborious and time-consuming. Therefore, a more efficient alternative approach can be applied, where the module-level RF immunity requirements are translated into equivalent IC-level requirements. These equivalent conditions can be used then by the chip designers to optimize the chip architecture and design for its intended application. Fig. 1 illustrates the procedure.

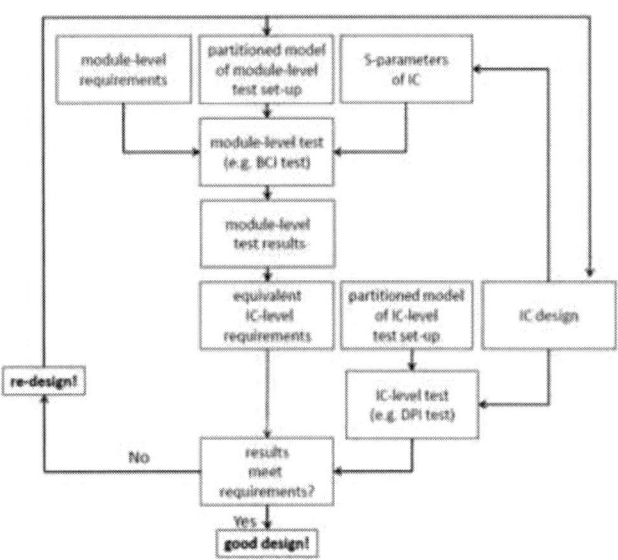

Fig. 1. Flowchart of module to IC level requirements conversion.

An IC design serves as a starting point in this flowchart. From the completed design of the IC its S-parameters are obtained by means of simulations (see Sec. III). These S-parameters are used then together with a partitioned model of the module-level test set-up and the EMC specifications to simulate the results of the module-level test, such as BCI. The results of this simulated test, being RF currents and voltages in case of the BCI test, are then used to calculate equivalent requirements for the IC-level test, which for the DPI test would be equivalent DPI power levels that the IC should be able to withstand. At the same time, the IC design is used together with the partitioned model of the IC-level set-up to simulate the results of the IC-level DPI test. If the obtained results meet the equivalent DPI levels (obtained from the BCI test results) it means that the IC is designed properly and would pass both IC and module-level tests. However, if the results do not comply with the equivalent IC-level requirements, the re-design of the IC and/or the module is needed.

Although this conversion has proven to be very useful and seems to be quite accurate, it is based on a few assumptions that have proven to be valid in many cases, but should be checked in each application of the proposed method. The first and most important assumption is that the IC is assumed to behave linearly as long as it meets the pass/fail criteria of the RF immunity test to be simulated. If that assumption holds,

the nonlinear large-signal analysis can be replaced by a linear small-signal AC analysis so that the IC can be modeled by its S-parameters.

The second assumption concerns the conversion to equivalent DPI requirements. As already explained in [7], the equivalent DPI levels at each port (i.e. at each pin referenced to ground) can be easily derived from the calculated small-signal RF currents and voltages at each port of the S-parameter model. However, this means that all these DPI levels should be injected at the same time to have a full equivalence with the module-level test whereas in a standard DPI test [5] the pins are tested one-by-one. Hence our second assumption is that this does not make a lot of difference in most practical cases. Obviously, this is a weakness of our method that needs further study, but fortunately, if the IC-level tests are simulated, it should be fairly straightforward to apply the different calculated pin injections at the same time.

III. SIMULATION OF S-PARAMETER IC MODELS

A method was developed to calculate the S-parameters of an automotive IC. As a case study an input stage of an automotive sensor interface was used, which was described in detail in [7]. The technique consists of three steps, namely, (1) determination of the operating point, (2) determination of the chip-level S-parameters in this operating point, and (3) incorporation of the package effects.

The first two steps deal with the determination of the chip-level parameters and are quite time-consuming as the full chip needs to be simulated. On the contrary, the third step was very quick, because simple lumped models were applied to simulate the package.

In order to perform steps (1) and (2), a full chip model needs to be constructed. To do so, RC back-annotated circuit models were extracted for the analog blocks whereas Verilog behavioral models were used for the digital blocks. To further improve the accuracy of the model, simple models for the inductance of long metal connections and the substrate coupling between the pins-under-test were added. These "manual" circuit additions were necessary as they were not taken into account by our RC back-annotated schematics.

To complete the chip model (to be used in steps (1) and (2)), the requested S-parameter ports need to be added but one has to make sure (by using DC blocks or switches) that the $50\,\Omega$ impedances of these ports do not load the circuits in the first step. For this first step, a real operating point analysis did not work very well in the investigated cases. Therefore, a transient analysis was used to bring all circuits in the correct operating conditions (e.g. capacitors charged to the right levels, no circuits clipping, sample & holds set correctly) so that step (2) could be carried out. As this step (2) was just a small-signal AC analysis, it proved to be much more straightforward than step (1).

In the final step (3), package effects were added to the chip-level S-parameters obtained in the previous step. Similar to step (2), this is a very straightforward step as the S-parameter

978-1-4799-5004-1/13 $31.00 © 2013 IEEE   250

model of the chip can be integrated in the package model very easily.

IV. MEASUREMENT OF S-PARAMETER IC MODELS

In order to measure S-parameters of ICs, a vector network analyzer is needed. In this work we used a 4-port Agilent E5071B ENA. To connect this instrument to the IC-under-test, our original idea was to use the DPI test board of the IC-under-test (see e.g. [12]), but we soon realized that it was better to use off-board DC blocks and bias tees instead of integrating them on the board such as in a DPI test board. In that way, the effect of these DC blocks and bias tees can be calibrated out and good $50\,\Omega$ connections can be made between the pins-under-test and their corresponding coaxial connectors.

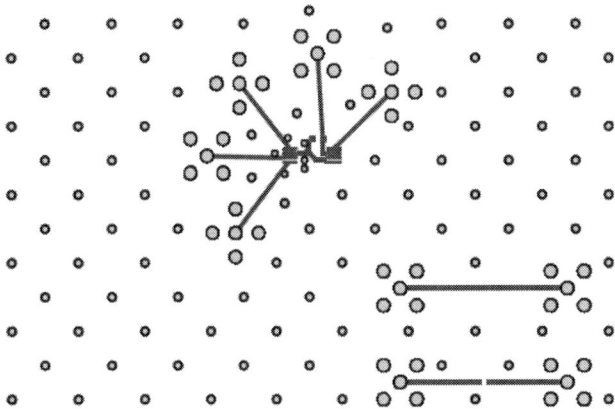

Fig. 2. Bottom layer of S-parameter test board.

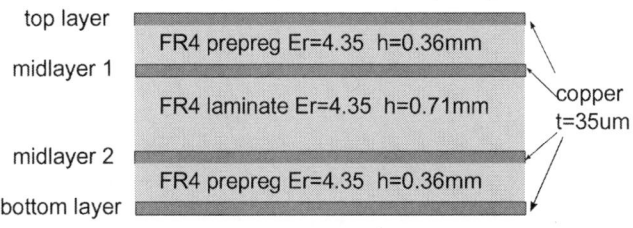

Fig. 3. Cross-section of the S-parameters board.

Fig. 2 shows the bottom side of the four-layer board, with a cross-section and substrate parameters depicted in Fig. 3, that was used to measure the results given in the next section. It is the same board that was also used to measure the S-parameters used in [7]. One can see five coaxial connectors with their connecting microstrip lines. This 5-port was measured by combining the results of three 4-port measurements into one 5×5 S-parameter matrix. However, only a 2×2 sub-matrix of it was used in the comparison with simulated results presented in this paper, because the other three pins were not really useful for this purpose as a decoupling capacitor (mounted on the top side of the board) was attached to them.

Prior to the actual measurements, the set-up was calibrated at the reference planes of the test board connectors. To eliminate the effect of the board, the reference planes were moved

to the IC pins using port extensions. These port extensions were determined using the test structures that are also shown in Fig. 2. In all cases the test frequency range was 300 kHz (lowest test frequency of the Agilent E5071B) to 3 GHz (considered to be the maximum usable test frequency of the test board).

V. COMPARISON OF MEASURED AND SIMULATED RESULTS

Figs. 4, 5 and 6 show both the measurement and simulation results for the two-port input stage of the ASIC described in [7]. Separate plots are shown for the amplitude (in dB) and phase (in degrees) of the s_{11}, s_{22} and s_{12} parameters. No results are shown for the s_{21} parameter as it was nearly identical to the s_{12} parameter.

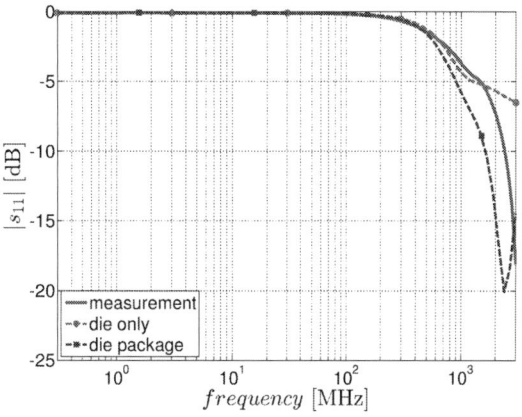

(a)

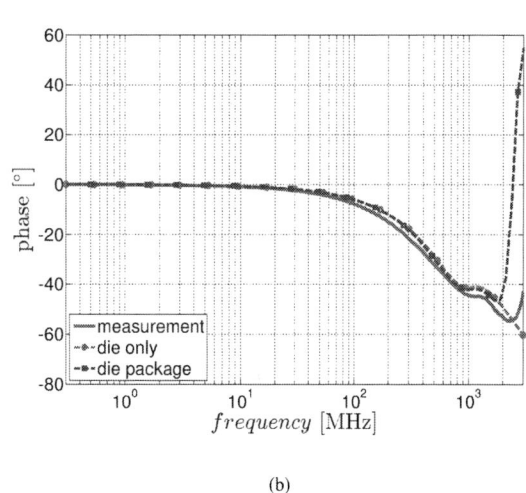

(b)

Fig. 4. Simulated and measured results for input reflection coefficient s_{11} (a) amplitude (b) phase.

The total calculation time of steps (1) and (2) of the S-parameter simulations was 16 hours. All simulations were performed in Cadence Spectre, using a server with a 6-core 2.4 GHz CPU and with 98 GB of RAM. These chip-level

978-1-4799-5004-1/13 $31.00 © 2013 IEEE

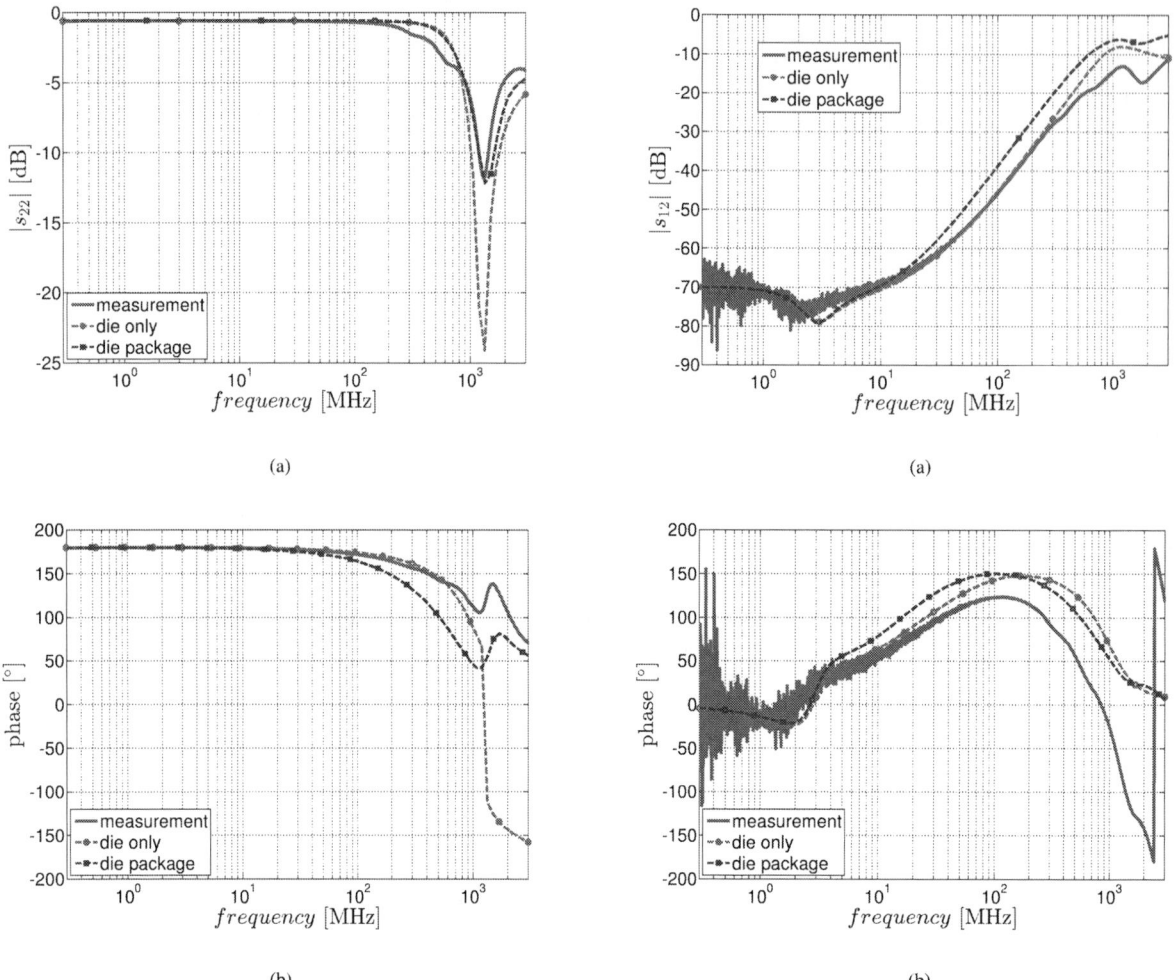

(a)

(b)

Fig. 5. Simulated and measured results for output reflection coefficient s_{22} (a) amplitude (b) phase.

(a)

(b)

Fig. 6. Simulated and measured results for transmission coefficient s_{12} (a) amplitude (b) phase.

results are represented in all figures by a green dashed line with circles (∘), labeled as "die only". They are compared with an additional calculation result that includes package effects, plotted as dashed blue line with squares (□) and labeled "die package", and with the measurement results drawn with solid red line and labeled "measurement". The package model consisted of two 2 nH inductances (connected to the two input bond pads) and one 5 nH inductance (connected to the ground bond pad located at the other side of the die).

From the results depicted in Figs. 4, 5 and 6 it is clear that a good agreement between simulations and measurements is obtained up to 1 GHz. Above 1 GHz the correspondence starts to deteriorate quickly but could probably be improved by using more accurate package models.

VI. CONCLUSIONS

In this paper, a method was proposed to transform the RF immunity requirements of automotive modules into equivalent IC-level requirements that can be applied to the ICs that are integrated in the modules. The conversion is based on a partitioned model of the full test set-up where the IC is modeled in its normal operating point by a linear S-parameter model obtained by means of simulations. In this way, the EMC behavior of the IC is validated when the circuit is still in its early design stage, before it has been taped-out. A good correspondence obtained between the simulated and measured S-parameters of the IC confirms the validity and usefulness of this approach.

REFERENCES

[1] *Electromagnetic compatibility specification for electrical/electronic components and subsystems*, Ford, EMC-CS-2009.1 Std., 2010.
[2] M. Coenen, H. Pues, and T. Bousquet, "Automotive RF immunity test set-up analysis: Why test results can't compare..." in *8th Workshop on Electromagnetic Compatibility of Integrated Circuits (EMC Compo)*, Dubrovnik, Croatia, 6-9 Nov. 2011, pp. 71–75.

978-1-4799-5004-1/13 $31.00 © 2013 IEEE

[3] *Road vehicles - Component test methods for electrical disturbances from narrowband radiated electromagnetic energy - Part 4: Bulk current injection (BCI)*, International Organization for Standardization, ISO 11452-4 Std., 2011.

[4] *Road vehicles - Component test methods for electrical disturbances from narrowband radiated electromagnetic energy - Part 2: Absorber-lined shielded enclosure*, International Organization for Standardization, ISO 11452-2 Std., 2004.

[5] *Integrated Circuits - Measurement of Electromagnetic Immunity, 150 kHz to 1 GHz - Part 4: Measurement of Conducted Immunity - direct RF power injection method*, International Electrotechnical Commission, IEC 62132-4 Std., 2006.

[6] S. Miropolsky and S. Frei, "Comparability of RF immunity test methods for IC design purposes," in *8th Workshop on Electromagnetic Compatibility of Integrated Circuits (EMC Compo)*, Dubrovnik, Croatia, 6-9 Nov. 2011, pp. 59–64.

[7] A. Durier, H. Pues, D. Vande Ginste, M. Chernobryvko, C. Gazda, and H. Rogier, "Novel modeling strategy for a BCI set-up applied in an automotive application: An industrial way to use EM simulation tools to help hardware and ASIC designers to improve their designs for immunity tests," in *8th Workshop on Electromagnetic Compatibility of Integrated Circuits (EMC Compo)*, Dubrovnik, Croatia, 6-9 Nov. 2011, pp. 41–46.

[8] A. Alaeldine, R. Perdriau, M. Ramdani, J.-L. Levant, and M. Drissi, "A direct power injection model for immunity prediction in integrated circuits," *IEEE Transactions on Electromagnetic Compatibility*, vol. 50, no. 1, pp. 52–62, Feb. 2008.

[9] I. Chahine, M. Kadi, E. Gaboriaud, A. Louis, and B. Mazari, "Characterization and modeling of the susceptibility of integrated circuits to conducted electromagnetic disturbances up to 1 GHz," *IEEE Transactions on Electromagnetic Compatibility,*, vol. 50, no. 2, pp. 285–293, May 2008.

[10] C. Gazda, D. Vande Ginste, H. Rogier, I. Couckuyt, T. Dhaene, K. Stijnen, and H. Pues, "Harmonic Balance surrogate-based immunity modeling of a nonlinear analog circuit," *IEEE Transactions on Electromagnetic Compatibility, early access*, vol. -, pp. –, 2013.

[11] V. Ceperic and A. Baric, "Modelling of electromagnetic immunity of integrated circuits by artificial neural networks," in *20th International Zurich Symposium on Electromagnetic Compatibility*, Zurich, Switzerland, 12-16 Jan. 2009, pp. 373 –376.

[12] H. Pues and D. Pissoort, "Design of IEC 62132-4 compliant DPI test boards that work up to 2 GHz," in *International Symposium on Electromagnetic Compatibility (EMC EUROPE)*, Rome, Italy, 17-21 Sep. 2012, pp. 1–4.

CURRAN ASSOCIATES INC.
proceedings
.com

9781479950041